PLANT BIOTECHNOLOGY

VOLUME 1

Principles, Techniques, and Applications

PLANT BIOTECHNOLOGY

VOLUME 1

Principles, Techniques, and Applications

Edited by

Bishun Deo Prasad, PhD

Sangita Sahni, PhD

Prasant Kumar, PhD

Mohammed Wasim Siddiqui, PhD

APPLE ACADEMIC PRESS

Apple Academic Press Inc.
3333 Mistwell Crescent
Oakville, ON L6L 0A2 Canada

Apple Academic Press Inc.
9 Spinnaker Way
Waretown, NJ 08758 USA

Printed in the United States of America on acid-free paper
Plant Biotechnology (2-volume set)
International Standard Book Number-13: 978-1-77188-580-5 (Hardcover)
International Standard Book Number-13: 978-1-315-21374-3 (eBook)
International Standard Book Number-13: 978-1-77188-582-9 (2-volume set)
International Standard Book Number-13: 978-1-315-21309-5 (eBook)

Library and Archives Canada Cataloguing in Publication

Plant biotechnology (Oakville, Ont.)
Plant biotechnology / edited by Bishun Deo Prasad, PhD, Sangita Sahni, PhD, Prasant Kumar, PhD, Mohammed Wasim Siddiqui, PhD.
Includes bibliographical references and index.
Contents: Volume 1. Principles, techniques, and applications
Issued in print and electronic formats.

ISBN 978-1-77188-580-5 (v. 1 : hardcover).--ISBN 978-1-315-21374-3 (v. 1 : PDF)
1. Plant biotechnology. I. Siddiqui, Mohammed Wasim, editor II. Prasad, Bishun Deo, editor III. Sahni, Sangita, editor IV. Kumar, Prasant, editor V. Title.

TP248.27.P55P63 2017 660.6 C2017-905057-5 C2017-905058-3

Library of Congress Cataloging-in-Publication Data

Names: Prasad, Bishun Deo, editor.
Title: Plant biotechnology. Volume 1, Principles, techniques, and applications / editors: Bishun Deo Prasad, Sangita Sahni, Prasant Kumar, Mohammed Wasim Siddiqui.
Other titles: Principles, techniques, and applications
Description: Waretown, NJ : Apple Academic Press, 2017. | Includes bibliographical references and index.
Identifiers: LCCN 2017034336 (print) | LCCN 2017044760 (ebook) | ISBN 9781315213743 (ebook) | ISBN 9781771885805 (hardcover : alk. paper)
Subjects: LCSH: Plant biotechnology.
Classification: LCC TP248.27.P55 (ebook) | LCC TP248.27.P55 P5545 2017 (print) | DDC 630--dc23
LC record available at https://lccn.loc.gov/2017034336

Apple Academic Press also publishes its books in a variety of electronic formats. Some content that appears in print may not be available in electronic format. For information about Apple Academic Press products, visit our website at **www.appleacademicpress.com** and the CRC Press website at **www.crcpress.com**

ABOUT THE EDITORS

Bishun Deo Prasad, PhD

Dr. Bishun Deo Prasad is an Assistant Professor and Scientist in the Department of Molecular Biology and Genetic Engineering, Bihar Agricultural University, Sabour, India. He has published several research papers in reputed peer-reviewed international journals which have been cited more than 100 times. He has also contributed to two authored book, has written several book chapters, and has submitted 10 sequences of different isolates to the National Center for Biotechnology Information (NCBI). He is a reviewer of the *International Journal of Agriculture Sciences* and *Journal of Environmental Biology*.

Dr. Prasad has received the DAE—Young Scientist Research award in 2013 and the Fast Track Scheme for Young Scientists award by the Department of Science and Technology (DST), India, in 2012. He has also been awarded with an Outstanding Achievement Award in 2014 from the Society for Scientific Development in Agriculture and Technology (SSDAT) and an Inventor of the Year Award, 2015 in the discipline of Molecular Biology and Genetic Engineering from the Society of Scientific and Applied Research Centre at an international conference (iCiAsT-2016) held at the Faculty of Science, Kasetsart University, Bangkok, Thailand in 2016.

Dr. Prasad acquired his BSc (Agriculture) degree from MPKV, Rahuri, Maharashtra, India, MSc (Agricultural Biotechnology) from Assam Agricultural University and his PhD from M. S. University from Baroda, Gujarat, India, with a thesis in the field of Plant Biotechnology. He also worked at the John Innes Centre (JIC), Norwich, UK, during his PhD. Subsequently, he worked as a postdoctoral research fellow at the University of Western Ontario University, London, Ontario, Canada. He also worked at V.M.S.R.F., Bangalore, as a Scientist and S. D. Agricultural University, Gujarat, as an Assistant Professor. He has received grants from various funding agencies to carry out his research projects. He is a member secretary of Biosafety Committee and member of different committees of Bihar Agricultural University, Sabour.

Dr. Prasad has been associated with biotechnological aspects of rice, *Brassica napus*, *Arabidopsis*, linseed, lentil, vegetable (bitter guard and pointed guard), and horticultural (mango, litchi, and banana) crops. He is also associated with host–pathogen interaction studies in rice, *B. napus*, and *Arabidopsis* as well as mutational breeding aspect in rice for abiotic stress tolerance. He is dynamically involved in teaching graduate and post-graduate courses of Biotechnology, Plant Breeding and Genetics, Vegetable Crops, and Horticultural Crops.

Sangita Sahni, PhD

Dr. Sangita Sahni is a Junior Scientist and Assistant Professor in the Department of Plant Pathology, Tirhut College of Agriculture, Dholi, Rajendra Agricultural University, Pusa, Samastipur, Bihar, India. She has published several research papers in reputed peer-reviewed national and international journals. She has published two authored book and several book chapters. She has isolated several bacterial isolates from different sources and submitted their sequences to the National Center for Biotechnology Information (NCBI).

Dr. Sahni acquired a BSc (Agriculture) degree from A.N.G.R.A.U, Hyderabad, India, and an MSc (Agriculture) in Mycology and Plant Pathology from Banaras Hindu University, Varanasi, India. She received her PhD (Agriculture) in Plant Pathology from the B.H.U, Varanasi. Subsequently, she worked as a postdoctoral research fellow at the University of Western Ontario University, London, Ontario, Canada.

Dr. Sahni has been awarded with the Dr. Rajendra Prasad National Education Shikhar Award for outstanding contribution in the field of education, a Young Scientist Award in 2014 from the Society for Scientific Development in Agriculture and Technology (SSDAT), and an Innovative Scientist of the Year Award, 2015, from the Scientific Education Research Society for outstanding contribution in the field of Plant Pathology. She is a Principal Investigator in All India Co-ordinated Research Programme at MULLaRP and Chickpea Pathology at T.C.A., Dholi. She is an officer in-charge of ARIS cell, TCA, Dholi, and a member of different committees of RAU, Pusa. She has been an active member of the organizing committees of several national and international seminars.

Dr. Sahni has been associated with molecular host–pathogen interaction studies in *Arabidopsis* and *B. napus*. She is also associated with pathological aspect of chickpea and MULLaRP. She is actively involved in teaching graduate and post-graduate courses in Plant Pathology and Biotechnology. She has proved herself as an active scientist in the area of Molecular Plant Pathology.

Prasant Kumar, PhD

Dr. Prasant Kumar is an Assistant Professor at the C. G. Bhakta Institute of Biotechnology, Department of Fundamental and Applied Science at Uka Tarsadia University, Surat, Gujarat, India, and is the author or co-author of several peer-reviewed journal articles and eight conference papers and a newsletter.

He is a reviewer and editorial board member of several peer-reviewed journals. He has been an active member of the organizing committees of several national and international seminars and conferences.

Dr. Kumar received a BSc (Agriculture) from Acharya N. G. Ranga Agriculture University through the all India combined entrance exam conducted by the Indian Council of Agriculture Research, India. After graduating from Acharya N. G. Ranga Agriculture University, he was selected for the MSc Biotechnology program of The Maharaha Sayajirao University of Baroda, Gujarat, through the all India combined biotechnology entrance exam conducted by Department of Biotechnology (Govt. of India) and Jawaharlal Nehru University, New Delhi. Along with completion of his postgraduation, with first class with distinction in Biochemistry, he qualified GATE, ICMR-JRF, UGC-NET exam of national repute. Later, he joined the PhD program in Biochemistry from The Maharaha Sayajirao University of Baroda. He was awarded an Indian Council of Medical Research Fellowship Award for the PhD from the Indian Council of Medical research, New Delhi, India. He worked as an Assistant Professor in Sardar Patel University, Anand, Gujarat, from August 2011 to June 2012.

Mohammed Wasim Siddiqui, PhD

Dr. Mohammed Wasim Siddiqui is an Assistant Professor and Scientist in the Department of Food Science and Post-Harvest Technology, Bihar Agricultural University, Sabour, India, and author or co-author of 34 peer-reviewed research articles, 26 book chapters, 2 manuals, and 18 conference papers. He has 11 edited and one authored books to his credit, published by Elsevier, USA; CRC Press, USA; Springer, USA; and Apple Academic Press, USA. Dr. Siddiqui has established an international peer-reviewed journal, *Journal of Postharvest Technology.*

He has been honored to be the Editor-in-Chief of two book series: "Postharvest Biology and Technology" and "Innovations in Horticultural Science," being published by Apple Academic Press, USA. Dr. Siddiqui is also a Senior Acquisitions Editor for Apple Academic Press, for Horticultural Science. He has been serving as an editorial board member and active reviewer of several international journals, such as *PLoS ONE,* (PLOS), *LWT—Food Science and Technology* (Elsevier), *Food Science and Nutrition* (Wiley), *Acta Physiologiae Plantarum* (Springer), *Journal of Food Science and Technology* (Springer), *Indian Journal of Agricultural Science* (ICAR), etc.

Recently, Dr. Siddiqui was conferred with the Best Citizen of India Award-2016; Bharat Jyoti Award, 2016; Outstanding Researcher Award, 2016; Best Young Researcher Award, 2015; and the Young Scientist Award, 2015. He was also a recipient of the Young Achiever Award, 2014, for outstanding research work by the Society for Advancement of Human and Nature (SADHNA), Nauni, Himachal Pradesh, India, where he is an honorary board member and lifetime author. He has been an active member of the organizing committee of several national and international seminars/conferences/summits. He is one of the key members in establishing the World Food Preservation Center (WFPC), LLC, USA. Presently, he is an active associate and supporter of WFPC, LLC, USA. Considering his outstanding contribution in science and technology, his biography has been published in *Asia Pacific Who's Who* and *The Honored Best Citizens of India.*

Dr. Siddiqui acquired his BSc (Agriculture) degree from Jawaharlal Nehru Krishi Vishwa Vidyalaya, Jabalpur, India. He received the MSc (Horticulture) and PhD (Horticulture) degrees from Bidhan Chandra Krishi Viswavidyalaya, Mohanpur, Nadia, India, with specialization in Postharvest Technology. He was awarded a Maulana Azad National Fellowship Award from the University Grants Commission, New Delhi, India. He is a member of Core Research Group at the Bihar Agricultural University (BAU) where he is providing appropriate direction and assistance to sensitizing priority of the research. He has received several grants from various funding agencies to carry out his research projects. Dr. Siddiqui has been associated with postharvest biotechnology and processing aspects of horticultural crops. He is dynamically indulged in teaching (graduate and doctorate students) and research, and he has proved himself as an active scientist in the area of postharvest biotechnology.

CONTENTS

LIST OF CONTRIBUTORS

Seyedeh Fatemeh Afzali
Department of Biological Science, Faculty of Science, Universiti Tunku Abdul Rahman, Malaysia

Samindra Baishya
Department of Biochemistry and Agricultural Chemistry, Assam Agricultural University, Jorhat, Assam, India

Akhil Ranjan Baruah
Department of Agricultural Biotechnology, Assam Agricultural University, Jorhat, Assam, India

Govinal Badiger Bhaskara
Department of Integrative Biology, University of Texas, Austin, Texas-78712, United States

Gaurav S. Dave
Department of Biochemistry, Saurashtra University, Rajkot 360005, India

Mitesh Dwivedi
C. G. Bhakta Institute of Biotechnology, Department of Fundamental and Applied Science, Uka Tarsadia University, Bardoli, Surat 394350, Gujarat, India

J. G. Hehir
Department of Crop Science, Oak Park Crops Research Centre, Teagasc, Carlow, Ireland

Vijay Kumar Jha
Department. of Botany, Patna University, Patna, Bihar, India

Aruna Joshi
Department of Botany, Faculty of Science, The Maharaja Sayajirao University of Baroda, Vadodra, Gujarat, India

Chandan Kishore
Department of Plant Breeding and Genetics, Bihar Agricultural University, Sabour 813210, Bihar, India

Devanshi Khokhani
Department of Plant Pathology, University of Wisconsin—Madison, 583 Russell Labs, 1630 Linden Dr. Madison, Wisconsin 53706, USA

Amarendra Kumar
Department of Plant Pathology, Bihar Agricultural University, Sabour 813210, Bihar, India

Anand Kumar
Department of Plant Breeding and Genetics, Bihar Agricultural University, Sabour 813210, Bihar, India

Jitesh Kumar
Department of Molecular Biology and Genetic Engineering, Bihar Agricultural University, Sabour, Bhagalpur, Bihar, India

Mahesh Kumar
Department of Molecular Biology and Genetic Engineering, Bihar Agricultural University, Sabour 813210, Bihar, India

Manoj Kundu
Department of Horticulture (Fruit and Fruit Technology), Bihar Agricultural University, Sabour, Bhagalpur, Bihar, India

Pankaj Kumar
Department of Molecular Biology and Genetic Engineering, Bihar Agricultural University, Sabour 813210, Bihar, India

Prasant Kumar
C. G. Bhakta Institute of Biotechnology, Department of Fundamental and Applied Science, Uka Tarsadia University, Bardoli, Surat 394350, Gujarat, India

Ravi Ranjan Kumar
Department of Molecular Biology and Genetic Engineering, Bihar Agricultural University, Sabour 813210, Bihar, India

Renu Kushwah
Department of Plant Molecular Biology and Biotechnology, IGKV, Raipur 492012, India

Sonam Kumari
Department of Molecular Biology and Genetic Engineering, Bihar Agricultural University, Sabour 813210, Bihar, India

Sunita Kumari
Krishi Vigyan Kendra, Kishanganj, Bihar Agricultural University, Sabour, Bhagalpur, Bihar, India

Vinod Kumar
Department of Molecular Biology and Genetic Engineering, Bihar Agricultural University, Sabour 813210, Bihar, India

Suhail Muzaffar
National Centre for Biological Sciences, GKVK Campus, Bellary Road, Bangalore 560065, India

Ram Balak Prasad Nirala
Department of Plant Breeding and Genetics, Bihar Agricultural University, Sabour 813210, Bihar, India

Awadhesh Kumar Pal
Department of Plant Breeding and Genetics, Bihar Agricultural University, Sabour 813210, Bihar, India

Ganesh Patil
Vidya Pratisthan's College of Agriculture Biotechnology, Vidyanagari, Baramati 413133, India

Mitesh B. Patel
C. G. Bhakta Institute of Biotechnology, Uka Tarsadia University, Tarsadi, Surat 394350, Gujarat, India

Ashutosh Pathak
Department of Botany, Faculty of Science, The Maharaja Sayajirao University of Baroda, Vadodra, Gujarat, India

Jayesh Pathak
Department of Silvicuture and Agroforestry, Navsari Agricultural University, Navsari, Gujarat, India

Chandra Prakash
Genome Research Centre, Department of Microbiology and Biotechnology Centre, Faculty of Science, The M. S. University of Baroda, Vadodara 390002, Gujarat, India

Bishun Deo Prasad
Department of Molecular Biology and Genetic Engineering, Bihar Agricultural College, Sabour 813210, Bihar, India

Ashish Ranjan
Department of Plant Pathology, University of Wisconsin—Madison, 583 Russell Labs, 1630 Linden Dr. Madison, Wisconsin 53706, USA

Kumari Rajani
Department of Seed Technology, Bihar Agricultural University, Sabour 813210, Bihar, India

Sunayana Rathi
Department of Biochemistry and Agricultural Chemistry, Assam Agricultural University, Jorhat, Assam, India

Tushar Ranjan
Department of Basic Science and Humanities Genetics, Bihar Agricultural University, Sabour 813210, Bihar, India

Nand K. Sah
Post-Graduate Studies and Research Centre, Department of Botany, T. N. B. College, Bhagalpur, T. M. B. University, Bhagalpur 812007, Bihar, India

Uday Sajja
Department of Biotechnology, Dr. B. R. Ambedkar University, Srikakulam, Andhra Pradesh, India

Sangita Sahni
Department. of Plant Pathology, Tirhut College of Agriculture, Dholi, Muzaffarpur, Bihar, India

Surojit Sen
Department of Zoology, Mariani College, Mariani 785634, Jorhat, Assam, India

Gaurav V. Sanghvi
Department of Botany, Faculty of Science, The Maharaja Sayajirao University of Baroda, Vadodara 390002, India

Shiv Shankar
Department of Food Engineering and Bionanocomposite Research Institute, Mokpo National University, 61 Dorimri, Chungkyemyon, Muangun, 534-729 Jeonnam, Republic of Korea

Md Shamim
Department of Molecular Biology and Genetic Engineering, Krishi Vigyan Kendra, Kishanganj, Bihar, India

Vaishali Sharma
DOS in Biotechnology, University of Mysore, Mysore, India

Sweta Sinha
Department of Molecular Biology and Genetic Engineering, Bihar Agricultural University, Sabour 813210, Bihar, India

G. Thapa
Molecular Plant Pathogen Interaction Group, Earth Institute, Science East, University College of Dublin, Belfield, Dublin 4, Ireland

Imran Uddin
Nanotechnology Innovation Centre, Department of Chemistry, Rhodes University, PO Box 94, Grahamstown, South Africa

LIST OF ABBREVIATIONS

2, 4-D	2, 4-dichloropheonoxyacetic acid
2D-GE	two-dimensional gel electrophoresis
ABA	abscisic acid
ACC	1-aminocyclopropane-1- carboxylic acid
AFLP	amplified fragment length polymorphism
AM	arbuscular mycorrhizal
AMP	adenosine monophosphate
AR	amplex red
ATMT	*Agrobacterium tumefaciens*-mediated transformation
AUR	amplex ultra red
BAP	6-benzylaminopurine
BC	backcross population
Bt	*Bacillus thuringiensis*
CIM	callus induction medium
CRISPR	clustered regularly interspaced short palindromic repeats
DAB	3,3′-diaminobenzidine
DAF	DNA amplification fingerprinting
DAGT	diacylglycerol acyltransferase
DHL	double haploid lines
DNA	deoxyribonucleic acid
ELISA	enzyme linked immunosorbent assay
EPR	electron paramagnetic resonance
ETS	expressed tagged sites
FACS	fluorescence activated cell sorting
GA3	gibberellic acid
GBS	genotyping-by-sequencing
GC-MS	gas-chromatography-mass-spectrometry
GC-TOF-MS	gas-chromatography-time-of-flight-mass-spectrometry
GM	genetic modification
GMOs	genetically modified organisms
IAA	indole-3-acetic acid
IBA	indole-3-butyric acid
ILs	introgression lines
ISSR	inter simple sequence repeats

KASPar	Kbioscience competitive allele specific PCR
Kn	kinetin
LC-MS	liquid-chromatography-mass-spectrometry
LD	linkage disequilibrium
MAGIC	multi-parent advanced generation inter crosses
MAS	marker-assisted selection
miRNA	microRNA
mRNA	messenger RNA
MS	mass spectrometry
MS	Murashige and Skoog
NAA	α-naphthalene acetic acid
NADP	nicotinamide adenine dinucleotide phosphate
NBT	nitroblue tetrazolium
NGS	next generation sequencing
NILs	near-isogenic lines
NMR	nuclear magnetic resonance
PAGE	polyacrylamide gels
PCD	programmed cell death
PCR	polymerase chain reaction
PEG	polyethylene glycol
PGPR	plant growth-promoting rhizobacteria
PHA	polyhydroxyalkanoate
PHB	poly (3-hydroxybutyrate)
PMC	pollen mother cell
PTGS	post-transcriptional gene silencing
PVP	polyvinylpyrrolidone
QRT-PCR	quantitative real-time PCR
QTL	quantitative trait locus
RAD	restriction site-associated DNA
RAPD	rapid amplified polymorphic DNA
REMI	restriction enzyme-mediated integration
RFLP	restriction fragment length polymorphism
RILs	recombinant inbred lines
RISC	RNA-induced silencing complex
RNAi	RNA interference
ROS	reactive oxygen species
RT-PCR	reverse transcription PCR
SCARs	sequence characterized amplified regions
SDS	sodium dodecyl sulfate
SIM	simple interval mapping

siRNAs	small interfering RNAs
SNP	single nucleotide polymorphism
SSR	simple sequence repeats
STMs	sequence-tagged microsatellites
STR	simple tandem repeats
STSs	sequence-tagged sites
TAIL-PCR	thermal asymmetric interlaced PCR
Ti	tumor-inducing
TILLING	targeting induced local lesions in genomes
T-RFLP	terminal restriction fragment length polymorphism
UV	ultraviolet
VIGS	virus-induced gene silencing
vir	virulence

ACKNOWLEDGMENT

At the end of editing this book, I close my eyes and remember the day when the idea of writing this book was seeded in my mind, followed by discussion about this with my other colleagues, which led to the foundation of this project. From that initial day to now, when we are finally publishing our book, there have been several ups and downs. However, with blessings of "Almighty God," we were able to convert our ideas, teaching, and research experiences to a logical end in the form of this book. Therefore, first of all, we would like to thank "Almighty God" from whom all blessings come. Further, I would like to express my gratitude to the many people who saw us through this book; to all those who provided support, talked things over, read, wrote, offered comments, allowed me to quote their remarks, and assisted in the editing, proofreading, and design. I would like to thank Dr. Tusar Ranjan and Dr. Mitesh Dwivedi for helping us in the process of editing this book.

With a profound and unfading sense of gratitude, I wish to express our sincere thanks to the Bihar Agricultural University, India, for providing me with the opportunity and facilities to execute such an exciting project and for supporting me toward research and other intellectual activities around the globe.

We feel privileged to acknowledge our immense sense of devotion to our parents and family members for their infinitive love, cordial affection, and incessant inspiration. Last not least: we beg forgiveness of all those who have been with us during the course of writing this book and whose names we have failed to mention.

PART I

History, Scope, and Importance of Plant Biotechnology

CHAPTER 1

HISTORY OF BIOTECHNOLOGY

SUHAIL MUZAFFAR[1*] and BISHUN DEO PRASAD[2*]

[1]*National Centre for Biological Sciences, GKVK Campus, Bellary Road, Bangalore 560065, India*

[2]*Department of Molecular Biology and Genetic Engineering, Bihar Agricultural College, Sabour, Bhagalpur, Bihar, India*

[*]*Corresponding author. E-mail: suhail.bt@gmail.com; dev.bishnu@gmail.com*

CONTENTS

ABSTRACT

The science of 'biotechnology' has received enormous attention recently due to its unlimited potential to benefit humanity. Biotechnology uses biological materials to create novel products for agricultural, pharmaceutical, medical, and environmental applications. The history of biotechnology begins with zymotechnology, which originated with a focus on brewing techniques. By the start of 20th century, zymotechnology began to expand and the concept of industrial fermentation gave rise to biotechnology. After domestication of crops and animals, humans began to make cheese, curd, and wine using simple fermentation techniques. Cheese is considered as one of the first products of biotechnology, as it was prepared by adding rennet (an enzyme) to sour milk. During the 1940s, the discovery of penicillin was a dramatic event. Although discovered in England, it was produced industrially in the U.S. using a deep fermentation process. Penecillin was one of the most important success stories of last century and doctors called it a "miracle drug". The introduction of principles of genetic engineering brought biotechnology to the forefront of science in society. With the development of synthetic human insulin, the biotechnology industry started to grow rapidly. Genetic engineering remains the centre of scientific discussion in modern world especially with ever emerging fields of gene therapy, stem cell technology, and genetically modified organisms. Although most of these scientific advancements are very recent but the service of biotechnology to society began centuries ago.

1.1 OVERVIEW

The term "Biotechnology" was first coined by a Hungarian agricultural engineer Károly Ereky in 1919. His scientific work laid the foundations of this new discipline and therefore he is regarded as "the father of biotechnology." The history of biotechnology started hundreds of years ago with the advent of fermentation when humans learned to brew beer. But, with the introduction of genetic engineering, biotechnology came to the forefront of modern day science. Discovery of DNA, development of synthetic human insulin, and genetically modified organisms were the milestones in biotechnology as it introduced genetic engineering in our day-to-day life. During the 1980s, biotechnology grew into one of the most promising industries of the time. Till this date, biotechnology has remained a hot topic among scientists, lawmakers, and the general public for both breakthrough technologies as well as various controversies like animal cloning, stem cell research, and gene therapy. The developments in the field of biotechnology can be divided into three different eras: (1) ancient biotechnology, (2) classical biotechnology, and (3) modern biotechnology (Verma et al., 2011).

1.2 BIOTECHNOLOGY TIME LINES

Timeline of key events occurred in biotechnology has been summarized in Table 1.1.

TABLE 1.1 Biotechnology Time Lines.

Periods	Important Discoveries/Events
6000 BC	• Sumerians and Babylonians used yeast to make beer.
4000 BC	• Baking leavened bread using yeast was discovered by the Egyptians.
320 BC	• Aristotle stated that all inheritance comes from the father.
1000	• Spontaneous generation hypothesis was proposed.
1673	• Anton van Leeuwenhoek described the role of microorganisms in fermentation.
1701	• Giacomo Pylarini practiced "inoculation" in which children were intentionally inoculated with smallpox to prevent a serious case later in life.
1809	• Heat sterilization of food was by devised by Nicolas Appert.
1856	• A technique for keeping animal organs alive outside the body, by pumping blood through them was discovered by Karl Ludwig.
	• Charles Darwin (1809–82) hypothesized that animal populations adapt their forms over time to best exploit the environment, a process he referred to as "natural selection."
1859	• Louis Pasteur (1859) asserted that microbes are responsible for fermentation.
	• Charles Darwin proposed theory of "natural selection."
1863	• Louis Pasteur invented the process of pasteurization.
1865	• Gregor Mendel presented his laws of heredity.
1870	• Walther Flemming discovered mitosis.
1871	• DNA was isolated from the sperm of trout found in the Rhine River.
1873–76	• Robert Koch investigated anthrax and developed techniques to view, grow, and stain organisms.
1880	• Louis Pasteur developed a method of attenuating or weakening pathogen agent of chicken cholera, so it would immunize and not cause disease.

TABLE 1.1 *(Continued)*

Periods	Important Discoveries/Events
1884	• Koch's postulates for testing whether a microbe is the causal agent of a disease.
	• Pasteur developed a rabies vaccine.
	• Christian Gram discovered Gram staining.
1900	• Rediscovery of Mendelian work by Hugo de Vries, Erich Von Tschermak, and Carl Correns.
1901	• Shigetane Ishiwatari, a Japanese biologist, first isolated the bacterium *Bacillus thuringiensis* (*Bt*) responsible for killing silkworms.
1902	• Human genetics born.
1905–08	• William Bateson and others demonstrated that some genes modify the action of other genes.
1907	• Researches on fruit flies, Thomas Hunt Morgan demonstrated that chromosomes have a definite function in heredity, establish mutation theory, and lead to a fundamental understanding of the mechanisms of heredity.
1909	• Wilhelm Johannsen coined the term "gene."
1910	• Thomas Morgan established that genes are carried on chromosomes.
1911	• Thomas Hunt Morgan began to map the positions of genes on chromosomes of the fruit fly.
	• Ernst Berliner isolated a bacterium that had killed a Mediterranean flour moth and rediscovered *Bt* and named it *Bacillus thuringiensis.*
1912	• Lawrence Bragg discovered that X-rays can be used to study the molecular structure of simple crystalline substances.
1915	• Berliner reported the existence of a crystal within *Bt.*
1926	• "The theory of the gene" published by Thomas Morgan.
1928	• Fredrick Griffiths noticed that a rough type of bacterium changed to a smooth type when an unknown "transforming principle" from the smooth type was present. Sixteen years later, Oswald Avery identified that "transforming principle" as DNA.
1938	• The term "Molecular Biology" was coined by Warren Weaver.
1941	• George Beadle and Edward Tatum discovered "one-gene-one-enzyme" hypothesis.
1943–53	• Cortisone (a 21-carbon steroid hormone), the first biotech product was manufactured in large amounts.
1944	• Waksman isolated streptomycin, an effective antibiotic for tuberculosis (TB).
1945–50	• Isolated animal cell cultures were grown in laboratories for the first time.

TABLE 1.1 *(Continued)*

Periods	Important Discoveries/Events
1947	• Barbara McClintock first reported on "transposable elements," known today as "jumping genes."
1950	• Discovery of Chargaff's Rules.
1953	• Double helix structure of DNA was published in Nature by James Watson and Francis Crick.
1953	• Gey developed the HeLa human cell line.
1957	• Francis Crick and George Gamov demonstrated "central dogma."
1962	• Watson and Crick shared the 1962 Nobel Prize for Physiology and Medicine with Maurice Wilkins for discovery of the double helical structure of DNA.
1966	• Genetic code cracked by Marshall Nirenberg, Heinrich Mathaei, and Severo Ochoa.
1967	• Arthur Kornberg and Dr. Severo Ochoa of New York University discovered "the mechanisms in the biological synthesis of deoxyribonucleic acid (DNA)."
1972	• Formation of first recombinant DNA molecule by Paul Berg.
1973	• Formation of world's first transgenic animal by Rudolf Jaenisch by introducing foreign DNA into its embryo created a transgenic mouse.
1978	• Herbert Boyer and his coworker constructed a synthetic version of the human insulin gene and transformed into *Escherichia coli*.
1980	• Discovery of polymerase chain reaction (PCR) by Kary Mullis and his coworker.
1982	• Human insulin, the first genetically engineered drug produced by bacteria was approved by the U.S. Food and Drug Administration.
	• Michael Smith at the University of British Columbia reported site-directed mutagenesis.
1983	• The first genetically engineered plant (tobacco) was created by Michael and his coworker.
1985	• Genetically engineered plants resistant to insects, viruses, and bacteria were field tested for the first time.
1986	• The Environmental Protection Agency (EPA) approved the release of the first genetically engineered crop, gene-altered tobacco plants.
1987	• Calgene, Inc. received a patent for the tomato polygalacturonase DNA sequence, used to produce an antisense RNA sequence that can extend the shelf life of fruit.
1990	• Human Genome Project, the international effort to map all of the genes in the human body, was launched.
	• Napoli and his coworkers demonstrated cosuppression of purple color in *Petunia* plants.

TABLE 1.1 *(Continued)*

Periods	Important Discoveries/Events
1993	• Kary Mullis won the Nobel Prize in Chemistry for inventing the technology of polymerase chain reaction (PCR).
1994	• The first genetically engineered food product, the Flavr Savr tomato, gained U.S. FDA approval.
1995	• Australian Genetic Manipulation Advisory Committee (GMAC) allows unrestricted, commercial release of a GM blue carnation in Australia.
1996	• The discovery of a gene associated with Parkinson's disease.
	• Ingard® insect-resistant (Bt) cotton is grown commercially in Australia.
1997	• Dolly—a cloned sheep from the cell of an adult ewe—was developed at Scotland's Roslin Institute.
	• Polly the first sheep cloned by nuclear transfer technology bearing a human gene was developed.
1998	• A rough draft of the human genome map is produced.
	• First complete animal genome of *Caenorhabditis elegans* worm was sequenced at the Sanger Institute, UK.
	• Forty million hectares of GM crops are planted globally, predominantly soy, cotton, canola, and corn.
	• Transgenic papaya cultivars namely SunUp and Rainbow resistant against PRS (papaya ring spot virus) were commercially released in Hawaii, USA.
2000	• Scientists at Celera Genomics and the Human Genome Project complete a rough draft of the human genome.
	• TILLING (targeting induced local lesions in genomes) was introduced in the model plant *Arabidopsis thaliana.*
	• The first entire plant genome of *Arabidopsis thaliana* was sequenced.
	• "Golden rice," a genetically modified variety with genes added which produce a vitamin A precursor, is created by Prof. Ingo Potrykus and coworkers.
2001	• Rice whole genome sequencing was completed.
2002	• Bollgard cotton became the first biotech crop technology approved for commercialization in India.
	• Sequencing of major genomes like mouse, chimpanzee, dog, and hundreds of other species were completed using SHOTGUN sequencing method.
2003	• Celera Genomics and National Institutes of Health (NIH) complete sequencing of the human genome.
2006	• The U.S. FDA approved a recombinant vaccine against human papillomavirus.
	• The 3-D structure of the human immunodeficiency virus, which causes AIDS was determined.

TABLE 1.1 *(Continued)*

Periods	Important Discoveries/Events
2008	• Dr. J. Craig Venter and his team replicate a bacterium's genetic structure entirely from laboratory chemicals.
2010	• Harvard researchers report building "lung on a chip" technology.
	• Dr. J. Craig Venter announces completion of "synthetic life" by transplanting synthetic genome capable of self-replication into a recipient bacterial cell.
	• Maize whole genome sequencing project was completed.
2011	• Advances in next generation sequencing enable human whole genome sequencing in less than 1 week for under $2000.
	• Tomato whole genome sequenced.
2012	• FDA issues draft rules for biosimilar drugs.
	• Barley and banana whole genome sequenced.
2013	• Bt brinjal was commercially released in Bangladesh.
	• Tobacco and chickpea whole genome sequenced.
	• Cisgenic apple and barley which confer scab resistance and improved phytase activity, respectively, were produced.
2014	• Whole genome sequencing of *Capsicum annuum* was completed.
2015	• Whole genome sequencing of wild potato.
	• "Magic" plant discovery could lead to growing food in space. Dr. Julia Bally and Prof. Peter Waterhouse have discovered a plant (ancient Australian native tobacco plant *Nicotiana benthamiana*) with huge genome properties that can have the potential to be the "laboratory rat" of the molecular plant world. This could open the door for things such as space-based food production.

1.3 PERIODS OF BIOTECHNOLOGY HISTORY

Ancient biotechnology (Pre-1800): Early applications and speculation

Classical biotechnology (1800–1950): Significant advances in the basic understanding of genetics

Modern biotechnology (1950 onward): Discovery of DNA, Recombinant DNA technology, genetically modified organisms, animal cloning, and stem cell research

1.3.1 ANCIENT BIOTECHNOLOGY (PRE-1800)

Most of the discoveries in biotechnology in the ancient period before 1800 were mainly based on the common observations of nature. The discovery of agriculture and the method of storing more viable and productive seeds for agricultural practices was possibly one of the first uses of biotechnology by humans. Ancient humans were hunters and food gatherers but agriculture made it possible for humans to settle at places where the farming conditions were optimum, e.g., availability of water, sunlight, and fertile land. Domestication of wild animals was a similar practice which made it possible for humans to quit hunting away from their homes. Domestication of plants started more than 10,000 years ago when humans started using plants and plant products as a reliable source of food. Rice, barley, and wheat were among the first domesticated plants. Selective domestication and breeding of wild animals were the beginning of observation and application of biotechnology principles. Around 250 BC, the Greeks started practicing crop rotation for maximum soil fertility and high agricultural yields (Wells, 1992).

Artificial selection for specific, desired traits has been used by humans for ages and has resulted in a variety of organisms like sweet corn, high milk-yielding cows, and hairless cats. These types of selections in which organisms with specific traits are chosen to breed for subsequent generations are dependent on naturally occurring traits. Corn is a remarkable example of a plant where the specific desired traits have been enhanced by selective breeding. Early teosinte plants (about 5000 BC) had small cobs with very few kernels but around 1500 AD, the corn cobs were more than five times the size and packed full highly nutritious kernels due to generations of selective breeding. Crossbreeding is also one of the earliest forms of biotechnological techniques used by humans to produce organisms with purebred parents of two different breeds or species. It is also called as designer crossbreeding intended to create an offspring that shares the traits of both parents, or producing an offspring with hybrid vigor. One of the earliest examples of crossbreeding in animals that benefitted humans is a mule, which is an offspring of a female horse and a male donkey. Mules contain 63 chromosomes while horse and donkey contain 64 and 62 chromosomes, respectively. Mules have been used for transportation, and farming for centuries due to their various positive characteristics like patience, long life span, faster speed, and more intelligence than donkeys.

After the discovery of agriculture and domestication of wild animals, humans came across new observations of food processing including the making of cheese and curd. In the course of development of biotechnology, yeast and bacteria have always been frontrunners among all organisms. Fermentation is thought to be discovered by an accident and since humans were not aware of how fermentation works, they thought it was a miracle or gift from their gods. Yeast is one of the oldest microorganisms that have been utilized by man for various purposes including the making of bread, alcohol, and vinegar. Due to its low pH, vinegar inhibits the growth of food-degrading microbes and therefore it has been used for food preservation in the ancient times.

In the ancient times, plants and plant parts were used as medicines for wound healing, fever, and infections. Ancient Egyptians were using honey for respiratory infections and wound healing. Honey is a natural antibiotic, so it can efficiently prevent wounds from being infected. In Ukraine, ancient people treated infected wounds with moldy cheese. These molds probably released natural antibiotics that destroyed the bacteria and prevented the infection. Around 600 BC, the Chinese people were using fungus-infested soybean curds to treat wounds and boils.

1.3.2 CLASSICAL BIOTECHNOLOGY

This phase started from the 1800s and extended through the first half of the 20th century. This is the phase of biotechnology when people started providing the scientific background to many of the common observations. A Dutch tradesman Antonie van Leeuwenhoek (1632–1723), while working in his draper's shop, observed minute organisms in the fabric using a simple microscope. His microscopic observations also included the microbes from the plaque between his own teeth and described his observations in a letter to the royal society, "I then most always saw, with great wonder, that in the said matter there were many very little living animalcules, very prettily a-moving. The biggest sort… had a very strong and swift motion, and shot through the water (or spittle) like a pike does through the water. The second sort oft-times spun round like a top… and these were far more in number." Although Leeuwenhoek did not have a formal education in science, he used to design magnifying lenses and microscopes. Using his microscopes, he discovered bacteria, protists, blood cells, sperm cells, and many other microscopic organisms (Gest, 2004). He is considered as the first microbiologist and widely known as the Father of Microbiology. His work opened a whole new world of microscopic life to the scientific community. Around the same time, an English natural philosopher Robert Hooke (1635–1703) discovered empty pores in a piece of cork and named them as cells. He designed different microscopes to observe microorganisms and recorded all his drawings and observations into a book *Micrographia,* which became an instant best seller (Hooke, 2003). The term "cell" got a wide recognition and Robert Hooke got the credit for discovering the building blocks of all life. Leeuwenhoek and Hooke independently laid the foundations of microbiology.

An English physician, Edward Jenner (1749–1823) successfully developed world's first vaccine for smallpox. His work is considered as one of the greatest for saving more lives than any other scientific work and therefore he is considered as the father of immunology. It was a common observation that the milkmaids who were infected with cowpox show immunity toward smallpox. Jenner postulated that the pus inside the blisters caused by the cowpox disease protected the milkmaids from smallpox. In 1796, Jenner proved his postulation by inducing immunity in an uninfected boy toward smallpox by exposing him to cowpox, a comparatively mild disease. Louis Pasteur (1822–95), a French microbiologist impacted the science of microbiology like no one else did. His pioneering work in the field of fermentation biology, vaccination, and pasteurization earned him the title "father of microbiology" (Feinstein, 2008). Before Pasteur, people used to believe in the obsolete doctrine of spontaneous generation or anomalous generation which states that living organisms are generated from nonliving matter or unrelated living. This theory was coherently synthesized by the Greek philosopher Aristotle and according to this theory, the insects arise from inanimate matter like soil and dust, and maggots arise from the dead flesh (Brack, 1998). Pasteur experimentally disproved the doctrine of spontaneous generation and showed that microorganisms could not generate in a sterilized flask without contamination.

Gregor John Mendel (1822–84), an Austrian Monk, was the first scientist to experimentally demonstrate that genetic information is essential for the development of phenotypic traits. Although the farmers were aware of the fact that crossbreeding can enhance certain desirable traits of different plants and animals, Mendel's experiments on pea plants (*Pisum sativum*) established the basic laws of heredity, now referred to as the laws of inheritance. Gregor Mendel, known as the "father of modern genetics," proposed that invisible internal factors of information are responsible for phenotypic traits and that these factors (later named as genes) are passed from one generation to another. However, Mendel did not get the due recognition for his groundbreaking work in his lifetime until other scientists like Hugo de Vries, Erich Von Tschermak, and Carl Correns validated his work years after his death. During the same time, a Scottish biologist Robert Brown discovered the nucleus in cells. Brown suggested that the nucleus plays a key role in fertilization and development of the embryo in plants. Brown not only named nucleus but also suggested that it may be the center of cellular development and creation (Oliver, 1913). In 1868, a Swiss biologist, Fredrich Miescher reported that white blood cells contain nuclein, a chemical compound made up of nucleic acids (Dahm, 2008). These two observations laid the foundations for molecular biology. In 1876, Robert Koch, a German physician discovered anthrax bacillus and elucidated its life cycle, thus launching the field of medical bacteriology. His work established that not only the bacteria but their spores are able to cause disease in healthy animals (Blevins and Bronze, 2010).

After Mendel, the principles of genetics and inheritance were redefined by an American geneticist T. H. Morgan whose discoveries formed the basis of the modern genetics. Morgan (1886–1945) studied the genetic characteristics of the fruit fly *Drosophila melanogaster* and showed that the genes are the mechanical basis of heredity which are carried on chromosomes. Morgan performed a test cross between the white-eyed male fly and red-eyed females. The resulting flies in F_1 generation had all red eyes. He then crossed males and females from the F_1 generation and observed a 3:1 ratio of red eyes to white eyes in the F_2 generation. This result was very similar to the results already shown by Mendel in peas. However, all the white-eyed F_2 flies were male and there were no white-eyed females at all. Correlation of genetic traits with male or female sexes had never been observed before (Miko, 2008). These results followed by many successive crosses led to the establishment of the chromosomal theory of inheritance.

"When I woke up just after dawn on September 28, 1928, I certainly didn't plan to revolutionise all medicine by discovering the world's first antibiotic, or bacteria killer. But, I suppose that was exactly what I did." These were the famous words by the British scientist Sir Alexander Fleming after the discovery of penicillin (Haven, 1994). Having witnessed the death of many soldiers from infected wounds in World War I, Fleming was actively searching for antibacterial agents. On 3 September 1928, Fleming returned to his laboratory after a long holiday with his family and noticed that one of the bacterial cultures in his laboratory was contaminated with a fungus and the bacterial colonies surrounding the fungus had been killed. Fleming grew the fungus in a pure culture and found that it produced a substance that killed a number of disease-causing bacteria. He identified the fungus as being from the *Penicillium* genus and named the antibacterial substance as penicillin. For his pioneering work in the discovery of antibiotics, Fleming shared Nobel Prize in Physiology or Medicine with Howard Florey and Ernst Boris Chain in 1945 (Hugh, 2002).

In 1941, two American geneticists George Beadle and Edward Tatum reported that genes regulate the biochemical events in cells (Beadle and Tatum, 1941). Beadle and Tatum induced mutations in the fungus *Neurospora crassa* by exposing to X-rays and demonstrated that these mutations caused changes in enzymes involved in metabolic pathways. Although the structure of genetic material was not discovered yet, these experiments led them to propose that genes directly control enzymatic processes in a cell. These findings led to "one gene-one enzyme hypothesis" and Beadle and Tatum were awarded Nobel Prize in Physiology or Medicine in 1958.

1.3.3 MODERN BIOTECHNOLOGY

After the World War II, many of the groundbreaking discoveries were reported which paved the path for modern biotechnology. The discovery that DNA (deoxyribonucleic acid) is the genetic material of most of the organisms and it can be isolated and manipulated has led to a new age of modern biotechnology. In 1953, Cambridge University scientists James D. Watson and Frances H. C. Crick for the first time cleared the mysteries around the genetic material, by proposing a structural model of DNA. According to this model, DNA is a double helix with two strands running in the antiparallel direction. Though DNA was discovered in 1869 by Friedrich Miescher, its role in determining the genetic inheritance was not studied (Dahm, 2008). Based on the X-ray diffraction image taken by Rosalind Franklin and Raymond Gosling in 1952, Watson and Crick determined that DNA is a double-helix polymer made of long chains of monomer nucleotides wound around each other (Watson and Crick, 1953a). According to this model, DNA replicated itself by separating into two individual strands, each of which serves as the template for a new DNA molecule. In 1962, James Watson and Francis Crick received the Nobel Prize in Physiology or Medicine for determination of the structure of DNA. After the structure of DNA was discovered, next question was that how the genetic material replicates itself and passes from one generation to another. Watson and Crick had already proposed the semiconservative model of replication but their hypothesis was not experimentally proved so far (Watson and Crick, 1953b). In 1958, two American scientists Matthew Meselson and Franklin Stahl allowed *Escherichia coli* to grow for several generations in a medium with ^{15}N. The DNA of the bacterial cells grown in ^{15}N medium had a higher density than the cells grown in the normal ^{14}N medium. After one replication, the density of the DNA was found to be of intermediate density, thus supporting the semiconservative mode of replication (Meselson and Stahl, 1958).

After elucidation of the structure of DNA, the next big question was how this double-stranded structure controls all the cellular processes like heredity and metabolism. Serious efforts were being made to understand the link between DNA and proteins. In 1961, Francis Crick, Sydney Brenner, Leslie Barnett, and R. J. Watts-Tobin demonstrated that three bases of DNA code for each amino acid in the genetic code (Crick et al., 1961). In the same year, Marshall Nirenberg and Heinrich J. Matthaei used a cell-free system to express a chain of phenylalanine amino acids by translation of poly-uracil RNA sequence, thus elucidating the nature of genetic codon (Nirenberg and Matthaei, 1961). Similarly, Severo Ochoa's laboratory demonstrated that the poly-adenine RNA sequence coded for the poly-lysine polypeptide (Speyer et al., 1963). Subsequent work by Har Gobind Khorana established the genetic codes for all the acids as well as stop codons. In 1964, an American biochemist Robert W. Holley determined the structure of transfer RNA, the adapter molecule that mediates the translation of RNA into protein (Holley et al., 1965). In 1968, Khorana, Holley, and Nirenberg were awarded the Nobel Prize in Physiology or Medicine for their work on genetic codes and elucidation of the structure of tRNAs. Khorana was the first scientist to chemically synthesize oligonucleotides and a full gene by assembled small fragments of oligonucleotides together with the help of DNA polymerase and ligase (Khorana, 1979).These custom-designed pieces of oligonucleotides are widely used in research labs for gene amplification, cloning, and sequencing, and have become an integral and indispensable part of molecular biology.

Restriction endonucleases are the basic molecular tools of modern biotechnology. The foundation for the discovery of restriction enzymes was laid by Luria, Anderson, Bertani, and Felix in the early 1950s (Luria and Human, 1952; Anderson and Felix, 1952; Bertani and Weigle, 1953). The term restriction enzyme originated from the studies of the phenomenon of host-controlled restriction and modification of a bacterial virus. Arber and Meselson discovered that the restriction is caused by an enzymatic cleavage of the phage DNA, and the enzyme was termed as restriction enzyme (Arber and Linn, 1969; Meselson and Yuan, 1968). HindII was the first type II restriction enzyme discovered, it was isolated from *Haemophilus influenzae* and characterized by Hamilton O. Smith and Thomas Kelly in 1970 (Smith and Welcox; 1970, Kelly and Smith, 1970). American scientist Daniel Nathans and his graduate student Kathleen Danna demonstrated that cleavage of Simian virus 40 DNA by restriction enzymes produces specific

DNA fragments that can be separated efficiently by gel electrophoresis (Danna and Nathans, 1971). For the discovery of restriction endonucleases, Werner Arber, Hamilton Smith, and Daniel Nathans were awarded the 1978 Nobel Prize in Physiology or Medicine.

In 1972, an American biochemist Paul Berg used restriction enzymes and DNA ligases to create the first recombinant DNA molecule. He generated recombinant DNA molecules from the Simian virus SV40 and lambda virus (Jackson et al., 1972). Herbert Boyer and Stanley N. Cohen were the first to introduce recombinant DNA molecules into a bacterial cell. They used a restriction enzyme to cut a plasmid and ligated a kanamycin resistance gene with the plasmid. They introduced this recombinant plasmid into a bacterial cell by transformation to create the kanamycin-resistant bacteria (Cohen and Chang, 1973). It was the first genetically modified organism ever produced. A German scientist, Rudolf Jaenisch created the first transgenic animal in 1973 by introducing SV40 viral DNA into mouse embryos but the transgene was not passed on to the offspring (Jaenisch and Mintz, 1974). In 1981, Frank Ruddle, Frank Constantini, and Elizabeth Lacy injected purified DNA into a mouse embryo and showed transmission of the transgenes to the subsequent progenies (Gordon and Ruddle, 1981). The first genetically engineered plant was developed by Michael W. Bevan, Richard B. Flavell, and Mary-Dell Chilton in 1983. They introduced a recombinant DNA containing antibiotic resistance gene into tobacco using agrobacterium-mediated transformation. These discoveries paved the way to a whole new area of biotechnology known as "genetically modified organisms."

Among the many discoveries of the 20th century in the field of biotechnology, polymerase chain reaction (PCR) has to be among the top due to its applications in molecular biology and medicine. It is a revolutionary method developed by the American biochemist Kary Mullis in the 1980s. PCR is an enzymatic process catalyzed by thermostable Taq DNA polymerase used to synthesize new DNA from the pre-existing template DNA (Mullis et al., 1986). It is used to amplify a few copies of DNA across several orders of magnitude, generating millions of copies. For this landmark technique of DNA synthesis, Kary Mullis was awarded Nobel Prize in Chemistry in 1993 Before Mullis, several scientific discoveries by many notable scientists paved the way to polymerase chain reaction. It includes chemical synthesis of oligonucleotides as well as fully functional human genes by H. Gobind Khorana (Khorana et al., 1976). Khorana used sequence-specific oligonucleotides as primers, templates, and building blocks for the gene in a reaction catalyzed by DNA polymerase. In 1969, an American microbiologist Thomas D. Brock isolated a new species of bacterium, *Thermus aquaticus*, from a hot spring in Yellowstone National Park (Brock and Freeze, 1969). This bacterium became a standard source of enzymes which are able to withstand higher temperatures. In 1971, one of the researchers in Khorana's lab named Kjell Kleppe envisioned a PCR-like process and described how a two-primer and polymerase system might be used to replicate a specific segment of DNA (Kleppe et al., 1971). In 1976, a thermostable DNA polymerase was isolated from the bacterium *T. aquaticus* and was found to be active at temperatures above 75°C (Chien et al., 1976). Discovered as a simple technique to amplify specific regions of DNA, a number of variants of this technique have evolved over the decades. These variants include assembly PCR, inverse PCR, and quantitative real-time PCR used for DNA assembly, identification of new genes, and studying the gene expression.

DNA sequencing is an indispensable technique for basic biological research and several applied fields like systemic and evolutionary biology, forensic medicine, and medical diagnosis. History of sequencing began when Sanger's research on insulin demonstrated the importance of sequence in biological macromolecules. It was the first time when proteins were shown to be composed of linear polypeptides formed by joining amino acid residues in a defined order (Sanger, 1949). In 1975, Sanger introduced his "plus and minus" method for DNA sequencing which led to the first modern generation method for DNA sequencing (Sanger and Coulson, 1975). The method involved the analysis of oligonucleotide products of DNA polymerase reactions that extended a specific primer annealed to a single-stranded DNA template. In 1977, Maxam and Gilbert developed a DNA sequencing method that was similar to the Sanger's method in using polyacrylamide gels to resolve the bands but was different in the way of generation of products ending in a specific base (Maxam and Gilbert, 1977). The complete sequence of bacteriophage ϕX was determined by the plus and minus method and was published in 1977. This was followed by the complete genome sequence of the Simian virus SV40 in 1978 (Fiers et al., 1978). In 1980, Frederick Sanger, Walter Gilbert, and Paul Berg were awarded the Nobel Prize in Chemistry for their pioneering work in DNA sequencing (Berg, 2010). This was the second Nobel Prize for Sanger as he had already received his first Nobel Prize in Chemistry in 1958 for his work in the structural analysis of proteins, especially insulin. In 1985, the laboratory of Leroy Hood at Caltech, in collaboration with Applied Biosystems (ABI), reported the automation of DNA sequencing which was easier and faster than the manual sequencing (Smith et al., 1985). In February 2001, the draft human genome sequences were published in the same week in *Science* and *Nature* journals using two different technologies and after that whole new world of genomics opened up for researchers (Venter et al., 2001; Lander et al., 2001).

Genetic engineering is the basis of modern biotechnology and it involves direct manipulation of genetic material of an organism, using molecular biology techniques. DNA may be inserted into the host genome by first isolating the genetic material of interest from an organism using molecular cloning methods or by synthesizing the DNA and inserting it into the host organism. Genes of interest can be manipulated, overexpressed, deleted, or silenced with different techniques available in the molecular biology. Genetic engineering has been used to express proteins from humans in organisms that normally cannot synthesize them. The expression of first human protein somatostatin in *E. coli* was carried out in Genentech, a company founded by Herbert Boyer and Robert Swanson. Human insulin was synthesized in bacteria in 1979 and used as a treatment for diabetes in 1982 (Ladisch and Kohlmann, 1992). Genetically engineered organisms have been produced by introducing the DNA modifications of the organism to enhance certain desired traits like disease resistance, nutrient quality, increased crop yield, and delayed ripening. Bt cotton is a genetically modified cotton variety, which has been genetically engineered to produce an insecticide for boll-worm. Bt cotton was created by expressing the Cry genes from a bacterium *Bacillus thuringiensis* which encode for Bt toxin. When insects consume the cotton plant, the Cry toxins get dissolved in the insect gut due to the high pH level and this leads to the death of the insects. Bt cotton has several advantages over non-Bt cotton, like the increase in the yield of cotton due to effective control of insects, early maturation of plants, and reduction in the environmental pollution due to minimal use of pesticides. One of the well-known genetically modified foods is "Flavr Savr tomato," which was the first commercially grown GM food to be granted a license for human consumption. To make it more resistant to rotting, an antisense gene which interferes with the production of the enzyme polygalacturonase was introduced into the tomato plant. This enzyme normally degrades pectin in the cell walls resulting in the softening of the fruit which in turn makes the fruit more susceptible to fungal and bacterial diseases. This genetic modification delayed the ripening of Flavr Savr tomato with changing the flavor and nutrient values of the fruit (Martineau, 2001). In 2000, vitamin A-enriched rice variety, golden rice, was the first food with improved nutrient value (Ye et al., 2000).

One of the recent advances in biotechnology is somatic cell nuclear transfer, which is a laboratory technique for creating a viable embryo from a somatic cell and an egg cell. The technique involves implanting a donor nucleus from a somatic cell into an enucleated oocyte (egg cell). Dolly the sheep became hugely famous for being the first successful mammal produced from reproductive cloning (Li et al., 2009). Somatic cell nuclear transplantation is currently a focus of study in stem cell research. Scientists aim to exploit this technique to obtain pluripotent cells from a cloned embryo. Since these cells are genetically similar to the donor organism, it creates an opportunity to generate patient-specific pluripotent cells, which can be used as potential therapies for different diseases (Lomax and DeWitt, 2013). Therapeutic cloning involves potential use of somatic cell nuclear transfer in regenerative medicine to cure the damaged tissues in a patient. Regenerative medicine uses tissue engineering and molecular biology techniques to replace, modify, or regenerate the cells, tissues, or organs in humans to restore the normal function. It involves stimulating the body's own repair mechanism to restore previously irreparable tissues and organs. This field holds the promise of engineering damaged tissues and organs via stimulating the body's own repair mechanisms to functionally heal previously irreparable tissues or organs (Lomax and DeWitt, 2013). Umbilical cord blood contains stem cells that can be used to treat hematopoietic diseases, genetic disorders, brain injury, and type 1 diabetes in humans (Lomax and DeWitt, 2013; Haller et al., 2008; Vendrame et al., 2006).

Functional genomics is a core branch of modern biotechnology involved in the study of gene and protein expression and functions mainly focusing on regulation of gene expression at transcription and translational levels, protein function, protein–protein interactions, etc. The primary goal of functional genomics is to understand the correlation between the genome and its phenotype in a given organism. Some of the recent approaches used to study a gene function include mutations, yeast-2-hybrid assay, quantitative real-time PCR, microarray, and gene silencing. Prior to the discovery of PCR, Southern blotting was being used to detect specific DNA sequences in DNA samples. Southern blotting involves the gel electrophoresis of the DNA sample and subsequent transfer of the DNA fragments to a filter membrane, followed by detection of the fragments by hybridization. This technique was developed by the British biologist E. M. Southern in 1975 (Southern, 1975). Other blotting methods like western blot and northern blot utilize similar principles to detect proteins and RNA (Towbin et al., 1979). One of the common techniques used in functional genomics is the yeast-2-hybrid assay used to study protein–protein interactions. This method was illustrated by American biologist Stanley Fields and Ok-Kyu Song in 1989 (Fields and Song, 1989). This system uses the GAL4 activated transcription of a protein involved in galactose metabolism, which is the basis of selection. Currently, the yeast-2-hybrid system is the most commonly used screening method for protein–protein interactions in eukaryotes (Gietz et al., 1997).

Serial analysis of gene expression (SAGE) is a technique used to study levels of mRNA in a sample of interest in the form of small tags that correspond to fragments of those transcripts. Although several variants have been developed over the decades, the original technique was developed by Victor Velculescu in the Johns Hopkins University in 1995 (Velculescu et al., 1995). SAGE is based on sequencing of mRNA output and not on the hybridization of mRNA, therefore the transcription levels are measured more quantitatively than the microarray technology. Massive parallel signature sequencing (MPSS) is another technique used to identify and quantify mRNA levels in a given sample. It produces data in a similar way like SAGE but employs a different method for analysis which involves a series of biochemical and sequencing steps. After Kary Mullis, several modifications were introduced into the basic PCR method to customize this standard technique for different purposes. Real-time quantitative PCR has become very accurate and highly sensitive method for the quantification of nucleic acids. In 1996, Applied Biosystems (ABI) made real-time PCR commercially available with the introduction of the 7700 instrument (Heid et al., 1996). Unlike conventional PCR, real-time PCR monitors the amplification of a targeted DNA during the PCR (in real-time), and not at the end of the reaction. Another technology devised for gene expression at mRNA level is microarray which involves the analysis of global gene expression in a given sample. A microarray (also known as a biochip) is a two-dimensional array of microscopic DNA spots attached to a solid surface used to measure the expression of a large number of genes simultaneously. The microarray is generally used for global gene expression analysis, the discovery of new genes, disease diagnosis, and drug discovery. The theory and methodology of microarrays were first introduced as antibody microarrays by Tse Wen Chang in 1983 in a scientific publication and various patents (Chang, 1983). Gene expression analysis using DNA microarray started gaining momentum after the method was reported by two Stanford scientists Ron Davis and Pat Brown in 1995 (Schena et al., 1995). Till this date, a number of companies including Affymetrix, Agilent, Applied Microarrays, and Illumina are continuously updating microarray technology for better efficiency and sensitivity.

RNA interference is considered a breakthrough technology of the recent times, which revolutionized the field of modern biotechnology. RNAi is a gene-silencing process where RNA molecules inhibit the gene expression either by causing the degradation of targeted mRNA molecules or by halting the translation. Early insights of RNAi-like phenomenon were discovered by

Napoli and Jorgensen in 1990 while working on anthocyanin biosynthesis in petunias (Napoli et al., 1990). They attempted to generate violet petunias by overexpression of chalcone synthase gene in petunias which unexpectedly resulted in the production of white petunias. It was found that the exogenously introduced gene caused the cosuppression of chalcone synthase gene. Two years later, Romano and Macino reported a similar phenomenon of gene silencing called quelling in *N. crassa* (Romano and Macino, 1992). In 1998, Fire and Mello demonstrated that double-stranded RNA could directly silence genes through RNA interference in the nematode *Caenorhabditis elegans* (Fire et al., 1998). The Nobel Prize in Physiology or Medicine 2006 was awarded jointly to Fire and Mello for the discovery of RNA interference. Earlier it was thought that the dsRNA had to unwind so that the antisense strand can bind to the target mRNA but the full-length antisense strand could never be detected by researchers. Hamilton and Baulcombe detected 25 nucleotides long antisense RNA fragments required for the target specificity during mRNA degradation (Hamilton and Baulcombe, 1999). To determine whether the 21–23 nt dsRNAs can cause the gene silencing via RNAi pathway, *Drosophila* cell extracts were incubated with chemically synthesized 21–22 nt dsRNAs, targeting a luciferase gene transcript. It was found that these siRNAs were efficiently able to mediate the cleavage of the target mRNA (Elbashir et al., 2001).

One of the advanced approaches used in the current molecular biology research is specific editing of the genome of an organism (or a single cell) to study the gene function. This mode of genetic engineering is known as "genome editing," which was selected by Nature Methods as the 2011 Method of the Year. Site-directed mutagenesis is one of the gene editing methods that uses either phage- or PCR-mediated methods to generate a desired mutation in a DNA sequence (Ling and Robinson, 1997). The basic procedure involves a synthetic primer that contains a desired point mutation. This single-strand primer is then extended using a DNA polymerase which replicates the rest of the gene and contains the mutated site. The mutated gene is then cloned in an expression vector and introduced into the desired host organism to study the effect of mutation on the phenotype of the organism. In the last decade, genetically engineered nucleases have emerged as promising tools for genome editing. Recently, zinc-finger nucleases (ZFNs), transcription activator-like effector nucleases (TALENs), and the clustered regularly interspersed short palindromic repeats (CRISPR)/Cas systems are being used for specific genomic editing to enhance various desired traits in animals and plants (Urnov et al., 2010). For example, ZFN system was used to precisely and site-specifically insert a transgene expression cassette to develop herbicide-tolerance in corn crop (Espinoza et al., 2013). Also, Shan et al. developed aromatic rice via TALEN, and developed wheat plants resistant to the powdery mildew via the CRISPR/Cas system (Telem et al., 2013; Shan et al., 2014).

KEYWORDS

- **biotechnology**
- **ancient biotechnology**
- **classical genetics**
- **discovery of DNA**
- **genetic engineering**

REFERENCES

Anderson, E.; Felix, A. Variation in Vi Phage II of *Salmonella typhi*. *Nature* **1952,** *170*, 492–494.

Arber, W.; Linn, S. DNA Modification and Restriction. *Ann. Rev. Biochem.* **1969,** *38*, 467–500.

Beadle, G. W.; Tatum, E. L. Genetic Control of Biochemical Reactions in *Neurospora*. *Proc. Natl. Acad. Sci. U. S. A.* **1941,** *27*, 499–506.

Berg, P. *The Nobel Prize in Chemistry 1980 Paul Berg, Walter Gilbert, Frederick Sanger*; 2010.

Bertani, G.; Weigle, J. Host Controlled Variation in Bacterial Viruses. *J. Bacteriol.* **1953,** *65*, 113.

Blevins, S. M.; Bronze, M. S. Robert Koch and the 'Golden Age'of Bacteriology. *Int. J. Infect. Dis.* **2010,** *14*, e744–e751.

Brack, A. *The Molecular Origins of Life: Assembling Pieces of the Puzzle*; Cambridge University Press: Cambridge, UK, 1998.

Brock, T. D.; Freeze, H. *Thermus aquaticus* gen. n. and sp. n., a Nonsporulating Extreme Thermophile. *J. Bacteriol.* **1969,** *98*, 289–297.

Chang, T.-W. Binding of Cells to Matrixes of Distinct Antibodies Coated on Solid Surface. *J. Immunol. Methods* **1983,** *65*, 217–223.

Chien, A.; Edgar, D. B.; Trela, J. M. Deoxyribonucleic Acid Polymerase from the Extreme Thermophile *Thermus aquaticus*. *J. Bacteriol.* **1976,** *127*, 1550–1557.

Cohen, S. N.; Chang, A. C. Recircularization and Autonomous Replication of a Sheared R-factor DNA Segment in *Escherichia coli* Transformants. *Proc. Natl. Acad. Sci. U. S. A.* **1973,** *70*, 1293–1297.

Crick, F.; Barnett, L.; Brenner, S.; Watts-Tobin, R. J. (1961). General Nature of the Genetic Code for Proteins. *Nature* **1961,** *192*, 1227–1232.

Dahm, R. Discovering DNA: Friedrich Miescher and the Early Years of Nucleic Acid Research. *Hum. Genet.* **2008,** *122*, 565–581.

Danna, K.; Nathans, D. Specific Cleavage of Simian Virus 40 DNA by Restriction Endonuclease of *Hemophilus influenzae*. *Proc. Natl. Acad. Sci. U. S. A.* **1971,** *68*, 2913–2917.

Elbashir, S. M.; Lendeckel, W.; Tuschl, T. RNA Interference Is Mediated by 21-and 22-Nucleotide RNAs. *Genes Dev.* **2001,** *15*, 188–200.

Feinstein, S. *Louis Pasteur: The Father of Microbiology;* Enslow Publishers, Inc.: Berkeley Heights, NJ, 2008.

Fields, S.; Song, O.-K. A Novel Genetic System to Detect Protein Protein Interactions. *Nature* **1989,** *340* (6230), 245–246.

Fiers, W.; Contreras, R.; Haegemann, G.; Rogiers, R.; Van De Voorde, A.; Van Heuverswyn, H.; Van Herreweghe, J.; Volckaert, G.; Ysebaert, M. Complete Nucleotide Sequence of SV40 DNA. *Nature* **1978,** *273*, 113–120.

Fire, A.; Xu, S.; Montgomery, M. K.; Kostas, S. A.; Driver, S. E.; Mello, C. C. Potent and Specific Genetic Interference by Double-stranded RNA in *Caenorhabditis elegans*. *Nature* **1998,** *391*, 806–811.

Gest, H. The Discovery of Microorganisms by Robert Hooke and Antoni Van Leeuwenhoek, Fellows of the Royal Society. *Notes Rec. Royal Soc.* **2004,** *58*, 187–201.

Gietz, R. D.; Robbins, A.; Graham, K. C.; Triggs-Raine, B.; Woods, R. A. Identification of Proteins that Interact with a Protein of Interest: Applications of the Yeast Two-hybrid System. In *Novel Methods in Molecular and Cellular Biochemistry of Muscle;* Springer, U.S. 1997; pp 67–79.

Gordon, J. W.; Ruddle, F. H. Integration and Stable Germ Line Transmission of Genes Injected into Mouse Pronuclei. *Science* **1981,** *214*, 1244–1246.

Haller, M. J.; Viener, H.-L.; Wasserfall, C.; Brusko, T.; Atkinson, M. A.; Schatz, D. A. Autologous Umbilical Cord Blood Infusion for Type 1 Diabetes. *Exp. Hematol.* **2008,** *36*, 710–715.

Hamilton, A. J.; Baulcombe, D. C. A Species of Small Antisense RNA in Posttranscriptional Gene Silencing in Plants. *Science* **1999,** *286*, 950–952.

Haven, K. F. *Marvels of Science* : 50 Fascinating 5-Minute Reads. Littleton, Colo: Libraries Unlimited, 1994, 182.

Heid, C. A.; Stevens, J.; Livak, K. J.; Williams, P. M. Real Time Quantitative PCR. *Genome Research* **1996,** *6*, 986–994.

Holley, R. W.; Apgar, J.; Everett, G. A.; Madison, J. T.; Marquisee, M.; Merrill, S. H.; Penswick, J. R.; Zamir, A. Structure of a Ribonucleic Acid. *Science* **1965,** *147*, 1462–1465.

Hooke, R. *Micrographia: or Some Physiological Descriptions of Minute Bodies Made by Magnifying Glasses, with Observations and Inquiries Thereupon.* Courier Corporation, Ney York, 2003.

Hugh, T. B. Howard Florey, Alexander Fleming and the Fairy Tale of Penicillin. *Med. J. Aust.* **2002,** *177*, 52.

Jackson, D. A.; Symons, R. H.; Berg, P. Biochemical Method for Inserting New Genetic Information into DNA of Simian Virus 40: Circular SV40 DNA Molecules Containing Lambda Phage Genes and the Galactose Operon of *Escherichia coli*. *Proc. Natl. Acad. Sci. U. S. A.* **1972,** *69*, 2904–2909.

Jaenisch, R.; Mintz, B. Simian Virus 40 DNA Sequences in DNA of Healthy Adult Mice Derived from Preimplantation Blastocysts Injected with Viral DNA. *Proc. Natl. Acad. Sci. U. S. A.* **1974,** *71*, 1250–1254.

Kelly, T. J.; Smith, H. O. A Restriction Enzyme from *Hemophilus influenzae*: II. Base Sequence of the Recognition Site. *J. Mol. Biol.* **1970,** *51*, 393–409.

Khorana, H. G. Total Synthesis of a Gene. *Science* **1979,** *203*, 614–625.

Khorana, H. G.; Agarwal, K.; Besmer, P.; Büchi, H.; Caruthers, M.; Cashion, P.; Fridkin, M.; Jay, E.; Kleppe, K.; Kleppe, R. Total Synthesis of the Structural Gene for the Precursor of a Tyrosine Suppressor Transfer RNA from *Escherichia coli*. 1. General Introduction. *J. Biol. Chem.* **1976,** *251*, 565–570.

Kleppe, K.; Ohtsuka, E.; Kleppe, R.; Molineux, I.; Khorana, H. Studies on Polynucleotides: XCVI. Repair Replication of Short Synthetic DNA's as Catalyzed by DNA Polymerases. *J. Mol. Biol.* **1971,** *56*, 341–361.

Ladisch, M. R.; Kohlmann, K. L. Recombinant Human Insulin. *Biotechnol. Prog.* **1992,** *8*, 469–478.

Lander, E. S.; Linton, L. M.; Birren, B.; Nusbaum, C.; Zody, M. C.; Baldwin, J.; Devon, K.; Dewar, K.; Doyle, M.; FitzHugh, W. Initial Sequencing and Analysis of the Human Genome. *Nature* **2001,** *409*, 860–921.

Li, J.; Liu, X.; Wang, H.; Zhang, S.; Liu, F.; Wang, X.; Wang, Y. Human Embryos Derived by Somatic Cell Nuclear Transfer Using an Alternative Enucleation Approach. *Cloning Stem Cells* **2009,** *11*, 39–50.

Ling, M. M.; Robinson, B. H. Approaches to DNA Mutagenesis: An Overview. *Anal. Biochem.* **1997,** *254*, 157–178.

Lomax, G. P.; DeWitt, N. D. Somatic Cell Nuclear Transfer in Oregon: Expanding the Pluripotent Space and Informing Research Ethics. *Stem Cells Dev.* **2013,** *22*, 25–28.

Luria, S.; Human, M. L. A Nonhereditary, Host-induced Variation of Bacterial Viruses. *J. Bacteriol.* **1952,** *64*, 557.

Martineau, B. *First Fruit: The Creation of the Flavr Savr Tomato and the Birth of Biotech Foods*; Schaum: New York, 2001; p 259.

Maxam, A. M.; Gilbert, W. A New Method for Sequencing DNA. *Proc. Natl. Acad. Sci. U. S. A.* **1977,** *74*, 560–564.

Meselson, M. and Stahl, F. W. The Replication of DNA in *Escherichia coli*. *Proc. Natl. Acad. Sci. U. S. A.* **1958,** *44*, 671–682.

Meselson, M.; Yuan, R. DNA Restriction Enzyme from *E. coli*. *Nature* **1968,** *217*, 1110.

Miko, I. Thomas Hunt Morgan and Sex Linkage. *Nat. Educ.* **2008,** *1*, 143.

Mullis, K.; Faloona, F.; Scharf, S.; Saiki, R.; Horn, G.; Erlich, H. Specific Enzymatic Amplification of DNA In Vitro: the Polymerase Chain Reaction. *Cold Spring Harbor Symp. Quant. Biol.* **1986,** *51*, 263–273.

Napoli, C.; Lemieux, C.; Jorgensen, R. Introduction of a Chimeric Chalcone Synthase Gene into Petunia Results in Reversible Co-suppression of Homologous Genes in Trans. *Plant Cell* **1990,** *2*, 279–289.

Nirenberg, M. W.; Matthaei, J. H. The Dependence of Cell-free Protein Synthesis in *E. coli* upon Naturally Occurring or Synthetic Polyribonucleotides. *Proc. Natl. Acad. Sci. U. S. A.* **1961,** *47*, 1588–1602.

Oliver, F. W. Makers of British Botany. In *JSTOR*, 1913.

Romano, N.; Macino, G. Quelling: Transient Inactivation of Gene Expression in *Neurospora crassa* by Transformation with Homologous Sequences. *Mol. Microbiol.* **1992,** *6*, 3343–3353.

Sanger, F. The Terminal Peptides of Insulin. *Biochem. J.* **1949,** *45*, 563.

Sanger, F.; Coulson, A. R. A Rapid Method for Determining Sequences in DNA by Primed Synthesis with DNA Polymerase. *J. Mol. Biol.* **1975,** *94*, 441–448.

Schena, M.; Shalon, D.; Davis, R. W.; Brown, P. O. Quantitative Monitoring of Gene Expression Patterns with a Complementary DNA Microarray. *Science* **1995,** *270*, 467–470.

Smith, H. O.; Welcox, K. A Restriction Enzyme from *Hemophilus influenzae*: I. Purification and General Properties. *J. Mol. Biol.* **1970,** *51*, 379–391.

Smith, L. M.; Sanders, J. Z.; Kaiser, R. J.; Hughes, P.; Dodd, C.; Connell, C. R.; Heiner, C.; Kent, S.; Hood, L. E. Fluorescence Detection in Automated DNA Sequence Analysis. *Nature* **1985,** *321*, 674–679.

Southern, E. M. Detection of Specific Sequences among DNA Fragments Separated by Gel Electrophoresis. *J. Mol. Biol.* **1975,** *98*, 503–517.

Speyer, J. F.; Lengyel, P.; Basilio, C.; Wahba, A. J.; Gardner, R. S.; Ochoa, S. Synthetic Polynucleotides and the Amino Acid Code. In *Cold Spring Harbor Symposia on Quantitative Biology;* Cold Spring Harbor Laboratory Press: Cold Spring Harbor, New York, 1963; pp 559–567.

Towbin, H.; Staehelin, T.; Gordon, J. Electrophoretic Transfer of Proteins from Polyacrylamide Gels to Nitrocellulose Sheets: Procedure and Some Applications. *Proc. Natl. Acad. Sci. U. S. A.* **1979,** *76*, 4350–4354.

Urnov, F. D.; Rebar, E. J.; Holmes, M. C.; Zhang, H. S.; Gregory, P. D. Genome Editing with Engineered Zinc Finger Nucleases. *Nat. Rev. Genet.* **2010,** *11*, 636–646.

Velculescu, V. E.; Zhang, L.; Vogelstein, B.; Kinzler, K. W. Serial Analysis of Gene Expression. *Science* **1995,** *270*, 484–487.

Vendrame, M.; Gemma, C.; Pennypacker, K. R.; Bickford, P. C.; Sanberg, C. D.; Sanberg, P. R.; Willing, A. E. Cord Blood Rescues Stroke-induced Changes in Splenocyte Phenotype and Function. *Exp. Neurol.* **2006,** *199*, 191–200.

Venter, J. C.; Adams, M. D.; Myers, E. W.; Li, P. W.; Mural, R. J.; Sutton, G. G.; Smith, H. O.; Yandell, M.; Evans, C. A.; Holt, R. A. The Sequence of the Human Genome. *Science* **2001,** *291*, 1304–1351.

Verma, A. S.; Agrahari, S.; Rastogi, S.; Singh, A. Biotechnology in the Realm of History. *J. Pharm. Bioallied Sci.* **2011,** *3*, 321.

Watson, J. D.; Crick, F. H. Molecular Structure of Nucleic Acids. *Nature,* **1953a,** *171*, 737–738.

Watson, J. D.; Crick, F. H. The Structure of DNA. In *Cold Spring Harbor Symposia on Quantitative Biology;* Cold Spring Harbor Laboratory Press: Cold Spring Harbor, New York, 1953b; pp 123–131.

Wells, B. (1992) Agriculture in Ancient Greece. In *Proceedings of the Seventh International Symposium at the Swedish Institute of Athens,* 16–17 May 1990.

Ye, X.; Al-Babili, S.; Klöti, A.; Zhang, J.; Lucca, P.; Beyer, P.; Potrykus, I. Engineering the Provitamin A (β-Carotene) Biosynthetic Pathway into (Carotenoid-free) Rice Endosperm. *Science* **2000,** *287*, 303–305.

CHAPTER 2

SCOPE AND IMPORTANCE OF PLANT BIOTECHNOLOGY IN CROP IMPROVEMENT

ASHISH RANJAN* and DEVANSHI KHOKHANI

Department of Plant Pathology, University of Wisconsin—Madison, 583 Russell Labs, 1630 Linden Dr. Madison, Wisconsin 53706, United States

**Corresponding author. E-mail: ashishranjan2005@gmail.com*

CONTENTS

ABSTRACT

Plant biotechnology involves breeding to improve plants for various reason such as increasing yield and quality, heat and drought resistance, resistance to phytopathogens, herbicide and insect resistance, increasing biomass for biofuel production, and enhancing the nutritional quality of the crops. This chapter presents a brief history of breeding, disadvantages of conventional breeding while advantages of non-conventional breeding techniques such as molecular marker-assisted breeding techniques and molecular farming. Tissue culture, as a form of large-scale plant micropropagation and its advantages along with the future of breeding program based on high throughput sequencing platform called genomic assisted molecular farming have been also discussed.

2.1 INTRODUCTION OF PLANT BREEDING

Plant breeding is a technique used for development of desired varieties of plants by manipulating its genetics. It is one of the oldest practices of human civilization. It started with domestication of plants approximately 10,000 years ago. Selection is an important step of breeding. In old times, it is believed that farmers used to select the best seeds from the plant which is more desirable and used for further propagation. Breeding is generally thought as a sexual way of developing new improved varieties but asexual reproduction is also used to develop the new varieties such as using tissue culture.

Breeders play an important role in breeding program. They are the professionals who conduct the plant breeding for the desired characters. Breeders follow different steps to come up with the desired plants. They are as follows:

1. Finding out a clear objective for the breeding program,
2. Collecting or creating the varieties (such as by mutation) from which trait of interest can be transferred called donor plant,
3. Introgression of the desired trait from donor to the recipient plant, this can be done by using conventional or nonconventional method,
4. Selection of the desired plants,
5. Evaluation of the varieties in fields at different locations for few years to find out the best suitable variety for commercial release,
6. Seed certification and approval of the same from the designated seed certifying agencies.

Work of Mendel and other scientists established that plant traits are transferable which were later on called genes. These studies helped breeder to transfer the traits in controlled manner. The variety produced through breeding is expected to have the stable traits and are transferable from one generation to other generation. A good breeder should have very good observation quality and sense of judgement while selecting the varieties.

The goal of the breeding is to perform focused and targeted gene manipulations. Breeding can have several objectives such as increasing yield and quality, resistance to lodging (bending or breaking over of plants before harvest), and shattering (fall out of seeds before harvest), heat and drought resistance, resistance to phytopathogens, herbicide and insect resistance, and increasing biomass for biofuel production. Plant breeding is needed to enhance the nutritional quality of the crops and for healthy living of humans. For example, golden rice, having high vitamin A content (Ye et al., 2000; Beyer et al., 2002). This variety might reduce vitamin A deficiency, a major problem in developing and the least developed country. Similarly, cereal varieties having higher level of amino acid such as lysine and threonine are also created for which they are normally deficient.

More than 50% of human populations of world use rice as staple food. Currently global climate change is also emerging as the major challenge for the crop productivity. Poor and developing countries have major challenges as the crop productivity is highly dependent on weather conditions, especially for irrigation. This situation might call for better drought and heat-resistant varieties. Ornamental plants have also become one of the important plants for human society. The demand of these plants include new colors and aesthetic variations. This encourages breeder to come up with new varieties.

Now with the advanced technology, quality traits are becoming more narrowly defined in breeding objectives. Rather than high protein or high oil, breeders are breeding for specifics, such as low linolenic acid which in low amount provides stability and enhanced flavor, and reduces the need for partial hydrogenation of the oil and production of trans-fatty acids, which is good for health.

While the world population is on increase, the land for cultivation is either not increasing or in some areas it is decreasing. The decrease in the farming land area is because of the residential and commercial needs of the increasing population. This makes yield increase necessary to sustain the growing population food demand. It is also needed to make plant products more digestible and safer to eat by reducing their toxic components and improving their texture.

There are various tools/technology used for plant breeding which has been described briefly in Table 2.1.

TABLE 2.1 Tools Used for Plant Breeding.

Conventional/Traditional Tools	Description
Emasculation	Removal of anther to prepare the plants for crossing
Male sterility	Eliminate needs of emasculation in plants for crossing
Hybridization	Crossing of unidentical plants to produce hybrid leading to transfer of genes into the recipient plants
Selection	A tool to select for the desired variability
Chromosome counting	Determination of ploidy level of the plants
Chromosome doubling	For making plants fertile by manipulating chromosome ploidy
Triploidy	Making fruits seedless
Linkage analysis	Determine association between genes
Statistical tools	For evaluation of germplasm
Relatively Advanced Tools	
Mutagenesis	For inducing mutations to create desired variability
Tissue culture	In vitro manipulation of genes at cellular or tissue level
Haploidy	For creating homozygous diploid
Isozyme markers	Uses enzyme-based marker to facilitate selection
In situ hybridization	To detect successful interspecific hybridization
Advanced Technology	
RFLP	Restriction endonulease enzyme-based molecular markers
RAPD	PCR-based molecular markers
Marker assisted selection	Marker-based selection process
DNA sequencing	Whole genome sequencing of the plant
Metabolomics	Study of a broad range of metabolites produced in plants
Transcriptomics	Study of expression of genes in plants
Plant transformation	Transfer of DNA to plant using recombinant technology

2.2 HINDRANCES AND SHORTCOMING OF THE CONVENTIONAL BREEDING PROGRAM

Conventional breeding involves crossing of plants having the desired traits, called donor plant to the variety not having those traits. But it is not easy to transfer all the desired characters into the variety of interest especially, quantitative traits which are contributed by many genes.

In conventional breeding method, it takes minimum of five to six generations to transfer a trait from one cultivar to other within a species. This method includes planting a large number of progenies to get the desired trait for the selection. The variety developed undergoes multilocation tests before being identified for use by farmers. This whole process can take 7–12 years. From the conventional breeding program, it is impossible to transfer characters from bacteria to plant such as *Bt* gene from bacteria to cotton. Early breeding program for selection of new varieties was more dependent on experience, and guess of the breeders while the new techniques are able explain more science for the decision-making. Selecting the best breed is an art in conventional breeding program. Disease resistance is one of the important traits for breeders. In most cases the disease resistance genes are recessive. During conventional breeding, transfer of the desired gene is also accompanied by undesirable genes around it, called linkage drag. Since recessive characters are tightly linked therefore several numbers of backcrossing can't remove it. This becomes the major disadvantage of conventional breeding.

2.3 NONCONVENTIONAL BREEDING AND ITS ADVANTAGES

The advancement in the technology might help us to improve the varieties in more focused and efficient ways. This so called nonconventional breeding uses recombinant DNA technology, a powerful tool to transfer gene of interest into the new varieties in a controlled manner. The advancement in technology has also helped to identify genes from the improved varieties much faster. Only the cloned gene(s) of agronomic importance is/are being introduced without co-transfer of other undesirable genes from the donor. The recipient genotype is least disturbed, eliminating the need for repeated backcrosses. Recombinant DNA technology helps to overcome natural barriers of breeding such as transferring genes from bacteria or genetically unrelated species. We have discussed the new advance technology such as molecular marker-assisted breeding, gene farming, and other techniques in later part of the chapter.

Using genetic markers in nonconventional breeding program can make selection process less art dependent. With the advancement of molecular biology techniques, the gene of interest can be transferred to the target variety in a single event. By following this, it takes 5–6 years to develop a new variety with stable gene expression. The variety produced can also be used as donor line to transfer the desired character to the elite varieties.

The conventional and nonconventional tools can be used as complementary to each other rather than exclusively. After developing varieties using nonconventional tools, the desired character can be transferred using conventional breeding methods.

2.4 MOLECULAR MARKERS AND ITS ADVANTAGE FOR THE BREEDING PROGRAM

Molecular markers are a set of DNA-based genetic markers that can detect DNA polymorphism both at the level of specific loci and at the whole genome level. They serve as the reference position on the chromosome to identify target gene location. The logic of using marker is to identify the more complicated target gene association with the easily verifiable marker genes. Markers help in selecting the desirable trait indirectly without affecting the phenotype of the trait. There are various different types of molecular markers. The comparisons of important markers are briefly given in Table 2.2.

In conventional breeding programs, selection is performed by selecting for the observable phenotype at the adult stage of the plants but with the advent of molecular marker, it can be done at the molecular level. This helps in selecting the plant for desirable trait at the early stages of the plants before they reach to adulthood. This is of tremendous help since it reduces the time of breeding program, especially the selection steps involved in the process. Molecular markers are usually chosen from the noncoding region as these regions are not affected by different plant tissue used or the developmental stages chosen to perform the assay. Markers are being applied for the selection of parental materials and accelerated selection of loci controlling traits that are difficult to select phenotypically.

Genetic/molecular markers can be used for the following several purposes:

1. For selecting the parental breeding line for crossing, especially in case of polygenetic inheritance where such a genetic marker can be selected, which cosegregate with the quantitative trait loci (QTL).

TABLE 2.2 Most Commonly Used Molecular Markers.

Marker	Principle	Advantage	Disadvantage
Restriction fragment length polymorphisms (RFLPs)	When homologous chromosomes are subjected to restriction enzyme digest, different restriction products varying in length are produced, generating variable electrophoretic mobility patterns (hence called restriction fragment length polymorphisms)	—Co-dominant in nature i.e. able to differentiate between homozygous and heterozygous. —Highly reproducible —Doesn't require sequence information	—Expensive and low throughput —Large amount of DNA needed for restriction digestion and Southern blotting —Not affective for detecting single nucleotide changes
Minisatellites/simple sequence repeats (SSRs)	A minisatellite is a section of DNA that consists of variant repeats of bases (di, tri, tetra, etc.), the repeat length varying between 10 bp and 60 bp, and sometimes over 100 bp	— Co-dominant in nature — Yield a high level of polymorphism, hence, useful as markers in linkage analysis and population studies	—Locus-specific probes required
Random amplified polymorphic DNA (RAPD)	The total genomic DNA is amplified using a single short (about 10 bp) random primer which amplifies specific DNA	—Yields high levels of polymorphism and is simple and quick to conduct —Doesn't require sequence information Require less amount of DNA (10-50 ng)	—Dominant markers, therefore it is impossible to distinguish between DNA amplified from a heterozygous locus or homozygous locus —Not reproducible.
Microsatellites/inter simple sequence repeats (ISSR)	Like minisatellite but requiring very small repetitive DNA Sequences, 2–5 bp long each having variable repeats depending on alleles	—Developed for one species, may be applied to another closely related species —Doesn't require sequence information as primers are designed based on repeats. —Reproducible compared to RAPD.	—Dominant in nature. —Too complex marker system for population genetic studies —Null allele arise due to non-amplification might complicate the calculation of allele frequency

TABLE 2.2 *(Continued)*

Marker	Principle	Advantage	Disadvantage
DNA amplification fingerprinting (DAF)	Variant of RAPD method having very small primers (5–8 bases)	—Best method to distinguish genetically closely related plants such as genetically modified plants which differ only for transgenes —More polymorphic compared to RAPD.	Dominant in nature Less effective for distinguishing among species of plants at a higher taxonomic level where genetic variation is already pronounced
Single nucleotide polymorphisms (SNPs)	It distinguishes a single base pair site in the genome that is different from one individual to another	—Co-dominant in nature —SNPs are often linked to genes, making them important in locating interesting genes such as disease genes. —Have low mutation rate —Abundant	—Costly and time intensive —Require sequence information —Low information content of single SNP
Amplified fragment length polymorphism (AFLP)	It uses 17–21 nucleotide long primers capable of annealing perfectly to their target sequences to amplify DNA restriction fragment	—It is very reliable and robust method —Does not require DNA sequence information —Very useful for detecting polymorphism between closely related genotypes	—Dominant in nature —Large amount of DNA is required —Methodology is complex

2. It can be helpful in decreasing the linkage drag (undesirable genes which comes with the gene of interest) during introgression of gene of interest from wild varieties.
3. Early detection of desirable traits.
4. Transferring multiple resistance genes (also called gene pyramiding) can be easily and quickly done by marker-assisted breeding (MAB) while it is difficult and time consuming using traditional breeding techniques.

Using DNA marker for trait(s) selection for crop improvement is called marker-assisted selection (MAS). Improvement in technology has led to pyramiding genes of interest. This is beneficial, especially in case of plant improved for resistance against a pathogen. Plants harboring single resistance gene against a pathogen can easily break down while if more resistance genes have been stacked, chances of breaking resistance can be delayed for long time until the pathogen evolves resistance against all the genes. In MAS, the elite variety is crossed with the variety having desired character leading to production of F_1 hybrid seeds. These F_1 hybrids are crossed with the elite variety which is called backcross 1 (BC_1). The crosses are performed for next few generations (BC_{2-5}) until the desired character is transferred into the elite variety (Acquaah, 2012). The marker-assisted breeding takes 3–6 years, making it much faster technique than the conventional breeding.

2.5 THE ROLE OF TISSUE CULTURE IN THE CROP IMPROVEMENT PROGRAM

Tissue culture broadly refers to the in vitro cultivation of plants from any part of plant (e.g., tissues, organ, embryos, protoplasts, single cells) on nutrient media under closely controlled and aseptic conditions. The plant parts used for such propagations are called explants. This leads to production of genetically identical progenies. But variations can also be selected by challenging with toxic agent and selecting such plants. Such variations are called somaclonal variations.

There are several advantages of using tissue culture in crop improvement:

1. Use of tissue culture can reduce the length of breeding programs of perennial species.

2. It is very useful for the cases where crosses of wild variety with the elite varieties are not very easy.
3. It can be used to rapidly multiply planting material and ensures true to type plants.
4. Tissue culture is known to cause early induction of flowering so it is useful in accelerating the breeding and testing time.
5. Tissue culture is useful in mass production of disease-free plants,
6. It is easy to maintain superior plants and genetic traits though tissue culture in comparison to sexual way of propagation where fixing particular traits require much more time.
7. As tissue culture has the ability to produce new plants from vegetative parts, breeders need to select only one plant to use as stock.
8. In crop breeding, the genetics of the crop is determined at the end of the process but in tissue culture genotype of the plant can be determined at first step and fixed for propagation.
9, Plant propagation using this method is independent of seasonal and raw material constraints.

Tissue culture products have witnessed a huge expansion globally, with an estimated global market of US$15 billion/annum. In vitro production of secondary metabolites from tissue culture has become one of the most important outcomes.

Plant propagation using tissue culture also called micropropagation involves four major steps—

1. Establishment of aseptic cultures,
2. Shoot bud/shoot multiplication,
3. Induction of rooting and hardening,
4. Transfer of plantlets to soil.

2.6 PLANT MOLECULAR FARMING

Plant molecular farming refers to the production of recombinant proteins (including pharmaceuticals and industrial proteins) and other secondary metabolites in plants. There are more than 100 proteins produced using various plant expression systems in last decade which signifies rapid progress in plant molecular farming (Fischer et al., 2000; Giddings, 2001).

2.6.1 ADVANTAGES OF USING PLANT MOLECULAR FARMING

2.6.1.1 LOW COST OF PRODUCTION

There is no need for fermenters or the skilled personnel for plant molecular farming. The recombinant proteins or metabolites can be produced in plants at 2–10% of the cost of microbial fermentation systems and at 0.1% of the cost of mammalian cell cultures. The low-cost production also depends on the product yield. The scale of plant-based production can be modulated rapidly in response to market demand. Transgenic plant lines can be stored indefinitely and inexpensively as seed (Schillberg et al., 2002). Several types of recombinant proteins can be used in unprocessed or partially processed material, therefore removing many of the downstream costs. The advantage of recombinant protein expression in the seeds of transgenic cereal plants is high levels of the product which can be accumulated in a small volume. This technique minimizes the cost associated with processing by using more or less land as required (Stoger et al., 2000).

2.6.1.2 DEVELOPMENT TIME SCALE

Plant cell suspension cultures have been used for molecular farming, especially where high containment is an advantage. The gene-to-protein time for transgenic plants encompasses the preparation of expression constructs, transformation, regeneration, and the production and testing of several generations of plant which might take up to 2 years.

2.6.1.3 PRODUCT AUTHENTICITY

As a production system for pharmaceutical proteins, plants are considered to be much safer than both microbes and animals because they generally lack human pathogens, oncogenic DNA sequences, and endotoxins. However, there is need to consider the structural authenticity of such proteins as they might influence their behavior *in vivo*. Since the protein synthesis pathways are conserved between plants and animals, plants are likely to fold and assemble recombinant human proteins efficiently. This is a great advantage over bacterial expression systems, in which many proteins either fail to fold properly resulting in accumulation as insoluble inclusion bodies or are degraded (resulting in low yields). Plants offer practical and safety advantages as well as lower production costs compared with traditional systems based on microbial or animal cells, or transgenic animals

2.6.1.4 HIGH YIELDS IN TRANSGENIC PLANTS

Following are the several ways of increasing yields in plants:

a. Increasing the efficiency of transcription and translation: Briefly, in case of dicotyledonous species, strong constitutive cauliflower mosaic virus 35S (CaMV35S) promoter (Tyagi, 2001) can be used to drive transgene expression while in case of cereals, the maize ubiquitin-1 (*Ubi*-1) promoter (Christensen and Quail, 1996) can be used to increase the rate of transcription. The rate of translation can also be optimized by making sure the translational start site matches the Kozak consensus for plants (Kawaguchi and Bailey-Serres, 2002) and by modifying codon usage in some transgenes (Koziel et al., 1996). Both transcription and translation can be increased by minimizing frequency of silencing (Voinnet, 2002; Kohli et al., 2003).
b. Facilitating isolation and purification of the proteins: Proteins can be targeted to plasma membrane or can be accumulated in a particular organelle (Schillberg et al., 2002). Antibodies targeted to the secretory pathway, using either plant or animal N-terminal signal peptides, usually accumulate to levels that are several orders-of-magnitude greater than those of antibodies expressed in the cytosol. It is advantageous to accumulate proteins in the seeds as this has been demonstrated that antibodies expressed in seeds remain stable for at least 3 years at ambient temperatures with no detectable loss of activity (Stoger et al., 2000).

2.6.2 DISADVANTAGES OF USING PLANT MOLECULAR FARMING

Plant molecular farming is one of the most rapidly growing industries but has to overcome following disadvantages—

a. Quality and homogeneity of the final product,
b. Concerns about biosafety and challenges in processing plant-derived pharmaceuticals under good manufacturing conditions. Pollen, seed,

or fruit dispersal may expose the transgene to nontargeted organisms such as birds, insects, and microbes (Commandeur et al., 2003).

c. Proper glycosylation of expressed foreign protein molecule as they lack few key enzymes to do so, which is also the case with microbial system.
d. Environmental impact: Contamination of food chain can happen when food crops are used for production of oral vaccines. This can be prevented by following strict containment strategy.
e. Risk of horizontal gene transfer. Till date, there is no report of horizontal gene transfer from plants to microbes. However, the regulatory policies and guidelines of different countries do consider this as biosafety issue.

2.7 INTEGRATION OF PHYSIOLOGICAL AND BIOCHEMICAL PATHWAY INTO THE DEVELOPMENT AND DESIGNING OF CROP IMPROVEMENT PROGRAMME

Yield is one of the most important aspects of crop breeding program. It also includes nutritional improvement of the edible plants. For example, introduction of provitamin A and *β*-carotene genes have resulted in the production of "golden rice" (Ye et al., 2000; Beyer et al., 2002). High-protein "phaseolin" and *AmA1* genes have been introduced into tobacco and potato (Chakraborty et al., 2000; Randhawa et al., 2009) leading to improvement in its yield, protein content, and quality.

It is important to note that how a biochemical step in starch biosynthesis if altered, can lead to increase in yield. For example, enhanced activity of an enzyme called ADP-glucose pyrophosphorylase, which has key function in starch biosynthesis in wheat endosperm, leads to 40% seed yield increase (Frey et al., 1999).

One of the important biochemical and physiological property of plant is fruit ripening. This is controlled by ripening genes which are activated by ripening phytohormone called ethylene. Cloning of ripening genes including polygalacturonase and pectin esterase involved in cell wall softening; phytoene synthase, required for carotenoid synthesis; and ACC synthase and ACC oxidase that catalyze the production of ethylene has been done. Gene silencing has been achieved by antisense or sense gene expression for these genes leading to delayed ripening, helping them to store for longer duration. For example, fruit ripening was delayed by controlling ethylene production in apple (Yao et al., 1995), and citrus (Wong et al., 2001).

Genetic engineering can be used to change the metabolic pathways to increase the amounts of various secondary metabolites, which play an important role in host-plant resistance to diseases, for example, medicarpin and sativan in alfalfa, deoxyanthocyanidin flavonoids (luteolinidin, apigenindin, etc.) in sorghum, and stilbene in chickpea and tobacco (Heller and Forkman, 1993). Designing crop breeding program keeping these things as target can add value to agri-foods.

2.8 GENOMIC ASSISTED MOLECULAR BREEDING: A USEFUL TOOL FOR FUTURE BREEDING

Advancement in genomics has led to enhanced understanding of functional aspects of plant which can help improve the crop plants. Genomic-assisted molecular breeding can be defined as a technique to develop new varieties of improved plants using information gained from molecular techniques such as genomics, proteomics, and metabolomics to manipulate DNA.

There are two ways in which advances in genomics can contribute to crop improvement. First, selection of superior genotypes can be done more efficiently by better understanding of the biological mechanisms. Second, decision-making process for breeding strategies can be improved more efficiently with new knowledge gained by genomics.

Different strategies and approaches can be applied to exploit genomic research for crop improvement. They are as follows.

2.8.1 TRANSCRIPTOMICS

The investigation of the total transcript content of any biological sample is known as transcriptomics. There are several methods to perform transcriptomics, such as cDNA array, microarray, and high-throughput RNA sequencing methods. Microarray and high-throughput RNA sequencing analysis have added tremendous speed to identification and characterization of different biological processes including physiological, developmental, stress, and resistance-related functional genes (Aharoni and Vorst, 2002; Potokina et al., 2004). These analyses give clue for functional association of level of gene expression to the trait. But these types of studies need further characterization of the genes. Following are limitations of using these techniques:

1. False positive signal from genes caused by low heritability of gene expression patterns and gene expression,
2. Small population size under study, the reason being higher cost of such studies,
3. Different microarray platforms (e.g., Affymetrix, Agilent, Amersham) with the same RNA sample or analysis of the same microarray gene expression data with different bioinformatics tools might not identify the same set of differently expressed genes for a given trait (Larkin et al., 2005).

These things suggest caution while using the technique(s) and interpreting their results.

2.8.2 ECO TILLING

It is desirable for the breeders to know relative value of all alleles for gene of interest since it will help them to decide which allele for a gene to be introgressed into the desirable variety. Local lesion in genome is called TILLING and the strategy of targeting induced local lesion in genomes is called Eco TILLING (Comai et al., 2004). It helps in identifying naturally present SNP and haplotyping leading to characterization of alleles. Eco TILLING can provide a series of alleles for the genes which are involved in important processes of the plant even though the known variants for these genes have not been observed through genetic studies.

2.8.3 METABOLOMICS

Metabolomics deals with study of a broad range of metabolites produced in plants associated with several different conditions including developmental changes, biotic and abiotic stresses. This study helps in parallel assessment of plant metabolites with other techniques such as transcriptomics. The metabolite profile including both primary and secondary metabolites can be assessed to understand their role in crop resistance to biotic and abiotic stress. This can also give information about nutritional status of the plant which directly affects human health.

Currently there are two major techniques called mass spectrometry (MS) and nuclear magnetic resonance (NMR) used for metabolite profiling. The three different variants of MS being used for metabolite analysis are:

Gas-chromatography-mass-spectrometry (GC-MS), gas-chromatography-time-of-flight-mass-spectrometry (GC-TOF-MS), and liquid-chromatography-mass-spectrometry (LC-MS).

In past decade, several metabolomics study have been carried out which has facilitated the identification of important sources of allelic variance for metabolic engineering (Harrigan et al., 2007; Roessner et al., 2001; Mercke et al., 2004).

The ongoing efforts to elucidate the metabolic response to biotic and abiotic stresses indicate that metabolomics-assisted breeding might be useful in the development of crops that are more resistant to stresses without yield losses.

These techniques will help breeders rapidly identify desirable genotypes and use them for breeding. In the post-genomics era, ultimate goal will be to use high-throughput approaches combined with automation, leading to increased amounts of sequence data in the public domain which will contribute to genomics research for crop improvement. Therefore in future marker-assisted breeding will gradually evolve into genomics assisted breeding.

KEYWORDS

- **plant breeding**
- **biotechnology**
- **TILLING**
- **molecular markers**
- **plant tissue culture**

REFERENCES

Acquaah, G. *History of Plant Breeding. Principles of Plant Genetics and Breeding*; John Wiley & Sons: UK, 2012; 22–39.

Aharoni, A.; Vorst, O. DNA Microarrays for Functional Plant Genomics. *Plant Mol. Biol.* **2002,** ***48***, 99–118.

Beyer, P.; Al-Babili, S.; Ye, X.; Lucca, P.; Schaub, P.; Welsch, R.; Potrykus, I. Golden rice: Introducing the β-Carotene Biosynthesis Pathway into Rice Endosperm by Genetic Engineering to Defeat Vitamin A Deficiency. *J. Nutr.* **2002,** *132*, 506–510.

Chakraborty, S.; Chakraborty, N.; Datta. A. Increased Nutritive Value of Transgenic Potato by Expressing a Nonallergenic Seed Albumin Gene from Amaranthus hypochondriacus. *Proc. Natl. Acad. Sci.* **2000,** *97*, 3724–3729.

Christensen, A. H.; Quail, P. H. Ubiquitin Promoter-based Vectors for High-level Expression of Selectable and/or Screenable Marker Genes in Monocotyledonous Plants. *Transgenic Res.* **1996,** *5*, 213–218.

Commandeur, U.; Twyman, R. M.; Fischer, R. The Biosafety of Molecular Farming in Plants. *AgBiotechNet* **2003,** *5*(110), 1–9.

Comai, L.; et al. Efficient Discovery of DNA Polymorphisms in Natural Populations by EcoTILLING. *Plant J.* **2004,** *37*, 778–786.

Fischer, R.; Emans, N. Molecular Farming of Pharmaceutical Proteins. *Transgenic Res.* **2000,** *9*, 279–299.

Frey, C.; Audran, E.; Marin, B.; Sotta, A.; Marion, P. Engineering Seed Dormancy by the Modification of Zeaxanthin Epoxidase Gene Expression. *Plant Mol. Biol.* **1999,** *39*,1267–1274.

Giddings, G. Transgenic Plants as Protein Factories. *Curr. Opin. Biotechnol.* **2001,** *12*, 450–454.

Harrigan, G. G.; et al. Impact of Genetics and Environment on Nutritional and Metabolite Components of Maize Grain. *J. Agric. Food Chem.* **2007,** *55*, 6177–6185.

Heller, W; Forkman, G. Biosynthesis of Flavonoids. In *The Flavonoids, Advances in Research Since 1986;* Harborne, J. B. Ed.; Chapman and Hall: London, UK, 1993, 499–535.

Kawaguchi, R.; Bailey-Serres, J. Regulation of Translational Initiation in Plants. *Curr. Opin. Plant Biol.* **2002,** *5*, 460–465.

Kohli, A.; et al. Transgene Integration, Organization and Interaction in Plants. *Plant Mol. Biol.* **2003,** *52*, 247–258

Koziel, M. G.; et al. Optimizing Expression of Transgenes with an Emphasis on Post-transcriptional Events. *Plant Mol. Biol.* **1996,** *32*, 393–405.

Larkin, J. E.; et al. Independence and Reproducibility Across Microarray Platforms. *Nat. Methods* **2005,** *2*, 337–343.

Mercke, P.; et al. Combined Transcript and Metabolite Analysis Reveals Genes Involved in Spider Mite Induced Volatile Formation in Cucumber Plants. *Plant Physiol.* **2004,** *135*, 2012–2024.

Potokina, E.; et al. Functional Association Between Malting Quality Trait Components and cDNA Array Based Expression Patterns in Barley (*Hordeum vulgare* L.). *Mol. Breed.* **2004,** *14*, 153–170.

Randhawa G. J.; Singh, M.; Sharma, R. Duplex, Triplex and Quadruplex PCR for Molecular Characterization of Genetically Modified Potato with Better Protein Quality. *Curr. Sci.* **2009,** *97*, 21–23.

Roessner, U.; et al. Metabolic Profiling Allows Comprehensive Phenotyping of Genetically or Environmentally Modified Plant Systems. *Plant Cell* **2001,** *13*, 11–29.

Schillberg, S.; et al. Antibody Molecular Farming in Plants and Plant Cells. *Phytochem. Rev.* **2002,** *1*, 45–54.

Stoger, E.; et al. Cereal Crops as Viable Production and Storage Systems for Pharmaceutical scFv Antibodies. *Plant Mol. Biol.* **2000,** *42*, 583–590.

Tyagi, A. K. Plant Genes and their Expression. *Curr. Sci.* **2001,** *80*, 161–169.

Voinnet, O. RNA Silencing: Small RNAs as Ubiquitous Regulators of Gene Expression. *Curr. Opin. Plant Biol.* **2002,** *5*, 444–451.

Wong, W. S.; Li, G. G.; Ning, W.; Xu, Z. F.; Hsiao, W. L. W.; Zhang, L. Y.; Li, N. Repression of Chilling Induced ACC Accumulation in Transgenic Citrus by Over Production of Antisense 1-minocyclopropane-1-carboxylate synthase RNA. *Plant Sci.* **2001,** *161*, 969–977.

Ye, X.; Al-Babili, S.; Klöti, A.; Zhang, J.; Lucca, P.; Beyer, P.; Potrykus, I. Engineering Provitamin A (β-carotene) Biosynthetic Pathway into (Carotenoidfree) Rice Endosperm. *Sci.* **2000,** *287*, 303–305.

Yao, J. L.; Cohen, D.; Atkinson, R.; Richardson, K.; Morris, B. Regeneration of Transgenic Plants from the Commercial Apple Cultivar Royal Gala. *Plant Cell Rep.* **1995,** *4*, 407–412.

CHAPTER 3

SCOPE OF PLANT BIOTECHNOLOGY IN THE DEVELOPING COUNTRIES

NAND K. SAH*

Post-Graduate Studies and Research Centre, Department of Botany, T. N. B. College, Bhagalpur, T. M. B. University, Bhagalpur 812007, Bihar, India

**Corresponding author. Email: nandksah1@gmail.com*

CONTENTS

ABSTRACT

By 2050, world human population is expected to go past 9.5 billion. To address the needs of the teeming billions is a huge challenge for the scientists. It is imperative that unequivocal plans are put in place to keep equilibrium between pace of growth of human population and production of food along with health care, energy and environmental concerns if the human race has to survive and flourish. Plant biotechnology promises lucrative scopes for food, health, energy, and environmental securities for mankind in future. Several recombinant plants have been cloned to address nutritional requirements along with environmental concerns. For example, Golden rice has been created with abundant vitamin A, Flavr Savr tomato ensures post-harvest quality for a long time, SUSIBA-2 rice boosts production along with cut down on global warming due to methane, Bt gene cloned in several plants including cotton, brinjal (egg plant), maize etc reduces damage due to worms. Along with intensive research on biodiversity, the traditional plant food including the coarse cereals, pulses, fruits and vegetables need to be revived, particularly those known in the developing countries like India, China etc. Post-harvest protection of the plant produces is another area that developing countries like India have to strengthen. In the recent years, it has been observed that tons of procured agricultural products are left uncared leading to biodeterioration jeopardizing food security. Traditional system of medicine including Unani, Chinese, and Ayurved has a bright future from the standpoint of reliable healthcare security at affordable costs. Intensive research and development is necessary also to modernize the traditional health care system that is often referred to as an alternative system of medicine. Plant and other biological resources are expected to play crucial role in the energy management. Some countries, like Brazil, have managed up to 60% of their energy requirement through these resources. Total energy demand of India is going to increase from 800 Mtoe at present to 1223 Mtoe by 2035. Multi-national projects should be launched for tapping abiotic and biotic sources of methane in a way that brightens the prospect of energy output with concomitant reduction in global warming.

3.1 INTRODUCTION

The estimated antiquity of green plants on earth is about 3 billion years. This is when oxygen is supposed to have evolved through photosynthesis

(Leslie, 2009). The green plants (autotrophs) evolved along with the non-green (heterotrophs) ones in parallel. Since then, evolution has brought about huge diversities in the plant kingdom. Of the 1.7 million existing species on silver earth, plants comprise about 15% (2.6 lacs) as per records. From the terrains at higher altitude of the Himalayas to the abysmal bottom of the Pacific Ocean, from the volcanic zones and springs at high temperature to the glaciers at ultralow temperatures, plants of diverse nature are observable. Knowledge about these diversities equipped inquisitive human minds with ideas to make use of them for food, health, energy, and environmental securities. A new era of research heralded in the late 1960s and early 1970s, when restriction enzymes were discovered and characterized by the molecular biologists (Werner Arber, Hamilton O. Smith, and Daniel Nathans). Herbert Boyer, who worked on restriction enzyme, and Stanley Cohen, who had an expertise in extra-chromosomal DNA (plasmid), shared their experience together and succeeded in cloning and transferring the first gene for frog ribosomal RNA into bacterial cells and examined its expression (Cohen et al., 1973). This technique has now matured as Recombinant DNA (rDNA) technology that forms the basis of modern vibrant biotechnology. rDNA technology got a shot in the arm with the discovery of thermocycler in 1983 (a device that performs polymerase chain reaction to produce large amount of DNA in a short time) by Kary Mullis (US Patent).

Great strides have been taken since the first plant cell was engineered with a foreign gene in 1983 by Herrera-Estrella and coworkers using *Agrobacterium tumefaciens* as a vector. *A. tumefaciens* is a pathogenic gram-negative soil bacterium that possesses a unique extra-chromosomal DNA, known as Ti-plasmid. T-DNA region of this plasmid is capable of integration into the plant genome. This property of the T-DNA is exploited to transfer foreign genes into a plant genome, especially of the dicotyledonous group. The first transgenic plant that hit the media headlines was a tomato, known commonly as "Flavr Savr" (pronounced "flavor saver") in 1994 (Keith et al., 1992). Its relatively stable rigid rind and ability to stay long even after ripening made it an instant favourite amongst the growers and consumers. The enzyme, polygalactouronase (PG) that softens tomato skin, particularly after ripening has been silenced by producing mRNA complementary to the PG gene-specific mRNA. So far, thousands of genetically modified plants have been developed in the laboratory throughout the world. The primary objectives of these endeavours include food, health, energy, and environmental securities for mankind. In all these areas, significant progress has been made.

There is a growing sense of relevance of GM crops and plants not only amongst the scientists but also agriculturists. This is corroborated by the fact (Table 3.1) that in 2014 GM plants were raised in about 4.5 billion acres (1.7 billion hectares). The area under cultivation for biotech (GM) plants has been growing at a pace of 3–4% every year. Nevertheless, there are some important apprehensions amongst the knowledgeable people at large that need to be addressed before universal acceptance. However, it depends also on the nature of GM plants. For example, Bt-cotton that is used for clothing and other nonfood requirements may be acceptable, if affordable for agriculturists. However, those that carry food value need to be discussed in detail before global acceptance.

TABLE 3.1 Plantation Area of Genetically Modified Crops (2013).

Countries	Crops	Million ha
Argentina	Cotton, maize, and soybean	24.4
Brazil	Cotton, maize, and soybean	40.3
Canada	Canola, cotton, maize, soybean, and sugarbeet	10.8
India	Cotton	11.0
USA	Alfalfa, canola, cotton, maize, papaya, squash, soybean, and sugarbeet	70.1

3.2 FOOD SECURITY

Professor M. S. Swaminathan, a great architect of successful green revolution in India during the 1960–70s, has put forward a novel concept of "EVERGREEN REVOLUTION," which entails food and nutritional security for all (Swaminathan, 2010). India became a food sovereign country, which peaked in cereals production (above 234 million tons) in 2008–09 (Swaminathan, 2013). It has now stabilized to around 230 million tons per annum that is sufficient to meet India's current food grain requirement. However, with the growing population at the rate of 1.9% per annum (1990–2007), it is imperative that food grain production also keeps pace with it (MSSRF report, 2008); but during this period, food grain production increased only at the rate of 1.2%. The scientists are aware that the production of food grain has almost reached plateau with the conventional methods of green revolution in most of the developed countries. Data show that India lags behind many countries in the rate of production of principal crops per hectare (Swaminathan and Bhavani, 2013). For example, Paddy, wheat, and maize production recorded in India in 2008 was 3370, 2802, and 2324 kg/ha as against 9731, 6501, and 7977 kg/ha, respectively in Egypt. There are ample avenues for boosting production here. Indiscriminate use of the production-boosters also needs to be guarded, otherwise serious threat to soil, environment, water resources, flora and fauna, and human health may appear. For a sustained and balanced growth, it is necessary that concepts of evergreen revolution are applied which largely embodies modern technology spearheaded by biotechnology with negligible or no harm to the environment ensuring food security for all. In 2011, Monsanto's DroughtGard™ maize became the first drought-resistant GM crop to merit approval for US marketing.

In India, about 60% families subsist on agricultural income. Their economic condition and access to the modern agricultural practices should be ensured. One way to achieve this goal is to engage them in subsidiary agriculture-based productions that have a ready market in the country and abroad (Table 3.2). The government has taken suitable measures in this direction. Special schemes such as "START UP INDIA" and "STAND UP INDIA" have been launched. The research on GM plants has not gathered as much pace in India as in other developed countries. One of the important reasons for this is paucity of adequate funds in research and development (R&D). However, recently scientists at CSMCRI, Bhavanagar Gujarat have shown that *SbASR-1* gene enhances the salinity and drought stress tolerance in transgenic groundnut by functioning as a late embryogenesis abundant (LEA) protein and a transcription factor (Tiwari et al., 2015). There are several areas that need to be geared-up to boost production and rev up agricultural income: (i) quality seeds; (ii) balanced manuring; (iii) raising coarse cereals, as shown in Table 3.3, that account for about 15% of total yield (2009–10); (iv) raising climate resilient plants to resist biotic and abiotic stresses; (v) nutritional quality improvement; (vi) soil bioremediation; (vii) post-harvest safety of food items; (viii) food processing units that make use of perishable farm products; and (ix) good marketing strategy that benefits all.

3.3 HEALTH SECURITY

Health and hygiene is another important area that has an immense scope of being strengthened by plant biotechnology. Commercial grade therapeutic biochemical, including enzymes, proteins, nutraceuticals, antigens (vaccines), plantibodies, alkaloids, and flavonoids, may be produced in plants (Pujari, 2015; Hatti-Kaul, 2015). The US-FDA approved the first plant-produced pharmaceutical, a treatment for Gaucher's Disease in 2012

TABLE 3.2 Some Agriculture-based Products That Have a Good Market in India.

Biochemicals/Enzymes	Gene Source	Name of the Genetically Engineered Plant	Application
α-Amylase	*Bacillus licheniformis*	Tobacco	Liquefaction of starch (industrial application).
1-3, 1-4-β-Glucanase	*Trichoderma reesei*	Tobacco and barley	Brewing (industrial application).
Phytase	*Aspergillus niger*	Increased utilization of phosphate from feed	
Avidin	Chicken	Maize	Research reagent and in industrial application.
β-Glucuronidase	*E. coli*	Maize	Research reagent and in industrial application.
Cyclodextrin glycosyltransferase	*Klebsiella pneuminae*	Potato	α and β-cyclodextrins produced.
Fructosyl transferase	*Bacillus subtilis*	Potato and tobacco	Fructan production.
Manitol-1phosphate dehydrogenase	*E. coli*	Tobacco	Increased tolerance to high salinity.
Tryptophan 2 oxygenase	*Pseudomonas ssyringae pv. savastanoi*	Tobacco and brinjal (egg plant)	Seedless fruit development (parthenocarpy) even under unfriendly conditions.
Biodegradable plastic PHB	*R. eutropha*	*Arabidopsis thaliana*	Gene targeted in chloroplast that produces PHB up to 14% of dry leaf mass.

TABLE 3.3 List of Coarse Cereals Grown in India.

S. No.	Common Name of Coarse Cereals	Botanical Name	Family
1	Jowar	*Sorghum vulgare*	Poaceae
2	Bajra (Pearl millet)	*Pennisetum glaucum*	Poaceae
3	Maize	*Zea mays*	Poaceae
4	Finger millets or ragi	*Eleusine coracana*	Poaceae
5	Small millets:		Poaceae
	Barnyard millet: Shyama/Sanwa	*Echinochloa frumentacea* (L.)	
	Proso millet: China	*Panicum miliaceum* (L.)	
	Little millets:	*Panicum sumatrance*)	
	Kodo millet: Kodo	*Paspalum scrobiculatum* (L.)	
	Foxtail millet: Kanguni/Kaoni	*Setaria italica* (L.)	
6	Barley	*Hordeum vulgare*	Poaceae
7	Khobi—subsidiary food	*Schoenoplectiella articulata*	Cyperaceae

(Lundmark, 2006; Manoj et al., 2012). Quality anti-oxidants that could be easily absorbable may be produced from plants. There is a long list of biopharmaceuticals produced from GM plants (Table 3.4). Important among them are anti-coagulant protein C, erythropoietin (treatment of anemia) obtainable from tobacco, thrombin inhibitor (hirudin) produced from Canola (*Brassica napus*), Interferon alpha and beta produced from rice, turnip, and tobacco, serum albumin from tobacco, lactoferrin, and insulin from potato and so on (Sharma et al., 1999). As per an estimate in 2010, market worth 48 billion US$ exists for monoclonal antibodies (mAb), which is expected to rise up to 86 billion US$ by 2015 (webpage pharmaceuticals and antibodies). mAbs are actually are excellent for passive immunization, particularly against specific cancer (Thomas et al., 2002). Biotech products from the indigenously engineered GM plants are yet to gather satisfactory momentum. Plants like tobacco, potato, tomato, and arabidopsis have been genetically engineered to produce antigens (vaccines) against various diseases (Table 3.5). Plants are suitable for this purpose because downstream processing becomes relatively easier due to extracellular export of the products after synthesis. Another important aspect is that these products are devoid of toxins that are often associated with bacterial products.

3.4 ENERGY SECURITY

Plants are in fact the primary and trusted sources of energy on earth. The petro products and coal (fossil fuel) found underneath earth are the photosynthetic products of the green plants that existed millions of years ago (Table 3.6). However, these reserves would not last for more than 200 years. It is, therefore, imperative that alternative sources of energy are discovered if we intend to save the human race. Scientists have discovered unique sources of energy including the renewable ones. The discovery of biodiesel, biogas, biowaste energy, ethanol, hydrogen, and methane gases have opened new streams of energy security. The National Biodiesel Board (USA) defines biodiesel as a mono alkyl ester. Biodiesel, actually, is fatty acid chain de-esterified from glycerol (Fatty acids esterified to glycerol are oils) that are produced by various types of green plants, including certain algal forms (Chlorella) and wild and cultivated seed plants. Ethanol is one of the cleanest fuels. It may be produced by fermentation of sugary substances. Brazil produces huge amount of ethanol that covers about 60% of its fuel needs. In India, biodiesel is considered as a renewable energy source of choice, because production of alcohols and products thereof is

TABLE 3.4 Production of Some Representative Therapeutics from GM Plants.

Therapeutics	Disease	Name of Host Plant	Botanical Name	Nature of Protein/Product
Anti-coagulant	Thrombosis	Tobacco	*Nicotiana tabacum*	Protein C
Thrombin inhibitor	Thrombosis	Canola	*Brassica napus*	Hirudin
GMCSF	Neutropenia	Tobacco	*Nicotiana tabacum*	GMCSF
Human growth hormone	Hypopituitarism	Tobacco	*Nicotiana tabacum*	Somatotropin (From chloroplast)
Blood protein	Anemia	Tobacco	*Nicotiana tabacum*	Erythropoetin
Wound repair and control of cell proliferation protein hormone	Wound	Tobacco	*Nicotiana tabacum*	EGF
Hepatitis B and C	Jaundice	Rice, turnip, and tobacco	*Oryza sativa, Nicotiana tobacum*	Interferon alpha and beta
Albumin	Liver cirrhosis, burns, and surgery	Tobacco	*Nicotiana tabacum*	Serum albumin
Antimicrobial agent	Infection	Potato	*Solanum tuberosum*	Lactoferrin
Hormone	Diabetes	Potato	*Solanum tuberosum*	Insulin

TABLE 3.5 Antigens (Vaccine) Produced from GM Plants.

Types of Antigens/Vaccine	Treatment	Name of the GM Plant	Family of the GM plant
Hepatitis B surface antigen	Hepatitis	Tobacco	*Nicotiana tabacum*
Rabies virus glycoprotein	Rabies	Tomato	*Solanum lycopersicum*
Cholera toxin B subunit	Cholera	Tobacco	*Nicotiana tabacum*
		Tomato	*Solanum lycopersicum*
VPI protein of foot and mouth disease virus	Foot and mouth disease of cattle	Tobacco	*Nicotiana tabacum*
Glycoprotein of swine-transmissible gastroenteritis cornavirus	Swine flu	Arabidopsis	*Arabidopsis thalliana*
Nowak virus capsid protein		Tobacco	*Nicotiana tabacum*
E. coli heat-labile enterotoxin B subunit		Potato	*Solanum tuberosum*

TABLE 3.6 Physical and Biological Hydrogen (H_2) and Methane (CH_4) Gas Production.

Fuel Gas	Physico-chemical Method	Biological Method	
Hydrogen	**Source: Water (H_2O)**	**Fermentation/anaerobic reactions**	**Photosynthesis/aerobic process**
	Thermolysis at 2500°C	*E. coli*	*Rhodospirillum rubrum.*
	$2H_2O = 2H_2 + O_2$		
	Electrolysis	*Citrobacteria intermedius*	*Rhodopseudomonas capsulate*
		Acetobacter sp	Thiocapsa roseopercisna
		Serratia sp	Chlorella sp
		Clostridium botulinum	Chlamydomonas sp
		Lactobacillus.	Cyanobacteria: Nostoc, Anabaena, Oscillatoria, Spirulina.
	Abiotic	**Fermentation/anaerobic reactions**	**Photosynthesis/aerobic process**
Methane	i) Destructive distillation of acetylene in presence of soda lime	Microbial sources: Methanococci, Methanospirillum, Methanosaracina.	Several green plants: under stress:
			Chlamydomonas,
	ii) Reaction of aluminium carbide with water or strong acids		*A. thaliana,*
			Typha latifolia,
	iii) Serpentinization involving water, carbon dioxide and the natural olivine, which is common on Mars		*Ocimum basilicum*, and
			Zea mays.
	iv) Carbon monoxide and hydrogen may react under hugh temperature and/or pressure to produce methane		Wetland soil of rice paddy.
			Fungi: basidiomycetous

TABLE 3.6 *(Continued)*

Fuel Gas	Physico-chemical Method	Biological Method	
		Organic feedstock: marine sea grass	*Pleurotus sapidus,*
		Eichhornia (aq. biomass)	*Laetiporus sulphureus,*
		Waste biomass, i.e., sugarcane bagasse, etc.	*Pycnoporus sanguineus*, and
		Beet pulp	*Lentinula edodes.*
		Doob grass pulp	Archaea
		Rice/wheat straw	
		Mirabilis leaf	
		Cauliflower leaf, etc.	
		Cattle waste (Dung gas)	
		Market organic waste	
		Human waste (Bio gas)	
		Whey (a fermented milk product)	
		Sewage sludge	

relatively costlier. Programmes have been chalked out to use nonedible oil from wild plant seeds for generation of biodiesel by treatment with alkali. The preferred plant for this is *Jatropha curcas*, which has energy as well as therapeutic values (Thomas et al., 2008). A national drive has been undertaken to maximize production of Jatropha seeds. In addition, the seeds of *Pongamia pinnata* (Karanz), *Calophyllum mophyllum* (Nag champa), *Havea brasiliensis* (Rubber), etc. have been identified that yield oil. From the seeds of Jatropha and of other plants, oil up to 40% may be obtained (Verma and Sah, 2013). But, in practice, only 30% oil from these seeds are extractable. Yield of oil from the seeds may be maximized by improvement in the expeller design and heat treatment. Attempt may be made to produce plants that produce seeds with more oil content and with relatively soft testa so that more oil may be extracted.

There are well studied microbial organisms that may produce liquid fuel and gases, such as ethanol, hydrogen, and methane (Table 3.6) (Handwerk, 2006). Ethanol may be produced by fermentation of cane and corn juice, mono-/di-saccharides, cellulose, lignocelluloses, hemicelluloses, etc. Cane-bagasse, straw, hay/dry grass, and wood may be used for producing these fuels on commercial scales. The best source of hydrogen is water, but it requires a temperature of 2500°C for thermolysis into H_2 and O_2. It may be produced by electrolysis of water and from fossil fuels. Biologically, it is producible with the proper use of bacteria, such as *Escherichia coli*, *Acetobacter, Serratia, Clostridium, Citrobacteria*, etc. in fermentation and *Rhodospirillum, Rhodopseudomonas, Thiocapsa,* Green algae (*Chlorella, Chlamydomonas*, etc.), and blue-green algae (*Anabaena, Nostoc, Oscillatoria, Spirulina*, etc.) in photosynthesis (Table 3.6). However, the major problem is how to tap this resource in a scientifically and technically viable way. Biomethane may be produced through microbial sources, saprophytic basidiomycetous fungi (Lenhart et al., 2012), green plants (Handwerk, 2006) organic feedstock, cattle waste, market organic waste, human waste, and so on (Table 3.6). Along with sound technical management of tapping fuel gases, it is necessary to develop suitable microbes that generate more fuel at affordable cost.

3.5 ENVIRONMENTAL SECURITY

Green plants are the best natural instruments that maintain a balance of 20% oxygen in the earth's biosphere that enables survival of aerobic living systems. If the ratio of oxygen gets disturbed, life on earth becomes perilous.

Clean air, water, and soil and ambient temperature are amongst those factors that keep the environment habitable. Undesirable changes in these factors cause environmental pollution posing threats to life on earth. Today, human race faces unprecedented intimidation from the electronic and plastic garbage, which is actually a necessary evil. The developed nations try to dump these wastes in and around the developing nations compounding the global environmental problem. Suitable scientific measures have to be undertaken with a global perspective if we sincerely wish to save the human race.

The latest Paris Summit in November, 2015 on global climate change has brought solace for mankind, as there was an agreement on maintaining sustainable low carbon emission (webpage Paris Summit on climate change, 2015). We need to work harder on how to reduce various types of environmental pollutions. Air pollution occurs due to release of sulphur oxides, carbon monoxide, chlorofluorocarbons (CFCs), nitrogen oxides, particulate matters of micrometer size (PM_{10} to $PM_{2.5}$), etc. by industries and motor vehicles. In addition, there is pollution due to noise by aeroplanes, loudspeakers, traffic, thunders, industry, high intensity sonar, and so on; photo-pollution due to over illumination and astronomical interference, littering by criminal throw of man-made inappropriate waste objects into the public places; soil pollution occurs by herbicide, pesticides, heavy metals, chlorinated and non-chlorinated hydrocarbons, radioactive contamination, plastic waste, etc. Water pollution occurs due to industrial effluents, discharge of untreated domestic sewage, chemical waste, leaking surface runoff into ground water, eutrophication, and littering.

3.5.1 PLANTS AS GUARDS AGAINST POLLUTION

In most of the cases of pollution, plants stand as natural vanguards of protection for mankind. Green plants not only reduce air pollution, but also soil and water pollution. Water of the holy Ganga is said to be clean by virtue of various types of herbs and plants that lie in its runoff from Gangotri to the Bay of Bengal. Plans are afoot to plant suitable trees and herbs that reduce pollution load due to absorption and transformation of chemical pollutants. Several plants including Arabidopsis and tobacco have been genetically engineered with microbial genes (a cassette of genes from *Pseudomonas oleovoram*, *Ralstonia eutropha*, *Alcaligenes eutrophus*, *Alcaligenes lotus*, and recombinant *E. coli*) to produce polyhydroxy alkanoate (PHA) and polyhydroxy butyrate (PHB) used for manufacturing various eco-friendly and biodegradable plastic materials (*Jacquel, 2008*). Production of these substances in plants may reduce its cost from Rs. 500/Kg to Rs. 150/Kg as against Rs. 50 Kg of the petroplastic. *Pseudomonas putida* was the first genetically engineered microbe engineered by A. M. Chakrabarty that was shown to reduce pollution level. But, due to immense pollution load in the sewage water this organism did not grow for long to reduce pollution adequately (Chakrabarty et al., 1974). Research that focuses on environmental safety along with boost in biomass production should be encouraged. Recently, scientists have developed a GM rice (SUSIBA-2) that recycles methane to boost starch production as well as to reduce carbon emission (CH_4), which accounts for 20% of global warming (Su et al., 2015). SUSIBA-2 is a transcription factor that modulates gene function related to synthesis of photosynthates.

3.6 METHODOLOGY

There are several methods of introduction of foreign genes into plants, but the most widely used one is through the natural or direct DNA transfer capacity of *A. tumefaciens*. This is a soil bacterium that causes tumor formation (called crown gall) largely in the dicotyledonous plant species. During this infection a part of the Ti-plasmid of *Agrobacterium*, called T-DNA, is integrated into the plant genome (De Cleene and De Ley, 1976). This fascinating natural capacity is exploited as a natural vector of foreign genes (inserted into the Ti-plasmid) into plant genomes. *Agrobacterium*-based and direct gene transfer techniques were developed in parallel, but the former is today the most widely-used method because of its simplicity and efficiency in many plants, albeit it still suffers from limitations in terms of the range of species (Herrera-Estrella et al., 1983). These limitations are due to the natural host range of *Agrobacterium*, which infects herbaceous dicotyledonous species most efficiently and is less effective on monocotyledonous and woody species (De Cleene and De Ley, 1976).

The development of novel direct gene transfer methodology, by-passing limitations imposed by *Agrobacterium*-host specificity and cell culture constraints, has allowed the engineering of almost all major crops, including non-amenable cereals, legumes, and woody species (Herrera-Estrella et al., 1983). Direct gene transfer methods are species and genotype-independent in terms of DNA delivery, but their efficiency is influenced by the type of target cell, and their utility for the production of transgenic plants in most cases depends on the ease of regeneration from the targeted cells, as most methods operate on cells cultured in vitro (Barcelo and Lazzeri, 1998). The direct gene transfer methods include particle bombardment, DNA uptake into protoplasts, treatment of protoplasts with DNA in the presence of polyvalent cations, fusion of protoplasts with bacterial spheroplasts, fusion of protoplasts with liposomes containing foreign DNA, electroporation-induced DNA uptake into intact cells and tissues, silicon carbide fiber-induced DNA uptake, ultrasound-induced DNA uptake, microinjection of tissues and cells, electrophoretic DNA transfer, exogenous DNA application and imbibitions, and macroinjection of DNA (Barcelo and Lazzeri, 1998; Walden and Schell, 1990). The major achievements of transgenic plant technology up to now concern tolerance to insect or disease pests, herbicide tolerance, and improved product quality.

3.6.1 GENOME EDITING

This is the latest, sound technique of genetic manipulation. Precise manipulation of genetic sequence of plants and organisms, including humans, in a lucid, handy, and error-free way is known as "genome editing." The classical processes of introducing changes in the genetic makeup, as mentioned above, include hybridization and mutagenesis and, in the modern age, rDNA technology. But, the latest spurt in enthusiasm in this field grew from a simple observation by the scientists of a Yogurt company in 2007 that their bacteria, when primed, apply a versatile defence strategy against their killer viruses (bacteriophage). This simple mechanism of bacterial defence to fight-off the bacteriophage has matured into a "molecular marvel" with tremendous promises in the areas including health, energy, food, and environmental securities for mankind (Travis, 2015; McNutt, 2015).

This technique got a shot in the arm by the discovery of engineered nucleases (enzymes which cut DNA/RNA) that have a recognition site of over 14 nucleotides in the genome. These molecular scissors are primarily of four types: (i) meganucleases; (ii) zinc finger nucleases (ZFNs); (iii) transcription-activator like effector nucleases (TALENs), and (iv) clustered regularly interspaced short palindromic repeats (CRISPR). These nuclease systems may act off-target and, therefore, have low specificity, stringent selection steps, and cumbersome handling. The latest engineered CRISPR-Cas system has overcome most of these limitations and has captivated global attention (Jinek et al., 2012; Hsu et al., 2013).

3.6.2 CRISPR-CAS SYSTEM

In 2007, *Streptococcus thermophilus* (a bacterial species with cells in chain that were used in making yogurt, which is a kind of curd) displayed fighting ability against the re-invading bacteriophage. Viral scraps left over during the first attack on the bacterial cell serve as a memory bank that creates guide ribonucleic acid (gRNA) to recognize the DNA of the returning viruses, and chop-off their genes with nucleases. Understanding of this mechanism furnished strong leads to a host of scientists, including Doudna of USA and Charpentier of Germany (2014), to adapt it to develop a CRISPR powered "Gene Drive" for much easier, low cost, and handy method of gene editing which also got recognized as mutagenic chain reaction (MCR) for work on the pigmentation trait in the laboratory grown Drosophila (fruit fly) to the next generation with 97% efficiency.

3.6.2.1 MECHANISM OF CRISPR-CAS SYSTEM

Using ingenuous tools and techniques of genetic engineering, scientists in many countries have been able to create a construct of type II CRISPR-Cas system in the form of circlets of DNA, i.e., plasmids. There are two primary components of CRISPR: (a) "gRNA" that targets and recognizes specific DNA sequences where manipulation is desired and (b) DNA cutting enzyme or nuclease that is commonly called Cas9. The design of CRISPR is such that permits high fidelity homology-directed repair (HDR) system to operate for desirable target gene manipulation. This technique has been refined several times in the past that helped genome editing evolve precisely, and has now become a convenient, cheap, and affordable technology.

A non-profit group "Addgene" has already distributed 50,000 copies of the plasmid carrying CRISPR-Cas9 clone to the needy groups of scientists throughout the world. The discovery of CRISPR eclipsed many important innovations in science in 2015. This technique has been widely used for targeting many important genes in cell lines and organisms, including humans, bacteria, zebrafish, *Caenorhabditis elegans*, plants, xenopus, yeast, drosophila, monkeys, rabbits, pigs, rats, and mice. It has become easy to induce large deletions and genomic rearrangement (such as inversion, translocation, etc.). Protein domains for transcription may be manipulated through the epigenetic pathways to facilitate effective modulation in gene function conducive to human welfare.

3.7 FUTURE PROSPECT AND CONCERNS

Plant biotechnology is to play a pivotal role in making India self reliant in food, health, energy, and environmental security. To achieve this, R&D has to be strengthened because, at present, we stand far behind the developed nations. Along with this, proper scientific culture and accountability has to be developed. Food production has to keep pace with the growth of population. Affluence of biodiversity has to be wisely exploited. Traditional food plants, particularly of cereals and herbs, have to be protected and exploited in a rational way. Green revolution has yielded desired results in terms of food sufficiency in India. But, this revolution has to be transformed into Evergreen revolution (Swaminathan, 2010) if India has to be amongst the developed nations of the 21st century. Energy needs are growing. In the coming 25 years, global energy demand is understood to grow by 60%. Of this, 85% will be provided by fossil fuel. The developing nations like India have to manage the needs wisely in an objective manner. Plants, particularly the genetically engineered ones, will play crucial roles. Alternative sources of energy, particularly, the renewable resources have to be tapped and enriched. Sound techniques should be developed to maximize production of fuel gases, as shown in Table 3.6.

Modern medical and paramedical research is very costly. Bioinformatics will play an important role in designing and developing targeted drugs. Cost and time of drug development will also come down with the use of knowledge of bioinformatics India is rich in herbal and alternative medicine. Ancient Indian texts, particularly the Atharvaveda, possess priceless treasure of knowledge of treatment of human ailments. It is necessary to strengthen R&D to substantiate our pristine health wisdom with the help of modern techniques as has been successfully launched by China (Sharma et al., 2008). No less important are the environmental concerns. All our efforts will go in vain for a secure future if our environment becomes inimical to healthy life. India is facing huge problems in cleaning natural rivers, particularly, the Ganga which is thc lifclinc of North India. Usc of suitable GM plants, genetically engineered organisms along with social awareness will play important role in controlling water pollution. The need of the hour, particularly for the developing nations, is to encourage research that addresses multiple problems in one go. Plants used for food should also carry therapeutics or nutraceuticals without harm to the environment. A lot of progress has been made in this direction. For example, GM potato may now induce or provide plantibodies (immunity booster molecules), rice may provide starch with vitamin A (golden rice), and SUSIBA 2 rice reduces global warming along with concomitant increase in yield (Su et al., 2015; Ye et al., 2000).

Recently, there has been a tendency to introduce *Bt* gene in almost every plant that are domesticated. Indiscriminate expression of *Bt* gene in every plant should be discouraged. Introduction of *Bt* gene in food items, particularly for human consumption, should be avoided as far as possible.

CRISPR-Cas9 is a simple yet highly efficient "gene drive" that promises overall human welfare in every field of biotechnology. This is indeed a high through-put technology with tremendous possibilities, where imagination is the limit, and that tends to lurk on "PLAYING GOD." This technique may be exploited to address the future needs.

3.8 CONCLUSION

Plant biotechnology is full of promise to enable mankind to achieve food, health, energy, and environmental securities in future. Well-planned R&D in this field may secure sound future for mankind. Multipurpose and multi-institutional research schemes should be encouraged to share knowledge and ideas in order to fulfil human needs with environmental security. Along with intensive research on biodiversity, the traditional plant food including the coarse cereals should be revived. Post-harvest protection of the plant produces is another area that countries like India have to strengthen. In the recent years, it has been observed that tons of procured agricultural products are left uncared leading to biodeterioration affecting food security. Intensive R&D is necessary also to modernize the traditional health care system here that is often referred to as an alternative system of medicine. It is expected to have a market of 14.3 billion US$ in the USA by the end of 2016, out of a global market of over 115 billion US$ (web page alternative med in US, 2016). In India, this system had a market of Rs. 2300 crore (about 350 million US$) in 2008, which has been growing at the rate of 15% per annum (Web page, Global alternative med). Developing countries, including India, have to look for alternative sources of energy for the future development. Plant and other biological resources are expected to play crucial role in the energy management. Some countries, like Brazil, have managed up to 60% of their energy requirement through these resources. Total energy demand of India is going to increase from 800 Mtoe at present to 1223 Mtoe by 2035 (Ahn and Graczyk, 2015). Fossil fuel energy will continue to dominate in meeting the demand (over 50%). But, over the years, contribution of this energy will decrease because of the increasing inputs from biomass and waste as well as other renewable resources. GM plants may play important role in meeting this demand. While meeting the above mentioned challenges, it is highly desirable that environmental concerns are made a part and parcel of the development plan; otherwise, the future of mankind will be jeopardized. Raising multipurpose herbs and plants through biotechnology are desirable for a balanced growth that ensures secured future of the developing countries and mankind. Multi-national projects should be launched for tapping abiotic (Althoff et al., 2014) and biotic sources of methane (Nisbet et al., 2009) in a way that brightens the prospect of energy output with resultant reduction in global warming (20%).

KEYWORDS

- **genome editing**
- **CRISPR-Cas system**
- **Ti-plasmid**
- **Flavr Savr**
- **transgenic plant**

REFERENCES

Ahn, S. J.; Graczyk, D. Understanding Energy Challenges in India. [online] 2015, 1–116. https://www.iea.org/publications/freepublications/publication/India_study_FINAL_WEB.pdf.

Barcelo, P.; Lazzeri, P. Direct Gene Transfer: Chemical, Electrical and Physical Methods. In *Transgenic Plant Research;* Lindsey, K., Ed.; Harwood Academic Publishers, **1998**.

Chakrabarty, A. M., Mylroie, J. R., Friello, D. A.; Vacca, J. G. Transformation of *Pseudomonas putida* and *Escherichia coli* with Plasmid-linked Drug-resistance Factor DNA. *Proc. Natl. Acad. Sci.* (USA), **1975,** *72*(9), 3647–3651.

Cohen, S.; Chang, A.; Boyer, H.; Helling, R. Construction of Biologically Functional Bacterial Plasmids In vitro. *Proc. Natl. Acad. Sci.* (USA), **1973,** *70*, 3240–3244.

De Cleene, M.; De Ley, J. The Host Range of Crown Gall. *Bot. Rev*. **1976,** *42*, 389–466.

Doudna, J. A.; Charpentier, E. The New Frontier of Genome Engineering with CRISPR-Cas9. *Science*, **2014,** *346*, 6213–6215.

Frederik, A.; Kathrin, B.; Peter, C.; Colin, M.; Derek, R. B.; Steffen, G.; Frank K. Abiotic Methanogenesis from Organosulphur Compounds Under Ambient Conditions. *Nature Commun.* **2014,** *5*, 4205.

Handwerk, B. Plants Exhale Methane, Add to Greenhouse Effect, Study Says. [Online] **2006**. http://news.nationalgeographic.com/news/2006/01/0111_060111_plant_methane.html

Hatti-Kaul, R. Enzyme Production Biotechnology UNESCO EOLSS Sample Chapters. [Online] **2015**. http://www.eolss.net/sample-chapters/c17/e6-58-05-01.pdf.

Herrera-Estrella, L.; Depicker, A.; Montagu, M. V.; Schell. J. Expression of Chimaeric Genes Transferred into Plant Cells Using a Ti-plasmid-derived Vector. *Nature*, **1983,** *303*, 209–213.

Hsu, P. D.; Scott, D. A.; Weinstein, J. A.; Ran, F. A.; Konermann, S.; Agarwala, V.; Zhang, F. DNA Targeting Specificity of RNA-guided Cas9 Nucleases. *Nat. Biotechnol.* **2013,** *31*(9), 827–832.

Jacquel, N.; Lo, C. W.; Wei, Y. H.; Wu, H. S.; Wang, S. S. Isolation and Purification of Bacterial Poly(3-hydroxyalkanoates). *Biochem. Eng. J.* **2008,** *39*(1), 15–27.

Jinek, M.; Chylinski, K.; Fonfara, I.; Hauer, M.; Doudna, J. A.; Charpentier, E. A Programmable Dual-RNA–Guided DNA Endonuclease in Adaptive Bacterial Immunity. *Science*, **2012,** *337*, 816–821.

Katharina, L.; Michael, B.; Stefan, R.; Thomas, R.; Neu, I. S.; Markus, G.; Claudia, K.; Sylvia, S.; Christoph, M.; Holger, Z.; Frank, K. Evidence for Methane Production by Saprophytic Fungi. *Nat. Commun.* **2012,** *3*, 1046.

Keith, R.; Hiatt, B.; Martineau, B.; Kramer, M.; Sheehy, R.; Sanders, R.; Houck, C.; Emlay, D. *Safety Assessment of Genetically Engineered Fruits and Vegetables: A Case Study of the Flavr Savr Tomato*. CRC Press, **1992**; p 288.

Leslie, M. On the Origin of Photosynthesis. *Science*, **2009,** *323*, 1286–1287.

Lundmark, C. Searching Evolutionary Pathways: Antifreeze Genes from Antarctic Hairgrass. *BioScience*. **2006,** *56*, 552.

Manoj, B.; Zhu, L.; Shen, G.; Payton, P.; Zhang, H. Expression of an Arabidopsis Sodium/Proton Antiporter Gene (AtNHX1) in Peanut to Improve Salt Tolerance. *Plant Biotechnol. Rep.* **2012,** *6*, 59–67.

McNutt, M. Breakthrough to Genome Editing. *Science*, **2015,** *350*, 1445.

Mullis, K. B. Process for Amplifying Nucleic Acid Sequences. U.S. Patent 4, **1985,** 683, 202.

Nisbet R. E. R.; Fisher, R.; Nimmo, R. H.; Bendall, D. S.; Crill, P.M.; Gallego-Sala, A.V.; Hornibrook, E. R. C.; López-Juez, E.; Lowry, D.; Nisbet, P. B. R.; Shuckburgh, E. F.; Sriskantharajah, S.; Howe, C. J.; Nisbet, E. G. Emission of Methane from Plants. *Proceed. Royal Soc. B.* **2009,** *276*, 1347–1354.

Pujari, S. Biochemicals Production in Transgenic Plants. [Online] **2015**. http://www.yourarticlelibrary.com/biotechnology/transgenic-plants/biochemical-production-in-transgenic-plants/33423/

Sharma, A. K.; Mohanty, A.; Singh, Y.; Tyagi, A. K. Transgenic Plants for the Production of Edible Vaccines and Antibodies for Immunotherapy. *Curr. Sci.* **1999,** *77*, 524–529.

Sharma, A.; Shanker, C.; Tyagi, L. K.; Singh, M.; Rao, C. V. Herbal Medicine for Market Potential in India: An overview. *Acad. J Plant Sci*. **2008,** 1, 26–36.

Su. J.; Hu, C.; Yan, X.; Jin, Y.; Chen, Z.; Guan, Q.; Wang, Y.; Zhong, D.; Jansson, C.; Wang, F.; Schnürer, A.; Sun, C. Expression of Barley SUSIBA2 Transcription Factor Yields High-starch Low-methane Rice. *Nature*, **2015,** *523*, 602–606.

Report on the State of Food Insecurity in Rural India. M. S. Swaminathan Research Foundation (MSSRF): Chennai, December **2008**.

Swaminathan, M. S. *From Green to Evergreen Revolution*. Academic Foundation: New Delhi, **2010**.

Swaminathan, M. S.; Bhavani, R. V. Food Production and Availability-Essential Prerequisites for Sustainable Food Security. *Indian J. Med. Res*. **2013,** *138*, 383–391.

Thomas, B. R.; Deynze, A. V.; Bradford, K. J. *Production of Therapeutic Proteins in Plants;* UCANR Publications, **2002**.

Pharmaceutical and Antibody Market. In *Agri Biotech in California Series (Publication No 8078) 1-9.7.* http://www.plantformcorp.com/pdf/plantform_backgrounder-market_overview forweb.pdf.

Thomas, R.; Sah, N. K.; Sharma, P. B. Therapeutic Biology of *Jatropha curcas*: A Minireview. *Curr. Pharmaceut. Biotech.* **2008,** *9*, 315–324.

Tiwari, V.; Chaturvedi, A. K.; Mishra, A.; Jha, B. Introgression of the *SbASR-1* Gene Cloned from a Halophyte *Salicornia brachiata* Enhances Salinity and Drought Endurance in Transgenic Groundnut (*Arachis hypogaea*) and Acts as a Transcription Factor. *Plos One,* **2015,** *10*(7), e0131567.

Travis, J. Making the Cut. *Science*, **2015,** *350*(6267), 1456–1457.

Verma, R.; Sah, N. K. National Policy on Biofuel in India. Jour. Management Value & Ethics (ISSN: 2249-9512), **2013,** *3*, 91–96.

Walden, R.; Schell, J. Techniques in Plant Molecular Biology-progress and Problems. *Eur. J. Biochem.* **1990,** *192*, 563–576.

Ye, X.; Al-Babili, S.; Klöti, A.; Zhang, J.; Lucca, P.; Beyer, P.; Potrykus, I. Engineering the Provitamin A (Beta-carotene) Biosynthetic Pathway into (Carotenoid-free) Rice Endosperm. *Science*, **2000,** *287*(5451), 303–305.

PART II
Plant Tissue Culture

CHAPTER 4

STERILIZATION TECHNIQUE

TUSHAR RANJAN[1*], SANGITA SAHNI[2], BISHUN DEO PRASAD[3], RAVI RANJAN KUMAR[3], KUMARI RAJANI[4], VIJAY KUMAR JHA[5], VAISHALI SHARMA[6], MAHESH KUMAR[3], and VINOD KUMAR[3]

[1]*Department of Basic Science and Humanities Genetics, Bihar Agricultural University, Sabour, Bhagalpur, Bihar, India*

[2]*Department of Plant Pathology, Tirhut College of Agriculture, Dholi, RAU, Pusa, Bihar, India*

[3]*Department of Molecular Biology and Genetic Engineering, Bihar Agricultural University, Sabour, Bhagalpur, Bihar, India*

[4]*Department of Seed Science and Technology, Bihar Agricultural University, Sabour, Bhagalpur, Bihar, India*

[5]*Department of Botany, Patna University, Patna, Bihar, India*

[6]*DOS in Biotechnology, University of Mysore, Mysore, Karnataka, India*

[*]*Corresponding author. E-mail: mail2tusharranjan@gmail.com*

CONTENTS

ABSTRACT

The main aspect of a tissue culture laboratory is the maintenance of aseptic working environment so as to avoid the contamination of growing cultures with microorganisms like bacteria, fungi, and their spores. The tissue culture laboratory has been divided into different work areas viz. (1) general laboratory comprising a washing area and a preparation area and (2) a clean, sterile laboratory comprising an inoculation room, also called as a transfer room and a culture room. The plant tissue culture techniques are set of experimental procedures, performed by various researchers, to grow a large number of cells or tissues under sterile and controlled growth conditions with certain objectives. These experiments essentially require a well-organized and well-equipped laboratory.

4.1 INTRODUCTION

The plant tissue culture techniques are nothing but a set of experimental procedures, performed by various researchers, to grow a large number of cells or tissues under sterile and controlled growth conditions with certain objectives. These experiments essentially require a well-organized and well-equipped laboratory. Before starting any work in the plant tissue culture laboratory, it is very essential to understand the organization of the laboratory and also to know the equipment and instruments required in the laboratory for different purposes (Bhojwani and Razdan, 1990; Misra and Misra, 2012). The main aspect of a tissue culture laboratory is the maintenance of aseptic working environment so as to avoid the contamination of growing cultures with microorganisms like bacteria, fungi, and their spores. The tissue culture laboratory has been divided into different work areas viz. (1) general laboratory comprising a washing area and a preparation area and (2) a clean, sterile laboratory comprising an inoculation room, also called as a transfer room and a culture room (Chawla, 2000). The design of the laboratory allows a workflow pattern that maintains maximum cleanliness and promotes minimal backtracking. The entry to the sterile area is restricted and the door is provided with a clean air curtain, which prevents the outside air form entering into the laboratory when the door is open. Moreover, the doors are opening inside the laboratory so as to maintain positive air pressure preventing the outside air from entering into the sterile area. The laboratory is also provided with a backup power generator set in order to assure the continuous power supply in case of power failure. Different equipment required for tissue culture work is installed in the respective laboratory areas. The tissue culture laboratory is also provided with a greenhouse facility. Every manipulation of tissue culture requires aseptic conditions (Misra and Misra, 2012). Therefore, culture vessels, media, aseptic manipulation aids, and explants to be used in tissue culture laboratories must be sterilized. Aseptic manipulation (inoculation and subculture) is carried out in the laminar airflow hood, the sterile cabinet. Infection may take place in three ways: (1) by air which contains fungal and bacterial spores, (2) explant itself is usually covered with pathogens on its surface, and (3) by user himself. Following sterilization procedures are adopted to keep off the microbial contaminations:

1. Culture media, culture vessels, and instruments used in tissue culture are generally sterilized by autoclaves or pressure cookers maintaining temperature and pressure 121°C and 15 psi, respectively, for at least 20 min. However the temperature, pressure, and duration of sterilization depend upon the nature and quantity of the materials.
2. Thermolabile plant growth regulator (PGR) stocks are sterilized by the filter sterilization.
3. The explants are surface sterilized by disinfectants such as $HgCl_2$ and NaOCl.

Sterilization (or aseptic) term in plant tissue culture laboratory refers to any process that eliminates (removes) or kills (deactivates) all forms of life and other biological agents (such as viruses which some do not consider to be alive but are biological pathogens nonetheless), including transmissible agents (such as fungi, bacteria, viruses, prions, spore forms, etc.) present in a specified region, such as a surface, a volume of fluid, or in a compound such as biological culture media. Sterilization can be achieved with one or more of the following: heat, chemicals, irradiation, high pressure, and filtration. Sterilization is distinct from disinfection, sanitization, and pasteurization in that sterilization kills, deactivates, or eliminates all forms of life and other biological agents. In a tissue culture laboratory, it is very essential to maintain aseptic conditions to avoid the contamination of plant cultures by the microorganisms and their spores. The plant cells, tissues, and organs are cultured in vitro on synthetic culture media (Bhojwani and Razdan, 1990). The culture medium, especially when it contains sugar, also supports the growth of microorganisms like bacteria and fungi. If these microorganisms either in cellular form or in spore form come in contact with the culture medium, due to their short life cycle, grow faster than the cultured plant tissue, subsequently killing the tissue. The culture vessels, instruments, culture medium, and the plant material itself may serve as a source of contamination (Fig. 4.1). Therefore, the surface of plant tissue and all nonliving articles including culture medium needs to sterilize before they are used for tissue culture purpose (Chawla, 2000). Different methods of techniques used in plant tissue culture laboratory are summarized in Table 4.1.

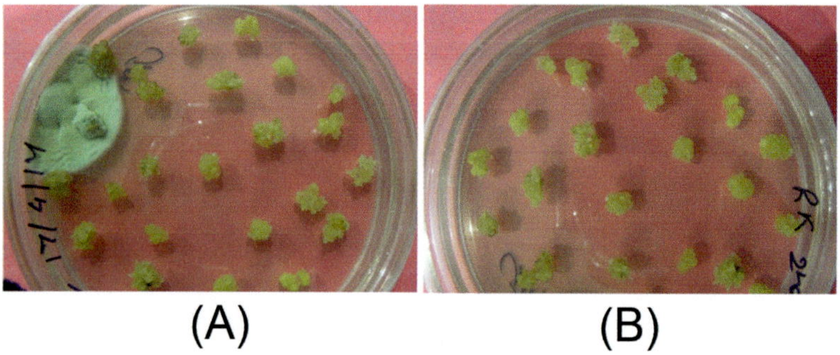

FIGURE 4.1 Contaminated culture (A) versus good culture (B).

TABLE 4.1 Sterilization Techniques Used in Plant Tissue Culture.

Techniques	Material Used
Steam sterilization/autoclaving (121°C at 15 psi for 20–40 min)	Nutrient media, culture vessels, glassware, and plasticwares
Dry heat (160–180°C for 3 h)	Instruments (needle, forceps, etc.), glasswares, tips, pipettes, and plasticwares
Flame sterilization	Instruments (needle, forceps, etc.), mouth of culture vessels
Filter sterilization (membrane filter made up of cellulose nitrate or cellulose acetate of 0.22 μm pore size)	Thermolabile substrates like growth factors, amino acids, vitamins, and enzymes
Alcohol sterilization	Laminar flow cabinet, worker's hand
Surface sterilization (sodium hypochlorite, hydrogen peroxide, etc.)	Explants

4.2 SURFACE STERILIZATION OF PLANT MATERIAL

Prior good care of stock plants may lessen the amount of contamination that is present on explants. Plants grown in the field are typically more "dirty" than those grown in a greenhouse or growth chamber, particularly in humid areas like Florida. Overhead watering increases contamination of initial explants. Likewise, splashing soil on the plant during watering will increase initial contamination. Treatment of stock plants with fungicides and/or bactericides is sometimes helpful. It is sometimes possible to harvest shoots and force buds from them in clean conditions. The forced shoots may then be free of contaminants when surface-sterilized in a normal manner. Seeds may be sterilized and germinated in vitro to provide clean material. Covering growing shoots for several days or weeks prior to harvesting tissue for culture may supply cleaner material. Explants or material from which material will be cut can be washed in soapy water and then placed under running water for 1–2 h (Kumar, 2003; Misra and Misra, 2012).

4.2.1 SODIUM HYPOCHLORITE

Sodium hypochlorite, usually purchased as laundry bleach, is the most frequent choice for surface sterilization. It is readily available and can be diluted to proper concentrations. Commercial laundry bleach is 5.25% sodium hypochlorite. It is usually diluted to 10–20% of the original concentration, resulting in a final concentration of 0.5–1.0% sodium hypochlorite. Plant material is usually immersed in this solution for 10–20 min. A balance between concentration and time must be determined empirically for each type of explant because of phytotoxicity.

4.2.2 ETHANOL/ISOPROPANOL

Ethanol is a powerful sterilizing agent but also extremely phytotoxic. Therefore, plant material is typically exposed to it for only few seconds or minutes. The more tender the tissue, the more it will be damaged by alcohol. Tissues such as dormant buds, seeds, or unopened flower buds can be treated for longer periods of time since the tissue that will be explanted or that will develop is actually within the structure that is being surface-sterilized. Generally 70% ethanol is used prior to treatment with other compounds (Street, 1973).

4.2.3 CALCIUM HYPOCHLORITE

Calcium hypochlorite is used more in Europe than in the United States. It is obtained as a powder and must be dissolved in water. The concentration that is generally used is 3.25%. The solution must be filtered prior to use since not the entire compound goes into solution. Calcium hypochlorite may be less injurious to plant tissues than sodium hypochlorite.

4.2.4 MERCURIC CHLORIDE

Mercuric chloride is used only as a last resort in the United States. It is extremely toxic to both plants and humans and must be disposed of with care. Since mercury is very phytotoxic, it is critical that many rinses with water must be used to remove all traces of the mineral from the plant material.

4.2.5 HYDROGEN PEROXIDE

The concentration of hydrogen peroxide used for surface sterilization of plant material is 30%, ten times stronger than that obtained in a pharmacy. Some researchers have found that hydrogen peroxide is useful for surface-sterilizing material while in the field.

4.2.6 ENHANCING EFFECTIVENESS OF STERILIZATION PROCEDURE

To increase the effectiveness of sterilization, surfactant (e.g., Tween 20) is frequently added to sodium hypochlorite. A mild vacuum may be used during the procedure. The solutions where explants are immersed are often shaken or continuously stirred for effective sterilization.

4.2.7 RINSING

After plant material is sterilized with one of the above compounds, it must be rinsed thoroughly with sterile water. Typically three to four separate rinses are done.

4.2.8 PLANT PRESERVATIVE MIXTURE

Plant preservative mixture (PPM) is a proprietary broad spectrum biocide, which can be used to control contamination in plant cell cultures, either during the sterilization procedure, or as a medium component. PPM comes in an acidic liquid solution (pH 3.8). The recommended dose is 0.5–2.0 mL of PPM per liter of medium. Higher doses are required to treat endogenous contamination and for *Agrobacterium*. Its makers say that PPM has several advantages over antibiotics as it is effective against fungi as well as bacteria, thus it can be substituted for a cocktail of antibiotics and fungicides. PPM is less expensive than antibiotics, which makes it affordable for wide and routine use. The formation of resistant mutants toward PPM is very unlikely because it targets and inhibits multiple enzymes. Many antibiotics adversely affect plant materials. If used as recommended, PPM does not adversely affect in vitro seed germination, callus proliferation, or callus regeneration. Seeds and explants with endogenous contamination can be sterilized with 5–20 mL/L of PPM. This is useful when routine surface sterilization is insufficient (Gautheret, 1983).

4.3 QUANTIFICATION OF CONTAMINATION

The aim of sterilization is the reduction of initially present microorganisms or other potential pathogens. The degree of sterilization is commonly expressed by multiples of the decimal reduction time, or D-value, denoting the time needed to reduce the initial number $\boldsymbol{N_0}$ to one tenth (10^{-1}) of its original value. Then the number of microorganisms $\boldsymbol{N}$ after sterilization time t is given by:

$$N / N_0 = 10^{(-t/D)}. \quad (4.1)$$

The D-value is a function of sterilization conditions and varies with the type of microorganism, temperature, water activity, pH, etc. For steam sterilization (see below) typically the temperature (in Celsius) is given as index. Theoretically, the likelihood of survival of an individual microorganism is never zero. To compensate for this, the overkill method is often used. Using the overkill method, sterilization is performed by sterilizing for longer time than it is required to kill the bioburden present on or in the item being sterilized. This provides a sterility assurance level (SAL) equal to the probability of a nonsterile unit (Smith, 2000).

4.4 DIFFERENT METHODS FOR STERILIZATION

Various methods or chemicals are used for sterilization purposes depending upon the constituents of nutrient medium (Biondi and Thorpe, 1981). Methods used for sterilization in plant tissue culture are discussed below.

4.4.1 MOIST HEAT STERILIZATION

A widely used method for heat sterilization is the autoclave, sometimes called a converter or steam sterilizer (Fig. 4.2). Autoclaves use steam heated to 121–134°C under pressure. To achieve sterility, the article is heated in a chamber by injected steam until the article reaches a time and temperature set point. The article is then held at that set point for a period of time which varies depending on the bioburden present on the article being sterilized and its resistance (D-value) to steam sterilization. A general cycle is 20 mins at 121°C at 100 kPa, which is sufficient to provide a sterility assurance level of 10^{-4} for a product with a bioburden of 10^{6} and a D-value of 2.0 min. Following sterilization, liquids in a pressurized autoclave must be cooled slowly to avoid boiling over when the pressure is released. This may be achieved by gradually depressurizing the sterilization chamber and allowing liquids to evaporate under a negative pressure, while cooling the contents.

FIGURE 4.2 An autoclave working on principle of moist heat sterilization.

Most autoclaves have meters and chart that record or display information, particularly temperature and pressure as a function of time. The information is checked to ensure that the conditions required for sterilization have been met. Indicator tape is often placed on packages of products prior to autoclaving, and some packaging incorporates indicators. The indicator changes color when exposed to steam, providing a visual confirmation. For autoclaving, cleaning is critical. Extraneous biological matter or grime may shield organisms from steam penetration. Proper cleaning can be achieved through physical scrubbing, sonication, and ultrasound or pulsed air (Razdan, 2003).

4.4.1.1 PRECAUTIONS

The following precautions must be taken before and during the use of autoclave.

1. Excessive autoclaving must be avoided as it may degrade media components, particularly sucrose and agar under high pressure and acidic environment. Few microorganisms may exist and can survive even at elevated temperature, but sterilization for 15–30 min may kill them.
2. The level of water must be checked as per the indicator before the operation.
3. The lid of the autoclave must be closed properly.
4. Air exhaust must be functioning normally.
5. Sudden reduction of pressure after autoclaving should be avoided. In doing so media begin to boil again and the media in the containers might burst out from their closures because of sudden drop in pressure.
6. Tight screwing of bottle should be avoided and their tops should be left loose during autoclave. After autoclaving these bottles should be kept in the laminar airflow and the tops of these bottles should be tightened just after cooling.

4.4.2 DRY HEAT STERILIZATION

Dry heat was the first method of sterilization, and is a longer process than moist heat sterilization. The destruction of microorganisms through the use of dry heat is a gradual phenomenon. With longer exposure to lethal temperatures, the number of killed microorganisms increases. Forced ventilation of hot air can be used to increase the rate at which heat is transferred to an organism and reduce the temperature and amount of time needed to achieve sterility (Fig. 4.3). At higher temperatures, shorter exposure times are required to kill organisms. This can reduce heat-induced damage to food products.

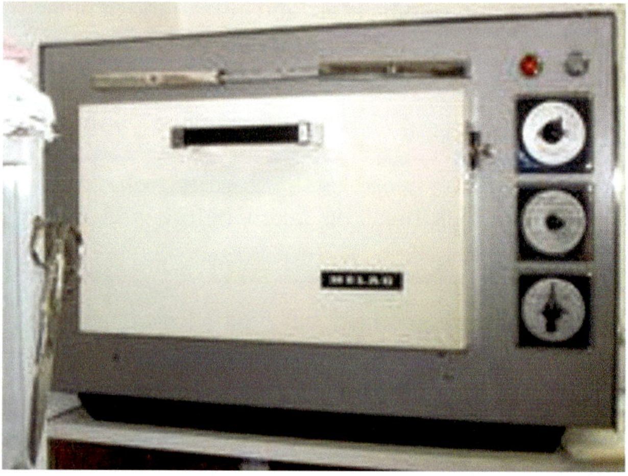

FIGURE 4.3 Hot air oven working on the principle of dry heat sterilization.

The standard setting for a hot air oven is at least 2 h at 160°C. A rapid method heats air to 190°C for 6 min for unwrapped objects and 12 min for wrapped objects. Dry heat has the advantage that it can be used on powders and other heat-stable items that are adversely affected by steam (e.g., it does not cause rusting of steel objects) (Casolari, 2004).

4.4.3 INCINERATION

Incineration is a waste treatment process that involves the combustion of organic substances contained in waste materials. This method also burns any organism to ash. It is used to sterilize biohazardous waste before it is discarded with nonhazardous waste. Bacteria incinerators are mini furnaces used to incinerate and kill off any microorganisms that may be on an inoculating loop or wire.

4.4.4 TYNDALLIZATION

Named after John Tyndall, tyndallization is an obsolete and lengthy process designed to reduce the level of activity of sporulating bacteria that are left by a simple boiling water method. The process involves boiling for a period (typically 20 min) at atmospheric pressure, cooling, incubating for a day, then repeating the process a total of three to four times. The incubation periods are to allow heat-resistant spores surviving the previous boiling period to germinate to form the heat-sensitive vegetative (growing) stage, which can be killed by the next boiling step. This is effective because many spores are stimulated to grow by the heat shock. The procedure only works for media that can support bacterial growth, and will not sterilize nonnutritive substrates like water. Tyndallization is also ineffective against prions (Thiel, 1999).

4.4.5 GLASS BEAD STERILIZATION

Glass bead sterilizers work by heating glass beads to 250°C. Instruments are then quickly doused in these glass beads, which heat the object while physically scraping contaminants off their surface. Glass bead sterilizers were once a common sterilization method employed in dental clinics as well as biologic laboratories, but are not approved by the U.S. Food and Drug Administration (FDA) and Centers for Disease Control and Prevention (CDC) to be used as sterilizers since 1997. They are still popular in European as well as Israeli dental practices although there are no current evidence-based guidelines for using this sterilizer (Zadik and Peretz, 2008).

4.5 STERILIZATION BY CHEMICALS

Chemicals are also used for sterilization. Heating provides a reliable way to rid objects of all transmissible agents, but it is not always appropriate if it will damage heat-sensitive materials such as biological materials, fiber optics, electronics, and many plastics. In these situations, chemicals, either as gases or in liquid form, can be used as sterilants. While the use of gas and liquid chemical sterilants avoids the problem of heat damage, users must ensure that article to be sterilized is chemically compatible with the sterilant being used. In addition, the use of chemical sterilants poses new challenges for workplace safety, as the properties that make chemicals effective sterilants usually make them harmful to humans. Table 4.2 indicates the range of percentage and treatment periods routinely used for sterilizing agents.

TABLE 4.2 A Comprehensive List of Chemicals, Concentration, and Treatment Time Period Range Used as Sterilizing Agent for Explants.

Chemicals	Concentration (% w/v)	Treatment Time (Min)
Sodium hypochlorite	1–1.4	5–30
Hydrogen peroxide	10–12	5–15
Calcium hypochlorite	9–10	5–30
Silver nitrate	1	5–30
Mercuric chloride	0.01–1	2–10
Bromine water	1–2	2–10

4.5.1 ETHYLENE OXIDE

Ethylene oxide gas is commonly used to sterilize objects that are sensitive to temperatures greater than 60°C and/or radiation such as plastics, optics, and electrics. Besides moist heat and irradiation, ethylene oxide is the most common sterilization method used for over 70% of total sterilizations, and for 50% of all disposable medical devices. Ethylene oxide treatment is generally carried out between 30°C and 60°C with relative humidity above 30% and a gas concentration between 200 mg/L and 800 mg/L, and typically lasts for at least 3 h. Ethylene oxide penetrates well, moving through paper, cloth, and some plastic films and is highly effective. Ethylene oxide can kill all known viruses, bacteria (including spores), and fungi, and is compatible with most materials even when repeatedly applied. However, it is highly flammable, toxic, and carcinogenic with a potential to cause adverse reproductive effects. Ethylene oxide sterilizers require biological validation after sterilization installation, repairs, or process failure. A typical process consists of a preconditioning phase, an exposure phase, and a period of post-sterilization aeration to remove ethylene oxide residues and by-products such as ethylene glycol and ethylene chlorohydrin. The two most important ethylene oxide sterilization methods are: (1) the gas chamber method and (2) the micro-dose method. To benefit from economies of scale, ethylene oxide has traditionally been delivered by flooding a large chamber with a combination of ethylene oxide and other gases used as diluents (usually CFCs or carbon). Drawbacks of this method include air contamination produced by CFC's and ethylene oxide residuals, operator exposure risks, training costs, and flammability issues requiring special handling and storage.

4.5.2 NITROGEN DIOXIDE

Nitrogen dioxide (NO_2) gas is a rapid and effective sterilant for use against a wide range of microorganisms, including common bacteria, viruses, and spores. The unique physical properties of NO_2 gas allow for sterilant dispersion in an enclosed environment at room temperature and ambient pressure. The mechanism for lethality is the degradation of DNA in the spore core through nitration of the phosphate backbone, which kills the exposed organism as it absorbs NO_2. This degradation occurs at even very low concentrations of the gas. NO_2 has a boiling point of 21°C at sea level, which results in a relatively high saturated vapor pressure at ambient temperature. Because of this, liquid NO_2 may be used as a convenient source for the sterilant gas. Liquid NO_2 is often referred to by the name of its dimer, dinitrogen tetroxide (N_2O_4). Additionally, the low levels of concentration required, coupled with the high vapor pressure, assure that no condensation occurs on the devices being sterilized. This means that no aeration of the devices is required immediately following the sterilization cycle. NO_2 is also less corrosive than other sterilant gases, and is compatible with most medical materials and adhesives (Gorsdorf et al., 1990).

4.5.3 GLUTARALDEHYDE AND FORMALDEHYDE

Glutaraldehyde and formaldehyde solutions (also used as fixatives) are accepted liquid sterilizing agents, provided that the immersion time is sufficiently long. To kill all spores in a clear liquid can take up to 22 h with glutaraldehyde and even longer with formaldehyde. The presence of solid particles may lengthen the required period or render the treatment ineffective. Sterilization of blocks of tissue can take much longer, due to the time required for the fixative to penetrate. Glutaraldehyde and formaldehyde are volatile, and toxic by both skin contact and inhalation. Glutaraldehyde has a short shelf life (<2 weeks), and is expensive. Formaldehyde is less expensive and has a much longer shelf life if some methanol is added to inhibit polymerization to paraformaldehyde, but is much more volatile. Formaldehyde is also used as a gaseous sterilizing agent; in this case, it is prepared on site by depolymerization of solid paraformaldehyde. Many vaccines, such as the original Salk polio vaccine, are sterilized with formaldehyde.

4.5.4 HYDROGEN PEROXIDE

Hydrogen peroxide, both as liquid and vaporized hydrogen peroxide (VHP), is another chemical sterilizing agent. Hydrogen peroxide is a strong oxidant, which allows it to destroy a wide range of pathogens. Hydrogen peroxide is used to sterilize heat- or temperature-sensitive articles such as rigid endoscopes. In medical sterilization, hydrogen peroxide is used at higher concentrations, ranging from around 35% up to 90%. The biggest advantage of hydrogen peroxide as a sterilant is the short cycle time. Whereas the cycle time for ethylene oxide may be 10–15 h, some modern hydrogen peroxide sterilizers have a cycle time as short as 28 min.

4.6 RADIATION STERILIZATION

Sterilization can be achieved using electromagnetic radiation such as electron beams, X-rays, gamma rays, or irradiation by subatomic particles. Electromagnetic or particulate radiation can be energetic enough to ionize atoms or molecules (ionizing radiation), or less energetic (nonionizing radiation) (Bauman et al., 2006).

4.6.1 NONIONIZING RADIATION STERILIZATION

Ultraviolet light irradiation (UV, from a germicidal lamp) is useful for sterilization of surfaces and some transparent objects. Many objects that are transparent to visible light absorb UV. UV irradiation is routinely used to sterilize the interiors of biological safety cabinets between uses, but is ineffective in shaded areas, including areas under dirt (which may become polymerized after prolonged irradiation, so that it is very difficult to remove). It also damages some plastics, such as polystyrene foam if exposed for prolonged periods of time. UV lights may be used to kill organisms in the experimental zone where aseptic manipulation and subculture is carried out. Nowadays every laminar airflow hood remains fitted with UV light source. It is, however, dangerous and should not be turned on while any other work is in progress because UV light of few wavelengths can damage even eyes and skin (Bauman et al., 2006).

4.6.2 IONIZING RADIATION STERILIZATION

The safety of irradiation facilities is regulated by the United Nations International Atomic Energy Agency and monitored by the different national Nuclear Regulatory Commissions. The incidents that have occurred in the past are documented by the agency and thoroughly analyzed to determine root cause and improvement potential. Such improvements are then mandated to retrofit existing facilities and future design. Gamma radiation is very penetrating, and is commonly used for sterilization of disposable medical equipment, such as syringes, needles, cannulas and IV sets, and food. It is emitted by a radioisotope, usually cobalt-60 (^{60}Co) or caesium-137 (^{137}Cs). Electron beam processing is also commonly used for sterilization. Electron beams use an on–off technology and provide a much higher dosing rate than gamma or X-rays. Due to the higher dose rate, less exposure time is needed and thereby any potential degradation to polymers is reduced. A limitation is that electron beams are less penetrating than either gamma or X-rays. Facilities rely on substantial concrete shields to protect workers and the environment from radiation exposure.

4.6.3 X-RAYS

High-energy X-rays (produced by bremsstrahlung) allow irradiation of large packages and pallet loads of medical devices. They are sufficiently penetrating to treat multiple pallet loads of low-density packages with very good dose uniformity ratios. X-ray sterilization does not require chemical or radioactive material: high-energy X-rays are generated at high intensity by an X-ray generator that does not require shielding when not in use. X-rays are generated by bombarding a dense material (target) such as tantalum or tungsten with high-energy electrons in a process known as bremsstrahlung conversion. These systems are energy inefficient, requiring much more electrical energy than other systems for the same result. Irradiation with X-rays or gamma rays, electromagnetic radiation rather than particles, does not make materials radioactive. Irradiation with particles may make materials radioactive, depending upon the type of particles and their energy, and the type of target material: neutrons and very high-energy particles can make materials radioactive, but have good penetration, whereas lower energy particles (other than neutrons) cannot make materials radioactive, but have poorer penetration. Sterilization by irradiation with gamma rays may, however, in some cases affect material properties (Bharati et al., 2009).

4.7 FILTER STERILIZATION

Fluids that would be damaged by heat, irradiation, or chemical sterilization, such as drug products, can be sterilized by microfiltration using membrane filters. This method is commonly used for heat-labile pharmaceuticals and protein solutions in medicinal drug processing. A microfilter with pore size 0.2 μm will usually effectively remove microorganisms. In the processing of biologics, viruses must be removed or inactivated, requiring the use of nanofilters with a smaller pore size (20–50 nm). Smaller pore sizes lower the flow rate, so to achieve higher total throughput or to avoid premature blockage, prefilters might be used to protect small-pore membrane filters. Membrane filters used in production processes are commonly made from materials such as mixed cellulose ester or polyethersulfone (PES). The filtration equipment and the filters themselves may be purchased as presterilized disposable units in sealed packaging, or must be sterilized by the user, generally by autoclaving at a temperature that does not damage the fragile filter membranes. To ensure proper functioning of the filter, the membrane filters are integrity tested post use and sometimes before use. The nondestructive integrity test assures the filter is undamaged, and is a regulatory requirement. Typically, terminal pharmaceutical sterile filtration is performed inside of a clean room to prevent contamination (Raju and Cooney, 1993).

4.8 LAMINAR AIRFLOW CABINET

This is the primary equipment used for aseptic manipulation. This cabinet should be used for horizontal airflow from the back to the front, and should be equipped with gas corks connected with gas burners. Air is drawn in electric fans and passed through the coarse filter and then through the fine bacterial filter. High efficiency particulate air filter (HEPA) is an apparatus designed in such a way that airflow through the working place flows in direct lines (i.e., laminar flow). Care is taken not to disturb this flow by putting anything in the path of airflow. Before commencing any experiment, it is desirable to clean the working surface with 70% alcohol. The air filters should be cleaned and changed periodically (Misra and Misra, 2012).

4.9 CONCLUSION

Despite the most stringent use of sterile techniques by the skilled person, however, the contamination of plant cultures remains a persistent problem that can result in losses ranging from small number of cultures to the catastrophic loss of whole batches of culture medium and tissue culture materials. Contamination by bacteria and fungi is an insidious process that continually threatens plant tissue cultures throughout the duration of culture period. Despite the fact that plant tissue cultures may be sterile when initiated, microorganisms can often contaminate cultures at any point during subsequent tissue culture manipulations. Current chapter covers on fundamental aspects of sterilization procedure. Here, we have discussed important techniques and methods for sterilization. However, there are several other methods which are also available for sterilizing tissue culture laboratory, equipment, and the plant materials.

KEYWORDS

- **aseptic**
- **sterilization**
- **disinfectants**
- **explant**
- **dry heat**
- **moist heat**

REFERENCES

Bauman, P. A.; Lawrence, L. A.; Biesert, L.; Dichtelmüller, H.; Fabbrizzi, F.; Gajardo, R.; Gröner, A.; Jorquera, J. I.; Kempf, C.; Kreil, T. R.; von Hoegen, I.; Pifat, D. Y.; Petteway, S. R. Jr.; MCai, K. Critical Factors Influencing Prion Inactivation by Sodium Hydroxide. *Vox. Sang.* **2006,** *91* (1), 34–40.

Bharati, S.; Soundrapandian, C.; Basu, D.; Datta, S. Studies on a Novel Bioactive Glass and Composite Coating with Hydroxyapatite on Titanium Based Alloys: Effect of γ-Sterilization on Coating. *J. Eur. Ceram. Soc.* **2009,** *29* (12), 2527–2535.

Bhojwani, S. S.; Razdan M. K. *Plant Tissue Culture: Theory and Practice*. Elsevier, 1990.

Biondi, S.; Thorpe, T. A. *Requirement for Tissue Culture Facility. Plant Tissue Culture: Application and Methods in Agriculture;* Academic Press: New York, 1981; pp 1–20.

Casolari, A. Food Sterilization by Heat. *Liberty Knowledge Reason*, **2004**.

Chawla, H. S. *Introduction to Plant Biotechnology* (2nd Edn.); Science Publishers: Enfield, NH, **2002**.

Gautheret, R. J. Plant Tissue Culture: A History. *Bot. Mag. (Tokyo)* **1983,** *96* (4), 393–410.

Görsdorf, S.; Appel, K. E.; Engeholm, C.; Obe, G. Nitrogen Dioxide Induces DNA Single-strand Breaks in Cultured Chinese Hamster Cells. *Carcinogenesis* **1990,** *2* (1), 3–6.

Kumar, U. *Methods in Plant Tissue Culture*. Agrobios, **2003**.

Misra, A. N.; Misra, M. *Sterilisation Techniques in Plant Tissue Culture,* **2012**.

Raju, G. K.; Cooney, C. L. Media and Air Sterilization. In *Biotechnology 2E, Bioprocessing;* Stephanopoulos, G., Ed.; Wiley-VCH: Weinheim, **1993**; Vol. 3, pp 157–184.

Razdan, M. K. *Introduction to Plant Tissue Culture* (2nd Edn.); Oxford Publishers: Enfield, NH, **2003**.

Smith, R. H. *Plant Tissue Culture: Techniques and Experiments* (2nd Edn.); Academic Press, **2000**.

Street, H. E. *Plant Tissue and Cell Culture*. Blackwell Scientific Publication: Oxford, **1973**; p 300.

Thiel, T. Sterilization of Broth Media by Tyndallization. In *Science in the Real World,* **1999**; pp 3–6.

Zadik, Y.; Peretz, A. The Effectiveness of Glass Bead Sterilizer in the Dental Practice. *J. Isr. Dent. Assoc.* **2008,** *25* (2), 36–39.

CHAPTER 5

BASIC PRINCIPLES AND RECENT ADVANCES IN ANTHER/POLLEN CULTURE FOR CROP IMPROVEMENT

GOVINAL BADIGER BHASKARA*

Department of Integrative Biology, University of Texas, Austin, Texas-78712, United States

**E-mail: bhaskartigp@gmail.com*

CONTENTS

ABSTRACT

Agricultural activity demands improved crop varieties with desirable traits such as quality, crop yield, and resistance to environmental stresses. Use of haploids has emerged as a key strategy for crop improvement. Haploids having a single set of chromosomes in the sporophytic phase have become a valuable source to screen for desired traits or to introduce a mutation in their genetic content. Furthermore, doubled haploids (DHs) can be obtained by spontaneous or induced chromosome doubling. DHs are homozygous at all loci, and they can be propagated through seed. DHs achieve complete homozygosity in a single generation. On the contrary, the conventional breeding method requires six to seven generations of self-crossing. In vitro production of haploids for crop improvement has been successfully achieved in many crops such as rice, wheat, barley, maize, tomato, potato, brassicas, grapes, sunflower and so on. In addition to crop improvement, DHs are an excellent source for gene mapping, cytogenetic research, and evolutionary studies. This chapter will focus on basic principles and recent advances in haploids production and their use for crop improvement.

5.1 INTRODUCTION

Crop improvement remains a major challenge for plant scientists. Agriculture productivity is constantly affected by abiotic and biotic stresses. Abiotic stresses such as drought, salinity, and extreme temperature and biotic stresses exerted by bacteria, virus, and fungi causes a severe loss in crop yield. It is, therefore, necessary to improve crops which possess desired traits such as increased resistance to biotic and abiotic stresses, increased tolerance to insect pests and herbicides, and increased nutrition and yield of the crops. In early years, farmers practiced crop improvement by selecting plants with desirable traits to collect seeds for subsequent farming. Since then, plant breedingis practiced in many ways to improve crops. In time periods, forced hybridization of plants is practiced in which two parent plants with desirable traits are identified and crossed each other to produce anew variety of crops with desirable traits termed as classical plant breeding.

Mendel's laws of genetics and advanced molecular biology studies revealed thatgenes controlled the traits of an organism. Genes are a segment of chromosomes, and, hence, crossing two plants to produce one individual plant is nothing but manipulation of the combination of chromosomes. The new combination of chromosomes may incorporate desired traits in the progeny. The alternative way for manipulation of chromosomes is to produce polyploidy plants. Plants are mostly diploid (2n), contain two sets of chromosomes. However, plants with three or more sets of chromosomes (polyploidy) are also common. Polyploidy in crops could be inducedby treating plants with chemical colchicine. Colchicine inhibits the spindle formation and arrests mitosis. Therefore, colchicine-treated cells contain double the number of actual chromosomes per cell. Due to its increase size in genetic content, polyploid plants may have desirable traits. In recent years, introducing the mutations in plants genetic content by treating plants with mutagenic chemicals, like ethyl methanesulfonate (EMS), or by radiations, such as gamma radiations, and select for the desired traits has been in practice. This method is popularly known as forward genetics. However, these mutagens often introduce many unwanted mutations, and sometimes they are deleterious to the progeny. Hence, progeny would then be crossed with one of the selected parental lines (backcross) to get rid of the unwanted mutations to obtain near-isogenic lines (NILs). NILs are homozygous for one of the parental lines but carry the desired mutation.

In practice of plant breeding, it is necessary to obtain homozygous lines to fix agronomic traits so that they donotlose the desired traits obtained from crossing event. Homozygosity could be achieved by crossing plants themselves (self-pollination or inbreeding) for several generations and plants/lines obtained through these crosses called"inbred"varieties. However, achieving homozygosity require several backcrosses or at least six generation of inbreeding. Therefore, these procedures are time-consuming and labor intensive. The quicker approach to achieve homozygosity is the use of haploid plants. Haploids contain a single set of chromosomes, and often they possess desirable traits. In diploid plants, the presence of the dominant allele may suppress these traits. Furthermore, we need to produce doubled haploids (DHs) for cultivation purpose. DHs can be obtained by spontaneous or by treating haploids with colchicine to induce the chromosome doubling. DHs are completely homozygous at all loci, and they can produce seeds via recombination. DHs achieve homozygosity in one generation which certainly cut down the significant amount of time to obtain complete homozygous lines compared to the conventional breeding method explained above.

5.2 MODE OF EMBRYOGENESIS IN PLANTS

Haploid occurs naturally, but the event of occurrence is very rare and also limited to few species. Hence, breeding requires *in vitro* (in flasks or test tubes) production of haploids from several crop species. A critical step in haploids production is selecting the haploidic tissue at a proper stage of development and, make the tissue embryogenic (capacity to produce embryos) by culturing it in proper tissue culture media with suitable growth parameters. Therefore, in vitro production of haploids is challenging and it also requires knowledge of embryogenesis. There are different modes of embryogenesis in plants as discussed below.

5.2.1 *ZYGOTIC EMBRYOGENESIS*

The process of embryo formation from zygote can be defined as zygotic embryogenesis. Plants undergo reproductive cycle to produce new individuals (offspring). A typical plant consists of roots, shoots, leaves, and reproductive organs (flower and fruit) called as sporophyte which is diploid (2n). During reproductive cycle, sporophyte produces multicellular male and female gametophytes within the flower. The gametophytes are haploid (n) as they are the result of meiotic cell division (Fig. 5.1). The gametophytes

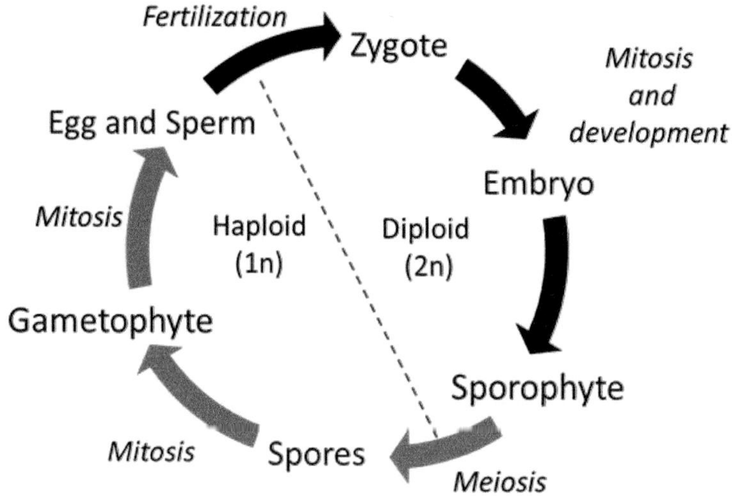

FIGURE 5.1 Plant life cycle shows an alternation of haploid and diploid generations. A diploid sporophyte produces haploid spores via meiosis. Spores further undergo several mitotic divisions to produce haploid gametophyte and eventually produce haploid gametes. Male (sperm) and female (egg) gametes undergo fertilization to produce Zygote which further divides to develop an embryo and a new individual plant (sporophyte) eventually. Grey and black arrows indicate the process of haploid and diploid generations, respectively. (Adapted from https://brainly.com/question/417970)

further undergo mitotic divisions to produce mature gametes. Both male and female gametes are haploid as they are produced from haploid gametophytes. The male gamete is mature pollen grain (sperm) produced by anthers and female gamete (egg) produced by the cells of the ovule in the ovary. The fusion of male and female gametes produces diploid zygote through a process known as fertilization. The zygote further undergoes several mitotic cell divisions to produce embryo and subsequent seed development. The embryo in the seed has all the basic features to produce acomplete plant. Thus, every plant has a life cycle of alternation between haploid and diploid generations.

5.2.2 APOMICTIC EMBRYOGENESIS

Production of embryo by apomix is is called apomictic embryogenesis. Apomixis is simply defined as, "the asexual formation of seed from the maternal tissues of the ovule, avoiding the processes of meiosis and fertilization, leading to embryo development"(Bicknell and Koltunow, 2004). There are two types of apomixis in plants: sporophytic and gametophytic (Fig. 5.2A and B for the comparison between zygotic and apomictic embryogenesis). In sporophytic apomixis, the embryo develops directly from the unreduced (diploid) somatic cell of the ovule (e.g., Integument, nucellus) also called as adventitious embryony. In gametophytic apomixis, the embryo develops from unreduced embryo sac. Unreduced embryo sac can originate from megaspore mother cell that failed to undergo reductional division, which is termed as diplospory, and, in some case, the somatic cell itself act as spore to produce embryo sac, which is termed as apospory. Unreduced embryo sac further develops and forms diploid egg cell and, diploid synergids and antipodals. The diploid egg cell eventually develops into an embryo without fertilization, which is called as parthenogenesis. Synergids and antipodals can also form embryos through a process called apogamy. In gametophytic apomixis, central cell develops autonomously to form functional endosperm; however, apomixis in some species required pollination to initiate endosperm development, which is called as pseudogamy. It is also reported in gymnosperm *Cupressus dupreziana* that embryos can also originate from unreduced pollen grain, which is known as paternal apomixis.

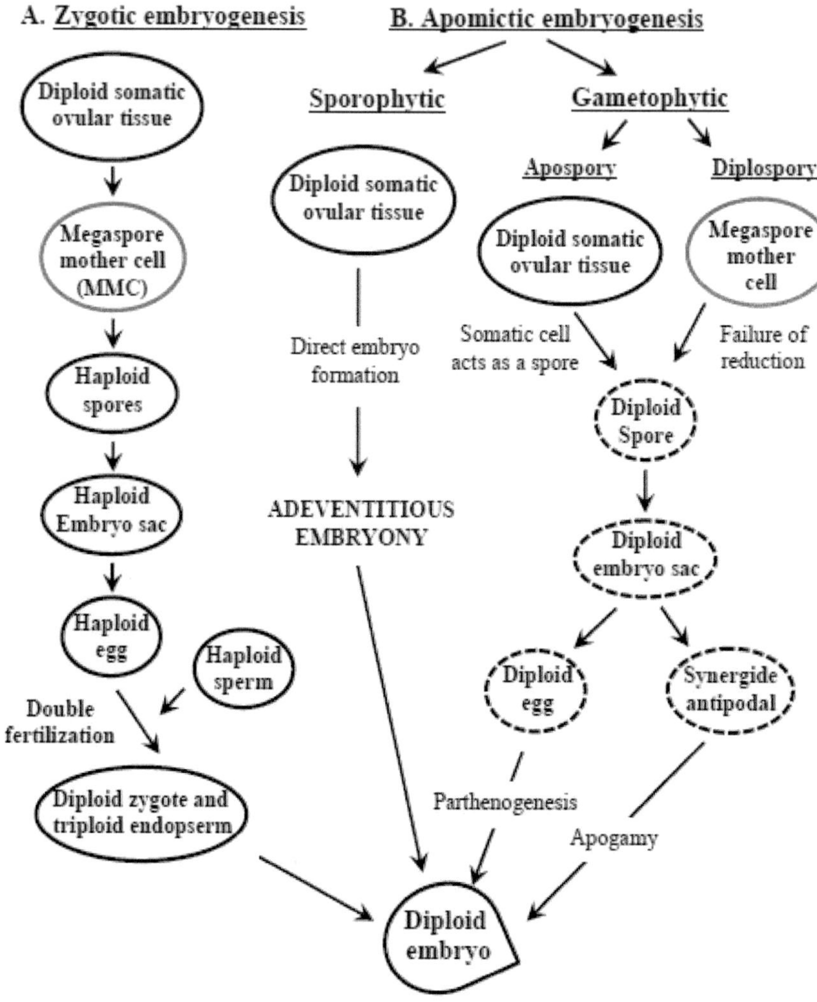

FIGURE 5.2 Comparison of zygotic (sexual) and apomictic (asexual) pathways for embryogenesis. (A) zygoticembryogenesis is showing gametes formation and fertilization to form zygote to produce anembryo. (B) Apomictic embryogenesis is showing sporophytic apomixis and gametophytic apomixis (apospory and diplospory). (Adapted from Koltunow et al., 1995)

5.2.3 SOMATIC EMBRYOGENESIS

Embryogenesis in plant kingdom is different from that of animal kingdom. Plant cells are totipotent in nature. Cellular totipotency can be defined as capacity of the plant cells to regenerate into whole plant. It is therefore, plants can initiate embryogenesis from many somatic cells other than a zygote which is termed as somatic embryogenesis, which produces offspring genetically identical (clonal propagation) to its parent (unless the somaclonal variation). Plant has unique cellular totipotency which allowed us to culture tissues from almost any part of the plant to generate new individual as a practice of plant tissue culture. The possibility of regenerating an entire plant from a single or few non-zygotic cells was proposed by Gottlieb Haberlandt (1854–1945) in 1902. Haberlandt is now popularly called the father of tissue culture. In 1958, Frederick C. Steward (1904–1993) and colleagues at Cornell University succeeded in producing plants from segments prepared from taproot of carrot using tissue culture techniques. These small explants consisted of differentiated secondary phloem cells.

5.2.4 GAMETIC EMBRYOGENESIS (ANDROGENESIS OR GYNOGENESIS)

Gametogenesis is an asexual mode of reproduction in the plant kingdom. Gametic cells produce gametes (pollens or embryo sac) under normal conditions via agametophytic pathway. However, under certain stages due to unfavorable environmental conditions, gametic cells directly produce embryos or embryo-like structure (ELS) instead of gametes. These embryos or ELS further develop into haploid sporophyte contains a single set of chromosomes. Haploids thus can be defined as "a sporophyte contain a single set of chromosomes (2n = x)."Haploids are sterile plants as they cannot undergo meiotic recombination to pair their chromosomes. Therefore, they no longer produce seeds andhaploids cannot be cultivatedthroughpropagation. However, this is not a limiting factor for the use of haploids in breeding as doubled haploids (DHs) can be generated by spontaneous or induced chromosome doubling (Burk et al., 1979; Kochhar et al., 1971; Sunderland and Wicks, 1971).

DHs are fertile, and they can propagate through seeds. Moreover, they are completely homozygous at all loci. DHs are of a significant advantage since they acquire complete homozygosity in one generation, which shortens the breeding time and allows an easy and quick selection of agronomic traits. On the other hand, producing homozygous lines by conventional breeding procedures are time-consuming and labor intensive as it demands several generations of backcrossing or self-pollination.

Gametic embryogenesis is of two kinds: Androgenic embryogenesis (androgenesis) and Gynogenic embryogenesis (Gynogenesis). During androgenesis, a male gametic cell (microspore) produces embryos directly instead of differentiating into mature pollen grains/gametes via a gametophytic pathway. Therefore, androgenesis can be defined as a capacity of a male gametic cell to switch completely from their gametophytic pathway of development to sporophytic growth and produce embryo (direct embryogenesis) or via callus formation (indirect embryogenesis) (Touraev et al., 1997). Gynogenic embryogenesis (gynogenesis) is the production of embryos exclusively from a female gametophyte (ovules, placenta attached ovules, ovaries or whole flower buds)

Androgenesis is extensively studied, and the technique has been used to produce haploids from several species. On the contrary, use of gynogenesis is limited due to lack of established protocols as well as the lack of detailed study on gametoclonal variation among gynogenic haploids.

5.3 CELLULAR ASPECTS OF ANDROGENESIS

During normal pollen development, pollen mother cell (PMC) undergoes meiosis to produce four haploid microspores (microsporogenesis). Microspores are uninucleated (single nucleus) and contain large vacuole. Microspores undergo asymmetric division (microspore mitosis) gives rise to small generative cell and a large vegetative cell, this bicellar product is called pollen. Generative cell within the young pollen divides again (second mitosis) to produce two sperm cells but the vegetative cell arrests in the G1 phase of the cell cycle and stops further dividing. The pollen grain at this stage accumulates starch granules and other storage products until the grain is entirely enlarged (Bedinger, 1992; Gresshoff and Doy, 1972; Maraschin et al., 2005; McCormick, 1993). Thus, pollen at this stage is highly differentiated gametophytic cell, and it will be ready to take process in sexual reproduction. In some cases, microspores produce embryo in an asexual mode, and these embryos can be called as androgenic embryos, and the process is known as androgenic embryogenesis. Therefore, androgenic embryo results from a complete reprogramming in the normal gametophytic development of microspore to sporophytic growth. Several lines of experimental evidences suggest that transition between microsporogenesis to microgametogenesis is critical point to determine the pathway because the disturbance at the first mitosis (microspore mitosis) is sufficient to push microspore toward sporophytic pathway to produce embryos instead of pollen.

There are three overlapping phases during the mode of action in androgenesis (Fig. 5.3) (Maraschin et al., 2005):

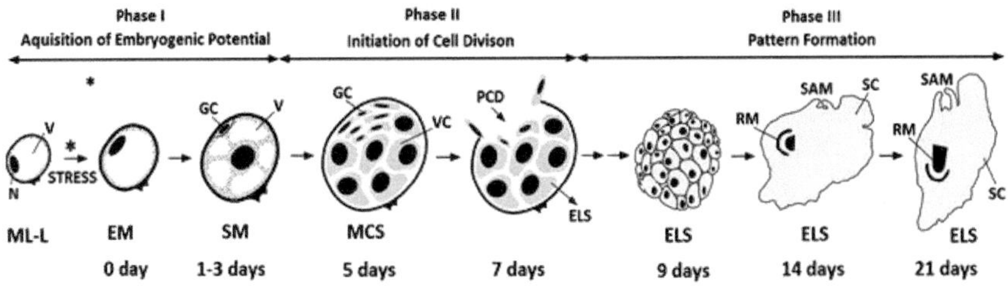

FIGURE 5.3 In vitro androgenic development in barley illustrating the three different phases of embryogenic development. ELS, Embryo-like structure; EM, enlarged microspore; GC, generative cell; MCS, multicellular structure; ML-L, mid-late to late uninucleate microspore; N, nucleus; PCD, programmed cell death; RM, root meristem; SAM, shoot apical meristem; SC, scutellum; SM, star-like microspore; V, vacuole; VC, vegetative cell. (Adapted from Maraschin et al., 2005)

Phase I: ***Acquisition of embryogenic potential.*** This phase involves in the repression of thecapacity of agametic cell to take part in the gametophytic development and leads to dedifferentiation of the cells to acquire an androgenic capacity. Various physiological and environmental factors influence this developmental phase.

PhaseII: ***Initiation of cell division.*** This phase involves early segmentation of microspores leads to the formation of multicellular structures (MCSs) within the pollen wall (exine).

Phase III: ***Pattern formation.*** In this phase, the release of MCSs by the bursting of pollen wall and pattern formation takes place to form embryo-like structures (ELS), also referred as androgenic embryoids.

5.4 DIFFERENT MODES OF ANDROGENESIS

Two steps are necessary for androgenesis: change of the gametophytic program to induce embryogenetic route and development of the embryoid (Sangwan and Camefort, 1983). Androgenetic embryogenesis has only been observed experimentally in vitro when microspores are cultivated at a specific developmental stage (Bhojwani et al., 1973). In vitro culture of anther at the tetrad stage or the binucleate stage can be used to induce androgenesis (Kameya and Hinata, 1970); however, microspores just before or at the time of first mitosis (microspore mitosis) were shown to be more androgenic (Nitsch and Nitsch, 1969). The androgenetic pathway in *Hyoscyamus niger* occurs in uninucleate pollen grains which synthesize mRNA during the first hour of culture. During anther culture, microspores exhibit different modes of development leading to androgenesis (Fig. 5.4).

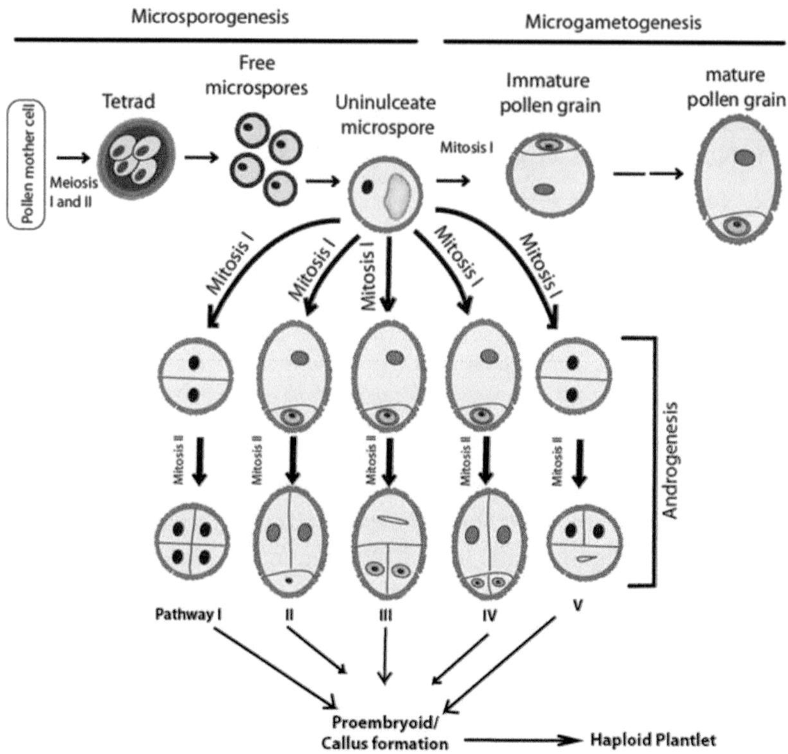

FIGURE 5.4 Different modes of androgenesis. Pollen mother cell produces mature pollen grain via gametophytic pathway under normal condition. The environmental stresses trigger the sporophytic development in uninucleate microspore just before or at the first mitosis. Uninucleate microspore then achieves embryogenic potential through different modes of action to produce embryos which subsequently develops in haploid plantlet. (Adapted from Bhojwani et al., 2015, and modified with references to Heberle-Bors, 1985 and Grando et al., 1997)

Pathway I. Uninucleate microspore, instead of the asymmetric division to form the generative and vegetative nucleus, divides symmetrically to yield two identical nuclei (Nitsch and Norreel, 1973) or there is direct segmentation (Rashid and Street, 1974). The symmetrical division was observed in *Brassica napus* (Indrianto et al., 2001; Zaki and Dickinson, 1991), where microspores divide symmetrically to produce two daughter cells, and both of them lead to a sporophytic development.

Pathway II. Uninucleate microspore divides asymmetrically to produce generative and vegetative cells, as it does during the normal pollen development. The generative cell degenerates after one or two division, but vegetative cell repeatedly divides to form haploid embryo/callus. The pathway

IIis commonlyobservedin *Nicotiana, Datura, Zea mays, Hordeum vulgare,* and *Triticum aestivum* (Clapham, 1971; Iyer and Raina, 1972; Pescitelli and Petolino, 1988; Sunderland and WICKS, 1971).

Pathway III. The Uninucleate microspore divides asymmetrically to form avegetative and generative cell, but the pollen embryos are mostly formed from generative cell alone; the vegetative cell either does not divide or divides only to a limited extent forming a suspensor-like structure, e.g., *H. niger.*

Pathway IV. The uninucleate microspore divides asymmetrically to form vegetative and generative cells. Both the cells divide repeatedly and produce developing embryos. This could be seen in anther culture of *Datura innoxia.*

Pathway V. In *B. napus*, the first division is symmetrical, and the pollen embryos develop exclusively from the vegetative cell.

Irrespective of the different modes of action that occurs in the division pattern of microspore, all these pollen grains must dedifferentiate to form a multicellular structure (MCS). These MCS eventually burst open from the pollen wall to undergo further differentiation to regenerate a complete plantlet. The regeneration of a plantlet from MCS can happen either by direct embryogenesis or indirect embryogenesis. During direct embryogenesis, MCS burst open at the globular stage and assume embryo-like structures from which plantlets emerge in 4–8 weeks. In case of indirect embryogenesis, MCSs divide a few times to form a callus, which bursts through a pollen wall. The callus further differentiates to form embryos or regenerate into roots and shoots. The callus derived plants exhibit genetic variations, and they are mostly undesirable for the homozygous haploid production.

5.5 CONTROVERSIES IN THE MODE OF ACTION IN ANDROGENESIS

It is widely accepted that androgenetic embryo formation happens due to shift in gametophytic development to sporophytic one but it is still controversial about the precise moment it occurs. The indeterministic theory argues that, shift occurs after the detachment of flower bud from the donor plant and culture conditions are responsible for it. According to this argument, every pollen grain cultured are capable of androgenic, if cultivated them before switching off of gametophytic development-determining genes (Vasil, 1973).

On the other hand, the deterministic theory states that environmentaffects male gamete differentiation during PMC meiosis. During normal meiosis, gametophytic determinants are maintained, and sporophytic determinants are eliminated. If the sporophytic determinants are maintained due to abnormal PMC meiosis, then cell become potentially androgenic. According to Heberle-Bors (1985), the androgenic capacity of the microspores is determined only at meiosis; after meiosis, it is only the viability of this pre-determined pollen that can be affected. It is proposed that after meiosis, pollen grain has one more chance to become embryogenic during uninucleate pollen stage.

5.6 FACTORS AFFECTING ANDROGENESIS

Androgenesis has been reported in more than 200 species, but in vitro production of haploids by this approach is limited to only few crops. Several endogenous and exogenous factors play a role in determining the embryogenecity in microspore. Most crucial factors are genotype of the donor plant and its growth conditions. In many cases, two different cultures of same genotype exhibit considerable variation in the same culture medium. Thus, it is often advised to modify the established protocols to deal with the new system.

5.6.1 GENOTYPE OF STOCK (DONOR) PLANTS

The success in androgenesis is highly dependent on the genetic potential of the donor plant. It could be intraspecific (within the species) or interspecific (across the species) variations. The observed intraspecific variation is so high that, some lines (cultivars) of a species exhibit good androgenesis; others are extremely poor or completely nonresponsive. For example, in *T. aestivum*, among 21 cultivars, only 10 cultivars can produce haploids. In rice, Japonica cultivars response for androgenesis is higher than Indica cultivars (Bajaj, 1990; Cho and Zapata, 1990; Miah et al., 1985; Raina and Zapata, 1997)

Genotypic variations could be minimized by developing the genotype-specific protocols, but it demands detailed study of physiological factors and extensive manipulation of in vitro culture conditions (Dunwell et al., 1987). As an alternative method, transfer the androgenic trait from the highly androgenic lines to breeding material which was originally recalcitrant (nonresponsive to anther culture). This approach has been successfully tried in potato, barley, and maize (Foroughi-Wehr et al., 1982; Jacobsen and Sopory, 1978; Petolino et al., 1988; Wenzel and Uhrig, 1981). Androgenic trait (or anther culturability) is a quantitative trait controlled by nuclear encoded genes (Miah et al., 1985). Therefore, selection of the quantitative trait loci (QTLs) responsible for anther culturability will be useful in selection of parents for haploid breeding. In rice, QTLs were identified for anther culturability in a population resulted from anther culture of DH line. Five QTLs were identified on chromosomes 6, 7, 8, 10, and 12 for callus induction frequency. Two QTLs for green plantlet differentiation on chromosome 1 and 9 and one QTL for albino plantlet differentiation and no independent QTL were found for green plantlet yield frequency. Therefore, anther culture responsiveness is not a simple, unique character;ratherit consists of several components, which correspond to independent and differently inherited mechanisms as follows:

1. Callus induction—the ability of microspores within the cultured anthers to startdivisions and give rise to proliferating cell accumulation.
2. Callus stabilization—the preservation of complete and fully functional cells within the callus.
3. Plantlet induction—the ability of cells within calluses to give rise to embryos and plants.
4. The green versus albino plantlets formation—the production of fully functional green haploid and DH plant.

5.6.2 PHYSIOLOGICAL STATE AND GROWTH CONDITIONS OF DONOR PLANT

The androgenic response is highly influenced by the physiology and the environmental condition of the donor plants. Anthers of the week donor plants produce only a few embryos or calli in vitro. However, an optimum physiological condition varies from species to species. It is observed thatfirst set of flower buds gives a better response than those born later (Sunderland and wicks, 1971; Sato et al., 1989). Therefore, anthers should be cultured from early buds during flowering. It is also observed that failure in plantlet formation from anthers in *D. metel* that have passed the peak stage of flowering (Narayanswamy and Chandy, 1971). On the contrary, microspores isolated from older plants of *B. napus* produced more embryos than younger plants (Takahata et al., 1996). It has been reported that anthers from the late sown plant of *B. juncea* had a higher androgenic response (Agarwal and Bhojwani, 1993) suggesting the seasonal changes influence the microspore embryogenesis.

Endogenous levels of hormones and nutritional status of the anther also affects androgenesis. Increased abscisic acid (ABA) during water stress increases the frequency of androgenic microspore in *H. Vulgare* (Wang et al., 1999). Plants were grown under nitrogen deficiency yield higher number of microspore embryos than those supplied with fertilizer (Sunderland, 1978). It is suggested that the characteristics of the embryogenic pollen, such as thinner exine structure, weak staining with acetocarmine, the presence of a vacuole, and absence of starch grains, are the phenomenon of nitrogen starvation (Heberle-Bors, 1982). It is also suggested to avoid a pesticide treatment during androgenic induction.

Environmental conditions, such as photoperiod, light quality and intensity, temperature, nutrition and fertilizers, CO_2 concentration, and several biotic and abiotic stresses, influence the process of androgenesis. Anthers collected from the field grown plants gave a better response than the greenhouse grown plants (Vasil, 1980). Day length and temperature during meiosis causes the deviation in the sexual balance and hence causes the increase in embryogenic pollen frequency in *N. tabacum* (Heberle-Bors, 1982). Some studies also showed that both photoperiod and light intensity influences the androgenesis (Dunwell, 1976). In *Brassica* species, the donor plants are initially grown at 20°C/15°C and the temperature will be reduced to 10°C/5°C just before the bolting. The buds collected at appropriate time from these stress-induced plants resulted in higher frequency of microspore embryos (Ferrie et al., 1995). The same approach of cold treatment however, did not work on other species such as asparagus (Wolyn and Nichols, 2003), pepper (Lantos et al., 2009) or *Saponaria vaccaria* L. (Kernan and Ferrie, 2006).

5.6.3 MICROSPORE DEVELOPMENTAL STAGE

The appropriate developmental stage of the microspore for androgenesis varies between species and also the treatment used for induction. Basically, haploids can be obtained by culturing the pollens at different developments stages (e.g., pollen tetrad, young-uninucleate, mid-uninucleate, late-uninucleate, binucleate, and mature pollen). Generally, microspores exhibit high androgenic competence around first mitosis (uninucleate stage or early, mid bicellular pollen grains); however, they lose androgenecity once they accumulate starch grains (Heberle-Bors, 1989). In *B. napus*, late uninucleate and early bicellular pollen grains were more responsive, whereas in *N. tabaccum,* younger uninucleate pollen grains responded better. However, in *Arabidopsis thaliana,* microspore mother cells at early meiosis were shown to be high embryogenic competent (Gresshoff and Doy, 1972). In *Lycopersicon esculentum* (tomato), microspores within the anthers at meiocyte, just before the compartmentalization of tetrad had shown better response (Seguí-Simarro and Nuez, 2005). The developmental stage of microspore can also affect ploidy level of the plant. The anthers with uninucleate microspores give rise to haploids, whereas, binucleate and older binucleate microspores generated diploids and triploids, respectively (Engvild et al., 1972). Therefore, collecting the anthers at the right stage to obtain pollen with embryogenic competence is crucial for androgenesis. The developmental stage of the pollen could be examined by acetocarmine staining (1% acetocarmine in 45% acetic acid) by squashing anthers into staining solution.

5.6.4 PRETREATMENTS

We have studied the different pathways involved in the induction of androgenesis process. It is evident from many studies that; uninucleate microspore just before the first mitosis is most favorable for androgenesis induction. During androgenesis, microspore undergoes several biochemical and physiological changes as studied before. Those changes have been used as markers to determine the microspore embryogenecity. However, these markers are not universal in all species, and different species exhibit different changes during androgenesis. Thestress-treateduninucleate microspore obtains a "star-like"structure with a centralized nucleus surrounded by star-like cytoplasmic strands (Touraev et al., 1996). Star-like microspore was shown as a sign of embryogenic microspores in number of different species such as wheat, tobacco, rice apples (Hofer, 2004).

Different kinds of stresses (cold, heat, starvation, and chemical) were applied as pretreatments for donor plants, excised spikes or flower buds, and cultured anthers. These stresses are divided into three categories (Shariatpanahi et al., 2006a) such as widely used stresses, neglected stresses, and novel stresses (Fig. 5.5).

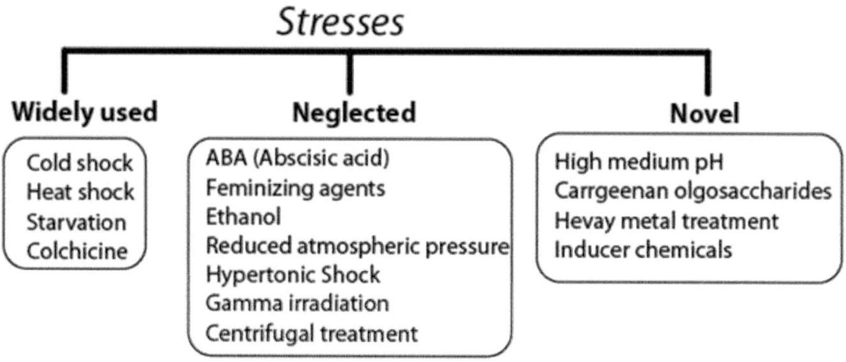

FIGURE 5.5 Different kinds of stresses used as pretreatments. (Adapted from Shariatpanahi et al., 2006)

5.6.4.1 WIDELY USED STRESSES

Cold, heat, starvation and colchicine stresses were extensively used to induce pollen embryogenesis.

5.6.4.1.1 Cold Pretreatment

Cold treatment is carried out at 4–10°C for few days to several weeks depend on the genotype of the donor plant. Cold treatment application was first reported in *Datura* anther/microspore cultures (Nitsch and Norreel, 1973). Later, several other studies reported low temperature induced microspore embryogenesis in barley, rice, bread wheat, durum wheat, oats, triticale, *Citrus clementina*, and flaxater (Shariatpanahi et al., 2006b). Cold treatment was also applied to tobacco anther culture to induce embryogenesis. Pretreatment of anther culture at 5°C for the first twodays promoted androgenesis in *Cyclamen persicum* Mill (Ishizaka and Uematsu, 1993). Several studies show that cold treated plants produceDHs spontaneously due to increased endoreduplication.

5.6.4.1.2 Heat Pretreatment

Heat pretreatment is veryeffective on pollen embryogenesis development. It is usually carried out at 33–37°C for a duration varying from several hours to several days. Heat treatment has been used in isolated microspores of rapeseed, wheat, tobacco, eggplant, and several other crops (Shariatpanahi et al., 2006b). The optimum temperature for heat treatment varies with the genotype. For example, cv. Duplo of spring rape yields highest embryos after 3-day treatment at 35°C, whereas other two cultivars were less responsive to 35°C pretreatment. Androgenesis was promoted in *B. compestris* L. by culturing the anthers at 35°C for 1–3 days beforeculture at 25°C.

5.6.4.1.3 *Starvation*

Nutrient starvation stress has been used to induce pollen embryogenesis in many crops. Sugar and nitrogen starvations have been used in tobacco. However, sugar starvation resulted in high microspore embryogenecity in isolated microspores of tobacco, rice, barley, apple and rye (Shariatpanahi et al., 2006a).

5.6.4.1.4 *Colchicine*

Colchicine treatment was used in isolated microspore cultures of *B. napus* and coffee. Microspore culture of *B. napus* treated with the concentration of 50 and 500 mg/l for 15 yielded high numbers of embryos and efficiency of chromosome doubling was 83–91%. Colchicine is a microtubule-disrupting drug. Drug treatment causes depolymerization of microtubule during cell division, thus a failure in spindle formation (a process necessary to pull chromosomes toward the opposite poles before cell enters to G2 phase) which results in endoreduplication or endomitosis.

The response of microspore to colchicine seems to be dependent on the developmental stage of microspores. In rapeseed, highest frequency of embryogenesis has been obtained in vacuolated microspores, somewhat earlier stage than the population responsive to heat induction (Zhao et al., 1996).

5.6.4.2 NEGLECTED STRESSES

Gamma irradiation was used to induce embryogenesis in anther cultures of rapeseed and tomato (Shtereva et al., 1998). The combination of radiation followed by cold treatment (10°C for nine days) increased the embryogenesis in tomato anther cultures. Radiations have also been used to obtain androgenic plants in recalcitrant rice barley, and *Solanum nigrum varieties*

(Aldemita and Zapata, 1991; Kopecký and Vagera, 2005; Vagera et al., 2004). Irradiation alters the auxin and cytokinin levels in somatic tissue, but it is not clear this change is the cause for androgenesis.

The other neglected stresses such as reduced atmospheric pressure, ABA, ethanol, hypertonic shock, centrifugal treatmenthave been tested only in few species (Imamura and Harada, 1980; Pechan and Keller, 1989; Tanaka, 1973; Wang et al., 1981). Hardly any recent reports widening the application of these stresses in other species are available. Recently, several novel stresses have been reported to induce microspore embryogenesis with considerable success in few species. High medium pH or medium containing heavy metals such as lithium was used to induce embryogenesis in *N. tabacum*microspores (Barinova et al., 2004; Zonia and Tupý, 1995). The microspores of *B. napus*were culturedin the presence of inducer chemicals (Liu et al., 2003). However, these all novel stresses to be tested in detail on other species.

5.7 ISOLATED MICROSPORE CULTURE

Basis of this method is to isolate microspores from the anthers and culture these microspores to produce haploid/DH plants. Most of the early studies preferred intact anther culture for the production haploid plants, but recently isolated microspore culture is preferred over anther culture as microspore culture brought several advantages to the field. The genetic purity of haploid callus (every cell in callus should be haploid) can be obtained by isolated microspore culture technique. This technique requires better equipment and skills to remove somatic anther wall and to release individual pollen grains. The removal of somatic tissue during pollen isolation is critical because presence of somatic tissue may produce diploid calli, embryos, and plants and result will be mixture of haploid and diploid calli. Isolated pollen culture was first performed in *Petunia hybrida* (Binding, 1972) by inoculating large number of pollen grains in a nutrient medium to obtain multicellular tissue masses from individual pollen grains. Later, microspore culture has been established for many species such as *Nictoianatabaccum, B. napus,* barley, and tobacco. In *B. napus*, the flowering buds incubated at 4°C several days before isolating the pollen. Pollen grains were then cultured in medium containing 1-naphthalene acetic acid (NAA) and 6-benzylaminopurine (BAP) as well as potato extracts to obtain embryoid and plants. This study successfully demonstrated the production of haploids by isolated microspore culture technique.

5.7.1 ISOLATION OF POLLEN

Several ways of isolating microspores were employed. Gradient centrifugation method, shed microspore method, and squeezing anther method were mainly used, and also many variations have been introduced to these basic methods to isolate clean microspore population.

5.7.1.1 SHEDDING METHOD

In this approach, anthers are directly placed in a liquid medium and allowed to shed their microspores. These microspores were then cultured for callus development and plant regeneration. Shedding method was first demonstrated in tobacco anther culture and later applied extensively to wheat (Wei, 1982) and barley (Ziauddin et al., 1990). It is recommended to pretreat the anthers with 0.3 M mannitol plus macronutrients for about a week to increase shedding during culture (Kasha et al., 2001). Shedding technique has been the most preferred method of haploid production in *Capsicum annum* L. (Pepper) (Supena et al., 2006). In pepper, selection of flower bud with more than 50% late unicellular microspores and pretreating them at 4°C for one day is critical. Anthers are then cultured in thedouble-layer medium system (discussed below) for one week at 9°Cand then incubate at 28°C in continuous darkness.

5.7.1.2 MAGENTIC-BAR STIRRING METHOD

It is just an extension of shedding method. Anthers cannot shed all the microspores to theliquidmedium, and aconsiderable amount of pollens are still enclosed within the anther wall. Applying the stirring force using magneticbar stirring will help to release the microspores completely, and, hence, it greatly affects the microspore yield from natural shedding (Cho and Zapata, 1990). The shedding and stirring procedures were not effective in wheat microspore cultures.

5.7.1.3 MECHANICAL ISOLATION METHOD

This method can be employed by maceration, blending, or squeezing.

5.7.1.3.1 Maceration

In this technique, anthers are pressed through a mesh filter using a glass or teflon rod to collect microspores. This technique was first established for isolation of microspores in *B. napus* (Lichter, 1982). Macerationfollowed by gradient centrifugation is practiced recently. In this method, surface sterilized buds were crushed by using mortar and pestle to release the microspores from anthers. The resulted homogenate is then passed through filter paper to separate microspores from somatic anther wall and bud tissues, and then centrifuge using gradient (Percoll, maltose, or sucrose gradient) to separate the developmental stages of microspores. Alternatively, crush the anthers directly (instead of flower buds) using mortar and pestle. This will reduce the amount of somatic tissue that has to be separate from the microspores.

5.7.1.3.2 Blending

Anthers suspended in 0.3 M mannitol were placed in a microblender and subjected to high speed (20,000 rpm) for 30 s. The microblender employs sharp blades at moderate velocity and, hence, is able to quickly cut anthers in pieces and release microspores. The micro blended suspension was then filtered through 100 μm mesh nylon filter screen to separated somatic tissue debris from microspores. The microblender system is composed of speed Waring base motor and rotor (microcontainer). Blending technique was first established to isolate microspores from barley (OLSEN, 1992) and later is adapted to wheat (Mejza et al., 1993). Microspores released through blending usually have less damages and higher embryogenic capacity and reproducibility than those obtained through maceration or any other above explained methods. However, blending isolation can also damage the microspores, especially when the blending speed and length is not optimized.

5.7.1.3.3 Squeezing

The pollens are extracted by pressing and squeezing the anthers with a glass rod against the sides of the beaker. The anther tissue debris is removed by filtering the pollen suspension, and large and healthy pollen are washed and collected. These pollens are cultured in a solid or liquid medium, and the callus or the embryo formed is transferred to a suitable medium to produce a haploid plant.

5.7.2 BASIC PROTOCOL FOR POLLEN CULTURE

1. Selection and surface sterilization of unopened flower bud.
2. Excision of anther without filament under aseptic conditions. Injured anthers should be removed as they result in callusing of anther wall tissue.
3. Crush anthers in 1% acetocarmine to test the pollen developmental stage.
4. Collect about 50–100 anthers having suitable microspores for androgenesis in a beaker containing liquid basal medium (MS or White or Nitsch and Nitsch).
5. Release microspores by pressing anthers against the side of the beaker with sterile glass piston of a syringe. Any other methods described above can be used for higher yield of microspores.
6. Remove anther tissue debris by filtering the homogenized anthers using a nylon sieve (pore diameter 40–60 μm)
7. Centrifuge the pollen filtrate at low speed (500–800 RPM) for 5 min. Suspend the pellet of pollen infresh liquid medium.
8. Wash the pollens with fresh liquid medium atleast twice by repeated centrifugation.
9. Prepare pollens with the density of 10^3–10^4 pollens/ml in liquid basal medium.
10. Pipette about 2–3 ml of pollens suspension intoa sterilepetri dish containing liquid medium or medium added with soft agar. Seal the petridishes with cellotape to avoid dehydration.
11. Incubatepetridishes at 27–30°C under cool white light (500 lux, 16 h).
12. Observe the development of young embryoids and subsequent development haploid plantlets after 30 days.
13. Maintain the haploid plantlets in alternating periods of light (2000 lux, 12–18 h) and darkness (6–12 h) at 27–50°C at about 2000 lux.
14. Transfer matured plantlets to soil.

5.7.3 NURSE CULTURE TECHNIQUE

The use of nurse culture technique was first demonstrated in *L. esculentum* (Sharp et al., 1972). In this method, intact anthers from the flower bud are placed horizontally on the top of the semisolid or solid basal medium within a conical flask (intact anthers act as a nurse tissue). A small filter paper disc is then placed over the top of the intact anthers. Meanwhile, a separate suspension of pollen from anthers of another similar bud in liquid medium (about ten pollen grains per 5 ml of medium) should be prepared. A drop of the pollen suspension is then pipetted onto a paper disc (transfer about10 pollen grains to each paper disc). In the case of *L. esculentum*, nurse culture produced numerous clones within a month, and they were uniformly haploids.

Other studies used different flower parts as nurse tissue. In *N. tabaccum* nurse culture technique, nurse tissue consisted of stamens, petals, sepals, and even adult leaves of the same plant at the stage of pollen mitosis. Addition of aqueous extract of anthers of *Nicotiana* to the microspore culture of *Datura* and *Lycopersicum* to nurse the culture is also practiced (Acharya and Ramji, 1977).

5.7.4 DOUBLE LAYER MEDIUM SYSTEM

This method is the derivative of shed-microspore culture, and it is the method of choice especially for microspore culture in *C. annuum* L. (Supena et al., 2006). Double layer system consists of two layers of the medium:under layer, and the upper layer. Under layer is the solid medium containing Nitsch components and 2% maltose, with 1% activated charcoal. The upper layer is the liquid medium containing 2.5 μM zeatin and 5 μM indole-3-acetic acid. Anthers are suspended in the liquid medium to shed their microspores. This system was tested for tengenotypes of hot pepper, and all of them responded to this protocol. This methodis also recommended for DHs production for hot pepper breeding.

5.8 ADVANTAGES OF MICROSPORE CULTURE OVER ANTHER CULTURE

Anther culture is a relatively straightforward technique for producing haploids, but its application is limited to few crops. Many crops are still recalcitrant and did not respond to anther culture. Often embryo yield is relatively poor from anther culture despite having thousands of microspores within the anthers. This failure of achieving significant yields from anther culture prompted to use isolated microspore culture for haploid production.

Microspore culture has several advantages over anther culture:

1. In microspore culture, isolated microspores will get uniform exposures to nutrient medium, which helps developing embryos to access to the better nutrient availability. In anther culture, microspores are overcrowded within the anther wall and, hence, do not expose uniformly to the nutrient medium.
2. In microspore culture, the genetic purity of haploid callus can be achieved. In anther culture, the anther wall may produce diploid somatic callus, which causes the genetic impurity of callus. Moreover, anther wall (tapetal layer) secretion negatively affects the microspore embryogenesis, and this secretion often disturbed the strict control of culture condition.
3. Microspore culture achieved a significant yield of embryos. It is possible to obtained hundreds or even thousands of embryos from the microspores isolated from single anther. Anther culture has a limited efficiency, producing only a few embryos per cultured anthers.
4. In microspore culture, in vitro factors can directly affect the microspore and microspore embryos. In anther culture, microspores areenclosed within the anthers, and, hence, they are not directly exposed to in vitro conditions, and the effect is mediated through anther wall.
5. In microspore culture, the androgenesis process, such as microspore maturation and embryo development, can be monitored starting from the single cell. This helps to regulate the various factors governing androgenesis.
6. Microspore culture is suitable to understand the cellular, physiological, biochemical, and molecular processes involved in androgenesis.
7. Microspore culture is ideal for mutagenic studies as single pollen grain can be uniformly exposed to chemical and physical mutagens.
8. Microspore culture is the only option for some species which are still recalcitrant for androgenesis.

Some of the disadvantages of microspore culture:

1. Isolated microspore culture requires better equipment and higher skills.
2. Viability and desiccation of the isolated microspore are crucial factors to be considered.
3. Yields are variable. In some cases, almost every microsporein the culture producedembryoids, and, in some cases, none of them had shown a response.

5.9 PRODUCTION OF HAPLOID AND DH PLANTS

The spontaneous development of haploid was first reported in Sea Island Cotton (Harland, 1920). Since then, the natural haploid embryos derived from the cells of embryosac apomictically have been reported in more than 71 species belongs to 39 genera in 16 families of angiosperms (Acharya and Ramji, 1977). However, natural haploids do not occur frequently, and, therefore, the practical importance of the natural haploids is less. This limitation prompted breeders to induce haploid production by several methods. Earlier (prior to 1960s), haploids were obtained through in vivo techniques, such as interspecific/wide hybridization, chromosome elimination, and pollination with irradiated pollen. These in vivo techniques could yield haploids infrequently and in very small numbers, and the use of haploids at that time was mainly for genetic studies than plant breeding.

5.9.1 *IN VIVO PRODUCTION OF HAPLOIDS*

In vivo production of haploids could be achieved by modifying the pollination method such as pollination with pollen of the same species (e.g., maize), pollination with irradiated pollen, pollination with pollen of a wide relative (e.g., barley and potato) or unrelated species (e.g., wheat). Pollination using these methods causes an elimination of the paternal chromosomes during early embryogenesis which results in haploid embryos. These pollination methods also causefertilization of the polar nuclei and development of functional endosperm that leads to the parthenogenicdevelopment of the haploid embryos. Parthenogenesis is the process of embryo formation without fertilization of the egg cell.

5.9.1.1 POLLINATION WITH POLLEN OF THE SAME SPECIES (INTRASPECIFIC CROSSING)

This method involves the legitimate crossing within the species with selected inducing genotypes. This intraspecific crossing results in mixture of normal hybrid embryos as well as haploid maternal embryos with normal triploid endosperms. The first identified haploid inducer line by this method was reported in maize genetic strain Stock 6. Kernels containing haploid embryos produce viable haploid seedlings; therefore, this method avoids in vitro culture of haploid embryos to regenerate an entire plant.

Pollination with irradiated pollen of the same species could also produce maternal haploids. Pollination in this method stimulates the embryo development, but irradiated pollen fails to fertilize the egg cell. This technique was successfully used in several species such as apple, blackberry, cucumber, onion, melon, sunflower, kiwifruit, and few others (Murovec and Bohanec, 2011). This method requires efficient emasculation, which is laborious, and a dose of irradiation is crucial to induce haploid production. Pollens exposed to lower doses can still fertilize the eggs, resulting inhybrid embryos and, subsequently, abnormal phenotype. Higher doses decrease the yield of embryos but originate mostly haploid embryos.

5.9.1.2 WIDE HYBRIDIZATION (INTERSPECIFIC OR INTERGENERIC CROSSING)

Crossing between the species is termed as wide hybridization, and it has been effectively used in several cultivated species for haploid production. Pollination (crossing) could be interspecific and intergeneric pollination. In wide hybridization method, there is a preferential elimination of paternal chromosomes, which results in haploids with maternal chromosomes.

5.9.1.2.1 Interspecific Pollination

Haploid barley was produced by wide hybridization between cultivated barley (*H. vulgare*, 2n = 2x = 14) as the female and wild *H. bulbosum* (2n = 2x = 14) as the male (Forster et al., 2007). The preferential elimination of the paternal chromosomes from wild *H. bulbosum* from the developing embryo results in haploid embryos. This method is also known as "bulbosum method."The haploid embryos obtained by bulbosumwere defective in the functional endosperm, and the embryos were rescued by growing them in vitro.

Haploids in wheat were also achieved by wheat-bulbosum wide hybridization, but it was only limited to few wheat (*T. aestivum*) cultivars which are crossing compatible with bulbos barley. Later, new hybridization method developed in which haploid wheat plantlets were obtained by interspecific crosses between wheat (*T. aestivum*) maize (*Z. mays*). The pollination developed the hybrid embryos, but in later stages of embryo development, preferential elimination of maize chromosomes resulted in wheat haploids. Later embryos were treated with 2,4-dichlorophenoxyacetic acid in planta, to rescue them. Later embryos were isolated at the suitable stage and cultured them in vitro to obtain plantlets. This technique was successfully applied to many commercial wheat cultivars and hybrids (Niu et al., 2014). Pollination with maize is successful in several other species such as barley, triticale (x Triticosecale), rye (*Secale cereal*), and oats (*Avena saliva*) (Wędzony et al., 2009).

Intergeneric pollination was used to produce haploid in cultivated potato (*S. tuberosum* L. sap. *tuberosum*, 2n = 4x; tetraploid) by crossing them with *S. phureja* (2n = 2x, diploid). After pollination, haploid embryos are obtained. Notice that haploid embryos obtained by this crossing are dihaploids (2n = 2x).

Wide hybridization was particularly helpful in theproduction of haploids from wheat cultivars, where anther culture did not respond very well due to high rates of albinism, the low response of some genotypes, and extended periods of inducing and regenerating process.

5.9.2 IN VITRO PRODUCTION OF HAPLOIDS

Haploids can be induced by in vitro culture of male (androgenesis) or female gametophyte (gynogenesis).

5.9.2.1 GYNOGENESIS

Gynogenesis is defined as in vitro induction of maternal haploids from the culture of female gametophyte. It can be induced by in vitro culture of various un-pollinated flower parts, such as ovules, placenta attached ovules, ovaries or whole flower buds. Gynogenesis is of two types, direct and indirect gynogenesis. Direct gynogenesis culture of female gametophyte directly produces embryoids. In indirect gynogenesis, culture of female gametophyte forms the callus which will eventually develop into embryoids. Gynogenesis is controlled by several factors such as developmental stages of gametes, the pretreatment of flower buds, culture media, and culture conditions. The female gametophyte develops into mature embryo sac during in vitro culture. This is in contrast to androgenesis, where microspores undergo saprophytic pathway. Mature microspores are usually not preferred for haploid production. Mature embryo sacs contain several haploid cells such as the egg cell, synergids, antipodal cells, and non-fused polar nuclei, which are capable forming of haploid embryos. The egg cells are the most gynogenetic among all other cells, and they can develop into haploid plants directly without the risk of gametoclonal variations or via callus formation.

Gynogenesis has been successfully used in several species such as onion, sugarbeet, cucumber, squash, sunflower, wheat barley, and few others (Murovec and Bohanec, 2011), but only onion and sugarbeet haploids are used in plant breeding.

Gynogenesis has several advantages. Gynogenic haploids display relatively higher genetic stability, and a lower rate of albinism. Gynogenesis of cereal species did not produce albino plants. However, this technique was always used as analternative approach to androgenesis or other induction techniques due to some of its limitations. The major limitation is a lack of established protocols for most species and production of diploid or mixoploid plants during gynogenesis. It could only produce few haploid plants as only one ovary per flower. The dissection of unfertilized ovaries and ovules requires special skills, and this made the gynogenesis more difficult.

5.9.2.2 ANDROGENESIS

Androgenesis is defined as in vitro production of haploids frommale gametic cells. Androgenesis can be induced by in vitro culture of immature anthers/ microspores. The first report of the isolation of haploid angiosperm tissue from anther culture was reported in *Tradescantia refle.* Guha and Maheshwari observed the embryo-like structures when the anthers of *D. innoxi-a*werecultured in vitro (androgenesis). Later, they confirmed the haploid nature of these embryoids and traced their origin to microspores (Guha and Maheshwari, 1967). This work was later expanded into *Nicotiana*and produced haploid embryos from different strains of *N. tabacum and N. Sylvestris* (Nitsch and Nitsch, 1969). To date, haploid production via anther/ microspore culture (androgenesis) was reported in over 250 plant species.

5.10 GENERAL APPROACH FOR ANTHER CULTURE FOR HAPLOID AND DOUBLE HAPLOID PRODUCTION

There is no thumb rule in designing the optimized condition for the anther/ microspore culture, and the protocols could be variedfor different species as well as different cultivars within species. In general, androgenesis steps includes, genotype selection, determination of pollen developmental stage, physiological state and growth conditions of donor plants, pretreatment, surface sterilization, anther excision, medium composition, culture condition and morphogenic development, reprogramming of gene expression and plant recovery, ploidy analysis, detection of homozygosity, and chromosome doubling if necessary. We have already discussed the genotype selection to pretreatmentearlier in this chapter. We will discuss remaining approaches here.

5.10.1 SURFACE STERILIZATION AND ANTHER EXCISION

After pretreatment of floral buds, surface sterilization of them will be carried out to obtain contaminant free anthers. In general, floral buds are immersed in 70% (v/v) ethyl alcohol for few minutes, followed by immersion in a sodium hypochlorite (about 1.5% active chlorine in water) containing few drops of mild detergent Tween 20 for 10–15 min. Then 3–5 min of wash with sterile distilled water followed by anther excision aseptically from the filaments and placed onto the medium.

5.10.1.1 MEDIUM COMPOSITION

Usually, basal medium such as N6 medium (Chu, 1978), MS medium (Murashige and Skoog, 1962), Nitsch and Nitsch medium (Nitsch and Nitsch, 1969), and B5 medium (Gamborg et al., 1968) are used in androgenesis. Half strength MS medium is suggested for the Solanaceae and N6 medium for the Cereals (Chu, 1978). Acarbohydrate source, such as high sucrose and in some species maltose, is preferred to induce embryo production (Powell, 1990). The exogenous application of plant growth regulators facilitated the androgenesis. Plant growth regulator, such as auxin, cytokinins, or a combination of both, is found to be crucial for androgenesis in many species. The addition of activated charcoal (0.5–2 g/l) to the medium favored the androgenesis in several species (Bajaj, 1990). The supplement of other substance, such as glutamine, casein, proline, biotin, inositol, coconut water, silver nitrate, polyvinylpyrrolidone (PVP), and aliphatic polyamines (PAs) facilitated the microspore embryogenesis (Reinert et al., 1977; Tiainen, 1992). pH of the media influences the androgenesis, and the pH is usually adjustedto 5.7–5.8. Medium Solidifying agents such agar is widely reported;however, other solidifying agents such as starch (potato, wheat, corn, or barley starch), gerlite, agarose, and ficoll have been used with additional benefits.

5.10.1.2 CULTURE CONDITIONS

The optimal condition for anther culture is to incubate them at 24–27°C and expose to light at an intensity of about 2000 lux for 14 h per day (Reinert et al., 1977). However, optimal conditions need to be fixed for each culture system. For example, Vasil (1973) reported using of alternating light periods (12–18 h; 5000–1000 lux/m^2).

5.10.1.3 PLOIDY LEVEL DETERMINATION AND HOMOZYGOSITY TESTING

Anther/microspore culture will not only yield pure haploid plants but also produces the undesired heterozygous plantlets (diploid, triploid, tetraploid, and so on) plants. Sometimes, the haploid culture will result in spontaneous doubling chromosomes gives rise to homozygous DHs which are more useful. Therefore, it is important to determine the ploidy level to evaluate the regenerants to separate haploid/DHs from undesired non-haploidregenerants. In early days, breeders used more indirect approach to determine ploidy levels, such as comparisons between regenerated and donor plants regarding plant morphology (plant height, leaf dimensions, and flower morphology), plant vigor and fertility; however, these are unreliable, and these markers can be influenced by environmental effects. The ploidy level can be determined by using conventional cytological techniques such as counting the chromosome number in root tip cells and measuring the DNA content using flow cytometry. In the early days, evaluation of regenerators was largely depending on phenotypic markers, progeny resting after self-pollination, and isozyme analysis. Recent development in biotechnology allowed the faster and more reliable techniques to test the homozygosity and assessment of plant origin. For example, amplified fragment length polymorphism (AFLP), random amplified polymorphic DNA (RAPD), sequence characterized amplified regions (SCAR), or simple sequence repeat (SST) has been commonly used.

5.10.1.4 CHROMOSOME DOUBLING

In haploids, meiosis cannot occur due to the absence of one set of homologous chromosomes, so there is no seed set. Therefore, most of the time haploids are sterile and often express reduced vigor. To restore the fertility in haploid plants, it is necessary that spontaneous doubling must occur during in vitro culture otherwise it has to be induced chemically. Spontaneous doubling of chromosomes has been reported in anther culture of barley (70–90%), bread wheat (25–70%), rice (50–60%), rye (50–90%), and maize (20%). Spontaneous doubling is highly preferred, as chemical induction may cause undesired mutation during the doubling process. However, for those species in which spontaneous doubling frequency is very low, it is necessary to induce the chromosome doubling. Various approaches have been used to induce chromosome doubling in vivo and in-vitro. Use of the microtubule disrupting drugs, such as colchicine and oryzalin, in millimolar concentrations have been reported for chromosome doubling (Castillo et al., 2009). Additionally, some herbicides, such as trifluralin and pronamide, have also been used. Antimicrotubule drugs can be applied at various stages of androgenesis. Drugs can be added in the microspore pretreating media or anther culture media. Drug treatments are applied after regeneration at either embryo, shoot, or plantlet level. Chromosome doubling was induced with nitrogen oxide (N_2O) treatment in maize seedlings (Kato and Geiger, 2002). Plantlets were treated at a high pressure of 600 KPa for two days at the six-leaf stage when plants develop flower primordia.

5.11 APPLICATION OF HAPLOIDS AND DHs IN CROP IMPROVEMENT

DHs brought several advantages and benefits to breeding programs. The rapid attainment of homozygosity in a single generation accelerates the breeding process. On the contrary, theconventional method requires several generations of self-pollinationto obtain homozygous true breeding cultivars with hybrid vigor. Therefore, DHs cut down the time needed for breeding and also minimize the labor cost. DHs are now anintegral part of plant breeding, and they have been used in commercial cultivar production of many species such as asparagus, barley, *B. juncea*, eggplant, melon, pepper, rapeseed, rice, tobacco, triticale, wheat, and so on. There is already more than 290 varieties that have been released (http://www.scri.ac.uk/assoc/COST851/COSThome.htm). Moreover, 50% of the available barley cultivars in Europe and three of the five Canada Western Red Spring (CWRS) wheat classesinCanadahavebeen produced via DH system (Dunwell, 2010). DH technology is useful in species suffering from inbreeding depression, in which it is hardto produce fertile homozygous lines by self-pollination, e.g., rye and grasses. Applying the DH technology to medicinal and aromatic plants is currently gaining much attention (Ferrie, 2009).

Producing double haploids in tree species is more useful. Trees species have long reproductive cycles, greater heterozygosity, large size, and self-incompatibility. Therefore, it is difficult to produce homozygous lines with better traits through conventional method even after several generations of selfing (Germana, 2006). The size reduction in the DH plants compared to diploid heterozygous may be of horticulture interest, e.g., ornamental plants or dwarfing rootstocks for fruit crops. Chinese researchers using haploid system obtained rubber tree taller by six meters, which could then be multiplied by asexual propagation to raise several clones. Another use of the double haploid system is producing triploid plants via anther culture. Triploid plants are for commercial purposes as they produce fruits without seeds. Triploids produced from anther culture has been reported in Apple (Hofer, 2004), *Pyrus pyrifolia* Nakai (Kadota and Niimi, 2004), *Carica papaya* L., and *C. clementine* Hort ex Tan (Germanà, 2009).

The haploid system provides a relatively easier system for induction of mutations, and, thereby, select for the desired mutant traits. For example, mutations were successfully introduced during microspore embryogenesis in rapeseed for herbicide resistance, disease resistance, salt tolerance, and also for seed quality traits. It saves time and space for the production of crops with desired traits. With the help of anther culture technique, tobacco mutants were selected for resistance to shank disease and wheat lines resistant to scab (*Fusarium graminearum*). Mutation breeding is another area of crop improvement in which double haploid system can help to accelerate the process. It is possible to obtain homozygous mutant lines in the first generation after mutagenic treatment. All mutated traits are immediately expressed, allowing screening for both recessive and dominant mutants in first generation without the need for self-pollination. There are two options for mutagenic treatment. The first option is treatment of dormant seeds that on germination and flowering produce M_1 gametes, which are used for haploid culture. This has been successfully used in barley, rice, and wheat, increasing the efficiency of selection for desired traits in a mutated double haploid population (Forster et al., 2007). The second option is the treatment of haploid cells in vitro. The mutagenic agent is usually applied to the microspore at the uninucleate stage before the first nuclear division. Treatment after the first nuclear division will generate chimeric and heterozygous plants, which are caused by spontaneous diploidization through nuclear fusion. In vitro mutagenic treatment can be followed by in vitro selection of desired traits such as disease and herbicide resistance.

In addition to the use of DHs in crop improvement, they have been useful in gene mapping and genomics and as targets for transformation. DHs are useful for genetic analysis such as QTL (Datta, 2005). QTLs affect important agronomic traits in several crops. The QTL study required recombinant inbred lines (RILs) to relocate phenotypic studies, but it will take a long time to develop RIL population. Recently, DHs are used to construct genetic maps and locate QTLs because DHs are homozygous, and they can be propagated without further segregation. Therefore, DHs offer the precise measurement of quantitative traits. DHs have been used for studies on QTLS for rice root characteristics, plant height and days to heading, grain yield, and cold tolerance. Based on the constructed linkage map of DH population from a female parent, which had a spreading plant type, and male parent, which had a compact plant type, two major QTL were detected on chromosomes 9 and 11 for tiller angle phenotype. Tiller angle has great significance in the high breeding of rice. DH rice populations have also been used in the QTL studies on rice grain quality, grain shape, aromatic traits, and brown planthopper resistance

The double haploid system is more advantageous to apply marker-assisted selection for finding a near-isogenic line defective in a particular trait; finding this by backcrossing combined with marker-assisted selection requires several generations of crosses. DHs play a major role in integrating genetic and physical maps, and, thereby, providing precision in targeting candidate genes (Künzel et al., 2000; Wang et al., 2001). DHs have been a key feature in establishing chromosomes maps in a range of species such as barley, rice, rapeseed, and wheat (Forster and Thomas, 2005).

DH system is particularly handy in generating stable homozygous transgenic lines. Transformation of the transgene to uninucleate microspores or haploid embryos will generate stable homozygous transgenic lines in one generation. On the contrary, heterozygous diploid cells/plants transformed with transgene will require atleast three generations of selfing to achieve homozygous for the transgene. Several transformation protocols have been employed such as microinjection, electroporation, particle bombardment, and *Agrobacterium tumefaciens* mediated transformation (Touraev et al., 2001). The first transgenic indica rice was reported to use such haploid microspore culture. The particle bombardment method has been used to regenerate fertile transgenic barley plants using microspore cultures (Datta, 2005).

KEYWORDS

- **embryogenesis**
- **androgenesis**
- **anther culture**
- **haploids**
- **crop improvement**

REFERENCES

Acharya, B. C.; Ramji, M. V. Experimental Androgenesis in Plants—A Review. In *Proceedings of the Indian Academy of Sciences-Section B;* Springer India, **1977**; pp 337–360.

Agarwal, P. K.; Bhojwani, S. S. Enhanced Pollen Grain Embryogenesis and Plant Regeneration in Anther Cultures of *Brassica Juncea* cv. PR-45. *Euphytica*, **1993,** *70*, 191–196.

Aldemita, R. R.; Zapata, F. Anther Culture of Rice: Effects of Radiation and Media Components on Callus Induction and Plant Regeneration. *Cereal Res. Commun.* **1991,** *19*(1), 9–32.

Bajaj, Y. P. S. In vitro Production of Haploids and Their Use in Cell Genetics and Plant Breeding. In *Haploids in Crop Improvement I;* Springer Berlin Heidelberg, 1990; pp 3–44.

Barinova, I.; Clément, C.; Martiny, L.; Baillieul, F.; Soukupova, H.; Heberle-Bors, E.; Touraev, A. Regulation of Developmental Pathways in Cultured Microspores of Tobacco and Snapdragon by Medium pH. *Planta*, **2004,** *219*, 141–146.

Bedinger, P. The Remarkable Biology of Pollen. *Plant Cell*, **1992,** *4*, 879.

Bhojwani, S.; Dunwell, J.; Sunderland, N. Nucleic-acid and Protein Contents of Embryogenic Tobacco Pollen. *J. Exp. Bot.* **1973,** *24*, 863–869.

Bicknell, R. A.; Koltunow, A. M. Understanding Apomixis: Recent Advances and Remaining Conundrums. *Plant Cell*, **2004,** *16*, S228–S245.

Binding, H. Nuclear and Cell Divisions in Isolated Pollen of Petunia Hybrida in Agar Suspension Cultures. *Nature*, **1972,** *237*, 283–285.

Burk, L.; Gerstel, D.; Wernsman, E. Maternal Haploids of *Nicotiana tabacum* L. from seed. *Science*, **1979,** *206*, 585–585.

Castillo, A.; Cistué, L.; Vallés, M.; Soriano, M. Chromosome Doubling in Monocots. In *Advances in Haploid Production in Higher Plants;* Springer, **2009**; pp 329–338.

Cho, M. S.; Zapata, F. J. Plant Regeneration from Isolated Microspore of Indica Rice. *Plant Cell Physiol.* **1990,** *31*, 881–885.

Chu, C.-C. The N6 Medium and Its Applications to Anther Culture of Cereal Crops. In *Proceedings of Symposium on Plant Tissue Culture;* Science Press: Beijing, China, **1978**; pp 43–50.

Clapham, D. In vitro Development of Callus From the Pollen of *Lolium* and *Hordeum*. *Z Pflanzenzucht*, **1971**.

Datta, S. K. Androgenic Haploids: Factors Controlling Development and Its Application in Crop Improvement. *Current Science-Bangalore*, **2005,** *89*, 1870.

Dunwell, J. A Comparative Study of Environmental and Developmental Factors Which Influence Embryo Induction and Growth in Cultured Anthers of *Nicotiana tabacum*. *Environ. Exper. Bot.* **1976,** *16*, 109–118.
Dunwell, J.; Francis, R.; Powell, W. Anther Culture of *Hordeum vulgare* L.: A Genetic Study of Microspore Callus Production and Differentiation. *Theor. Appl. Genet.* **1987,** *74*, 60–64.
Dunwell, J. M. Haploids in Flowering Plants: Origins and Exploitation. *Plant. Biotech. J.* **2010,** *8*, 377–424.
Engvild, K. C.; Linde-Laursen, I.; Lundqvist, A. Anther Cultures of *Datura innoxia*: Flower Bud Stage and Embryoid Level of Ploidy. *Hereditas*, **1972,** *72*, 331–332.
Ferrie, A. Current Status of Doubled Haploids in Medicinal Plants. In *Advances in Haploid Production in Higher Plants*; Springer, **2009**; pp 209–217.
Ferrie, A.; Epp, D.; Keller, W. Evaluation of *Brassica rapa* L. Genotypes for Microspore Culture Response and Identification of a Highly Embryogenic Line. *Plant Cell Rep.* **1995,** *14*, 580–584.
Foroughi-Wehr, B.; Friedt, W.; Wenzel, G. On the Genetic Improvement of Androgenetic Haploid Formation in *Hordeum vulgare* L. *Theor. Appl. Genet.* **1982,** *62*, 233–239.
Forster, B. P.; Heberle-Bors, E.; Kasha, K. J.; Touraev, A. The Resurgence of Haploids in Higher Plants. *Trends Plant Sci.* **2007,** *12*, 368–375.
Forster, B. P.; Thomas, W. T. Doubled Haploids in Genetics and Plant Breeding. *Plant Breed Rev.* **2005,** *25*, 57–88.
Gamborg, O. L.; Miller, R. A.; Ojima, K. Nutrient Requirements of Suspension Cultures of Soybean Root Cells. *Exp. Cell. Res.* **1968,** *50*, 151–158.
Germanà, M. Haploids and Doubled Haploids in Fruit Trees. In *Advances in Haploid Production in Higher Plants;* Springer, **2009;** pp 241–263.
Germana, M. A. Doubled Haploid Production in Fruit Crops. *Plant Cell Tiss. Org.* **2006,** *86*, 131–146.
Gresshoff, P. M.; Doy, C. H. Haploid *Arabidopsis thaliana* Callus and Plants from Anther Culture. *Aust. J. Biol. Sci.* **1972,** *25*, 259–264.
Guha, S.; Maheshwari, S. Development of Embryoids from Pollen Grains of *Datura* In vitro. *Phytomorphology*, **1967,** *17*, 454–461.
Harland, S. A Note on a Peculiar Type of "Rogue" in Sea Island Cotton. *Agr. News, Barbados* **1920,** *19*, 29.
Heberle-Bors, E. On the Time of Embryogenic Pollen Grain Induction During Sexual Development of *Nicotiana tabacum* L. Plants. *Planta*, **1982,** *156*, 402–406.
Heberle-Bors, E. Isolated Pollen Culture in Tobacco: Plant Reproductive Development in a Nutshell. *Sex Plant Reprod.* **1989,** *2*, 1–10.
Hofer, M. In vitro Androgenesis in Apple—Improvement of the Induction Phase. *Plant Cell Rep.* **2004,** *22*, 365–370.
Imamura, J.; Harada, H. Stimulatory Effects of Reduced Atmospheric Pressure on Pollen Embryogenesis. *Naturwissenschaften*, **1980,** *67*, 357–358.
Indrianto, A.; Barinova, I.; Touraev, A.; Heberle-Bors, E. Tracking Individual Wheat Microspores In vitro: Identification of Embryogenic Microspores and Body Axis Formation in the Embryo. *Planta*, **2001,** *212*, 163–174.
Iyer, R.; Raina, S. K. The Early Ontogeny of Embryoids and Callus from Pollen and Subsequent Organogenesis in Anther Cultures of *Datura metel* and Rice. *Planta*, **1972,** *104*, 146–156.
Jacobsen, E.; Sopory, S. The Influence and Possible Recombination of Genotypes on the Production of Microspore Embryoids in Anther Cultures of Solanum Tuberosum and Dihaploid Hybrids. *Theor. Appl. Genet.* **1978,** *52*, 119–123.

Kadota, M.; Niimi, Y. Production of Triploid Plants of Japanese Pear (*Pyrus pyrifolia* Nakai) by Anther Culture. *Euphytica*, **2004,** 138, 141–147.
Kameya, T.; Hinata, K. Induction of Haploid Plants from Pollen Grains of *Brassica*. *Jpn J. Breed.* **1970,** *20*, 82–87.
Kasha, K.; Hu, T.; Oro, R.; Simion, E.; Shim, Y. Nuclear Fusion Leads to Chromosome Doubling During Mannitol Pretreatment of Barley (*Hordeum vulgare* L.) Microspores. *J. Exp. Bot.* **2001,** *52*, 1227–1238.
Kernan, Z.; Ferrie, A. Microspore Embryogenesis and the Development of a Double Haploidy Protocol for Cow Cockle (*Saponaria vaccaria*). *Plant Cell Rep.* **2006,** *25*, 274–280.
Kochhar, T.; Sabharwal, P.; Engelberg, J. Production of Homozygous Diploid Plants by Tissue Culture Technique. *J. Hered.* **1971,** *62*, 59–61.
Koltunow, A. M.; Bicknell, R. A.; Chaudhury, A. M. Apomixis: Molecular Strategies for the Generation of Genetically Identical Seeds Without Fertilization. *Plant Physiol.* **1995,** *108*, 1345.
Kopecký, D.; Vagera, J. The Use of Mutagens to Increase the Efficiency of the Androgenic Progeny Production in *Solanum nigrum*. *Biologia plantarum*, **2005,** *49*, 181–186.
Künzel, G.; Korzun, L.; Meister, A. Cytologically Integrated Physical Restriction Fragment Length Polymorphism Maps for the Barley Genome Based on Translocation Breakpoints. *Genetics*, **2000,** *154*, 397–412.
Lantos, C.; Juhász, A. G.; Somogyi, G.; Ötvös, K.; Vági, P.; Mihály, R.; Kristóf, Z.; Somogyi, N.; Pauk, J. Improvement of Isolated Microspore Culture of Pepper (*Capsicum annuum* L.) Via Co-culture with Ovary Tissues of Pepper or Wheat. *Plant Cell Tiss. Org.* **2009,** *97*, 285–293.
Lichter, R. Induction of Haploid Plants from Isolated Pollen of *Brassica napus*. *Zeitschrift für Pflanzenphysiologie*, **1982,** *105*, 427–434.
Liu, X.-P.; Liu, Z.-W.; Tu, J.-X.; Chen, B.-Y.; Fu, T.-D. Improvement of Microspores Culture Techniques in *Brassica napus* L. *Hereditas (Beijing)*, **2003,** *25*, 433–436.
Maraschin, S. D. F.; De Priester, W.; Spaink, H. P.; Wang, M. Androgenic Switch: An Example of Plant Embryogenesis from the Male Gametophyte Perspective. *J. Exp. Bot.* **2005,** *56*, 1711–1726.
McCormick, S. Male Gametophyte Development. *Plant Cell*, **1993,** *5*, 1265.
Mejza, S. J.; Morgant, V.; DiBona, D. E.; Wong, J. R. Plant Regeneration from Isolated Microspores of *Triticum aestivum*. *Plant Cell Rep.* **1993,** *12*, 149–153.
Miah, M.; Earle, E.; Khush, G. Inheritance of Callus Formation Ability in Anther Cultures of Rice, *Oryza sativa* L. *Theor. Appl. Genet.* **1985,** *70*, 113–116.
Murashige, T.; Skoog, F. A Revised Medium for Rapid Growth and Bio Assays with Tobacco Tissue Cultures. *Physiologia Plantarum*, **1962,** *15*, 473–497.
Murovec, J.; Bohanec, B. Haploids and Doubled Haploids in Plant Breeding. *Plant Breeding. Rijeka, Croatia: In Tech*, **2011,** 87–106.
Narayanswamy, S.; Chandy, L. P. In vitro Induction of Haploid, Diploid, and Triploid Androgenic Embryoids and Plantlets in *Datura metel* L. *Ann. Bot.* **1971,** *35*, 535–542.
Nitsch, C.; Norreel, B. Factors Favoring the Formation of Androgenetic Embryos in Anther Culture. In *Genes, Enzymes, and Populations;* Springer, **1973;** pp 129–144.
Nitsch, J.; Nitsch, C. Haploid Plants from Pollen Grains. *Science*, **1969,** *163*, 85–87.
Niu, Z.; Jiang, A.; Abu Hammad, W.; Oladzadabbasabadi, A.; Xu, S. S.; Mergoum, M.; Elias, E. M. Review of Doubled Haploid Production in Durum and Common Wheat Through Wheat × Maize Hybridization. *Plant breed.* **2014,** *133*, 313–320.

Olsen, F. Isolation and Cultivation of Embryogenic Microspores from Barley (*Hordeum vulgare* L.). *Hereditas*, **1992,** *115*, 255–266.

Pechan, P.; Keller, W. Induction of Microspore Embryogenesis in *Brassica napus* L. by Gamma Irradiation and Ethanol Stress In vitro. *Cell Dev. Biol.* **1989,** *25*, 1073–1074.

Pescitelli, S.; Petolino, J. Microspore Development in Cultured Maize Anthers. *Plant Cell Rep.* **1988,** *7*, 441–444.

Petolino, J.; Jones, A.; Thompson, S. Selection for Increased Anther Culture Response in Maize. *Theor. Appl. Genet.* **1988,** *76*, 157–159.

Raina, S.; Zapata, F. Enhanced Anther Culture Efficiency of Indica Rice (*Oryza sativa* L.) Through Modification of the Culture Media. *Plant Breed.* **1997,** *116*; 305–315.

Rashid, A.; Street, H. Segmentations in Microspores of *Nicotiana sylvestris* and *Nicotiana tabacum* Which Lead to Embryoid Formation in Anther Cultures. *Protoplasma*, **1974,** *80*, 323–334.

Reinert, J.; Bajaj, Y. P. S.; Nitsch, C.; Clapham, D.; Jensen, C. *Haploids;* Springer, 1977.

Sangwan, R.; Camefort, H. The Tonoplast, A Specific Marker of Embryogenic Microspores of *Datura* Cultured In vitro. *Histochemistry*, **1983,** *78*, 473–480.

Sato, T.; Nishio, T.; Hirai, M. Plant Regeneration from Isolated Microspore Cultures of Chinese Cabbage (*Brassica campestris spp. pekinensis*). *Plant Cell Rep.* **1989,** *8*, 486–488.

Seguí-Simarro, J. M.; Nuez, F. Meiotic Metaphase I to Telophase II as the Most Responsive Stage During Microspore Development for Callus Induction in Tomato (*Solanum lycopersicum*) Anther Cultures. *Acta Physiologiae Plantarum*, **2005,** *27*, 675–685.

Shariatpanahi, M. E.; Bal, U.; Heberle-Bors, E.; Touraev, A. Stresses Applied for the Re-programming of Plant Microspores Towards In vitro Embryogenesis. *Physiologia Plantarum*, **2006a,** *127*, 519–534.

Sharp, W.; Raskin, R.; Sommer, H. The Use of Nurse Culture in the Development Haploid Clones in Tomato. *Planta,* **1972,** *104*, 357–361.

Shariatpanahi, M. E.; Belogradova, K.; Hessamvaziri, L.; Heberle-Bors, E.; Touraev, A. Efficient Embryogenesis and Regeneration in Freshly Isolated and Cultured Wheat (*Triticum aestivum* L.) Microspores Without Stress Pretreatment. *Plant Cell Rep.* **2006b,** *25*; 1294–1299.

Shtereva, L.; Zagorska, N.; Dimitrov, B.; Kruleva, M.; Oanh, H. Induced Androgenesis in Tomato (*Lycopersicon esculentum* Mill). II. Factors Affecting Induction of Androgenesis. *Plant Cell Rep.* **1998,** *18*, 312–317.

Spillane, C.; Steimer, A.; Grossniklaus, U. Apomixis in Agriculture: The Quest for Clonal Seeds. *Sex Plant Reprod.* **2001,** *14*, 179–187.

Sunderland, N. Strategies in the Improvement of Yields in Anther Culture. In *Proceedings of Symposium on Plant Tissue Culture;* Science Press: Peking, **1978;** pp 65–86.

Sunderland, N.; Wicks, F. M. Embryoid Formation in Pollen Grains of *Nicotiana tabacum*. *J. Exp. Bot.* **1971,** *22*, 213–226.

Supena, E. D. J.; Suharsono, S.; Jacobsen, E.; Custers, J. Successful Development of a Shed-microspore Culture Protocol for Doubled Haploid Production in Indonesian Hot Pepper (*Capsicum annuum* L.). *Plant Cell Rep.* **2006,** *25*, 1–10.

Takahata, Y.; Komatsu, H.; Kaizuma, N. Microspore Culture of Radish (*Raphanus sativus* L.): Influence of Genotype and Culture Conditions on Embryogenesis. *Plant Cell Rep.* **1996,** *16*, 163–166.

Tanaka, M. Effect of Centrifugal Treatment on the Emergence of Plantlet from Cultured Anther of Tobacco. *Jpn. J. Breed.* **1973,** *23*, 171–174.

Tiainen, T. (1992). The Role of Ethylene and Reducing Agents on Anther Culture Response of Tetraploid Potato (*Solanum tuberosum* L.). *Plant Cell Rep.* **1973,** *10*, 604–607.

Touraev, A.; Ilham, A.; Vicente, O.; Heberle-Bors, E. Stress-induced Microspore Embryogenesis in Tobacco: An Optimized System for Molecular Studies. *Plant Cell Rep.* **1996,** *15*, 561–565.

Touraev, A.; Tashpulatov, A.; Indrianto, A.; Barinova, I.; Katholnigg, H.; Akimcheva, S.; Ribarits, A.; Voronin, V.; Zhexsembekova, M.; Heberle-Bors, E. Fundamental Aspects of Microspore Embryogenesis. In *Biotechnological Approaches for Utilisation of Gametic Cells,* COST 824: Final Meeting, Bled, Slovenia, July 1–5, 2000 (Office for Official Publications of the European Community); **2001,** pp 205–214.

Touraev, A.; Vicente, O.; Heberle-Bors, E. Initiation of Microspore Embryogenesis by Stress. *Trends In Plant Sci.* **1997,** *2*, 297–302.

Vagera, J.; Novotný, J.; Ohnoutková, L. Induced Androgenesis In vitro in Mutated Populations of Barley, *Hordeum vulgare*. *Plant Cell Tiss. Org.* **2004,** *77*, 55–61.

Vasil, I. Androgenetic Haploids [Flowering plants]. *Int. Rev. Cytol. Suppl.* **1980,** *11*, 195–223.

Vasil, I. K. The New Biology of Pollen. *Naturwissenschaften*, **1973,** *60*, 247–253.

Wang, J.; Hu, D.; Wang, H.; Tang, Y. Studies on Increasing the Induction Frequency of Pollen Callus in Wheat. *Acta. Genet. Sin.* **1981,** *2*, 71–77.

Wang, M.; Hoekstra, S.; van Bergen, S.; Lamers, G. E.; Oppedijk, B. J.; van der Heijden, M. W.; de Priester, W.; Schilperoort, R. A. Apoptosis in Developing Anthers and the Role of ABA in This Process During Androgenesis in *Hordeum vulgare* L. *Plant Mol. Biol.* **1999,** *39*, 489–501.

Wang, Z.; Taramino, G.; Yang, D.; Liu, G.; Tingey, S.; Miao, G.; Wang, G. Rice ESTs with Disease-resistance Gene or Defense-response Gene-like Sequences Mapped to Regions Containing Major Resistance Genes or QTLs. *Mol. Genet. Genomics.* **2001,** *265*, 302–310.

Wędzony, M.; Forster, B.; Żur, I.; Golemiec, E.; Szechyńska-Hebda, M.; Dubas, E.; Gotębiowska, G. Progress in Doubled Haploid Technology in Higher Plants. In *Advances in Haploid Production in Higher Plants;* Springer, **2009;** pp 1–33.

Wei, Z. M. Pollen Callus Culture in *Triticum aestivum*. *Theor. Appl. Genet.* **1982,** *63*, 71–73.

Wenzel, G.; Uhrig, H. Breeding for Nematode and Virus Resistance in Potato Via Anther Culture. *Theor. Appl. Genet.* **1981,** *59*, 333–340.

Wolyn, D.; Nichols, B. Asparagus Microspore and Anther Culture. In *Doubled Haploid Production in Crop Plants;* Springer, **2003;** pp 265–273.

Zaki, M.; Dickinson, H. Microspore-derived Embryos in *Brassica*: The Significance of Division Symmetry in Pollen Mitosis I to Embryogenic Development. *Sex Plant Reprod.* **1991,** *4*, 48–55.

Zhao, J.; Simmonds, D. H.; Newcomb, W. High Frequency Production of Doubled Haploid Plants of *Brassica napus* cv. Topas Derived from Colchicine-induced Microspore Embryogenesis Without Heat Shock. *Plant Cell Rep.* **1996,** *15*, 668–671.

Ziauddin, A.; Simion, E.; Kasha, K. Improved Plant Regeneration from Shed Microspore Culture in Barley (*Hordeum vulgare* L.) cv. Igri. *Plant Cell Rep.* **1990,** *9*, 69–72.

Zonia, L.; Tupý, J. Lithium Treatment of *Nicotiana tabacum* Microspores Blocks Polar Nuclear Migration, Disrupts the Partitioning of Membrane-associated Ca^{2+}, and Induces Symmetrical Mitosis. *Sex Plant Reprod.* **1995,** *8*, 152–160.

CHAPTER 6

EMBRYO CULTURE AND ENDOSPERM CULTURE

MANOJ KUNDU[1*], JAYESH PATHAK[2*], and SANGITA SAHNI[3]

[1]*Department of Horticulture (Fruit and Fruit Technology), Bihar Agricultural University, Sabour, Bhagalpur, Bihar, India*

[2]*Department of Silvicuture and Agroforestry, Navsari Agricultural University, Navsari, Gujarat, India*

[3]*Tirhut College of Agriculture, Dholi, Dr. Rajendra Prasad Central Agricultural University, Pusa, Samastipur, Bihar, India*

[*]*Corresponding author. E-mail: manojhorti18@gmail.com*

CONTENTS

ABSTRACT

The vulnerability of cultivated crops to different environmentally stresses have increased immensely. Improvement of crops against these stresses lies in the introduction of natural variability through conventional and biotechnological interventions. Intervarietal and interspecific crosses, followed by selection resulted overall improvement in the quality and yield potential of different crops. The abortions of hybrid embryos at early stage of their development are the major limiting factor. Alternatively, embryo culture, sometimes called embryo rescue, is the method of choice to save the early abortion of hybrid embryos obtained through intervarietal and interspecific crosses. It involves isolating and growing of an immature or mature zygotic embryo under sterile conditions in nutrient medium to obtain a viable plant. The basic premise for this technique is that the integrity of the hybrid genome is retained in a developmentally arrested or an abortive embryo and that its potential to resume normal growth may be realized if supplied with the proper growth substances. In this present chapter a detailed discussion was made on the development of embryo culture, technique of embryo culture, factor affecting embryo culture, etc. At the end of this chapter endosperm culture technique has been discussed. Endosperm is a triploid tissue formed by fusion of one male and two female nuclei (double fertilization) and it is present in developing seeds of more than 80% of angiosperms. In vitro culture of endosperm enables regeneration of triploid plants, which are of considerable commercial importance in many species.

6.1 INTRODUCTION

Plant is the primary source of food, feed, and energy for life existing on Earth. Without plants, life cannot be envisaged on the earth. But with the ever increasing population, one of the major challenges for agriculture is to increase the productivity of each and every crop in an environmentally sustainable manner. Moreover, in the recent situations of climate change, the vulnerability of different crops to biotic and abiotic stresses increased many folds. These conditions help to realize the importance of different wild species in different crops. Hybridization between cultivated species and wild relative may increase the resistance of the crop to different biotic and abiotic stress with higher yield potential. But it has been observed in many cases that in such crosses hybrid embryos are aborted at early stage of their development. Moreover, instead of availability of resistant sources in nature, sometimes it is very difficult to use them in conventional breeding because of cross incompatibility problem between cultivated species and the resistant sources. For example, in papaya, *Carica cauliflora* shows resistant to viruses while *C. candamarcensis* and *C. pentagona* are resistant to frost. But these species could not be utilized in papaya improvement program because crossing between cultivated papaya (*C. papaya*) and these resistant sources do not form mature seeds (Ray, 2002) due to the cross incompatibility problem among these species. This again raises the question what will be the possible way to overcome from these problems. Embryo culture is the only possible alternate technique to solve the problem.

6.2 MEANING OF EMBRYO CULTURE

In angiosperms, embryo represents the beginning of sporophyte. Normally, the fertilized egg or zygote undergoes embryogenesis in the postfertilization stage within the ovule and thus embryo is formed inside the seed. The typical seed embryo is a bipolar structure consisting of contrasting meristem at each pole—the primordial shoot or the plumule and the primordial root or radicle and one or two lateral appendages, the cotyledons. The mature embryo, therefore, possesses the basic organization of the adult plant. During seed germination, a plant is produced through progressive and orderly changes in embryo. Like many other plant organs, embryo can be used as explant and cultured aseptically in the test tube containing medium.

However, the technique of in vitro culture of mature or immature zygotic embryo, isolated from seed, in suitable culture medium under aseptic condition to regenerate into whole plantlet is called as embryo culture.

6.3 EMBRYO CULTURE

The embryo of different developmental stages, formed within the female gametophyte through sexual process, can be isolated aseptically from the bulk of maternal tissues of ovule, seed, or capsule and cultured in vitro under aseptic and controlled physical in the glass vials containing solid or liquid nutrient medium to grow directly into plantlet.

6.3.1 DEFINITION

Embryo culture can be defined as the in vitro culture either of the polarized egg, zygote, pre-embryo, or mature embryo.

6.3.2 HISTORY

Embryo culture, sometimes also called embryo rescue has been practiced by plant breeders for over half a century. However, the first systematic attempt to grow the embryos of angiosperms in vitro, under aseptic conditions was made by Hanning (1904) who cultured mature embryos of *Raphanus* and the conifers *Cochlearia* on a mineral salt medium supplemented with sugar (Norstog, 1979) and obtained viable plant from this culture. Subsequently, many workers raised plants by cutting embryos excised from mature seeds. In the year 1924, to evaluate the efficacy of embryo culture technique to curtail the dormancy period of different plant species, Dietrich cultured mature and immature embryos of various plant species. From this experiment, he reported that the mature embryos grew immediately, circumventing dormancy while precocious germination was observed from immature embryos without further embryo development. Further progress in the field of embryo culture was provided by Liabach (1925) who demonstrated the most important practical application of this technique. He crossed *Linum perenne* L. with *Linum austriacum* L. but obtained hybrid seeds of very light and shrivelled nature without any germinability. The excised embryos from such seeds were cultured on moist filter paper dipped in sucrose solution. This led to the regeneration of hybrid plants. Since then, the technique of embryo culture has been widely used to produce hybrids which otherwise was not possible due to postfertilization barriers to crossability leading to embryo abortion. Later on, while working with Datura embryo, van Overbeek et al. (1941) discovered that embryos could be grown in culture on media containing coconut milk (CM). This discovery ultimately led to understand the importance of reduced nitrogen in the form of amino acids for embryo culture. Since then, embryo culture technique has widely been used to understand the physical and nutritional requirements, embryonic development, by passing seed dormancy, shortening the breeding cycle, test seed viability, provide material for micropropagation, and rescue immature hybrid embryos from incompatible crosses (Hu and Wang, 1986). And it has become a routine operation for improvement of almost all agricultural as well as horticultural crops.

6.3.3 TYPES OF EMBRYO CULTURE

According to Pierik (1989), there are in principle two types of embryo culture.

6.3.3.1 CULTURE OF IMMATURE EMBRYO

This type of embryo culture is mainly used to grow immature embryos originating from unripe or hybrid seeds which fail to germinate. Excising such embryos is difficult and generally a complex nutrient medium is required to raise them to produce plants.

6.3.3.2 CULTURE OF MATURE EMBRYOS

Mature embryos are excised from ripe seeds and cultured mainly to avoid inhibition in the seed for germination. This type of culture is relatively easy as embryo requires simple nutrient medium containing mineral salts, mineral salts, sugar, and agar for growth and development.

6.3.4 PHASES OF EMBRYO DEVELOPMENT

Raghavan (1966) recognized two phases of embryo developments.

1. **Heterotrophic phase:** This phase lasts up to the globular stage. Here, the pro-embryos are solely depends on the endosperm and surrounding tissues for their nutrition. They are not competent to synthesize required substances.
2. **Autotrophic phase:** This phase starts at the late heart shaped stage. The embryos are competent to synthesize required substances for growth from mineral salts and sugar.

6.3.5 TECHNIQUE OF EMBRYO CULTURE

6.3.5.1 SURFACE STERILIZATION

Embryos of seed plants normally develop inside the ovule which in turn is covered by ovaries. Since they already exist in a sterile environment, disinfection of the embryo surface is unnecessary unless the seed coats are injured or systemic infection is present. Instead, mature seeds, entire ovule, or fruits are surface sterilized. Surface sterilization is carried out by immersing the material in hypo chorine-containing commercial bleach (5–10% clorox, 0.45% sodium or calcium hypochlorite) for 5–10 min or ethanol (70–75%) for 5 min. A small amount (0.01–0.1%) of a surfactant may be added to disinfection solution. In case of infected seeds, the excised embryos may be immersed in 70% alcohol plus 5–10 min exposure to 2.6% sodium hypochlorite solution.

6.3.5.2 EXCISION OF EMBRYO

Embryo excision operation is carried out aseptically in a laminar airflow hood. A stereomicroscope equipped with cool-ray fluorescent lamp is required for excision of small embryo. The commonly used dissecting tools are forceps, dissecting needles, scalpels, razor blades, and Pasteur pipettes. Mature embryo can be isolated with relative ease by splitting open the seeds. Soaking a hard-coat seeds for few hours to a few days before sterilization makes its dissection easier. In case of embryos embedded in liquid endosperms, the incision is made at micropolar end of young ovule and pressure applied at opposite end to force the embryo out through the incision.

6.3.5.3 EMBRYO-ENDOSPERM TRANSPLANT

It is very difficult to culture embryo in vitro abort at very early stages of development because of lack of knowledge of nutritional requirements. The chances of development of immature or abortive embryos increases if they are surrounded by endosperm tissue excised from another seed of same species. Generally, endosperm older than the embryo by 5 days was more efficient as a nurse tissue than one of the same age as the embryo.

6.3.5.4 NUTRITIONAL REQUIREMENT

The nutritional requirements of an embryo during its development in vitro consists of two phases: (1) Heterotropic phase—an early phase wherein the embryo is dependent and draws upon the endosperms and material tissues and (2) The autotrophic phase—a later phase in which the embryo is metabolically capable of synthesizing substances required for its growth, thus becoming fairly independent for nutrition.

The media constituents for in vitro growth of young or immature embryos also differ from those of mature embryos. This often necessitates the transfer of embryos from one medium to another for their orderly growth. Hanning (1904) used a minimal salt–sucrose solution to culture excised embryos of Raphanus and Cochlearia and grow them into plantlets. Since the beginning, a number of nutritional formulations have been used. It has been noted that nutritional requirements for pro-embryo and hybrid embryos are very critical. Continuous efforts are on to improve the culture media by the addition and alteration of inorganics, organics, and plant extracts as well as regulators. Detailed studies on nutritional requirements have been done in two angiospermic species namely *Capsella* and *Hordium*.

The most important aspect in culturing immature embryo is to define a culture medium that can sustain their growth and development. In a less than optimum medium, the immature embryos may fail to survive, turn into undifferentiated callus, or germinate prematurely (precocious germination), to give rise to weak seedlings.

So, isolated mature embryos can grow on a simple inorganic medium supplemented with an energy source, whereas younger embryos require an elaborate medium for their successful culture (Fregene et al., 1998; Asif et al., 2001).

1. Mineral Salts

Inorganic nutrients of MS, B5, and White's media with certain degree of modification are the most widely used basal media for embryo culture. Monnier (1978) modified the MS medium for immature embryo culture of Capsella which contains higher levels of potassium and calcium and reduced levels of ammonium (NH_4NO_3) and FeEDTA and double concentration of MS micronutrients.

2. Carbohydrates

Sucrose is the most commonly used source of energy for embryo culture. Addition of maltose, lactose, raffinose, or mannitol may be required in embryo culture of some species. In some cases glucose is found to be better than sucrose. Mature embryos grow fairly well at low sucrose concentration but younger embryos require higher level of carbohydrates.

3. *Nitrogen and Vitamins*

Ammonium nitrate is better than KNO_3, $NaNO_3$, and $(NH_4)_2$ HPO_4 especially the presence of NH^{4+} in the medium has been found essential for proper growth and differentiation of embryos. Various amino acids and their amides like aspargine, glutamine, and casein hydrolysate have been widely used in embryo culture.

4. *Natural Plant Extract*

The CM effectively stimulates the growth of excised young embryos of sugarcane, barley, tomato, carrot, interspecific hybrids of Vigna and fern species. Van Overbeek et.al. (1941) suggested that the CM contains some "Embryofactor" which presumably makes up for deficiencies of certain sugars, amino acids, growth hormones, and other critical metabolites of the culture medium.

In addition to CM, water extracts from dates, bananas, hydrolysate of wheat gluten and tomato juice were also effective.

5. *Growth Regulators*

Auxin and cytokinins are not generally used in embryo culture since they induce callus formation. At very low concentration, GA promotes embryogenesis of young barley embryos without inducing precocious germination. ABA also has a similar effect on barley and Phaseolus embryos.

6. *pH of Medium*

Excised embryos grow well in a medium with a pH 5–7.5. Generally, the medium pH is adjusted 0.5 units higher than the desired pH in order to compensate for uncontrollable change in its value during the autoclaving process.

7. *Incubation Conditions*

The embryo cultures are incubated at 25 ± 2°C whereas in case of species to warm temperature requires 27–30°C incubation temperature and species occurring in cold regions or seasons require incubation temperature of 17–22°C. Embryo of most of the plants grow well at temperatures between 25°C and 30°C. The optimum temperature for *Datura tatula* is reported to be 35°C.

Generally, an initial dark incubation (4 days) of embryo in culture is essential, following which they can grow to a mature stage even under continuous light regime. According to Narayanaswawmy and Norstog, light is not critical for embryo growth.

6.3.6 ROLE OF SUSPENSOR IN EMBRYO CULTURE

Suspensor is actively involved in embryo development. The suspensor is an ephemeral structure found at the radicular end of the pro-embryo and attains maximum development by the time embryo reaches globular stage. In cultures, the presence of a suspensor is critical, particularly for the survival of young embryos. The requirement of the suspensor may be substituted by the addition of GA or ABA to the culture medium.

6.3.7 EXCISION OF EMBRYO

According to Raghavan (1983), the term pro-embryo refers to those developmental stages of the embryo that precede cotyledon initiation. Embryos from mature seeds can be easily removed by splitting the seeds. Seeds with hard coats should be soaked in water (1 h to few days) so that the embryos can be removed easily. Excision of older embryos requires one excision in the ovule on the side lacking the embryo. Embryos excised from dry seeds can be put into culture tube with a pair of forceps. But if the embryos are too small, a dissecting needle will be required. Before excision, the tip of the needle should be moistened with the culture media/sterile water. Dehydrated embryos will easily stick onto the tip of the needle and can be then be transferred to the culture medium.

Excision of pro-embryos requires an incision at the micropilar end of the young ovule and application of pressure from the opposite side to force the embryo out. When the heart-shaped and younger embryos are excised, it is important to keep the suspensor intact. Utmost care must be taken to avoid injury. Excision of embryos may be done in a dish or watch glass with drops of liquid media to avoid injury. Excised small embryos/pro-embryos can also be picked up with the help of the Pasteur pipette.

6.3.8 PROTOCOL FOR EMBRYO CULTURE

The following protocol for embryo culture is based on the method used for *Capsella bursapastoris*. With modification, this basic protocol should be applicable to embryo culture in general.

1. The capsules in the desired stages of development are surface sterilized for 5–10 min in 0.1% $HgCl_2$ in a laminar air flow.
2. Wash repeatedly in sterile water.
3. Further operations are carried out under a special design dissecting microscope at a magnification of about 90×. The capsules are kept in a depression slide containing few drops of liquid medium.
4. The outer wall of capsule is removed by a cut in the region of the placenta; the halves are push apart with forceps to expose the ovules.
5. A small incision in the ovule followed by slight pressure with a blunt needle is enough to free the embryos.
6. The excised embryos are transferred by micropipette or small spoon headed spatula to standard 10 cm petri dishes containing 25 mL of solidified standard medium. Usually 6–8 embryos are cultured in petri dish.
7. The petri dishes are sealed with cello tape to prevent desiccation of the culture.
8. The cultures are kept in culture room at 25 ± 1°C and given 16 h illumination by cool white fluorescent tube.
9. Subcultures into fresh medium are made at approximately 4 weeks interval. In case of fresh seed or dry and imbibed seeds, the schedule is slightly changed. Seeds are cleaned by 5% Teepol for 10 min and dipped in 70% ethyl alcohol for 60 s. Surface sterilization in 0.1% $HgCl_2$ is followed by washing in sterile water, then the seeds are decotylated using a sharp scalpel and embryos are transferred to solid nutrient medium.

6.3.8.1 STEPS INVOLVED IN EMBRYO CULTURE

Dip the seeds in 70% alcohol for 1 min and then sterilize with 20% sodium hyprochloride solution for 10 min.

Rinse 3–5 times with sterile distilled water.

Take capsule in depression slide. Give cut at placenta, it will be divided into two halves.

With the help of forceps, expose the ovule. Give small incision on ovule give pressure with a needle.

Transfer it with micropipette in 25 mL of solid medium in petri dish. Place 6–8 embryos in petri dish.

Seal with cello tape or petri dish put on petri dish. The globular stage embryos were inoculated on the culture medium.

The culture was incubated at 25+/−2°C in dark for 30 days. After this rapid period, the cultures were transferred to a growth chamber with 16/8 h photoperiod (200lux) at 25°C.

The embryos were developed into seedlings of about 5 cm height within 3–4 weeks.

After 4 weeks subculturing was going on.

6.3.8.2 STEPS INVOLVED IN EMBRYO CULTURE IN CULTIVATED PAPAYA

1. C. papaya var. "Wahington" was used as the female parent and C. Caullflora as the male parent. C. papaya C. Caullflora
2. Flower of female and male parents were covered with butter paper bags (8cm × 12cm) before anthesis of flower.
3. When the female flowers had fully opened, pollen from the male parent was collected in a watch glass (between 9 and 11a.m.) and dusted on the stigma of the female flower with a fine brush. This was repeated 2–3 times a day for 2–3 days. After pollination, the female flowers were covered again with butter paper bags to avoid contamination.
4. The fruits were collected 89 days after fertilization and the seeds removed. The seeds were surface sterilized.

6.3.9 APPLICATION OF EMBRYO CULTURE

1. Rescuing Embryos from Incompatible Crosses

In interspecific and intergeneric hybridization programs, incompatibility barriers often prevent normal seed development and production of hybrids. Although there may be normal fertilization in some incompatible crosses, embryo abortion results in the formation of shrivelled seeds. Poor and abnormal development of the endosperm caused embryo starvation and eventual abortion. Isolation of hybrid embryos before abortion and their in vitro culture may prevent these strong postzygotic barriers. The most useful and popular application of embryo cultures is to raise rare hybrids by rescuing embryos of incompatible crosses.

2. Overcoming Dormancy and Shortening Breeding Cycle

Long periods of dormancy in seeds delay breeding works, especially in horticultural and crop plants. Using embryo culture techniques, the breeding cycle can be shortened in these plants. For example, the life cycle of Iris was reduced from 2 to 3 years to less than 1 year. Similarly, it was possible to obtain two generations of flowering against one in Rosa sps. Germination of excised embryo is regarded as a more reliable test for rapid testing of viability in seeds, especially during dormancy period.

3. Overcoming Seed Sterility

In early ripening fruit cultivars, seeds do not germinate because their embryos are still immature. Using the embryo culture method, it is possible to raise seedling from sterile seeds of early ripening stone fruits, peach, apricot, and plum.

"Makapuno" coconuts are very expensive and most relished for their characteristic soft fatty endosperms in place of liquid endosperm. Under normal conditions, the coconut seeds fail to germinate. Guzman et.al. (1971) obtained 85% success in raising field-grown makapuno trees with the aid of embryo cultures.

4. *Production of Monoploid*

An embryo culture has been used in the production of monoploids of barley. With the cross Hordeum vulgare, fertilization occurs normally but thereafter chromosomes of *H. bulbosum* are eliminated, resulting in formation of Monoploid *H. vulgare* embryo which can be rescued by embryo cultures.

5. *Clonal Micropropagation*

The regenerative potential is an essential prerequisite in nonconventional methods of plant genetic manipulations. Because of their juvenile nature, embryos have a high potential for regeneration and hence may be for in vitro clonal propagation. This is especially true of conifers and graminaceous members.

Both organogenesis and somatic embryogenesis have been induced in major cereals and forage grasses form embryonic tissues. Generally, callus derived from immature embryos of cereals has the desired morphogenetic potential for regeneration and clonal propagation.

6.4 ENDOSPERM CULTURE

Tissue culture methods are also used for culturing endosperm, which is unique, first, in its function of supplying nutrition to the developing embryo and second, in being triploid in its chromosome constitution. Triploid plants are useful for production of seedless fruits (e.g., apple, banana, watermelon, etc.) and for production of trisomics for cytogenetic studies.

In angiosperm, double fertilization gives rise to the embryo and endosperm while in gymnosperms, the enlarged female gametophyte acts as the endosperm. In angiosperm, the endosperm forms the immediate environment of the embryo and it is rich in nutrition and growth substances. It is a genetically unique triploid tissue that arises from the fusion of one male gamete with two polar nuclei (Fahn, 1974) an event of double fertilization. However, in gymnosperm, the endosperm is a haploid tissue.

About 81% of angiospermic families are known to possess triploid endosperms in their developing seeds. The presence of the endosperm in the seeds up to the germination stage designates them as endospermous seeds, for example, cereals, coconut, coffee, magnolia, rhododendron while seeds who have endosperms at the beginning but are consumed during maturity are called nonendospermous seeds, for example, legumes, cucurbits, etc.

Angiospermic endosperms being triploid in nature and gymnosperms being of haploid nature, they are appropriate for the in vitro experimental system. Lack of vascular elements and any degree of differentiation also make it a system for morphogenetic studies.

In vitro culture of the endosperm tissue begins in the 1930s with *Zea mays* but the materials did not respond properly (Lampe and Mills, 1933; LaRue, 1949; Tamoki and Ullstrup, 1958; Sehgal, 1969). In the process of culturing endosperm tissue of Hordium and Triticum, Sehgal (1974)

6.4.1 STEPS INVOLVED IN ENDOSPERM CULTURE

The technique of endosperm culture involves the following steps: (1) The immature seeds are dissected under aseptic conditions and endosperms along with embryos are excised; sometimes mature seeds can also be used. (2) The excised endosperms are cultured on a suitable medium and embryos are removed after initial growth. (3) The initial callus phase is followed by "shoot bud differentiation" or "embryogenesis." (4) The shoots and roots may subsequently develop and complete triploid plants that can be established for further use.

Isolation of the uninjured cellular endosperm is a prerequisite for endosperm culture. Selection of a particular age of responding endosperm tissue is a very important factor and it may vary from cellular endosperm to mature endosperm (4–7 days in rice, 7–10 days in lolium, more than 12 days in maize, wheat, fully matured emdosperm in santtalum). The embryo–endosperm relationship is another factor for proliferation of endosperm tissues in vitro. Any basal medium with auxin and cytokinin as growth regulators and tomato juice, yeast extract, casein hydrolysate, CM (liquid endosperm), young corn juice, as undefined substances are some general requirements of endosperm culture.

6.4.2 APPLICATIONS OF ENDOSPERM CULTURE

The regeneration of triploid plants from the endosperm of citrus by Wang and Chang (1978) is a landmark achievement though they failed to transfer the plants to the field. Later on Gmitter et al. (1990) successfully transferred triploid hybrid plants of citrus from endosperm tissues.

Seed sterility is the prime character of triploid plants. Triploidy is not desirable where seeds are important for agriculture. But there are some plants, some tropical fruits where seedlessness caused by triploidy may have some commercial value. Triploid production in tropical fruits by the conventional methods is laborious and time consuming. It is hoped that scientists will utilize the cellular totipotency of triploid endosperm, as an alternative method for crop improvement.

KEYWORDS

- **embryo culture**
- **endosperm culture**
- **monoploids**
- **crops**
- **plant**

REFERENCES

Aleza, P.; Juárez, J.; Cuenca, J.; Ollitrault, P.; Navarro, L. Recovery of Citrus Triploid Hybrids by Embryo Rescue and Flow Cytometry from 2x × 2x Sexual Hybridisation and Its Application to Extensive Breeding Programs. *Plant Cell Rep.* **2010,** *29*, 1023–1034.

Arditti, J. Factors Affecting the Germination of Orchid Seeds. *Bot. Rev.* **1967,** *33,* 1–97.

Bharathy, P. V.; Karibasappa, U. S.; Patil, S. G.; Agrawal, D. C. *In ovulo* Rescue of Hybrid Embryos in Flame Seedless Grapes—Influence of Pre-Bloom Sprays of Benzyladenine. *Sci. Hortic.* **2005,** *106*, 353–359.

Chawla, H. S. *Introduction of Plant Biotechnology* (3rd Edn.); Oxford and IBH Publishing Co. Pvt. Ltd.: New Delhi, **2012;** p 40.

Dietrich, K. Über Kultur von Embryonen ausserhalb des Samens. Flora (Jena) **1924,** 117, 379–417.

Dunwell, J. M. Pollen, Ovule and Embryo Culture as Tools in Plant Breeding. In *Plant Tissue Culture and Its Agricultural Applications*; Withers, L. A., Alderson, P.G. Eds.; Butterworths: London, **1986;** pp 375–404.

Froelicher, Y.; Bassene, J.; Jedidi-Neji, E.; Dambier, D.; Morillon, R.; Bernardini, G.; Costantino, G.; Ollitrault, P. Induced Parthenogenesis in Mandarin for Haploid Production: Induction Procedures and Genetic Analysis of Plantlets. *Plant Cell Rep.* **2007,** *26*, 937–944.

Godbole, M.; Murthy, H. N. *In vitro* Production of Haploids *via* Parthenogenesis in Culinary Melon (*Cucumis melo* var. *acidulous*). *Indian J. Biotechnol.* **2012,** *11*, 495–497.

Grouh, M. S. H.; Vahdati, K.; Lotfi, M. Production of Haploids in Persian Walnut Through Parthenogenesis Induced by Gamma-Irradiated Pollen. *J. Am. Soc. Hortic. Sci.* **2011,** *136*, 198–204.

Hanning, E. Zur physiologic pflanzicher embryonen, I. Uber die culture von crucifer. Enbroner assuerhalb des Embryosaks. *Botanical Gazette* **1904,** *62*, 45–80.

Hu, C.; Wang, P. Embryo Culture: Technique and Applications. In *Handbook of plant cell culture*; Evans, D. A., Sharp, W. R., Ammirato P. V, Eds.; Macmillan: New York, **1986;** Vol. 4, pp 43–96.

Ji, W.; Li, Z. Q.; Zhou, Q.; Yao, W. K.; Wang, Y. J. Breeding New Seedless Grape by Means of *in vitro* Embryo Rescue. *Genet. Mol. Res.* **2013,** *12*, 859–869.

Kasha, K. J.; Kao, K. N. High Frequency Haploid Production in Barley (*Hordium vulgare* L.). *Nature* **1970,** *225*, 874–876.

Laibach, F. Das Taubwerden von Bastardsamen und die künstliche Aufzucht früh absterbender Bastardembryonen. *Zeitschrift fur Biologie* **1925,** *17*, 417–459.

Momotaz, A.; Kato, M.; Kakihara, F. Production of Intergeneric Hybrids between Brassica and Sinapis Species by Means of Embryo Rescue Techniques. Euphytica **1998,** *103*(1), 123–130.

Nasertorabi, M.; Madadkhah, E.; Moghbeli, E.; Grouh, M.S.H.; Soleimani, A. Production of Haploid Lines from Parthenogenetic Iranian Melon Plants Obtained of Irradiated Pollen (*Cucumis melo* L.). *Int. Res. J. Appl. Basic Sci.* **2012,** *3,* 1585–1589.

Norstog, K. Embryo Culture as a Tool in the Study of Comparative and Developmental Morphology. In *Plant Cell and Tissue Culture*; Sharp, W. R., Larsen, P. O., Paddock, E. F., Raghavan, V., Eds. Ohio State Univ. Press: Columbus, **1979;** p 179–202.

Plapung, P.; Khamsukdee, S.; Potapohn, N.; Smitamana, P. Screening for Cucumber Mosaic Resistant Lines from the Ovule Culture Derived Double Haploid Cucumbers. *Am. J. Agric. Biol. Sci.* **2014,** *9*(3), 261–269.

Ponce, M. T.; Agüero, C. B.; Gregori, M. T.; Tizio, R. Factors Affecting the Development of Stenospermic Grape (*Vitis vinifera* L.) Embryos Cultured *in vitro*. *Acta Hortic.* **2000,** *528*, 667–671.

Ramming, D. W; Emershad, R. L. *In ovulo* Embryo Culture of Seeded and Seedless *Vitis vinifera* L. *Hort Scince* **1982,** *17*, 487.

Ray, P. K. *Breeding Tropical and Subtropical Fruits*. Narosa Publishing House: New Delhi, **2002;** p 107.

Reforgiato-Recupero, G.; Russo, G.; Recupero, S. Mandarin Tree Named 'Top Mandarin Seedless'. U.S. Patent US PP18,568 P, 2008.

Roy, A. K.; Malaviya, D. R.; Kaushal, P.; Kumar, B.; Tiwari, A. Interspecific Hybridization of *Trifolium alexandrinum* with *T. constantinopolitanum* using Embryo Rescue. *Plant Cell Rep.* **2004,** *22*(9), 705–710.

Singh, F. In vitro Propagation of Orchids 'state of the art'. J. Orchid Soc. India **1992,** *6*, 11–14.

Singh, N. V.; Singh, S. K.; Singh, A. K. Standardization of Embryo Rescue Technique and Bio-Hardening of Grape Hybrids (*Vitis vinifera* L.) using Arbuscular Mycorrhizal Fungi (AMF) under Sub-Tropical Conditions. *Vitis* **2011,** *50*, 115–118.

Tang, D. M.; Wang, Y. J.; Cai, J. S.; Zhao, R. H. Effects of Exogenous Application of Plant Growth Regulators on the Development of Ovule and Subsequent Embryo Rescue of Stenospermic Grape (*Vitis vinifera* L.). *Sci. Hortic.* **2009,** *120*, 51–57.

Tang, F.; Chen, F.; Teng, N.; Fang, W. Intergeneric Hybridization and Relationship of Genera within the Tribe Anthemideae Cass. (I. *Dendranthema crassum* (kitam.) kitam. × *Crossostephium chinense* (L.) Makino). *Euphytica* **2009,** *169*(1), 133–140.

Tian, L. L.; Wang, Y. J.; Niu, L.; Tang, D. M. Breeding of Disease-Resistant Seedless Grapes using Chinese Wild *Vitis* spp. I. *In vitro* Embryo Rescue and Plant Development. *Sci. Hortic.* **2008,** *117*, 136–141.

Tokunaga, T.; Nii, M.; Tsumura, T.; Yamao, M. Production of Triploids and Breeding Seedless Cultivar 'Tokushima 3X No.1' from Tetraploid × Diploid Crosses in Sudachi (*Citrus sudachi* Shirai). *Jpn Soc Hortic Sci.* **2005,** *4*(1), 11–15.

Toolapong, P.; Komatsu, H.; Iwamasa, M. Triploids and Haploid Progenies Derived from Small Seeds of 'Banpeiyu' Pummelo, Crossed with 'Ruby Red' Grapefruit. *J. Jpn Soc. Hortic. Sci.* **1996,** *65*, 255–260.

van Overbeek, J.; Conklin, M. E.; Blakeslee, A.F. Factors in Coconut Milk Essential for Growth and Development of Very Young Datura Embryos. *Sci.* **1941,** *94*, 350–351.

Williams, T. E.; Roose, M. L. 'TDE2' Mandarin hybrid (Shasta Gold® Mandarin), 'TDE3' Mandarin hybrid (Tahoe Gold® Mandarin) and 'TDE4' Mandarin hybrid (Yosemite Gold® Mandarin): Three new mid and late-season triploid seedless mandarin hybrids from California. In: Proceedings of 10th international citrus congress, vol 1, International Society of Citriculture, Agadir, Marruecos, **2004,** pp 394–398

Yahata, M.; Yasuda, K.; Nagasawa, K.; Harusaki, S.; Komatsu, H.; Kunitake, H. Production of Haploid Plant of 'Banpeiyu' Pummelo [Citrus maxima (Burm.) Merr.] by Pollination with Soft X-Ray-Irradiated Pollen. *J. Jpn. Soc. Hortic. Sci.* **2010,** *79*, 239–245.

Yeung, E. C.; Thorpe, T. A.; Jensen, C. J. In vitro Fertilization and Embryo Culture. In *Plant Tissue Culture: Methods and Applications in Agriculture*; Thorpe, T. A. Ed.; Academic: New York, **1981;** pp 253–271.

Zhang, Y. X.; Lespinasse, Y.; Chevreau, E. Obtention de plantes haploides de pommier (*Malus × domestica* Borkh) issues de parthenogenese induite *in situ* par du pollen irradie et culture *in vitro* des pepins immature. *CR Acad. Sci.* **1988,** *307*, 451–457.

CHAPTER 7

CALLUS INDUCTION

TUSHAR RANJAN[1], BISHUN DEO PRASAD[2*], SUNITA KUMARI[3], RAM BALAK PRASAD NIRALA[4], RAVI RANJAN KUMAR[2], VIJAY KUMAR JHA[5], VAISHALI SHARMA[6], MD SHAMIM[2], and ANAND KUMAR[4]

[1]*Department of Basic Science and Humanities Genetics, Krishi Vigyan Kendra, Kishanganj, Bihar, India*

[2]*Department of Molecular Biology and Genetic Engineering, Krishi Vigyan Kendra, Kishanganj, Bihar, India*

[4]*Department of Plant Breeding and Genetics, Bihar Agricultural University, Sabour, Bhagalpur, Bihar, India*

[5]*Department of Botany, Patna University, Patna, Bihar, India*

[6]*DOS in Biotechnology, University of Mysore, Mysore, Karnataka, India*

[*]*Corresponding author. E-mail: dev.bishnu@gmail.com*

CONTENTS

ABSTRACT

Tissue culture is an important biotechnological tool applied for crop improvement, where in vitro aseptic culture of cells, tissues, organs, or whole plant takes place under controlled nutritional and environmental conditions to generate the clones of particular plants. The controlled conditions could be nutrients, pH, temperature, and proper gaseous and liquid environment which provide conductive environment for their growth and multiplication. It can be used in fundamental research to study cell division, plant growth, plant propagation, elimination of plant diseases, plant improvement, and production of secondary metabolites as well. The key to the successful use of tissue culture is the manipulation of medium compositions to achieve desired outcomes. The most common application of tissue culture is micropropagation, which usually involves growing of plants in vitro (agar-solidified nutrient medium). Micropropagation facilitates production of virus-free planting material and propagation of plant species from undifferentiated callus. Micropropagation technology has a vast potential to produce plants of better quality, isolation and characterization of new useful variants in well-adapted high yielding genotypes with better disease resistance, and stress tolerance capacities.

7.1 INTRODUCTION AND HISTORICAL BACKGROUND

Plant callus (or *calli*) is a mass of unorganized or undifferentiated parenchyma cells derived from plant tissue (also referred as explants), which has full potential to be matured into whole plantlets. The plant callus is basically used in the biological research and has paved a revolution for biotechnologists. Generally, in plant biology, callus cells are those cells that spread over a plant wound to protect them. Callus formation is induced from plant tissues after their surface sterilization and plating onto nutrient medium in vitro. Phytohormones or plant growth regulators (PGRs), such as auxins, cytokinins, and gibberellins, are supplemented into the medium to induce callus induction or somatic embryogenesis. When the plant biotechnologists move the callus onto a fresh new medium with different growth regulators, the cells in the callus are induced to organize and form new plant organs. Recently, callus initiation has been described for all major groups of land plants. Callus formation is pivotal to many research and applied tissue culture protocol. Callus can be multiplied and later used to clone many whole plantlets. Apart from that, various genetic engineering approaches employ callus initiation procedures followed by genetic transformation. Then, callus is regenerated in whole plant, which leads to formation of transgenic plants (Razdan, 2003; White, 1939). Callus is also being generated for use in biotechnological procedures such as the formation of suspension cultures from which invaluable plant products could be collected and harvested. To express its cellular totipotency, the differentiated plant cells in the tissue first undergo dedifferentiation, i.e., the highly differentiated cells are reverted to the meristematic state, followed by repeated cycles of cell division and elongation to produce unorganized proliferative mass of cells. This unorganized proliferative mass of cells is known as callus. The repeated cell divisions followed by cell expansion from the cultured explants leads to an irreversible change in the size, shape, symmetry, and structural organization of the explant. The endogenous production in combination with exogenously supplied hormone makes a threshold level and their interaction results in the formation of unorganized cellular mass at the cut ends of the explant. Later on, gradually the whole tissue is involved in callus formation. The callus culture can be continuously maintained by serial subcultures (Chawla, 2002).

Henri-Louis Duhamel du Monceau investigated wound healing responses in elm trees, and reported formation of callus on live plants for the first time. In 1908, E. F. Simon was able to induce callus from poplar stems that also produced roots and buds. Later on in 1939, the first reports of callus induction in vitro came from three independent researchers. P. White induced callus obtained from tumor developing procambial tissues of hybrid *Nicotiana glauca* that did not require hormone supplementation. Gautheret and Nobecourt were able to maintain callus cultures of carrot using auxin hormone additions (Gautheret, 1983; White, 1939).

Tissue culture is an important biotechnological tool applied for crop improvement, where in vitro aseptic culture of cells, tissues, organs, or whole plant takes place under controlled nutritional and environmental conditions to generate the clones of particular plants. The controlled conditions could be nutrients, pH, temperature, and proper gaseous and liquid environment which provide conductive environment for their growth and multiplication. It can be used in fundamental research to study cell division, plant growth, plant propagation, elimination of plant diseases, plant improvement, and production of secondary metabolites as well. The key to the successful use of tissue culture is the manipulation of medium compositions to achieve desired outcomes (Touchel et al., 2008). The most common application of tissue culture is micropropagation, which usually involves growing of plants in vitro (agar-solidified nutrient medium). Micropropagation facilitates production of virus-free planting material and propagation of plant species from undifferentiated callus. Micropropagation technology has a vast potential to produce plants of better quality, isolation and characterization of new useful variants in well-adapted high yielding genotypes with better disease resistance, and stress tolerance capacities (Hussain et al., 2012).

7.2 CALLUS INDUCTION

Almost all plant land species have been shown to be capable of producing callus in tissue culture. Callus induction propagates on the nutrient medium consisting of agar and a mixture of micro- and macronutrients for the particular cell type. There are several types of basal salt mixtures used in tissue culture which are basically modified Murashige and Skoog (MS) medium, White's medium, and woody plant medium. Vitamins are basically known to provide or enhance growth, for example, Gamborg B5 vitamins (Murashige and Skoog, 1962). Enrichment of medium with nitrogen, phosphorus, and potassium is critically important for plant cells. Plant callus is usually derived from somatic tissues. The tissues used to initiate callus formation depend on plant species and which tissues are available for explant culture. The cells that initiate callus generation and somatic embryos usually undergo rapid division and these cells are partially undifferentiated such as meristematic plant tissue. In alfalfa, *Medicago truncatula*, however, callus and somatic embryos are obtained from mesophyll cells which undergo dedifferentiation. Explants from different parts of plant could be used for callusing. The most successful explants are often young tissues. Study indicates that pith cells from young stem are usually a good source of explant material. Initially, callus cells multiply without differentiation, but eventually later on differentiation occurs within the mass of tissue. Actively dividing cells are present around periphery in the callus at uppermost. The degree of overall differentiation usually depends on the hormonal balance of the nutrient medium and physiological status of the tissue (Burris et al., 2009; Chen et al., 2000; O'Dowd et al., 1993).

In theory, any part obtained from any plant species can be employed to induce callus tissue; however, the successful induction of callus not only depends upon plant species, but also does depend on their qualities. Dicotyledons are rather amenable for callus tissue induction compared to monocotyledons because the callus of woody plants generally grows slowly. Although, stems, leaves, roots, flowers, seeds part of plants are used, but younger and fresh explants are recommended or preferable as explanting materials. Explants are sterilized with 2% sodium hypochlorite solution and/

or 70% ethanol solution. The period of time for sinking the plant materials in sterilant solution depends upon plant species, their parts, and age. For example, a piece of tobacco stem of 3 cm in length is generally submerged in 70% ethanol solution for 2–3 min, which is followed by exposing with 1.2% sodium hypochlorite solution for 10 min. The explants should be rinsed with sterilized water. Thus, explant (such as stem or other part of plants) sterilized is trimmed to approximately 1 cm in length using a cleaned scalpel and each sterilized piece is now transferred with a tweezers to a solid nutrient medium in a flask or even a petri dish. The plant material is incubated aseptically at 25°C on the solid nutrient medium for several weeks and a callus is produced. The callus is subcultured by transferring a piece to new and fresh solid medium. After many subsequent transfers, the callus becomes soft and fragile (Lloyd and Mc Cown, 1981).

7.3 TECHNIQUES IN CALLUS INDUCTION

7.3.1 PREPARATION OF THE NUTRIENT MEDIUM

One of the most frequently used medium for plant tissue cultures is developed by Murashige and Skoog (MS) for tobacco tissue culture for the first time. To induce a callus from an explant and to cultivate the callus and cells in suspension, different kinds of medium (inorganic salt medium) have been designed according to our convenience. Agar is commonly added to the medium to develop solid medium for callus induction. MS basal medium is named after the name of scientists Murashige and Skoog who formulated this medium. Murashige and Skoog (1962) medium (MS) is the most suitable and commonly used basic tissue culture medium for the plant regeneration. The basal medium is formulated to provide essential nutrients required for the growth and development of explants or subculture. The tissue culture medium consists of 95% water, macro- and micronutrients, vitamins, amino acids, and sugars. These nutrient components in the medium are used by the plant cells for the synthesis of organic molecules while cofactors for the synthesis of enzyme. The macronutrients are required in millimolar (mM) quantities while micronutrients in much lower (micromolar, μM) concentrations. Vitamins which are organic substances and parts of enzymes or cofactors for essential metabolic functions are also included in the medium. Sugar is essential for growth and development in vitro because plant cultures are basically unable to photosynthesize effectively especially in the beginning. PGRs at a very low concentration (0.1–100 μM) regulate the initiation and development of shoots and roots on explants either in semisolid or in liquid medium. The auxins and cytokinins are the two most critical PGRs used in tissue culture. However, other PGRs may also be added as per the requirement and mode of tissue culture. The relative effects of auxin and cytokinin ratio determine the morphogenesis (rooting and shooting) of cultured tissues. Nutrient salts and vitamins are prepared as stock solutions (20X concentrations of that required in the medium) as specified and stored at 4°C to avoid frequent repetition during the optimization of the media and continuous culture. The required amount of concentrated stocks is mixed to prepare 1 L of medium. The quantity of component of the salts and vitamins required to make the stock solution is listed in the table given below (Tables 7.1–7.5) (Murashige and Skoog, 1962). The heat-labile PGRs are filtered through the bacteria-proof membrane (0.22 μm) filter and added to the autoclaved medium after being cooled enough (less than 60°C).

TABLE 7.1 Preparation of Callus Induction Medium or MS Medium.

MS Major Salts	mg/L Medium	500 mL Stock (20X)
NH_4NO_3	1650 mg	16.5 g
KNO_3	1900 mg	19 g
$CaCl_2.2H_2O$	440 mg	4.4 g
$MgSO_4.7H_2O$	370 mg	3.7 g
KH_2PO_4	170 mg	1.7 g

TABLE 7.2 Components of MS Medium.

MS Minor Salts	mg/L Medium	500 mL Stock (200X)
H_3BO_3	6.2 mg	620 mg
$MnSO_4.4H_2O$	22.3 mg	2230 mg
$ZnSO_4 4H_2O$	08.6 mg	860 mg
KI	0.83 mg	83 mg
$Na_2MoO_4.2H_2O$	0.25 mg	25 mg
$CoCl_2 6H_2O$	0.025 mg	2.5 mg
$CuSO_4.5H_2O$	0.025 mg	2.5 mg
MS Vitamins	**mg/L Medium**	**500 mL Stock (200X)**
Thiamine (HCl)	0.1 mg	10 mg
Niacine	0.5 mg	50 mg
Glycine	2.0 mg	200 mg
Pyrodoxine (HCl)	0.5 mg	50 mg

TABLE 7.3 Plant Growth Regulator.

Plant Growth Regulator	Nature	Mol. Wt.	Solvent
Benzyl amino purine	Autoclavable	225.2	1 N NaOH
Naphthalene acetic acid	Heat labile	186.2	Ethanol

TABLE 7.4 Preparation of MS Nonsulphates (10X). Preparation Should Be Done in Double Distilled or RO Water Autoclave for 15 psi for 15 min and Stored in the Refrigerator.

Name of Chemical	For 1 L Medium
NH_4NO_3	16.50 g
KNO_3	19 g
KH_2PO_4	1.70 g
$CaCl_2.2H_2O$	4.40 g
KI	8.30 mg
H_3BO_3	62 mg
$CoCl_2.6H_2O$	0.25 mg
$Na_2MoO_4.2H_2O$	2.50 mg

TABLE 7.5 Preparation of Iron-EDTA Stock (200X) (100 mL): Heat Both Solutions with 40 mL Double Distilled or RO H_2O with Stirrer to Dissolve Separately. Add $FeSO_4$ Solution Slowly to the EDTA Solution by Warming it up and Mix Well. Make it up to 100 mL and Store in the Refrigerator.

Name of Chemical	For 100 mL Solution
$FeSO_4.7H_2O$	557 mg
Na_2EDTA	745 mg

7.3.1.1 INORGANIC SALT

Most commonly used MS medium has very high concentration of inorganic salt such as nitrate, potassium, and ammonia, which plays a critical role in optimal growth. Another medium known as B5 medium established by Gamborg et al. is also being used by many researchers. But comparatively the levels of inorganic nutrients in the B5 medium are lower than in MS medium. Now scientists have developed many other media by modifying the nutrient compositions of some of the typical media. Although it is not necessary to test different kinds of basal medium when a callus is induced, it is better to use only one or hardly two kinds of basal medium in the combination of different kinds and concentrations of phytohormones. Optimization for composition of the most suitable medium composition should be done afterward in order to obtain higher level of products as well as higher growth rate (Ramsay and Galitz, 2003).

7.3.1.2 CARBON SOURCE

Sucrose (or glucose) ranging from 2% to 4% is best suitable carbon sources which are typically added to the basal medium. Other sugars such as fructose and maltose also support the growth of various plant cells. However, the most suitable source of carbon and their optimal concentration should be optimized to establish the efficient production process of useful metabolites. It is necessary to optimize the medium compositions including carbon sources in each case because it varies from plant to plant species and products as well.

7.3.1.3 VITAMINS

The most suitable basal medium such as MS medium mainly includes myo-inositol, nicotinic acid, pyridoxine HCl, and thiamine HCl. Out of all these vitamins, thiamine is an essential one for many plant cells whereas other vitamins help in growth stimulation. The recommended level of myo-inositol in the MS medium is 100 mg/L which is actually very high, but generally, high level of the vitamin is not required.

7.3.1.4 PHYTOHORMONES

Phytohormones also called growth regulators are required mainly for the induction of callus tissues and to enhance/promote the growth of many cell lines. Most frequently used phytohormones in plant tissue culture labs are auxin, 2,4-dichlorophenoxyacetic acid (2,4-D) and naphthaleneaceic acid (NAA). The recommended concentration of auxins in the medium generally lies between 0.1 μM and 50 μM. Interestingly, cytokinin (kinetin or benzyladenine) is also required sometimes together with auxins for callus induction at concentrations of 0.1 μM and 10 μM (Table 7.3). Other derivatives of auxin and kinetin are also used during callus induction. Since each plant species requires different levels of phytohormones for their callus induction, it is important to select the most appropriate growth regulators and to determine their optimal concentrations. Gibberellic acid is also supplemented to the medium if it is necessary.

7.3.1.5 ORGANIC SUPPLEMENTS

In order to promote the optimal growth of cells, organic supplements are added to the medium. These organic supplements may include casamino acid, peptone, yeast extracts, malt extracts, and coconut milk. Yeast extract supplies some of the unknown factors critical for the induction of callus. Coconut milk is also known as a supplier of growth regulators.

7.3.2 SURFACE STERILIZATION OF THE EXPLANTS

The explants may be opted from any part of the plant such as shoot tip, apical buds, roots, leaf, endosperm, seed, cotyledons, embryo, and male flower. The choices of explants depend upon the type and physiological condition of the explants. Whatever may be the explants, it is essential to treat with the disinfectants, e.g., sodium or calcium hypochlorite solution 0.3–0.6%, mercuric chloride, and alcohol followed by washing the explants with sterile distilled water. Other disinfectants and detergents may also be used as per the standard protocol of the particular plant from the particular explants. But the exposure and duration of the disinfectants is compromised with the vitality of the explants. Aseptic manipulation is the technique of culturing different types of explants in microorganism-free environment. Aseptic technique is surely required for the successful foundation and maintenance of plant cell, tissue, and organ cultures as well. The in vitro condition in which the plant cells are grown also provides optimum situation for the proliferation of microorganisms. Generally, microorganisms outgrow the plant tissues, leading to the death of plant cells. Contamination can also spread from one culture to another culture. The main motto of aseptic technique is to minimize the possibility that microorganisms remain in or enter the cultures. Explants require surface sterilization before they can be placed in culture on the nutrient agar. This is generally obtained by using diluted chlorine bleach. Some explants, such as very small seeds or spores are surface sterilized in a capped centrifuge tubes and require centrifugation to pellet the seeds and pouring off the solutions using pipette. Explants that float in the disinfectant can be wrapped in clean cloth to prevent their floating out of the test tube as the solutions are changed.

7.3.3 INOCULATION OF THE EXPLANTS

Once surface sterilized, the explants are transferred onto the suitable nutrient medium (sterilized by autoclaving) in culture vessels under sterile conditions. Both steps, surface sterilization and inoculation, are carried out in the laminar flow hood to avoid the risk of contamination. A general procedure for preparing the explant is as follows:

1. Wash the explants in warm, soapy water and rinse in tap water. This procedure is very beneficial for stem, leaf, and shoot tip explants from the field or greenhouse because it removes surface contaminants.
2. Sometimes a brief alcohol rinse or swabbing with alcohol-wetted cheesecloth is appropriate especially with surfaces that are hairy or coated with thick wax.
3. Immerse the explants in the chlorine bleach solution which should always be made up fresh. Always add 1–2 drops of Tween 20, detergent, or other wetting agent per 100 mL of bleach solution. A 10% bleach solution is prepared by adding chlorine bleach of 10mL to a measuring cylinder and diluting with water to 100 mL. When disinfesting the culture test tubes, spill the bleaching solution into the culture tube cap and then into the culture tube; this helps to disinfest the cap. Mix the tube by placing your hand over the tube. Agitation occasionally recommended for 5–30 min for disinfestations. Commercial available bleach once opened or exposed can lose their effectiveness.
4. Decant the bleach solution and rinse the explant in sterile water three to five times. This step is carried out in a transfer hood.

Surface sterilization protocol for paddy seeds

1. Dehusk the selected healthy seeds. Wash them twice with sterile distilled water.
2. Wash the seeds with 70% ethanol for 2 min.
3. Wash the seeds once with distilled water.
4. Add sodium hypochlorite (4% active chlorine) and a drop of Tween 80 to the seeds and wash them with continuous shaking for 15 min.
5. Wash the seeds with sterile distilled water 2–3 times.
6. Add 0.1% mercuric chloride solution to the seeds and wash them for 5 min.
7. Wash the seeds thoroughly (4–5 times) with sterile distilled water.
8. Place the surface sterilized rice seeds in an appropriate medium for germination or callus.

9. Wrap the petri dishes and the test tube with the parafilm.
10. Place the cultures on the culture shelf at appropriate temperature and light for growth and incubate at appropriate temperature in dark for callus induction.

7.3.4 INCUBATION OR GROWING THE CULTURES OR CALLUS INDUCTION IN PLANT TISSUE CULTURE ROOM AT OPTIMUM

After inoculation, the culture vessel is wrapped with the paraffin wax and allowed to incubate the explants in the culture room under optimum photoperiod, humidity, and light intensity condition. The typical photoperiod required in most of the crops is 16 h day and 8 h night with temperature 25°C and relative humidity in the range of 50–60%. Protocol for callus induction has been described below by considering rice as a typical example:

1. Dehusk selected healthy seeds. Wash them twice with sterile distilled water.
2. Wash the seeds with 70% ethanol for 2 min.
3. Wash the seeds once with distilled water.
4. Add sodium hypochlorite (4% active chlorine) and a drop of Tween 80 to the seeds and wash them with continuous shaking for 15 min.
5. Wash the seeds with sterile distilled water 2–3 times.
6. Add 0.1% mercuric chloride solution to the seeds and wash them for 5 min.
7. Wash the seeds thoroughly (4–5 times) with sterile distilled water.
8. Place the surface sterilized rice seeds in to callus induction medium (CIM; Medium composition mentioned below).
9. Incubate the petri dishes for 21 days in darkness at 28°C in an incubator.
10. Subculture the calli on a fresh CIM at 2-week interval.

7.3.5 REGULAR SUBCULTURE OF THE EXPLANTS LEADING TO REGENERATION OF THE PLANTLETS

The explants grown on the medium in the culture vessel consumes nutrients causing nutrient deficiency and due to growth of the explants there is scarcity of the space in the vessel as well. These necessities the need of the transfer in another culture vessel supplemented with the medium. In addition, modification in the media components is required during different stages of the plant. Therefore, subculture is also required to meet the different media requirement during the regeneration into the plantlet. Multiplied callus can be stimulated to form shoots by increasing the cytokinin concentration and decreasing auxin concentration of the culture medium. Shoot masses can be cut apart and transferred to rooting medium. Once rooted, regenerated plants can be acclimatized to natural rather than "in vitro" growth conditions. Regenerated plants are valuable if the parent plant is itself unique or the regenerated plants were genetically engineered. If multiplied callus used to form suspension cultures on which genetic engineering or cell selection was imposed, resultant regenerated plants using tissue culture could possess special traits or capabilities (Wang et al., 2011). Protocol for regeneration from rice calli is being discussed below. Figure 7.2 describes steps involved in regeneration of rice from calli.

1. Transfer embryogenic calli on shoot regeneration medium (MS medium with 3 mg/L kinetin, 2 mg/L NAA, 6 g/L phytagel, pH 5.8) and keep in darkness for 1 week.
2. Transfer the calli to light for 1 week.
3. Subculture the calli to fresh medium for 3 weeks.
4. Transfer the shoots to half strength MS medium for rooting for 3 weeks.
5. Transfer the plants to pots and incubate in the growth chamber for approximately 2 weeks at 30°C and 65% humidity.
6. Transplant into big pots and transfer to poly house/greenhouse till maturity.

7.3.6 ACCLIMATIZATION OF THE PLANTLETS

The in vitro raised plantlets are not in position to survive under natural condition directly. These plantlets have to be planted under the condition and the soil enriched with nutrients. Therefore, in vitro raised plantlets are planted under greenhouse condition to make it hardened before planting into the natural environments. This is the typical tissue culture technique; medium, the method of surface sterilization, and the subculture duration varies from species to species and the source and physiological condition of explants.

7.4 CALLUS MULTIPLICATION

Actively growing callus can be initiated on culture medium with physiological balance of cytokinin and auxin ratio. After increasing of callus biomass two to four times (after 2–4 weeks of growth), callus can be divided and placed on fresh callus initiation medium for the callus multiplication. Multiplication procedures can be repeated several times (up to eight sequential transfers) before chromosome instability occur. Specific auxin to cytokinin ratios in plant tissue culture medium leads to growth of unorganized and proliferating mass of callus cells. Callus cultures are often broadly classified into either compact or friable. Friable calluses fall apart easily and can be used to generate cell suspension cultures. Callus (Fig. 7.1) can directly undergo direct organogenesis and/or embryogenesis where the cells will form entirely new plants.

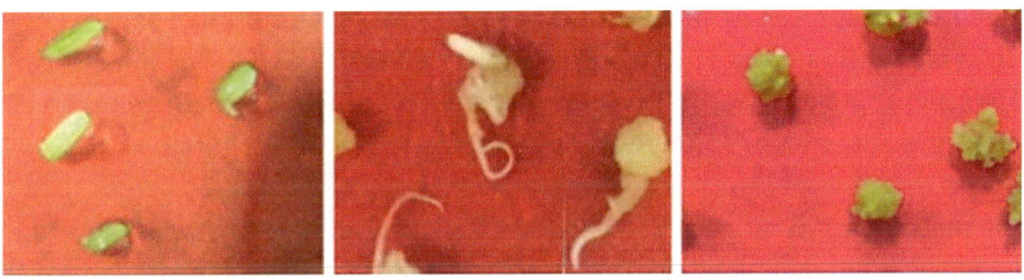

FIGURE 7.1 Induction and subculturing of embryonic calli derived from rice seeds.

FIGURE 7.2 Shoot regeneration, rooting, and primary hardening of plants obtained from rice calli.

7.5 SUSPENSION CULTURE

Callus multiplicate as an undifferentiated or unorganized mass of cells. So it is very difficult to follow many cellular events during their growth and developmental stages. To overcome such limitations of callus culture, the cultivation of free cells as well as small cell assemblage in a liquid medium which is chemically known as a suspension was initiated to study the morphological and biochemical advances during the different developmental stages. Suspension culture is a type of culture in which small aggregates or may be single cell multiply while suspended in agitated liquid medium. The growth rate of the suspension cultured cells is normally more than that of the solid cultured cells. Suspension culture is more desirable in production of very useful metabolites up to the large scale. A piece of the callus is transferred to a liquid medium in an Erlenmeyer flask and the flask is placed on a rotary or reciprocal shaker. The culture conditions also relay upon species of plant and other unknown factors, but generally the cells are cultivated at 100 rpm on a rotary shaker at 25°C. By subculturing for several generations, a fine cell suspension culture containing small cell aggregates could be established. The time required to maintain the cell suspension culture varies greatly and depends on the source tissue of the plant species and the medium composition. The cells in suspension are also used for a large-scale culture with jar fermenters and tanks. For commercialization, it is necessary to progress through several stages by increasing the volume at each step until the required bioreactor size is attained. Ideally, it is likely that large-scale suspension cultures will be relevant for industrial production of many precious phytoproducts, for example, pharmaceuticals and food additives as well in a manner akin to that of microbial fermentation (Lamport, 1964).

Suspension cultures are suspensions of individual plant cells; cell clusters also called microcalli are grown in liquid broth medium. Suspension cultures are maintained by bringing pieces of callus to liquid broth medium which are subsequently placed on a shaker. Within a few days, microcalli should be detached from the original inoculum and grown in the regularly agitated medium. Suspension cultures mainly grow at its best if the larger pieces of callus are taken off after the initiation of culture. Stepwise procedure for suspension culture has been described below:

1. Rice suspension cell culture will be generated from actively growing rice callus (callus induction protocol for rice seeds mentioned in Experiment No. 6).

2. After three subcultures, inoculate friable calli into a liquid callus induction medium (composition of CIM mentioned in Experiment No. 6).
3. Gently break apart the calli using a sterilized scalpel.
4. Transfer aseptically 1–2 g calli in 25 mL liquid CIM medium kept into 250 mL conical flask.
5. Transfer the flask on rotary shaker at 140 rpm at 25°C.
6. Serial subculture is done every 3 days by adding fresh liquid CIM medium for 1 week and then subculture weekly.
7. Determine the growth of cell suspension by measuring fresh weight every 2 days for 2 weeks.

7.6 CALLUS CELLS DEATH

Callus could become brown and may die during culture, but the detailed mechanism for callus browning is not well understood till date. In *Jatropha curcas* callus cells, small organized callus cells became disorganized and varied in size after browning occurred. Studies suggest that browning has also been correlated with oxidation in explant tissues and explant secretion of phenolic compounds (He et al., 2009).

7.7 APPLICATION OF CALLUS INDUCTION

Cells in callus are not necessarily genetically homogeneous because a callus is often made from structural tissue, not from an individual cell. Despite that, callus cells are generally considered similar enough for standard scientific analysis. Plant calli can differentiate into a whole plant, a process called regeneration, through addition of plant hormones in culture medium. This ability is known as totipotency of unorganized cells. Regeneration of a whole plant from a single cell allows researchers to recover whole plants that have a copy of the transgene in every cell. Regeneration of a complete plantlet that has some genetically transformed cells and untransformed cells are called a chimera. Insertion of gene into callus cells through biolistic bombardment, also known as a gene gun, or *Agrobacterium tumefaciens*. Cells harboring the gene of interest can then be recovered into whole plants using a combination of PGRs. The whole newly recovered plants can be used to determine gene function(s) experimentally. Whole recovered plantlets could also be used to enhance crop traits for modern agriculture.

Callus is of particular use in micropropagation where it can be used to produce genetically similar copies of plants having desirable characteristics. Meristemic callus leads to the regeneration of virus- or disease-free whole plantlets.

Suspension cultures are utilized during various biotechnological procedures such as inoculum for plant bioreactors, which mirror biofermentor. Several economically important plant products can be produced from bioreactors while proliferation of plant cells as cell suspensions. Such type of biotechnology is still underway of development and industrial scale-up. Suspension cultures may be treated with cellulase enzymes that break cell walls and produces protoplast cultures that can be utilized in DNA transformation experiments and in protoplast fusion (somatic hybridization) experiments as well.

KEYWORDS

- **callus induction**
- **dedifferentiation**
- **regeneration**
- **plant growth regulators**
- **2,4-dichlorophenoxyacetic acid (2,4-D)**

REFERENCES

Burris, J. N.; Mann, D. G. J.; Joyce, B. L.; Stewart, C. N. An Improved Tissue Culture System for Embryogenic Callus Production and Plant Regeneration in Switchgrass (*Panicum virgatum L.*). *Bioenergy Res.* **2009,** *2*(4), 267–274.

Chawla, H. S. *Introduction to Plant Biotechnology* (2nd Edn.); Science Publishers: Enfield, NH, **2002**.

Chen, Y.-C.; Chang, C.; Chang, W.-C. A Reliable Protocol for Plant Regeneration from Callus Culture of Phalaenopsis. *In Vitro Cell. Dev. Biol. Plant* **2000,** *36*(5), 420–423.

Gautheret, R. J. Plant Tissue Culture: A History. *The Bot. Mag. Tokyo* **1983,** *96*(4), 393–410.

He, Y.; Guo, X.; Lu, R.; Niu, B.; Pasapula, V.; Hou, P.; Cai, F.; Xu, Y.; Chen, F. Changes in Morphology and Biochemical Indices in Browning Callus Derived from *Jatropha curcas* Hypocotyls. *Plant Cell, Tissue Organ Cult. (PCTOC)* **2009,** *98*(1), 11–17.

Hussain, A., Qarshi, I.A., Nazir, H. and Ullah, I. (2012). Plant Tissue Culture: Current Status and Opportunities. *Agricultural and Biological Sciences.* "Recent Advances in Plant in vitro Culture", book edited by Annarita Leva and Laura M. R. Rinaldi, **2012**.

Lamport, D. T. A. Cell Suspension Cultures of Higher Plants: Isolation and Growth Energetics. *Exp. Cell Res.* **1964,** *33*(1), 195–206.

Lloyd, G.; McCown, B. Commercially-feasible Micropropagation of Mountain Laurel, Kalmia Latifolia, by use of Shoot-tip Culture. *Comb. Proc. Int. Plant Propag. Soc.* **1981,** *30*, 421–427.

Murashige, T.; Skoog, F. A Revised Medium for Rapid Growth and Bio Assays with Tobacco Tissue Cultures. *Physiol. Plant.* **1962,** *15*(3), 473–497.

O'Dowd, N. A.; McCauley, P. G.; Richardson, D. H. S.; Wilson, G. Callus Production, Suspension Culture, and In vitro Alkaloid Yields of Ephedra. *Plant Cell Tissue Organ Cult.* **1993,** *34*(2), 149–155.

Razdan, M. K. *Introduction to Plant Tissue Culture* (2 Edn.); Oxford Publishers: Enfield, NH, **2003**.

Touchell, D.; Smith, J.; Ranney, T. G. Novel Applications of Plant Tissue Culture. *Com. Proc. Int. Plant Propag. Soc.* **2008,** *22*(58), 22–25.

Wang, X. D.; Nolan, K. E.; Irwanto, R. R.; Sheahan, M. B.; Rose, R. J. Ontogeny of Embryogenic Callus in *Medicago truncatula*: The Fate of the Pluripotent and Totipotent Stem Cells. *Ann. Bot.* **2011,** *107*(4), 599–609.

White, P. R. Potentially Unlimited Growth of Excised Plant Callus in an Artificial Nutrient. *Am. J. Bot.* **1939,** *26*(2), 59–4.

CHAPTER 8

PROTOPLAST ISOLATION AND FUSION

UDAY SAJJA[1], TUSHAR RANJAN[2], and BISHUN DEO PRASAD[3*]

[1]*Department of Biotechnology, Dr. B. R. Ambedkar University, Srikakulam, Andhra Pradesh, India*

[2]*Department of Basic Science and Humanities Genetics, Bihar Agricultural University, Sabour, Bhagalpur, Bihar, India*

[3]*Department of Molecular Biology and Genetic Engineering, Bihar Agricultural University, Sabour, Bhagalpur, Bihar, India*

[*]*Corresponding address. E-mail: dev.bishnu@gmail.com*

CONTENTS

ABSTRACT

Plant protoplasts provide a unique single-cell system to underpin several aspects of modern biotechnology. Major advances in genomics, proteomics, and metabolomics have stimulated renewed interest in this osmotically fragile wall-less cells. Reliable procedures are available to isolate and culture protoplasts from a range of plants, including both monocotyledonous and dicotyledonous crops. Importantly, novel approaches to maximize the efficiency of protoplast-to-plant systems include techniques already well established for animal and microbial cells, such as electrostimulation and exposure of protoplasts to surfactants and respiratory gas carriers, especially perfluorochemicals and hemoglobin. However, despite at least four decades of concerted effort and technology transfer between laboratories worldwide, many species still remain recalcitrant in culture. In the context of plant genetic manipulation, somatic hybridization by protoplast fusion enables nuclear and cytoplasmic genomes to be combined, fully or partially, at the interspecific and intergeneric levels to circumvent naturally occurring sexual incompatibility barriers. Uptake of isolated DNA into protoplasts provides the basis for transient and stable nuclear transformation, and also organelle transformation to generate transplastomic plants. Isolated protoplasts are also exploited in numerous miscellaneous studies involving membrane function, cell structure, synthesis of pharmaceutical products, and toxicological assessments.

8.1 INTRODUCTION

The plant cell consists of outer cellulosic wall with a pectin-rich matrix. Two adjacent cells are joined by middle lamella. Inner to the outer cellulosic wall, the cytoplasm of each cell is bound by a plasma membrane. The living cytoplasm of a plant cell surrounded by a thin plasma membrane constitutes the protoplast. In hypertonic solutions when the plasma membrane shrinks, a little manipulation results in the release of spherical, osmotically fragile protoplasts. Loosely protoplasts are also called naked cells. In a protoplast, plasma membrane is the only barrier between the cytoplasm and the immediate environment.

The term protoplast was introduced in 1880 by Hanstein. The first isolation of protoplasts was achieved by Klercker (1892) employing a mechanical method. A real beginning in protoplast research was made in 1960 by Cocking, who used an enzymatic method for the removal of cell wall.

Rakabe and his associates successfully achieved the regeneration of whole tobacco plant from protoplasts in 1971. Since the first successful isolation of protoplasts by Cocking (1960), substantial progress has been made toward improving the technology. Attempts have also been made to isolate protoplasts from several crop species and protoplast-based plant regeneration systems are made available for a great number of species (Maheshwari et al., 1986). The improvements that have occurred include modification of protoplast isolation procedures (Mei-Lei et al., 1987), media composition (Kao and Michayluk, 1975), preconditioning of protoplast donor tissues (Shahin 1985), utilization of conditioned media or feeder cells (Bellincampi and Morpugo, 1987; Kyozuka et al., 1987; Lee et al., 1990), and manipulation of culture environment (d'UtraVaz et al., 1992). Rapid progress occurred after 1980 in protoplast fusion to improve plant genetic material, and the development of transgenic plants.

Sexual incompatibility is the biggest barrier to obtain full hybrids between desired individuals. This is a big handicap in crop improvement program through hybridization. In early 1960s, Cocking for the first time isolated protoplast from plant cells using cell wall-degrading enzymes. This breakthrough has paved the way for increased interest in the possibility of genetic modification of somatic cells in higher plants. Later Takebe et al. in 1971 reported complete regeneration of plants form leaf protoplasts of tobacco. The above two pathbreaking studies have increased the potential of protoplast culture techniques.

Further studies by other researchers have resulted in standardizing the protoplast isolation and culture protocols across many plant species. Given a combination of chemical and physical stimuli to the protoplasts, it is possible to make protoplast potentially totipotent resulting in repeated mitotic divisions producing daughter cells. Further using tissue culture techniques, the daughter cells can be made to regenerate into a complete fertile plant. Two genetically different protoplasts isolated from somatic cells can be fused experimentally to produce somatic hybrids with fused nuclei. These fused protoplasts called somatic hybrids can also be regenerated into a whole plant. Protoplasts fusion has now become a versatile technique to producc interspecific or even intergeneric hybrids in which sexual incompatibility is a problem. The protoplasts to plant systems are now standardized in many plants as the procedure of isolating protoplasts has undergone little change over decades.

Plant protoplasts provide a unique single cell system to underpin several aspects of modern biotechnology. Major advances in genomics, proteomics, and metabolomics have stimulated renewed interest in this osmotically fragile wall-less cells. Reliable procedures are available to isolate and culture protoplasts from a range of plants, including both monocotyledonous and dicotyledonous crops. Importantly, novel approaches to maximize the efficiency of protoplast-to-plant systems include techniques already well established for animal and microbial cells, such as electrostimulation and exposure of protoplasts to surfactants and respiratory gas carriers, especially perfluorochemicals and hemoglobin. However, despite at least four decades of concerted effort and technology transfer between laboratories worldwide, many species still remain recalcitrant in culture. In the context of plant genetic manipulation, somatic hybridization by protoplast fusion enables nuclear and cytoplasmic genomes to be combined, fully or partially, at the interspecific and intergeneric levels to circumvent naturally occurring sexual incompatibility barriers (Hain et al., 1985). Uptake of isolated DNA into protoplasts provides the basis for transient and stable nuclear transformation, and also organelle transformation to generate transplastomic plants. Isolated protoplasts are also exploited in numerous miscellaneous studies involving membrane function, cell structure, synthesis of pharmaceutical products, and toxicological assessments (Davey et al., 2005; Helgeson et al., 1986).

8.2 ISOLATION OF PROTOPLASTS

The essential step in the isolation of protoplast is the removal of the cell wall without damaging the cell or protoplasts. The plant cell is an osmotic system. The cell wall exerts the inward pressure upon the enclosed protoplasts. Likewise, the protoplast also puts equal and opposite pressure upon the cell wall. Thus, both the pressures are balanced. Now if the cell wall is removed, the balanced pressures will be disturbed. As a result, the outward pressure of protoplast will be greater and at the same time in absence of cell wall, irresistible expansion of protoplast takes place due to huge inflow of water from the external medium. Greater outward pressure and the expansion of protoplast cause it to burst. The isolated protoplast is an osmotically fragile structure at its nascent stage. Therefore, if the cell wall is to be removed to isolate protoplast, the cell or tissue must be placed in a hypertonic solution of a metabolically inert sugar such as mannitol at higher concentration (13%) to plasmolyze the cell away from the cell wall. Mannitol, an alcoholic sugar, is easily transported across the plasmodesmata, provides a stable osmotic environment for the protoplasts, and prevents the usual expansion and bursting of protoplast even after loss of cell wall. That is why, this hypertonic solution is known as osmotic stabilizer or plasmolyticum or osmolyticum. Once the cells are stabilized in such a manner by plasmolysis, the protoplasts are released from the containing cell wall either mechanically or enzymatically.

In plant breeding program, much desirable combination of characters could not be transmitted through the conventional method of genetic manipulation. The success of a protoplast culture system primarily lies with consistent yields of a large population of uniform and highly viable protoplasts. Several protoplast isolation and purification protocols have been published with optimized yield and reproducibility. They are often procedures of elaborate nature, labor intensive involving too many explant or protoplast handling steps, and require extended exposure of explant to digestion environment. Further, the efficacy of such protocols or that of enzyme combinations used therein could be limited to a few plant species. These restrictions must be overcome by improvement of the existing conditions and methods. A number of commercial cellulases and pectinases which allow protoplast release are available. By manipulating the source and concentrations of these, protoplasts may be released from most tissues; however, generalizations cannot be made. The enzymes and techniques used for isolation of protoplasts have a bearing on their subsequent behavior and development. Methods with too many steps often result in the introduction of cell contamination at some stage or the other.

8.2.1 SOURCE MATERIAL FOR PROTOPLAST ISOLATION

The physiological status of the source tissue influences the release of viable protoplasts. Furthermore, seasonal variation, which affects the reproducibility of protoplast isolation from glasshouse-grown plants, can be effectively eliminated using in vitro-grown (axenic) shoots, seedlings, and embryogenic cell suspensions. Nevertheless, Keskitalo (2001) reported that protoplast isolation from cultured shoots of *Tanacetum vulgare* and *T. cinerariifolium* was most successful during winter and spring (December to April). In contrast, other workers have not observed seasonal variation in vitro. Mliki et al. (2003) isolated protoplasts from Tunisian varieties of grape (*Vitis vinifera*) and concluded that the highest yields were from leaves of cultured shoots 4–5 weeks after transfer of the shoots to new medium. The essential components of nutrient medium are discussed in Table 8.1. An advantage of seedlings is that protoplasts can be isolated from radicles, hypocotyls, cotyledon tissues, roots, and root hairs within a few days of seed germination. For example, Dovzhenko et al. (2003) reported a reproducible and rapid cotyledon-based protoplast system for *Arabidopsis thaliana*, which will facilitate molecular studies with this model species. Similarly, Sinha et al. (2003) found that cotyledons from in vitro-grown seedlings of white lupin gave higher yields compared to leaves, hypocotyls, and roots. Although protoplast yield from cotyledons increased with seedling age, viability declined. Concurrent investigations by Sinha et al. (2003) optimized protoplast isolation from cotyledons of this legume.

TABLE 8.1 Critical Components of Medium Used for the Isolation and Purification of Protoplasts.

Media Components	Washing Media mg/L	Flotation Media mg/L	Culture Media mg/L
MS salts without NH_4NO^3			
KNO_3	1900	1900	1900
Myoinositol	2000	2000	2000
L-Glutamine	100	100	100
Casein hydrolysate	20	20	20
Thiamine HCl	10	10	10
Glycine	2	2	2
Nicotinic acid	0.5	0.5	0.5
L-Serine	0.1	0.1	0.1
Sucrose	–	17%	1.4%
Mannitol	9%	–	8.2%
pH	5.7	5.7	5.7

8.2.1.1 EXPLANT PREPARATION PRIOR TO ENZYME EXPOSURE

Leaves of field-grown plants were thoroughly washed in running water for about10 min, cleaned gently with dilute Labolene of Glaxo India Ltd., and quickly rinsed with tap water first and then with deionized water. Sterilization was carried out with 2% sodium hypochlorite for 5 min followed by several rinses with sterile distilled water. Surface decontaminated leaves of greenhouse plants as well as those excised from shoot cultures were chopped into 2–3 mm strips in most cases. But leaves after peeling of the epidermal layer were also used except in *Basella, Centella, Amaranthus,* tomato, and chili where peeling is not very efficient. Cotyledons of sunflower, niger, chili seedlings and placental tissue of tomato fruits were similarly chopped into segments for enzymatic digestion. The embryogenic callus of sandalwood was lightly squashed with a spatula and dropped into protoplast isolation solution. All tissues were weighed in sterilized petri plates for fresh weight determination before they were subjected to enzyme digestion (Rao and Prakash, 1995).

8.2.2 PROCEDURE FOR PROTOPLASTS ISOLATION

Mechanical procedures, involving slicing of plasmolyzed tissues, are now rarely employed for protoplast isolation, but are useful with large cells and when limited (small) numbers of protoplasts are required. Recently, this approach has been used successfully to isolate protoplasts of the giant marine alga, *Valonia utricularis*, for patch clamp analyses of their electrical properties, including physiological changes of the plasma membrane induced by exposure of isolated protoplasts to enzymes normally used to digest cell walls (Binder et al., 2003). When large populations of protoplasts are required, which is the norm, enzymatic digestion of source tissues is essential (Davey and Kumar, 1983). Interestingly, it was the release of protoplasts by natural enzymatic degradation of cell walls during fruit ripening that stimulated investigations, more than four decades ago, of protoplast isolation from roots of tomato seedlings. Subsequently, cellulase and pectinase enzymes became available commercially for routine use. An additional major advance in protoplast isolation involved treatment of tobacco leaves with pectinase to separate the cells, followed by cellulase to remove their walls. The procedure was further simplified by a single treatment with a mixture of enzymes (Power and Cocking, 1970). Basically, plant protoplasts can be isolated from cells by two methods.

8.2.2.1 MECHANICAL METHOD

Protoplast can be isolated from almost all plant parts: roots, leaves, fruits, tuber, root nodules, pollen mother cell, etc. Protoplast isolation by mechanical method is a crude and tedious procedure. Cells are plasmolyzed causing the protoplast to shrink from the cell wall (Fig. 8.1). The protoplast obtained from this method is then cultured on suitable culture medium. The overall mechanical methods have been divided into three major steps: (1) a small piece of epidermis from a plant is selected; (2) the cells are subjected to plasmolysis, this causes protoplasts to shrink away from the cell walls; and

(3) the tissue is dissected to release the protoplasts. The principal deficiency of this approach is that the protoplasts released are few in number. Due to difficulty in protoplast isolation by mechanical methods, very small number of protoplasts gets isolated using this protocol.

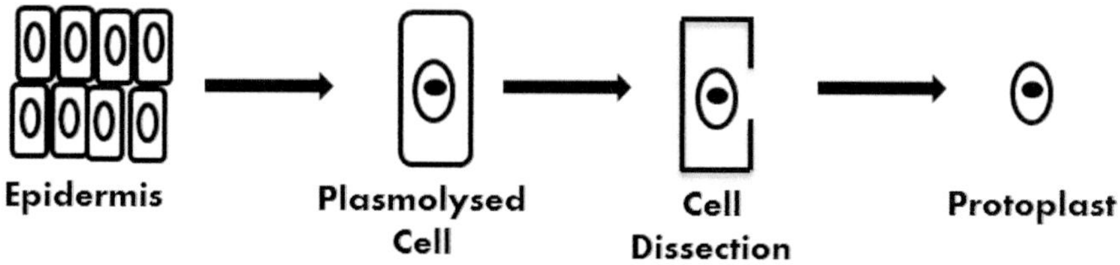

FIGURE 8.1 Isolation of protoplasts by mechanical methods.

8.2.2.1.1 Protocol for Protoplast Isolation by Mechanical Method

1. Young leaves are obtained from plants growing outdoors and are initially washed with tap water to remove any dust particles.
2. The leaves are washed with phosphate buffer and homogenized gently with the mortar and pestle.
3. The crude protoplast suspension is centrifuged at very low 50–100 rpm for 10 min.
4. The supernatant containing intact protoplast is carefully pipetted out and the pellet containing cell debris and other cell organelles is discarded.
5. Small volume of supernatant is placed on the slide and covered with cover slip.
6. The slide is observed under light microscope to find out viable protoplast.

8.2.2.2 Enzymatic Method

Enzymatic method is a very widely used technique for the isolation of protoplasts. The advantages of enzymatic method include good yield of viable cells, and minimal or no damage to the protoplasts. Protoplasts can be isolated from a wide variety of tissues and organs that include leaves, roots, shoot apices, fruits, embryos, and microspores. Among these, the mesophyll tissue of fully expanded leaves of young plants or new shoots is most frequently used. In addition, callus and suspension cultures also serve as good sources for protoplast isolation. Digestion is usually carried out after incubation in an osmoticum (a solution of higher concentration than the cell contents which causes the cells to plasmolyze). This makes the cell walls easier to digest. Debris is filtered and/or centrifuged out of the suspension and the protoplasts are then centrifuged to form a pellet. On resuspension, the protoplasts can be cultured on media which induce cell division and differentiation. A large number of plants can be regenerated from a single experiment—a gram of potato leaf tissue can produce more than a million protoplasts, for example. Protoplasts can be isolated from a range of plant tissues: leaves, stems, roots, flowers, anthers, and even pollen. The isolation and culture media used vary with the species and with the tissue from which the protoplasts were isolated. Protoplasts are used in a number of ways for research and for plant improvement. They can be treated in a variety of ways (electroporation, incubation with bacteria, heat shock, high pH treatment) to induce them to take up DNA. The protoplasts can then be cultured and plants regenerated. In this way, genetically engineered plants can be produced more easily than is possible using intact cells/plants. The enzymes that can digest the cell walls are required for protoplast isolation (Fig. 8.2). Chemically, the plant cell wall is mainly composed of cellulose, hemicellulose, and pectin which can be respectively degraded by the enzymes cellulase, hemicellulase, pectinase, and sometimes xylanase. In fact, the various enzymes for protoplast isolation are commercially available. The enzymes are usually used at a pH 4.5–6.0, temperature 25–30°C with a wide variation in incubation period that may range from half an hour to 20 h. To maintain the osmotic pressure inside the cells, osmoprotectant/stabilizer is most routinely used during protoplast culture. Osmotic stabilization should be manipulated by adding to the protoplast isolation and culture solutions, mannitol as the sole osmoticum or mannitol in combination with either sorbitol or magnesium sulfate. For tobacco leaves, niger cotyledons,

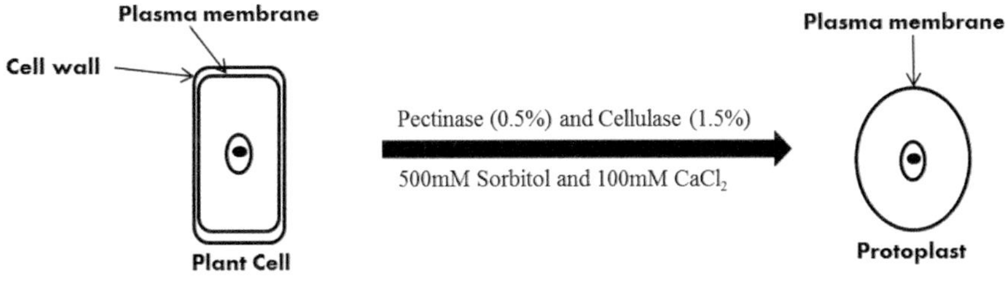

FIGURE 8.2 Isolation of protoplasts by enzymatic methods.

and sandalwood callus, mannitol alone was finally used as the osmoticum. Frequently used enzymes in plant protoplasts isolation and purification are listed in Table 8.2. The enzymatic isolation of protoplasts can be carried out by using two approaches:

1. *Two-step or sequential method:* The tissue is first treated with pectinase (macerozyme) to separate cells by degrading middle lamella. These free cells are then exposed to cellulose to release protoplasts. Pectinase breaks up the cell aggregates into individual cells while cellulase removes the cell wall proper.
2. *One-step or simultaneous method:* This is the preferred method for protoplast isolation. It involves the simultaneous use of both the enzymes—macerozyme and cellulase.

TABLE 8.2 Popularly Used Enzymes in Plant Protoplasts Isolation and Purification with Their Source.

Enzyme	Source
Cellulase	
Cellulase Onozuka R-10	*Trichoderma viride*
Cellulase Onozuka RS	*T. viride*
Cellulase YC	*T. viride*
Cellulase CEL	*T. viride*
Cellulysin	*T. viride*
Driselase	*Irpex lacteus*
Hemicellulase	
Helicase	*Helix pomatia*
Hemicellulase	*Aspergillus niger*
Rhozyme HP-150	*A. niger*
Pectinase	
Macerase	*Rhizopus arrhizus*
Macerozyme R-10	*R. arrhizus*
Pectinol	*Aspergillus* sp.
Pectolyase Y23	*A. japonicus*
Zymolyase	*Arthrobacter luteus*

8.2.2.2.1 Protocol for Protoplast Isolation by Enzymatic Method

1. Place 12.5 cm^3 of 13% sorbitol solution into a petri dish.
2. Place the squares of lettuce leaf bottom side down in the petri dish.
3. Cut up the squares into pieces approximately 5 mm × 5 mm or less. It is very important that the pieces are as small as possible.
4. Place the lid on the dish and incubate in the oven at 35°C for 5 min.
5. Add the Viscozyme and cellulase enzymes to the dish. Swirl gently for proper mixing of enzymes.
6. Incubate for a further 20 min; swirl gently at regular intervals.
7. Place the 60-mm mesh into the funnel and secure with tape.
8. Hold the funnel over the lid of the petri dish.
9. Pour the digested lettuce into the filter funnel. Try to minimize the distance the protoplasts have to fall.
10. Place the funnel over a centrifuge tube and gently pour the contents of the petri dish lid through the funnel again.
11. Balance your tube with another and centrifuge for 5 min at 2000 rpm.
12. The protoplasts will have formed a pellet at the bottom of the tube. Use the dropper to remove the liquid supernatant from the tube, without disturbing the pellet. Remove all but 0.5 cm^3 of the liquid and then resuspend the pellet by gently tapping the tube.
13. Place a drop of the suspension on a slide, cover with a cover slip and examine under high power.
14. If your suspension is too concentrated to see the protoplasts clearly you can add several drops of the sucrose solution to the suspension in the centrifuge tube. Critical steps involving plant protoplasts are schematically represented in Figure 8.3.

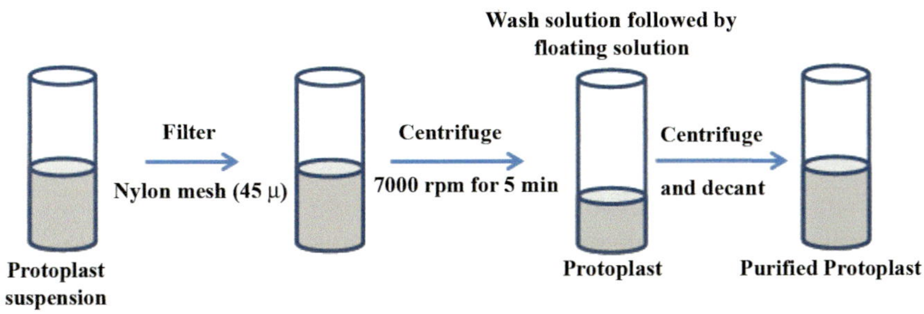

FIGURE 8.3 Schematic representation of steps involved in protoplasts isolation. (Adapted fromhttp://www.biocyclopedia.com/index/biotechnology/plant_biotechnology/in_vitro_culture_techniques/biotech_protoplast_fusion_hybridization.php)

8.2.3 FACTORS AFFECTING PROTOPLAST ISOLATION

A high quality and high yield of protoplasts can be achieved by standardizing the extraction protocol by manipulating various components during the extraction procedure. The various components which are used in the extraction process are explant, enzymes, osmoticum, and physical conditions like temperature and pH. It is highly desirable to take a disease-free young explant to achieve a high yield of protoplasts. In other cases, leaves of in vitro-cultured seedling are also an excellent source of good quality protoplasts. The enzymes which are used in the extraction process also play an important role in achieving high-quality protoplasts. Usually cellulase, pectinase, hemicellulase, pectinol, and meicellase are frequently used. One has to standardize the combination and concentration of the enzymes to be used. Whether they have to be used in combination at a time, or sequentially is to be standardized. The concentration and type of osmoticum (sucrose, mannitol, sorbitol, and glucose) to be used has to be standardized for the species from which we are extracting the protoplasts. Most widely used osmoticum is mannitol. Most importantly, the incubation temperature and the pH of the isolation solution should be standardized for each plant species. Usually, a low concentration of enzyme mixture at low temperature incubation and high pH (5–8) for a short incubation time works wonderfully for majority of the plant species. The conditions usually vary from one plant species to another and also from one plant part to another from which the explant is taken.

8.3 PROTOPLAST CULTURE

Once the protoplast are isolated using enzymatic isolation or mechanical methods as described above, the protoplast are cultured using different methods after purification. Some of the prominent methods are described below.

8.3.1 SUSPENSION CULTURE

In this method of protoplast culture, the protoplast are isolated and cultured in a liquid medium containing a suitable concentration of osmoticum. The osmoticum is required to prevent the bursting of protoplast. A 2 mL protoplast suspension containing approximately 10^5 protoplasts are used as an initial inoculum. The inoculum is added to the culture medium in an Erlenmeyer flask (25 mL) and is incubated in a shaking incubator at a very low rpm. The temperature of the incubator is maintained at 25°C under 200 lux light intensity. The cultures thus produced can be used for callus production.

8.3.2 HANGING DROP METHOD

Hanging drop method was first developed by Kao and his group in 1970 and was subsequently well standardized and modified. In this method, the isolated protoplasts at a concentration of 10^4–10^5 which are in a culture media are used. The suspension is taken in a micropipette and 25–50 µL of suspension is pipetted out on to a plastic petri plate. The plates are sealed with a parafilm and incubated in an inverted position, hence the name hanging drop. The incubator temperature is maintained at 25°C under 200–300 lux. Some researchers incubate in dark.

8.3.3 AGAR PLATE METHOD

In agar plate method, the protoplasts are kept in fixed position. By doing so we can avoid formation of clumps and the immobilized protoplasts can then be subcultured as required. In practice the protoplasts are suspended in a molten agar medium (1.2% w/v) at 40°C and are dispensed into small plates (3.5–5 cm diameter) and are allowed to solidify. The solidified layer is cut into small blocks and is transferred into a large petri plate (9 cm diameter) containing liquid medium of same composition.

8.4 PROTOPLAST VIABILITY

The most suitable protoplasts for further studies and work are those which are spherical in shape when observed under light microscope. The viable protoplasts can be identified using the following stains:

Fluorescein diacetate (FDA) staining method: When viable protoplasts are treated with FDA at a final concentration of 0.01% in acetone, the stain accumulates inside the plasmalemma of the protoplasts. Viable protoplast contains esterases which cleave FDA to release fluorescein. Presence of fluorescein can be observed under a fluorescence microscopy within 5 min of staining. FDA dissociates from membrane after about 15 min.

Calcofluor white (CFW) staining: This staining method assures protoplast viability by detecting onset of cell wall formation. Calcofluor binds to beta-linked glucosides in newly synthesized cell wall which can be observed as a fluorescent ring around the membrane. Optimum staining is achieved when 0.1 mL of protoplast is mixed with 5.0 μL of 0.1% w/v solution of CFW.

Protoplast viability can also be detected by monitoring oxygen uptake of cells by oxygen electrode, which shows respiration. Variation of protoplast size with changing osmotic concentration also enables to determine viability of protoplast.

8.5 APPLICATIONS OF PROTOPLAST CULTURE

1. Study of cell wall formation: The early deposition of cellulosic microfibrils and their orientation at the protoplast surface can be followed using light and electronic microscopy.
2. Study of cell division and organogenesis: The isolated protoplasts provide an ideal system for the study of cell division. Under an inverted microscope, one can study the process of cell division and organogenesis.
3. Study of photosynthesis: Isolated protoplasts provide an excellent system for the study of biochemical and biophysical aspects of photosynthesis in both C3 and C4 plants.
4. Somatic hybridization: Conventional hybrid breeding programs in certain species have the limitations of incompatibility. This is overcome by protoplast fusion process. Using protoplast fusion, it is possible to manipulate hybrid production such that desirable characters which are genetically controlled can be transferred from one species to another. In certain cases, certain characters which are controlled by the cytoplasmic genome can also be transferred between different species.

8.6 PROTOPLAST FUSION (OR SOMATIC HYBRIDIZATION)

Protoplast fusion is a type of genetic modification in plants by which two distinct species of plants are fused together to form a new hybrid plant with the characteristics of both, a somatic hybrid. Hybrids have been produced either between different varieties of the same species (e.g., between nonflowering potato plants and flowering potato plants) or between two different species (e.g., between wheat *Triticum* and rye *Secale* to produce *Triticale*). Plants from distantly related or unrelated species are unable to reproduce sexually as their genomes/modes of reproduction, etc. are incompatible. Protoplasts from unrelated species can be fused to produce plants combining desirable characteristics such as disease resistance, good flavor, and cold tolerance. Before fusing the protoplasts, they have to align properly (Fig. 8.4). Fusion is carried out by application of an electric current or by treatment with chemicals such as polyethylene glycol (PEG). Fusion products can be selected for on media containing antibiotics or herbicides. These can then be induced to form shoots and roots and hybrid plants can be tested for desirable characteristics.

FIGURE 8.4 Aligning of protoplasts (A) before their fusion (B).

After the cell wall barrier is removed, the isolated protoplasts seem to fuse spontaneously under certain conditions. The process of somatic hybridization involves the following steps: (1) protoplast isolation, (2) protoplast fusion, (3) selection of somatic hybrids, and (4) culture of somatic hybrids to regenerate complete plants.

Protoplast fusion can be broadly classified into two categories (Fig. 8.5):

1. *Spontaneous fusion*: During the isolation process, it is frequently observed that, the protoplasts are fusing spontaneously. This spontaneous fusion of adjacent protoplasts results in a multinucleated protoplast. The giant protoplast of *Acetabularia* has been fused mechanically by pushing together two protoplasts. This fusion does not require the presence of fusion-inducing agents. However, in this procedure protoplasts are likely to get injured. Protoplast released from meiocytes in enzyme solutions readily fuse by gentle tapping in depression slide.
2. *Induced fusion*: In induced fusion, fusion-inducing chemicals are used to fuse isolated protoplasts. The isolated protoplasts normally carry negative charge on their surface (−10 mV to −30 mV) and hence there is a strong tendency to repel. The fusion-inducing chemicals generally reduce the electronegativity of the isolated protoplasts and allow them to fuse with each other (Narayanaswamy, 1994). As described by Bengochea and Dodds (1986), three important methods are generally used for protoplast fusion: (1) PEG-induced fusion,

(2) electrofusion, and (3) high Ca^{++}- and high pH-induced protoplast fusion. For all practical purposes, the first two methods are more commonly used than the third method.

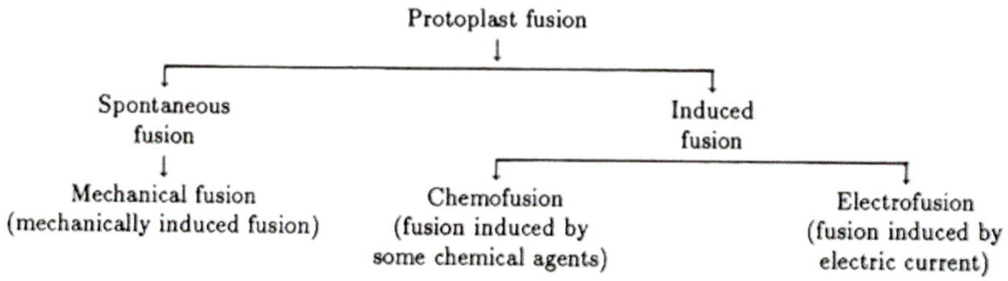

FIGURE 8.5 Methods of plant protoplasts hybridization.

8.6.1 *PROTOPLAST FUSION METHODS*

8.6.1.1 PEG TREATMENT

Protoplasts from different genera can be induced to fuse by high molecular weight PEG (M. W. 1500–6000) (Fig. 8.6). PEG appears to act as a molecular bridge between the surfaces of adjacent protoplasts either directly or indirectly through Ca^{2+}. Fusion presumably results from disturbance and redistribution of electric charges when the PEG molecules are washed away. If this assumption is correct, one should be able to increase the frequency of fusion by increasing the degree of charge disturbance. Since exposure of animal cells and plant protoplasts to solutions containing a high concentration of Ca^{2+} at a high pH has also been shown to induce fusion, an investigation into the combined effects of PEG and high pH–high calcium solutions on fusion is of particular interest (Kao and Michayluk, 1974; Keller and Melchers, 1973).

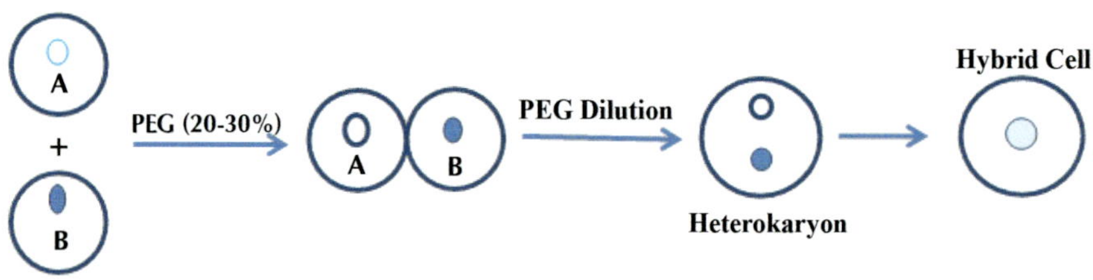

FIGURE 8.6 Represents the mechanism of PEG-induced protoplasts fusion.

8.6.1.2 ELECTROFUSION

Recently, mild electrical stimulation is being used to fuse protoplasts. This technique is known as electrofusion of protoplasts. Two glass capillary microelectrodes are placed in contact with the protoplasts. An electric field of low strength (10 kV m^{-3}) gives rise to dielectrophoretic pole generation within the protoplast suspension. This leads to pearl chain arrangement of protoplasts. The number of protoplasts within a pearl chain depends upon the population density of the protoplast and the distance between the electrodes. Subsequent application of high-intensity electric impulse (100 kV m^{-3}) for some microseconds results in the electric breakdown of membrane and subsequent fusion (Fig. 8.7) (Abe and Takeda, 1986).

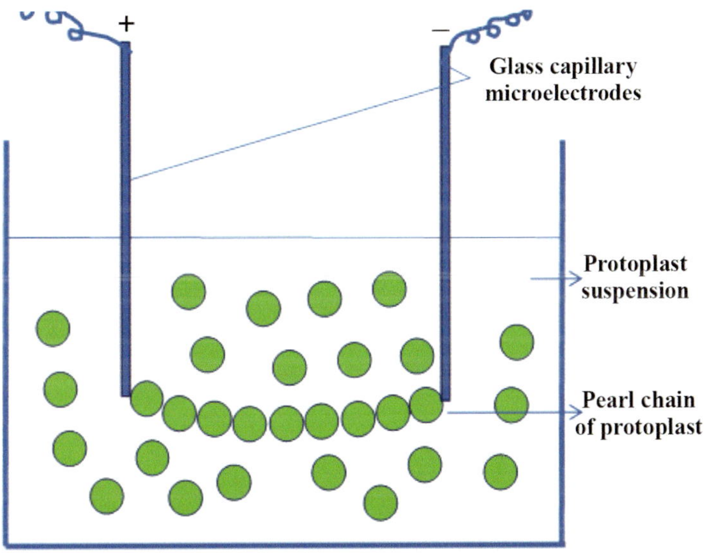

FIGURE 8.7 Electrofusion method of protoplasts fusion. (Adapted from http://www.biologydiscussion.com/plants/plant-protoplast/methods-of-protoplast-fusion-spontaneous-and-induced-fusion/14799)

8.6.1.3 HIGH PH OR CA^{++} TREATMENT

Kelier and Melchers (1973) developed a method to effectively induce fusion of tobacco protoplasts at a high temperature (37°C) in media containing high concentration of Ca^{2+} ions at a highly alkaline condition (pH 10.5). Equal densities of protoplasts are taken in centrifuge tube and protoplasts are spun at 100 g for 5 min. The pellet is suspended in 0.5 mL of medium. A 4 mL of 0.05 M $CaCl_2$, $2H_2O$ in 0.4 M mannitol at pH 10.5 is mixed with the protoplast suspension. The centrifuge tube containing protoplast at high pH or Ca^{2+} is placed in water bath at 30°C for 10 min and is spun at 50 g for 3–4 min. This is followed by keeping the tubes in water bath (37°C) for 40–50 min (Fig. 8.8). About 20–30% protoplast are involved in this fusion experiment (Toister and Loyter, 1971).

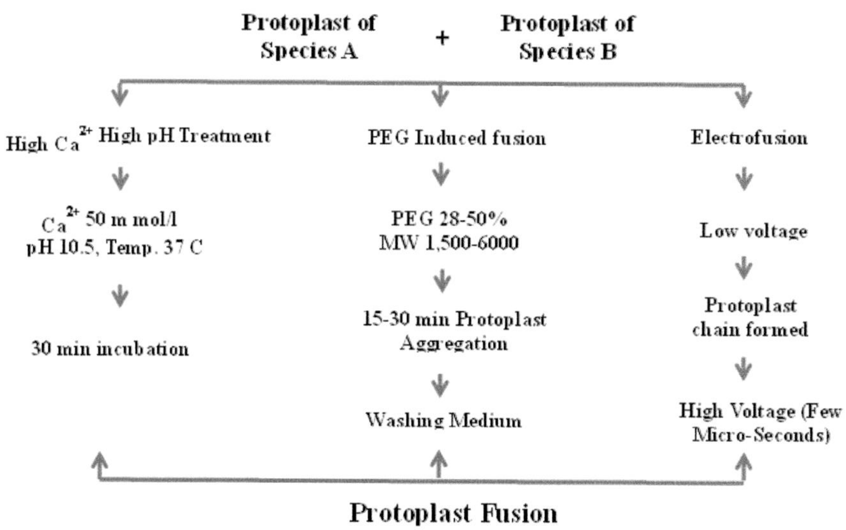

FIGURE 8.8 High pH or calcium method of protoplasts fusion.

8.6.1.4 $NANO_3$ TREATMENT

Equal densities of protoplasts from two different sources are mixed and then centrifuged at 100 g for 5 min to get a dense pellet. This is followed by addition of 4 mL of 5.5% sodium nitrate in 10.2% sucrose solution to resuspend the protoplast pellet. The suspended protoplasts are kept in water bath at 35°C for 5 min and again centrifuged at 200 g for 5 min. The pellet is once again kept in water bath at 30°C for 30 min. The fusions of protoplasts take place at the time of incubation. Finally, the protoplasts are plated in semisolid culture medium. The frequency of fusion is not very high in this method.

8.6.1.5 FUSION INDUCED BY OTHER CHEMICALS

Some other chemicals have also been observed to promote protoplast fusion: (1) 15% solution of polyvinyl alcohol (PVA) in combination with 0.05 $CaCl_2$ and 0.3 M mannitol are used to fuse plant protoplasts, (2) lectins are also known to agglutinate protoplasts, and (3) various proteins are also used for agglutination of protoplast.

8.6.2 MECHANISM OF PROTOPLAST FUSION

Several explanations were given by researchers regarding the mechanism of protoplast fusion; however, two convincing explanations are mentioned below. It is observed that when the protoplasts adhere to each other, the external fusagen causes configurational changes to the membrane proteins and lipids, which increase the fluidity of the membrane proteins and lipids thereby increasing the fluidity of the membrane. As the membranes of the adjacent protoplasts adhere to each other, intermixing of lipid molecules occur resulting in the fusion of the adjacent membrane. Most commonly used fusagen is PEG which has a high molecular weight. PEG acts as a bridge between two adjacent plasma membranes (Fig. 8.7). The elution of PEG subsequently causes changes in surface potential leading to contact and fusion of adjacent membranes. In addition, strong affinity of PEG for water may cause local dehydration upon removal, which increases the fluidity of the membrane thereby resulting in the fusion of protoplast subsequently.

8.7 REGENERATION OF PROTOPLAST

The first success of regeneration of plants from protoplasts cultures of *Nicotiana tabacum* was achieved by Takebe et al. (1971). Regeneration of protoplasts into whole plant occurs in two stages: (1) formation of cell wall and (b) development of callus. The newly formed cell wall is composed of loosely bound microfibrils and this formation requires continuous supply of nutrients. After the formation of cell wall, the protoplast loses its characteristic spherical shape. After the formation of cell wall, the protoplast undergoes normal cell division. Before the cell division, the protoplast increases in size and undergoes general division within 2–7 days. By the end of 3 week, visible colonies are formed and now they are inoculated into a mannitol-free medium for callus development. Once the callus is formed with appropriate manipulations, the callus will undergo organogenesis to form a whole plant.

8.8 APPLICATIONS OF PROTOPLAST AND THEIR FUSION

Plant protoplasts can be fused and the fusion products can be cultured to produce somatic hybrid plants. This technique could be used in studying physiology and dynamics of cell membrane, which include the uptake of macromolecules and viruses during infection. Interestingly, we can use this technique to produce germplasm, previously unavailable to the plant breeder. Production of hybrids between distantly related and sexually incompatible species is possible due to advancement in protoplast fusion. These naked cells are also responsible for somaclonal variations observed during plant tissue culture, which are beneficial for mankind. Characters affected include both qualitative and quantitative traits. Protoplasts are widely used for DNA transformation (for making beneficial genetically modified organisms), since the cell wall would otherwise block the passage of DNA into the cell. Protoplasts regenerate into whole plants first by developing into an undifferentiated callus and then by regeneration of shoots (caulogenesis) from the callus using plant tissue culture methods. Growth of protoplasts into callus, shooting, rooting, and regeneration into whole plantlets requires the optimal balance of plant growth hormones in the tissue culture medium that vary from species to species of plant. Some of the protoplasts, for example, protoplasts from mosses (*Physcomitrella patens*) do not require phytohormones for regeneration and even during callus induction. Interestingly, they regenerate directly into the filament-like appendages called protonema, mimicking a germinating moss spore. Protoplasts of different origin (inter species) are induced to fuse by using an electric field, PEG, and several chemicals as well. Additionally, protoplasts of plants expressing fluorescent proteins in certain cells may be used for fluorescence activated cell sorting (FACS), where only cells fluorescing a chosen wavelength are retained. Protoplasts have a wide range of applications (Table 8.3) and some of the major points are listed below:

1. The protoplast in culture can be regenerated into a whole plant.
2. Hybrids can be developed from protoplast fusion.
3. It is easy to perform single cell cloning with protoplasts.
4. Genetic transformations can be achieved through genetic engineering of protoplast DNA.

5. Protoplasts are excellent materials for ultrastructural studies.
6. Isolation of cell organelles and chromosomes is easy from protoplasts.
7. Protoplasts are useful for membrane studies (transport and uptake processes).
8. Isolation of mutants from protoplast cultures is easy.
9. Study of virus uptake and their replication.
10. Study of morphogenesis.

TABLE 8.3 Recent Advancement in Transfer of Important Agronomic Traits through Protoplast Fusion.

Species	Useful Traits Transferred	Reference
B. napus (+) *B. rapa*	Increased biomass and yield	Qian et al. (2003)
B. napus (+) *Crambe abyssinica*	Increased erucic acid content in seeds	Wang et al. (2003)
B. napus (+) *Sinapis arvensis*	Enhanced resistance to blackleg	Hu et al. (2002)
B. oleracea (+) *Moricandia arvensis*	Introduction of the C3–C4intermediate trait	Ishikawa et al. (2003)
Citrus limonia (+) *C. sunki* cv. *Tanaka*	Tolerance to citrus blight, tristeza virus, and *Phytophthora*	Costa et al. (2003)
C. reticulata cv. *Blanco* (+) *C. paradisi*	Production of mixoploid plants tolerant to citrus exocortis virus (CEV)	Liu and Deng (2002)
C. reticulata cv. *Blanco* (+) *C. volkameriana*	Tolerance to citrus blight, tristeza virus, and *Phytophthora*	Costa et al. (2003)
C. sinensis cv. *Ruby Blood* (+) *C. volkameriana*	Tolerance to citrus blight, tristeza virus, and *Phytophthora*	Costa et al. (2003)
C. sinensis (+) *Fortunella crassifolia*	Increased plant vigor	Cheng et al. (2003)
C. unshiu cv. *Guoqing* No. 1 (+) *C. grandis* cv. *Buntan Pink*	Generation of seedless cybrids	Guo et al. (2004)
Solanum tuberosum (+) *S. stenotomum*	Resistance to bacterial wilt (*R. solanacearum*)	Fock et al. (2001)
S. tuberosum (+) *S. nigrum*	Resistance to potato blight (*Phytophthora infestans*)	Szczerbakowa et al. (2003)
S. melongena (+) *S. sisymbrifolium*	Resistance to bacterial and fungal wilts	Collonnier et al. (2003)

8.9 CONCLUSIONS

It is now over 100 years since Klercker (1892) first made crude preparations of plant protoplasts. Since that time, enormous progress has been made in refining the methodologies for protoplast isolation, culture, and genetic manipulation through somatic hybridization and transformation. Currently, protoplasts provide systems for investigating most aspects of plant cell physiology and genetics, including proteomic and genomic studies. Despite the enormous progress achieved in this area, several important challenges remain. These include the recalcitrance of some protoplast systems to express their totipotency, with leaf protoplasts of cereals being a classic example. The advent of the new millennium has seen a marked resurgence of interest in protoplast technology with a particular focus on the generation of novel somatic hybrid and cybrid plants that cannot be produced through conventional breeding. Protoplasts are naked cells that are the equivalent, in general terms, to cultured animal cells. However, protoplast exhibits the unique property of totipotency. Consequently, protoplasts provide a cell system that can be manipulated readily, using physiological and pharmacological perturbations, and such experimentation can be followed through differentiation pathways to the whole (plant) organism and subsequent generations.

KEYWORDS

- **plant protoplasts**
- **somatic hybridization**
- **totipotency**
- **PEG**
- **cybridization**

REFERENCES

Abe, S.; Takeda, J. Possible Involvement of Calmodulin and the Cytoskeleton in Electrofusion of Plant Protoplasts. *Plant Physiol.* **1986,** *81*, 1151–1155.

Bellincampi, D.; Morpugo, G. Conditioning Factor Affecting Growth in Plant Cells in Culture. *Plant Sci.* **1987,** *51*, 83–91.

Bhatla, S. C.; Kiessling, J.; Reski, R. Observation of Polarity Induction by Cytochemical Localization of Phenylalkylamine-binding Receptors in Regenerating Protoplasts of the Moss *Physcomitrella patens*. *Protoplasma* **2002,** *219*, 99–105.

Binder, K. A.; Wegner, L. H.; Heidecker, M.; Zimmermann, U. Gating of Cl− Currents in Protoplasts from the Marine Alga *Valonia utricularis* Depends on the Transmembrane Cl− Gradient and Is Affected by Enzymatic Cell Wall Degradation. *J. Membr. Biol.* **2003,** *191*, 165–178.

Davey, M. R.; Kumar, A. Higher Plant Protoplasts—Retrospect and Prospect. *Int. Rev. Cytol. Suppl.* **1983,** *16*, 219–299.

Davey, M. R., Anthony, P., Power, J. B., Lowe, K. C. Plant Protoplasts: Status and Biotechnological Perspectives. *Biotechnol. Adv.* **2005,** *23 (2), 131–171.*

Dovzhenko, A.; Koop, H. U. Sugarbeet (*Beta vulgaris* L): Shoot Regeneration from Callus and Callus Protoplasts. *Planta* **2003,** *217*, 374–381.

d'UtraVaz, F. B.; Slamet, I. H.; Khatum, A.; Cocking, E. C.; Power, J. B. Protoplast Culture in High Molecular Oxygen Atmosphere. *Plant Cell Rep.* **1992,** *11*, 416–418.

Hain, R.; Czernilofsky, A. P. Uptake, Integration, Expression and Genetic Transmission of a Selectable Chimaeric Gene by Plant Protoplasts. *Mol. Gen. Genet.* **1985,** *199*, 161–168.

Helgeson, J. P.; Hunt, G. J.; Haberlach, G. T.; Austin, S. Somatic Hybrids between *Solanum brevidens and Solanum tuberosum: Expression of a Late Blight Resistance Gene and Potato Leaf Rolls Resistance. Plant Cell Rep.* **1986,** *5 (3), 212–214.*

Kao, K. N.; Michayluk, M. R. A Method for High-frequency Intergeneric Fusion of Plant Protoplasts. *Planta (Berl)* **1974,** *115*, 355–367.

Kao, K. N.; Michayluk, M. K. Nutritional Requirements for Growth of *Vinca hajastana* Cells and Protoplasts at a Very Low Population Density in Liquid Media. *Planta* **1975,** *126*, 105–110.

Keller, W. A.; Melchers, G. The Effect of High pH and Calcium on Tobacco Leaf Protoplast Fusion. *Z. Naturforsch.* **1973,** *28*, 737–774.

Kyozuka, J.; Hayashi, Y.; Shinomoto, K. High Frequency Plant Regeneration from Rice Protoplasts by Novel Nurse Culture Methods. *Mol. Gen. Genet.* **1987,** *206*, 408–413.

Lee, L.; Schroll, R. E.; Grimes, H. D.; Hodges, T. K. Plant Regeneration from Indica Rice (*Oryza sativa* L.) Protoplasts. *Planta* **1990,** *178*, 325–333.

Maheshwari, S. C.; Gill, R.; Maheshwari, N.; Gharyal, P. K. The Isolation and Culture of Protoplasts. In *Differentiation of Protoplasts and Transformed Plant Cells;* Reinert, J., Binding, H., Eds.; Springer-Verlag: New York, 1986; pp 20–48.

Mei-Lei, M. C. T.; Reitveld, E. M.; Van Marrewijk, G. A. M.; Kool, A. J. Regeneration of Leaf Mesophyll Protoplasts of Tomato Cultivars (*L. esculentum*)*:* Factors Important for Efficient Protoplast Culture and Plant Regeneration. *Plant Cell Rep.* **1987,** *6*, 172–175.

Power, J. B.; Cocking, E. C. Isolation of Leaf Protoplasts: Macromolecular Uptake and Growth Substance Response. *J. Exp. Bot.* **1970,** *21*, 64–70.

Rao, K. S.; Prakash, A. H. A Simple Method for the Isolation of Plant Protoplasts. *J. Biosci.* **1995,** *20* (5), 645–655.

Shahin, E. A. Totipotency of Tomato Protoplasts. *Theor. Appl. Genet.* **1985,** *69*, 235–240.

Sinha, A.; Wetten, A. C.; Caligari, P. D. S. Effect of Biotic Factors on the Isolation of *Lupinus albus* Protoplasts. *Aust. J. Bot.* **2003,** *51*, 103–109.

Thorpe, T. A. History of Plant Tissue Culture. *Mol. Biotechnol.* **2007,** *37 (2), 169–180.*

Toister, Z.; Loyter, A. Ca2+-induced Fusion of Avian erythrocytes. *Biochim. Biophys. Acta (Amst.)* **1971,** *241, 719–724.*

CHAPTER 9

SOMACLONAL VARIATION

ASHUTOSH PATHAK and ARUNA JOSHI*

Department of Botany, Faculty of Science, The Maharaja Sayajirao University of Baroda, Vadodara-390002, Gujarat, India.

**Corresponding author. E-mail: drarunajoshi@yahoo.co.in*

CONTENTS

ABSTRACT

Propagation through plant tissue culture has gained importance as it is rapid with no seasonal constrains and plants can be multiplied throughout the year. These in vitro plants are derived through asexual mode of regeneration and may develop as somaclonal variants. The reason may be due to preexisting variation in the explants or are generated during culture. These somaclones can be detected by different techniques like morphological, physiological, biochemical, protein and isozyme or molecular markers. Somaclonal variations in form of off-types are undesirable in micropropagation program, as the main goal is to generate true-to-type progeny. But on the other hand it provides genetic variability which has applications in plant breeding as well as in genetic improvement of the plant species.

9.1 INTRODUCTION

Plant cells are totipotent in nature which means that they have the ability to regenerate into whole plant. This nature of plant cells is being used in in vitro micropropagation technique which has now become one of the most popular means of mass propagation. The commercialization of micropropagation technology in horticultural crops began with orchids in 1970s, which was later witnessed in many ornamentals, fruits, spices, plantation crops, and medicinal plants as well. But many a times, off-types or variants are observed among clonally propagated plants, which are known as "somaclonal variants". One of the reasons is, during regeneration, the cells enter a dedifferentiation and redifferentiation program which differs from the natural growth pattern and hence can act as a mutagenic system. This reprogramming of cells can create a wide range of epigenetic variation in newly regenerated plants (Jain, 2000).

9.2 WHAT IS SOMACLONAL VARIATION?

The growth of plant cells under in vitro conditions is asexual, which involves only mitotic division of the cell which usually generates genetically uniform plants. But in tissue culture sometimes uncontrolled and random spontaneous variations occur, which is unexpected and undesired (Karp, 1994; Larkin, 1998). Variant formation in tissue culture generated plant was first reported by Braun in 1959. Earlier this variation was referred to as *calliclones* (Skirvin and Janick, 1976), *phenovariants* (Sibi, 1976), and *protoclones* (Shepard et al., 1980). But in 1981, Larkin and Scowcroft coined the term "somaclonal variation" which was then used for all type of variations derived through tissue culture. It is defined as "Quantitative or qualitative alterations among the clonally propagated somaclones". However, other names such as protoclonal, gametoclonal, and mericlonal variation are also used sometimes to describe the origin of variants, such as from protoplast, anther, and meristem cultures, respectively (Karp, 1994; Chen et al., 1998). Somaclonal variants may differ from the source plant permanently or temporarily and the variation in regenerated plants may be either genetic or epigenetic. Temporary changes result from epigenetic or physiological effects and are nonheritable and reversible (Kaeppler et al., 2000). On the other hand, permanent variants referred to somaclonal variants are heritable and often represent an expression of preexisting variation in the source plant or are due to the de novo variation via an undetermined genetic mechanism (Larkin and Scowcroft, 1981).

9.3 BASIS OF SOMACLONAL VARIATION

Somaclonal variations appear to result from preexisting genetic variations in the explant and are also induced through tissue culture (Fig. 9.1) (Evans et al., 1984).

9.3.1 PREEXISTING VARIATION

Heritable cellular variation could result from mutations and/or epigenetic changes (Kaeppler et al., 2000). Generally, to test for preexisting somaclonal variation, somaclones may be subjected to another round of in vitro regeneration. Clones with preexisting variation should yield more variability in the first generation than in the second and thereafter variation should be eliminated or stabilized. Subsequent variation is more likely to be tissue culture derived (Skirvin et al., 1994). There are some factors which cause preexisting variations and they are:

- ***Chimeras***

Many plants are known to have chimeras and they are one of the main sources of preexisting variation (George, 1993). Chimeras consist of tissues that differ in genetic constitutions which are developed from a meristem

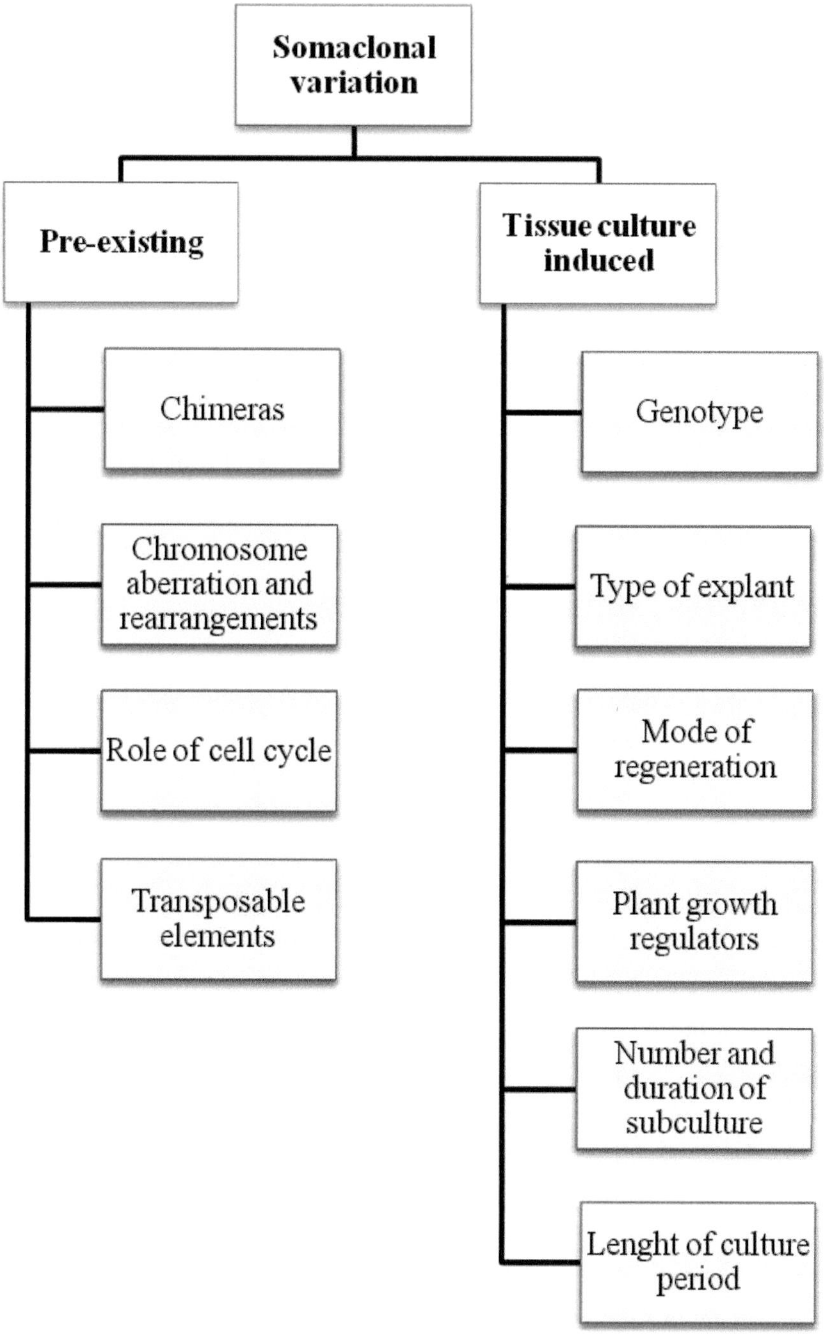

FIGURE 9.1 Overview of factors responsible for development of somaclonal variants.

containing layers/sectors of mutated tissue (Hartmann and Kester, 1983). When more than one explant is taken from one plant, it could cause variation (Kunitake et al., 1995). Hence, it is important to assess the entire plant for genetic uniformity before using it for tissue culture.

- ***Chromosomal aberration and rearrangement***

In tissue-cultured cells, the predominant type of aberration is the result of changes in chromosome structure. Therefore events leading to chromosome breakage, and in some instances subsequent exchange or reunion of fragments, appear to be of fundamental importance (Lee and Phillips, 1988). Late replicating heterochromatin and nucleotide pool imbalance are two possible origins of chromosome rearrangement in tissue culture (Bryant, 1976). The former involves the mitotic cell cycle of higher organisms. This cell cycle consists of four phases, each with a species specific and cell type specific duration. Any perturbation affecting the synchrony between chromosome replication during S phase and cell division would likely result in chromosome aberration. Because heterochromatic regions replicate later than euchromatic segments, their integrity may be particularly vulnerable to fluctuations in the cycle (Lee and Phillips, 1988). Some examples of somaclonal variation such as morphology of regenerated plants, position effects, qualitative variation and chromosome rearrangements, changes in sequence copy number, and gene amplification are involved with chromosome rearrangements (Lee and Phillips, 1988).

- ***Role of cell cycle***

Cell cycle plays an important role in controlling the growth and morphogenesis of plants and changes in it can create errors during tissue culture may alter the normal life (Larkin and Scowcroft, 1981). Lee and Phillips (1988) suggested that duration of four stages (G_1, S, G_2, and M) of cell cycle has varied at species or cell level, and irregularities in replication stage at S phase induce chromosomal aberration. Karp (1994) suggested that during protoplast cultures, errors at stage M of cell resulted in variation in chromosome number and structure. This suggests that any alteration in the normal process of the cell cycle can cause somaclonal variation. However, mitotic recombination including somatic crossing over and sister chromatid exchange which produce several types of chromosome rearrangements observed in tissue culture, especially if the exchanges were symmetric or between nonhomologous chromosomes (Larkin and Scowcroft, 1981).

- ***Transposable elements***

Transposons/transposable elements were first discovered in maize culture by McClintock in 1950. They are mobile DNA sequences that can induce gene mutations and contribute to genome rearrangements. Activation of these transposable elements is another source of chromosome-based somaclonal variation. Chromosome breakage is a means for initiating activity of maize transposable elements (Peschke et al., 1987). The discovery of activation of maize transposable elements in tissue culture suggested a possible relationship between somaclonal variation and mobile elements. Larkin and Scowcroft (1981) suggested that tissue culture environment probably provides a favorable environment for DNA sequence transposition. In many cases, the regeneration is indirect and induction of callus followed by subsequent development of shoot and root would disrupt normal cell function, which may activate transposable elements, stress-induced enzymes, or other products (Pietsch and Anderson, 2007). Gao et al. (2009) observed that the new insertions of transposons in a rice cultivar regenerated through tissue culture were responsible for somaclonal variation. Therefore, it has been suggested that transpositional events such as activation of transposable elements and the putative silencing of genes, and a high frequency of methylation pattern variation of single-copy sequences play a major role in somaclonal variation (Hirochika, 1993; Barret et al., 2006).

9.3.2 TISSUE CULTURE-INDUCED VARIATION

During in vitro culture, the propagation methods, genotype, nature of tissue used as starting material, type, and concentration of growth regulators, number as well as the duration of subcultures are some of the factors that determine the frequency of variation (Pierik, 1987). They are discussed below:

- ***Genotype***

Genotype is an important variable used for somaclonal variation and it can influence frequency of regeneration as well as frequency of somaclones.

Many times, some genotypes are more susceptible to produce somaclonal variants than other genotypes (Karp, 1991; Kaeppler and Phillips, 1993). For example, in modern bananas (*Musa* spp.) plants derived from ancestors *Musa acuminata* showed higher variability than plants derived from the ancestor *M. balbisiona*. Different genomes respond differently to stress-induced variation indicating that somaclonal variation has genotypic components. The differences in stability are related to differences in genetic makeup, whereby some components of the plant genome make them unstable during the culture process. Embryogenic cell suspension age and genotype affected the frequency and phenotype of variants produced significantly in *Coffea arabica* (Etienne and Bertrand, 2003). Similarly, the genotype and type of explant strongly influenced occurrence of somaclonal variation in callus cultures of strawberry (Popescu et al., 1997). In *Musa* spp., the type and rate of variation was specific to the genotype (Stover, 1987; Israeli et al., 1991), interaction between the genotype and the tissue culture environment (Martin et al., 2006), and genome composition (Sahijram et al., 2003).

- ***Type of explant***

Highly differentiated tissues such as roots, leaves, and stems generally produce more variants than explants from axillary buds and shoot tips which have preexisting meristems (Sharma et al., 2007). This is because the propagules develop from axillary buds maintaining the integrity of their histogenic layers. However, there are exceptions in this; for example, in banana shoot tip exhibited somaclonal variation (Israeli et al., 1991). Whereas the use of undifferentiated tissue such as the pericycle, procambium, and cambium as starting material for tissue culture reduces the chance of variation (Sahijram et al., 2003). Gross changes in the genome including endopolyploidy, polyteny, and amplification or diminution of DNA sequences could also occur during somatic differentiation in normal plant growth and development (D'Amato, 1977). Tissue source therefore can affect the frequency and nature of somaclonal variation (Kawiak and Lojkowska, 2004; Chuang et al., 2009).

- ***Mode of regeneration***

In vitro growth conditions can be extremely stressful on plant cells and may instigate highly mutagenic processes (Kaeppler and Phillips, 1993; Shepherd and Dos Santos, 1996). The processes of dedifferentiation and redifferentiation in in vitro cultures may involve both qualitative and quantitative changes in the genome, and different DNA sequences may be amplified or deleted during these changes in the state of the cell that is related to the original tissue source and regeneration system (Lee and Phillips, 1988). Disorganized growth phase in tissue culture is considered as one of the factors that cause somaclonal variation (Rani and Raina, 2000). In tissue culture, the organized structure of the plant is changed which is one of the reason for mutations to occur (Araujo et al., 2001; Cooper et al., 2006). It is well known that shoots of preformed origin (e.g., meristems, shoot tips, and axillary buds) show much less variation than those that arise from adventitious budding systems, such as organogenesis or embryogenesis (Karp, 1989). Although the direct formation of plant structures from cultured plant tissue, without any intermediate callus phase, minimizes the chance of instability; the stabilizing influence of the meristem is usually lost when plants are grown in culture (Karp, 1994). McClintock (1984) reported that callus induction occurs when cut surfaces comes in contact with medium which is analogous to wound responses observed in vivo. This may activate the transposable elements and stimulate the appearance of stress-induced enzymes. Systems subject to instability and disorganized growth demonstrated that cellular organization is a critical feature and that somaclonal variation is related to disorganized growth (Karp, 1994; Sivanesan, 2007). However, variation is not limited to callus regenerants and Evans (1988) reported considerable variation for morphological traits (flower shape, leaf shape, plant height), pollen viability, and chromosome number among adventitious *Nicotiana alata* plants are regenerated directly from leaf explants without a callus intermediate. Pathak et al. (2013) also reported variation in shoot cultures derived from both direct and indirect organogenesis in *Bacopa monnieri* and *Tylophora indica*, respectively.

- ***Plant growth regulators***

Plant Growth Regulators (PGRs) are one of the key factors for plant regeneration and they trigger the morphogenesis via cell-cycle disturbance which might induce variability (Peschke and Phillips, 1992). The genetic composition of a cell population can therefore be influenced by the relative levels of both auxins and cytokinins, and the levels of PGRs in the culture media have been linked with somaclonal variation to a greater extent (D'Amato, 1975; Vidal and De Garcia, 2000; Martin et al., 2006). In 1991, Swartz reported the possibility of unbalanced concentrations of auxins and cytokinins in inducing polyploidy. Rice and banana cultivar "Williams" reported genetic variability in cultures grown in high concentrations (>15 mg/L) of N^6-benzyladenine (Oono, 1985; Gimenez et al., 2001). Whereas diphenylurea derivatives induced somaclonal variation in banana (Roels et al., 2005) and soybean (Radhakrishnan and Ranjitha Kumari, 2008). LoSchiavo et al. in 1989 reported that auxins used during cultures of callus or cell suspension increased the rate of DNA methylation and in turn genetic variation. Many authors have reported that the synthetic auxins induce genetic variation and polyploidy by affecting normal DNA replication and postreplication mechanism (Bouman and De Klerk, 2001; Ahmed et al., 2004; Mohanty et al., 2008). Induction of callus using 2,4-dichlorophenoxyacetic acid (2,4-D) at high concentration has been implicated as cause of somaclonal variation in strawberry (Nehra et al., 1992), soybean (Gesteira et al., 2002), and cotton (Jin et al., 2008). Sometimes the type of auxin used in the media affects the generation of somaclone; for example, in leaf-color variants observed in Caladium somaclones varied with auxin type and the number of variants was higher on media containing either 2,4,5-trichlorophenoxyacetic acid (2,4,5-T) or 2,4-dichlorophenoxyaceticacid (2,4-D) than 1-naphtalene acetic acid (NAA) or indole-3-butyric acid (IBA) (Ahmed et al., 2004).

- ***Number and duration of subcultures***

During micropropagation, a high rate of proliferation is achieved in relatively shorter periods and leads to more frequent subculturing. The rapid multiplication of a tissue may affect genetic stability leading to somaclonal variation (Israeli et al., 1995). For example, the long period in culture increased the number of somaclonal variants observed in wheat (Hartmann et al., 1989). Similarly, Bairu et al. (2006) observed an increase in the rate of occurrence of variants with progressive subculturing of micropropagated bananas. Two possibilities have been hypothesized to explain an increased rate of mutation occurring in long-term cultures: a sequential accumulation of mutations over time and an augmented rate of mutation over time (Duncan, 1997).

- ***Length of culture period***

Total time period for which the culture has been maintained under in vitro condition is one of the most important and critical factors for induction of somaclonal variation. To maintain clonal stability, many commercial tissue culture laboratories regularly sample their plants ex vitro in a field or greenhouse to make sure that the characteristics of the particular plant is maintained. If the aim is to obtain true-to-type plants, one should minimize the amount of time that a culture is maintained in vitro, whereas to obtain variants, cultures should be maintained in vitro for a long time.

Somaclonal variation can be genotypic or phenotypic in nature, which can be either genetic or epigenetic in origin in the latter case. Typical genetic changes are: changes in chromosome numbers (polyploidy and aneuploidy), chromosome structure (translocations, deletions, insertions, and duplications), and DNA sequence (base mutations). Typical epigenetic changes are: gene amplification and gene methylation. The nature of somaclonal variation is of two types: heritable and epigenetic; heritable variation is stable through the sexual cycle or repeated asexual propagation. Whereas epigenetic variation (also known as developmental variation) may be unstable even when asexually propagated and it includes persistent changes in phenotype that involve the expression of particular genes (Hartmann and Kester, 1983).

9.4 SELECTION METHODS FOR SOMACLONAL VARIANTS

Selection of somaclonal variants can be done through two ways: (1) in vivo selection of somaclonal variants (Fig. 9.2) and (2) in vitro selection of somaclonal variants (Fig. 9.3) (Chawla, 2013).

Explant

↓

Shoot regeneration via direct or indirect organogenesis

↓

Multiplication of shoots

↓

Rooting and hardening of plants into field

↓

Screening for desirable traits

↓

Agronomic trails

FIGURE 9.2 Steps for in vivo selection of somaclonal variants. (Modified from Chawla, 2013)

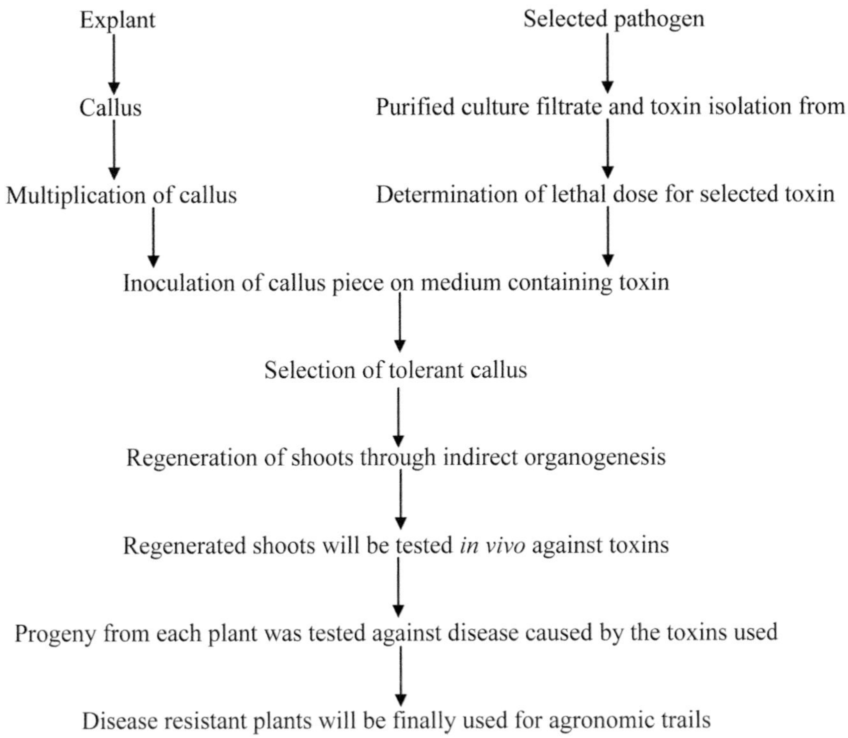

FIGURE 9.3 Steps for in vitro selection of somaclonal variants. (Modified from Chawla, 2013)

9.4.1 *IN VIVO* SELECTION OF SOMACLONAL VARIANTS

In this technique, the approach is to find somaclonal variants among the regenerated plants for various characters. In this, plants are regenerated through normal regeneration process—the explant will be inoculated into regeneration medium followed by rooting and hardening process. Then they were acclimatized and transferred to field and allowed to grow. These plants will be screened for desired traits/variation(s). In this, both dominant and homozygous recessive traits can be directly selected. If the regenerants are heterozygous, then recessive traits can be selected in the progenies of regenerants. Epigenetic variation can also be avoided when progenies are used. Thus, in case of self-fertilizing crops, it is recommended to screen progenies of the regenerants. Major disadvantage of this method is that it is time consuming and it requires screening of many plants.

9.4.2 *IN VITRO SELECTION OF SOMACLONAL VARIANTS*

In addition to the variants/mutants obtained as a result of the application of a selective agent(s) in the presence or absence of a mutagen, many variants have been obtained through the tissue culture cycle. For this any type of explant can be used but leaf explant will be more suitable for callus induction. Then the explant will be inoculated on medium that gives rise to organogenic callus which can differentiate and induce shoots when subcultured. Simultaneously, the pathogen (example, tobacco mosaic virus) was selected against which the resistant species needs to be developed. The pathogen will be grown and then filtrate was prepared for toxin isolation and then its lethal dose was determined for particular tissue. This toxin will be added in medium (in a concentration which will be decided in early step) and callus will be transferred to that medium and plants will be allowed to grow. Then they were tested for resistance against that toxin in vivo followed by agronomic values for first and second generation.

But the major difficulty in this technique is due to the instability of the resistant characteristic when removed from selective media. Many different reasons are there for this instability, one of which is the genetic lesion responsible may help a high-reversion rate or the alteration may simply be a transient genetic or epigenetic adaptation. The most common explanation is the presence of wild type escapes in the cell population and under nonselective conditions; this cell would rapidly proliferate eventually taking over the culture. In order to test the stability of the resistance trait, establish at least two cultures for each selected liner—one subculture routinely in the presence and one in the absence of the selective conditions. At various intervals over a period of several months perform dose response test. Further, the R_0 plants which were found to be resistant should be analyzed for resistance to that character in R_1 and subsequent generations.

There are many plant species in which somaclonal variants are induced/detected, and the traits developed are desired one; some of the plant species are listed in Table 9.1.

9.5 DETECTION METHODS FOR SOMACLONAL VARIANTS

Early detection of variants is essential to reduce the losses especially in large-scale production. There are mainly five different types of markers which can be used to detect somaclones (Fig. 9.4):

TABLE 9.1 Plants in Which Somaclonal Variations Are Generated Through Tissue Culture.

Plant species		Source of Variation	Reference
Scientific Name	**Common Name**		
Allium sativum L.	Garlic	Callus culture, genotype	Al-Zahim et al. (1999)
Beta vulgaris L.	Sugar beet	Callus culture, explant	Zhong et al. (1993)
Cajanus cajan (L.) Millsp.	Pigeon pea	Activation of transposable elements	Chintapalli et al. (1997)
Camellia sinensis (L.) O. Kuntze	Tea	Genotype	Thomas et al. (2006)
Capsicum annuum L.	Chili pepper	Genotype and BA	Hossain et al. (2003)
Curcuma aromatica Salisb.	Wild turmeric	Callus culture, 2,4-D, duration in culture	Mohanty et al. (2008)
Daucus carota L.	Carrot	Callus culture	Dugdale et al. (2000); Cooper et al. (2006)
Eleusine coracana (L.) Gaertn.	Finger millet	Callus culture	Baer et al. (2007)
Fragaria L.	Strawberry	BA	Biswas et al. (2009)
Fragaria × *ananassa* Duch.	Strawberry	Explant, genotype, duration in culture	Popescu et al. (1997)
Glycine max (L.) Merr.	Soybean	TDZ, number of subcultures	Radhakrishnan and Ranjitha Kumari (2008)
Hordeum vulgare L.	Barley	Callus culture, *Helminthosporium sativum* culture treatment	Kole and Chawla (1993)
Musa acuminata L.	Dessert banana	Genotype, explant source, number of subcultures	Sheidai et al. (2008)
Musa spp.	Plantain	Explant, chimeric effect	Krikorian et al. (1993)
Musa acuminata L.	Brasilero bananas	BA	Gimenez et al. (2001)

TABLE 9.1 *(Continued)*

Plant species		Source of Variation	Reference
Scientific Name	**Common Name**		
Oryza sativa L.	Rice	Callus culture	Araujo et al. (2001)
Oryza sativa L.	Rice	Callus culture, genotype, duration in culture	Sultana et al. (2005)
Pyrus pyraster Burgsd	Wild pear	Treatment with $FeSO_4$ + 6 mM $KHCO_3$	Palombi et al. (2007)
Saccharum L.	Sugarcane	Callus culture, salt, polyethylene glycol	Yadav et al. (2006)
Saccharum L.	Sugarcane	*Colletotrichum falcatum* culture filtrate	Sengar et al. (2009)
Saccharum L.	Sugarcane	Callus culture	Singh et al. (2000); Singh et al. (2008)
Solanum tuberosum L.	Potato	Embryogenic culture, genotype	Rietveld et al. (1993)
Sorghum bicolor L.	Sorghum	Explant	Zhang et al. (2010)
Triticum aestivum L.	Wheat	Callus culture, 2,4-D, genotype	Arun et al. (2007)
Triticum aestivum L.	Wheat	Embryogenic culture, genotype	Ahmed and Abdelkareem (2005)
Triticum aestivum L.	Wheat	Nicotinic acid	Bogdanova (2003)
Zea mays L.	Maize	Embryogenic culture, 2,4-D	Dennis et al. (1987)
Zea mays L.	Maize	Callus culture, mannitol, polyethylene glycol	Matheka et al. (2008)

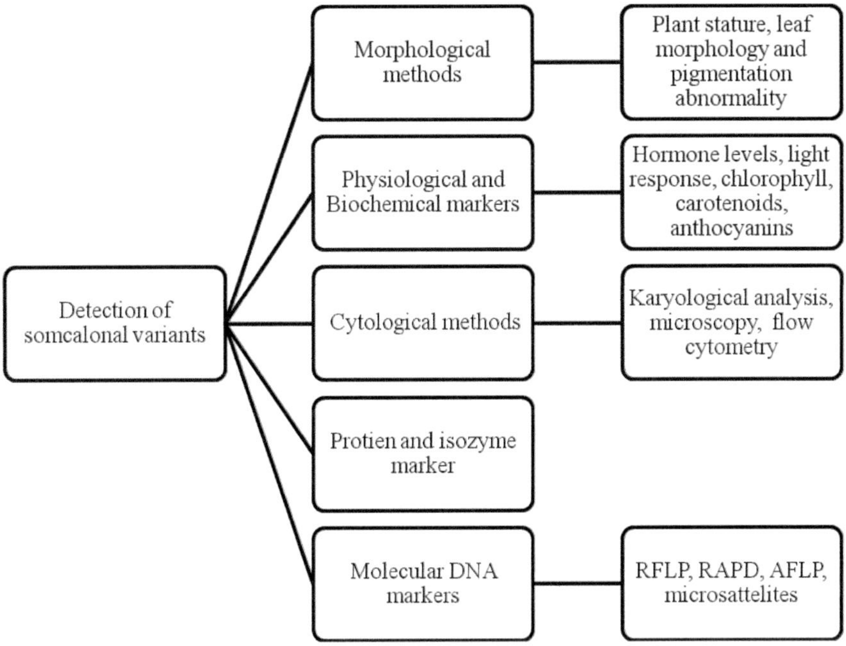

FIGURE 9.4 Detections methods for somaclonal variation.

- ***Morphological methods***

One of the oldest and most extensively used methods is to detect variants based on characters such as difference in plant stature, leaf morphology, and pigmentation (Israeli et al., 1991). It has advantages such as it does not require laboratory; one can observe the variants with naked eye and is suitable for preliminary detection. However, morphological traits are often strongly influenced by environmental factors and may not reflect the true genetic composition of a plant (Mandal et al., 2001). Also this marker used for phenotypic characters is limited in number and often developmentally regulated (Cloutier and Landry, 1994). For example, major changes to the genome, as a result of in vitro manipulation may not be expressed as an altered phenotype and vice versa. Furthermore, the detection of variants using morphological features is often mostly feasible for perennial plants. Variations observed could also be due to physiological changes induced by in vitro culture environment, which are temporary and may disappear once the culture conditions are withdrawn. Sometimes the regenerated plants did not present somaclonal variations, which could be improper due to the fact that the recessive mutations are not expressed in the R_0 plants (plants regenerated from tissue culture), which are generally heterozygous for the mutation. Therefore, plants which look normal could segregate abnormal plants in the R_1 or R_2 generations (successive sexual generations of R_0).

- ***Physiological and Biochemical markers***

Physiological and/or biochemical markers have advantages over morphological markers as they are comparatively faster and can be carried out at early stages of plant growth (Israeli et al., 1995). One of the important physiological parameters for plant growth is light, which is an energy source for photosynthesis and temperature is essential prerequisites. Photoinhibition is physiological stress, which is expressed as a decline in photosynthetic capability of oxygen-evolving photosynthetic organisms due to excessive illumination (Adir et al., 2003). Damasco et al. (1997) evaluated the responses of tissue cultured normal and dwarf off-type Cavendish bananas to suboptimal temperatures under field and controlled environmental conditions and observed that the dwarf off-types showed improved tolerance to low temperature and light compared to the normal one. Another physiological factor which can be used is application of gibberellic acid which is known to regulate the developmental processes like stem elongation and enzyme induction. Sandoval et al. (1995) and Damasco et al. (1996) used gibberellic acid to identify the variants of *Musa* spp. and banana, respectively. For biochemical marker, analysis of different pigments such as chlorophyll, carotenoids, anthocyanins can be used (Shah et al., 2003). Mujib (2005) and Wang et al. (2007) used chlorophyll and carotenoid content to identify the variants in pineapple and sweet potato variants, respectively. Another way to identify the variants is to check the carbon dioxide assimilating potential of the plants (Saradhi and Alia, 1995). But the major drawback of biochemical tests is due to their complex nature and one should have high expertise.

- ***Cytological methods***

Genetic composition of a plants changes with the chromosomal variations and with the changes in content of RNA/DNA. Karyological analysis and observation of chromosomal aberration using light microscopy, oil immersion, and other complex microscopy techniques have been used for detecting somaclonal variation in in vitro plants (Al-Zahim et al., 1999; Raimondi et al., 2001; Mujib et al., 2007). But this technique is time consuming, tedious, and cumbersome especially when chromosome number is high or difficult to observe due to their small size; hence as a better alternative, flow cytometry is used (Rani and Raina, 2000; Dolezel et al., 2004). This technique was used to identify somaclonal variants in many plants like strawberry (Nehra et al., 1992) and potato (Sharma et al., 2007). Advantages of flow cytometry are: convenient sample preparation and analysis, efficient method for large-scale studies of ploidy level, and even smallest modifications in chromosome number can be detected. But sometimes the cytosolic compounds interfere with quantitative DNA staining in flow cytometry, it is time consuming and absence of DNA reference standards cause limitations in the use of flow cytometry.

- ***Proteins and isozyme marker***

Any morphological variation is a result of biochemical variation, which is expressed as variation amongst the proteins. Proteins are the most abundant organic molecules in cells having diverse functions and isozymes are multiple molecular forms of enzymes. Isozymes were one of the most widely used markers for identifying the genetic variation in many organisms; and depending on the number of loci, their state, and the specific isozymes used, one or several bands are visualized and the polymorphism of the bands reveals variation (Weising et al., 2005). The variation characterized has been summarized into three categories: (1) altered electrophoretic mobility, (2) loss/gain of protein bands, and (3) altered level of specific protein.

Proteins and isozymes like peroxidase, malate dehydrogenase, and superoxide dismutase have been extensively used to study variation in sugarcane (Srivastava et al., 2005) and beans (Gonzalez et al., 2010). Total protein and isozymes markers did not reveal variations in closely related cultivars of *Musa* spp. (Bhat et al., 1992). Whereas Mandal et al. (2001) used both salt-soluble polypeptide and isozymes for variant identification in various banana cultivars, but they reported limitation of salt soluble peptide as markers for variant identification. Biggest advantages of this marker are due to its codominant expression and are easy to perform. Major drawback of them is it has tissue-specific expression. Other disadvantages are: they are influenced by environmental factors and the numbers of isozymes are limited and only DNA regions coding for soluble proteins can be sampled (Venkatachalam et al., 2007).

- ***Molecular DNA markers***

In tissue culture-derived plants, the variations arise from changes in chromosome number and/or structure, or from more subtle changes in the DNA (Gostimsky et al., 2005). Hence, it is necessary to examine the variation at the molecular level in order to determine locations and extent of deviance from the true-to-type clone plant (Cloutier and Landry, 1994). Nowadays most of the analysis of variants is done using molecular markers due to its advantages over other techniques like the variation at morphological level occurs at a much lower frequency than at the DNA level (Evans et al., 1984). Other advantages are: many of these markers are codominant in nature, DNA from any tissue can be used for analysis, they are phynotypically neutral and identification can be done when the plant is still in growth stage in culture media. There are different types of molecular markers which differ in their principles, methodologies, and applications; hence, one requires careful consideration in choosing one or more of such methods. No molecular markers are available till date which fulfils all requirements.

- ***Restriction Fragment Length Polymorphism (RFLP)***

Restriction enzymes/endonucleases are the enzymes produced by a variety of prokaryotes and are naturally used to destroy invading foreign DNA molecules (Weising et al., 2005). In RFLP, extracted DNA is digested with restriction enzymes and the resultant fragments are separated by gel electrophoresis (Karp et al., 1996). Molecular variations among the organisms are revealed by change in sites of restriction enzymes and variation in length of the fragments is observed after digestion with the restriction enzymes. This differential profile is generated due to nucleotide substitutions or DNA rearrangements like insertion, deletion, single nucleotide polymorphisms, etc. (Agarwal et al., 2008). RFLP markers are relatively highly polymorphic, codominantly inherited, and highly reproducible (Agarwal et al., 2008). But the technique has some limitations as it is time consuming, costly, large amount of plant tissue is required, and analyses involves the use of radioactive/toxic reagents (Karp et al., 1996; Piola et al., 1999) which limits its uses.

- ***Random Amplified Polymorphic DNA (RAPD)***

RAPD involves the use of single short primers of arbitrary sequence to amplify segments of target genomic DNA. These short primers referred to as "random/universal" primers are used to reveal polymorphisms among the amplification products which are seen as bands on agarose gel (Williams et al., 1990). Biggest advantage of this method is that no prior knowledge on genome sequence is required and also it can be done using crude DNA with very less quantity; hence, it is widely used for tissue culture-induced variation in many plants (Hashmi et al., 1997; Bairu et al., 2006; Pathak et al., 2013). The technique has certain limitation due to lower reproducibility and reliability; it is a dominant marker and hence is less informative. Although it has limitations, it remained a favorite technique as it is cheaper compared to other techniques. In addition, the problem of RAPD reliability and transferability among laboratories could be minimized and eliminated by following a standard protocol, replication of amplification reactions, and a conservative criterion of band selection (Belaj et al., 2003).

- ***Amplified Fragment Length Polymorphism (AFLP)***

AFLP was developed in early 1990s to overcome the limitation of reproducibility associated with RAPD and it combines the power of RFLP and flexibility of PCR-based technology. The fingerprints are produced, without any prior knowledge of sequence, using a limited set of generic primers. The number of fragments detected in a single reaction can be "tuned" by selection of specific primer sets. AFLP technique is reliable since stringent reaction conditions are used for primer annealing. AFLP has now become a preferred technique as it combines the reliability of RFLP with the efficiency of RAPD. It is a very sensitive and reliable marker technique that could be useful for detecting specific genomic alterations associated with tissue culture variation and identifying slightly different genotypes. AFLP analysis has been used to study tissue culture-induced somaclonal variation in many plant species (Sanchez-Teyer et al., 2003; James et al., 2004; Chuang et al., 2009). However, this technique requires more optimization and is more expensive than RAPD (Weising et al., 2005). It also requires pure DNA samples for analysis and expertise to perform, which has limited its use.

- ***Microsatellite markers***

Microsatellite markers are simple sequence repeats (SSRs), inter simple sequence repeats (ISSRs), short tandem repeats (STRs), sequence-tagged microsatellite sites (STMS), and simple sequence length polymorphisms (SSLP). Microsatellites consist of tandemly reiterated short DNA (one to five) sequence motifs which are abundant and occur as interspersed repetitive elements. Polymorphism results from differences in the number of repeat units between individuals at a particular microsatellite locus and is believed to be due to unequal crossing over or slippage of DNA polymerase during replication of repeat tracts (Levinson and Gutman, 1987; Coggins and O'prey, 1989). Microsatellite markers are inherited in a Mendelian fashion but they tend to be hypervariable as slippage in replication is more likely than point mutations, and are implicated for the extensive interindividual length polymorphisms observed in microsatellite assays (Litt and Luty, 1989; Agarwal et al., 2008). For the microsatellite assay, the sequence information of repeat-flanking regions is required to design locus-specific PCR primer pairs and they are unlabeled or radio-labelled or fluoro-labelled. Subsequently, amplified PCR products are separated on polyacrylamide gel and visualized by autoradiography, fluorometry, or staining with silver or ethidium bromide (Weising et al., 2005). Microsatellite markers have been used to check genetic stability of several micropropagated plants (Rahman and Rajora, 2001; Khlestkina et al., 2010; Zhang et al., 2010). In comparison to AFLP and RFLP, the microsatellite marker technique such as ISSR is cost efficient, overcomes hazards of radioactivity, and requires lesser amounts of DNA (Zietkiewicz et al., 1994). It combines the benefits of AFLP and microsatellite analysis with the universality of RAPD. It is highly reproducible and became popular due to their codominant inheritance nature, high abundance in organisms, enormous extent of allelic diversity as well as the ease of assessing microsatellite size variation using PCR with pairs of flanking primers (Li et al., 2002; Weising et al., 2005; Agarwal et al., 2008). However, a major drawback for the use of microsatellites is that the development of primers is time consuming (Squirrell et al., 2003).

9.6 ADVANTAGES

- Progenies of R_0 plants (R_1) will be screened for desired characteristics and the most suitable one will be selected for subsequent generations (R_2, R_3).
- Seeds will be mutated and used for mutation breeding experiment.
- New types of genetic variation arise through tissue culture regeneration such as changes in copy number of repeated DNA sequences,

activation of transposons which have not been induced by conventional mutagenesis.

- High frequency of chromosome rearrangements and cytoplasmic DNA changes and the tissue specificity of certain mutations offer an opportunity to target the technique to generate specific variants.
- Using specific selection agent in tissue culture, plant species having resistance to different stress (salt, draught, etc.) and diseases (fungal, herbicide, etc.) can be regenerated.
- Cell lines can be produced which will have increased secondary metabolite content.
- This is a more simple technique for genetic variability than recombinant DNA, and they result in a rich source of genetic variability. Also somaclonal variation has no regulatory hurdles like products of recombinant DNA.

9.7 DISADVANTAGES

- If the main aim is micropropagation, induction of somaclonal variants is not desirable.
- Variation generated by somaclonal technique is random and hence necessary to screen large number of progeny of regenerants.
- The reproducibility of variation depends on many factors and hence optimization should be done for desired variant recovery.
- R1 progenies are composed of segregating populations and with mutation breeding, for quantitative traits, it is impossible to select individual with improvements in R1 generation.

9.8 APPLICATIONS OF SOMACLONAL VARIANTS

- The technique could be used to uncover new variants that retain all the favorable qualities of an existing variety while adding one additional trait.
- Induced somaclonal variation can be used for genetic manipulation of crops with polygenic traits.
- Plants with desired characteristics such as bigger fruit size, more interesting flowers, as well as improved quality with better yield of metabolites can be produced.

- It can also be an important tool for plant breeding via generation of new varieties that could exhibit disease resistance.
- New plant varieties with resistance to different herbicide can be produced using somaclonal variation.
- New varieties can be developed which will have resistance to different abiotic stress (cold, salt, draught, aluminium, etc.)-tolerant plants can be produced.
- In vitro selection technique can be used to produce insect-resistant plants.
- Agriculturally useful variants have been identified for several traits in crop plants, for example, solids in tomato, male sterility in tomato and rice, yield in rice, earliness in maize, freezing tolerance in wheat, and disease resistance in wheat, sugar cane, potato, celery, and tomato.
- For ornamental plants, different varieties with leaf and flower morphology and color variants have been used.

KEYWORDS

- **chimeras**
- **molecular markers**
- **plant tissue culture**
- **somaclonal variation**
- **genome**

REFERENCES

Adir, N.; Zer, H.; Shochat, S.; Ohad, I. Photoinhibition—A Historical Perspective. *Photosynth. Res.* **2003,** *76*, 343–370.

Agarwal, M.; Shrivastava, N; Padh, H. Advances in Molecular Marker Techniques and their Applications in Plant Sciences. *Plant Cell Rep.* **2008,** *27*, 617–631.

Ahmed, E. U.; Hayashi, T.; Yazawa, S. Auxins Increase the Occurrence of Leaf-Colour Variants in Caladium Regenerated from Leaf Explants. *Sci. Hortic.* **2004,** *100*, 153–159.

Ahmed, K.; Abdelkareem, A. Somaclonal Variation in Bread Wheat (*Triticum aestivum* L.). II. Field Performance of Somaclones. *Cereal Res. Commun.* **2005,** *33*, 485–492.

Al-Zahim, M. A.; Ford-Lloyd, B. V.; Newbury, H. J. Detection of Somaclonal Variation in Garlic (*Allium sativum* L.) using RAPD and Cytological Analysis. *Plant Cell Rep.* **1999,** *18*, 473–477.

Araujo, L. G.; Prabhu, A. S.; Filippi, M. C.; Chaves, L. J. RAPD Analysis of Blast Resistant Somaclones from Upland Rice Cultivar IAC 47 for Genetic Divergence. *Plant Cell Tissue Org. Cult.* **2001,** *67*, 165–172.

Arun, B.; Singh, B. D.; Sharma, S.; Paliwal, R.; Joshi, A. K. Development of Somaclonal Variants of Wheat (*Triticum aestivum* L.) for Yield Traits and Disease Resistance Suitable for Heat Stressed and Zero-Till Conditions. *Field Crops Res.* **2007,** *103*, 62–69.

Baer, G.; Yemets, A.; Stadnichuk, N.; Rakhmetov, D.; Blume, Y. Somaclonal Variability as a Source for Creation of New Varieties of Finger Millet (*Eleusine coracana* (L.) Gaertn.). *Cytol. Genet.* **2007,** *41*, 204–208.

Bairu, M. W.; Fennell, C.; van Staden, J. The Effect of Plant Growth Regulators on Somaclonal Variation in Cavendish Banana (*Musa* AAA cv. 'Zelig'). *Sci. Hortic.* **2006,** *108*, 347–351.

Barret, P.; Brinkman, M.; Beckert, M. A Sequence Related to Rice Pong Transposable Element Displays Transcriptional Activation by in vitro Culture and Reveals Somaclonal Variations in Maize. *Genome* **2006,** *49*, 1399–1407.

Belaj, A.; Satovic, Z.; Cipriani, G.; Baldoni, L.; Testolin, R.; Rallo, L.; Trujillo, I. Comparative Study of the Discriminating Capacity of RAPD, AFLP and SSR Markers and of their Effectiveness in Establishing Genetic Relationships in Olive. *Theor. Appl. Genet.* **2003,** *107*, 736–744.

Bhat, K. V.; Bhat, S. R.; Chandel, K. P. S. Survey of Isozyme Polymorphism for Clonal Identification in *Musa*. I. Esterase, Acid Phosphate and Catalase. *J. Hortic. Sci.* **1992,** *67*, 501–507.

Biswas, M. K.; Dutt, M.; Roy, U. K.; Islam, R.; Hossain, M. Development and Evaluation of in vitro Somaclonal Variation in Strawberry for Improved Horticultural Traits. *Sci. Hortic.* **2009,** *122*, 409–416.

Bogdanova, E. D. Epigenetic Variation Induced in *Triticum aestivum* L. by Nicotinic Acid. *Russ. J. Genet.* **2003,** *39*, 1029–1034.

Bouman, H.; De Klerk, G. J. Measurement of the Extent of Somaclonal Variation in Begonia Plants Regenerated under Various Conditions. Comparison of three assays. *Theor. Appl. Genet.* **2001,** *102*, 111–117.

Braun, A. C. A Demonstration of The Recovery of the Crown-Gall Tumor Cell with the use of Complex Tumors of Single-Cell Origin. *Proc. Natl. Acad. Sci.* U.S.A. **1959,** *45*, 932–938.

Bryant, J. A. The Cell Cycle. In *Molecular Aspect of Gene Expression in Plants*; Bryant J. A., Ed.; Academic Press: New York, **1976;** pp 117–216.

Chawla, H. S. *Introduction to Plant Biotechnology* (3rd Edn); Oxford and IBH publishing Co. Pvt. Ltd.: New Delhi, India, **2013;** pp 110–122.

Chen, W. H.; Chen, T. M.; Fu, Y. M.; Hsieh, R. M.; Chen, W. S. Studies on Somaclonal Variation in *Phalaenopsis*. *Plant Cell Rep.* **1998,** *18*, 7–13.

Chintapalli, P.; Moss, J.; Sharma, K.; Bhalla, J. In vitro Culture Provides Additional Variation for Pigeonpea. *In Vitro Cell. Dev. Biol. Plant* **1997,** *33*, 30–37.

Chuang, S. J.; Chen, C. L.; Chen, J. J.; Chou, W. Y.; Sung, J. M. Detection of Somaclonal Variation in Micro-Propagated *Echinacea purpurea* using AFLP Marker. *Sci. Hortic.* **2009,** *120*, 121–126.

Cloutier, S.; Landry, B. Molecular Markers Applied to Plant Tissue Culture. *In Vitro Cell Dev. Biol. Plant* **1994,** *30*, 32–39.

Coggins, L. W.; O'Prey, M. DNA Tertiary Structures Formed in vitro by Misaligned Hybridization of Multiple Tandem Repeat Sequences. *Nucleic Acids Res.* **1989,** *17*, 7417–7426.

Cooper, C.; Crowther, T.; Smith, B. M.; Isaac, S.; Collin, H. A. Assessment of the Response of Carrot Somaclones to *Pythium violae*, Causal Agent of Cavity Spot. *Plant Pathol.* **2006,** *55*, 427–432.

D'Amato, F. The Problem of Genetic Stability in Plant Tissue and Cell Cultures. In *Crop Genetic Resources for Today and Tomorrow*; Frankel O. H., Hawkes J. G., Eds.; Cambridge University Press: New York, **1975;** pp 333–348.

D'Amato, F. Cytogenetics of Differentiation in Tissue and Cell Culture. In *Applied and Fundamental Aspects of Plant Cell, Tissue and Organ Culture*; Reinert J., Bajaj Y. P. S., Eds.; Springer: New York, **1977;** pp 343–464.

Damasco, O. P.; Godwin, I. D.; Smith, M. K.; Adkins, S. W. Gibberellic Acid Detection of Dwarf Off-Types in Micropropagated Cavendish Bananas. *Aust. J. Exp. Agric.* **1996,** *36*, 237–341.

Damasco, O. P.; Smith, M. K.; Godwin, I. D; Adkins, S. W.; Smillie, R. M.; Hetherington, S. E. Micropropagated Dwarf Off-Type Cavendish Bananas (*Musa* spp., AAA) Show Improved Tolerance to Suboptimal Temperatures. *Aust. J. Agric. Res.* **1997,** *48*, 377–384.

Dennis, E. S.; Brettell, R. I. S; Peacock, W. J. A Tissue Culture Induced Adh1 Null Mutant of Maize Results from a Single Base Change. *Mol. Gen. Genet.* **1987,** *210*, 181–183.

Dolezel, J.; Valarik, M.; Vrana, J.; Lysak, M. A.; Hribova, E.; Bartos, J.; Gasmanova, N.; Dolezelova, M.; Safar, J.; Simkova, H. Molecular Cytogenetics and Cytometry of Bananas (*Musa* spp.). In *Banana Improvement: Cellular, Molecular Biology, and Induced Mutations*; Jain S.M. and Swennen R., Eds.; Science Publishers, Inc.: Enfield, **2004;** pp 229–244.

Dugdale, L. J.; Mortimer, A. M.; Isaac, S.; Collin, H. A. Disease Response of Carrot and Carrot Somaclones to *Alternaria dauci*. *Plant Pathol.* **2000,** *49*, 57–67.

Duncan, R. R. Tissue Culture-Induced Variation and Crop Improvement. *Adv. Agron.* **1997,** *58*, 201–240.

Etienne, H.; Bertrand, B. Somaclonal Variation in *Coffea arabica*: Effects of Genotype and Embryogenic Cell Suspension Age on Frequency and Phenotype of Variants. *Tree Physiol.* **2003,** *23*, 419–426.

Evans, D. A. Applications of Somaclonal Variation. In *Biotechnology in Agriculture*; Mizrahi A., Ed.; Allan R. Liss: New York, **1988;** pp 203–223.

Evans, D. A.; Sharp, W. R.; Medina-Filho, H. P. Somaclonal and Gametoclonal Variation. *Am. J. Bot.* **1984,** *71*, 759–774.

Gao, D. Y.; Vallejo, V.; He, B.; Gai, Y. C.; Sun, L. H. Detection of DNA Changes in Somaclonal Mutants of Rice using SSR Markers and Transposon Display. *Plant Cell Tissue Org. Cult.* **2009,** *98*, 187–196.

George, E. F. *Plant propagation by tissue culture, Part 1: The Technology* (2nd Edn.); Exegetics Ltd.: Edington, Wilts., England, 1993.

Gesteira, A. S.; Otoni, W. C.; Barros, E. G.; Moreira, M. A. RAPD Based Detection of Genomic Instability in Soybean Plants Derived from Somatic Embryogenesis. *Plant Breed.* **2002,** *121*, 269–271.

Gimenez, C.; de Garcia, E.; de Enrech, N. X.; Blanca, I. Somaclonal Variation in Banana: Cytogenetic and Molecular Characterization of the Somaclonal Variant CIEN BTA-03. *In Vitro Cell Dev. Biol. Plant* **2001,** *37*, 217–222.

Gonzalez, A.; De la Fuente, M.; De Ron, A.; Santalla, M. Protein Markers and Seed Size Variation in Common Bean Segregating Populations. *Mol. Breed.* **2010,** *25*, 723–740.

Gostimsky, S. A.; Kokaeva, Z. G.; Konovalov, F. A. Studying Plant Genome Variation using Molecular Markers. *Russ. J. Genet.* **2005,** *41*, 378–388.

Hartmann, C.; Henry, Y.; Buyser, J.; Aubry, C.; Rode, A. Identification of New Mitochondrial Genome Organizations in Wheat Plants Regenerated from Somatic Tissue Cultures. *Theor. Appl. Genet.* **1989,** *77*, 169–175.

Hartmann, H. T.; Kester, D. E. *Plant Propagation: Principles and Practices.* Prentice-Hall: Englewood Cliffs, N.J., **1983,** p 727.

Hashmi, G.; Huettel, R.; Meyer, R.; Krusberg, L.; Hammerschlag, F. RAPD Analysis of Somaclonal Variants derived from Embryo Callus Cultures of Peach. *Plant Cell Rep.* **1997,** *16*, 624–627.

Hirochika, H. Activation of Tobacco Transposons during Tissue Culture. *EMBO J.* **1993,** *12*, 2521–2528.

Hossain, A. M.; Konisho, K.; Minami, M.; Nemoto, K. Somaclonal Variation of Regenerated Plants in Chili Pepper (*Capsicum annuum* L.). *Euphytica* **2003,** *130*, 233–239.

Israeli, Y.; Lahav, E.; Reuveni, O. In vitro Culture of Bananas. In *Bananas and plantains*; Gowen S., Ed.; Chapman and Hall: London, **1995;** pp 147–178.

Israeli, Y.; Reuveni, O.; Lahav, E. Qualitative Aspects of Somaclonal Variations in Banana Propagated by in vitro Techniques. *Sci. Hortic.* **1991,** *48*, 71–88.

James, A. C.; Peraza-Echeverria, S.; Herrera-Valencia, V. A.; Martinez, O. Application of the Amplified Fragment Length Polymorphism (AFLP) and the Methylation-Sensitive Amplification Polymorphism (MSAP) Techniques for the Detection of DNA Polymorphism and Changes in DNA Methylation in Micropropagated Bananas. In *Banana Improvement: Cellular, Molecular Biology, and Induced Mutations,* Jain S. M., Swennen R., Eds.; Science Publishers, Inc.: Enfield, **2004;** pp 287–306.

Jain, S. M. Tissue Culture-Derived Variation in Crop Improvement. *Euphytica* **2000,** *118*, 153-166.

Jin, S.; Mushke, R.; Zhu, H.; Tu, L.; Lin, Z.; Zhang, Y.; Zhang, X. Detection of Somaclonal Variation of Cotton (*Gossypium Hirsutum*) using Cytogenetics, Flow Cytometry and Molecular Markers. *Plant Cell Rep.* **2008,** *27*, 1303–1316.

Kaeppler, S. M.; Kaeppler, H. F.; Rhee, Y. Epigenetic Aspects of Somaclonal Variation in Plants. *Plant Mol. Biol.* **2000,** *43*, 179–188.

Kaeppler, S.; Phillips, R. DNA Methylation and Tissue Culture Induced Variation in Plants. *In Vitro Cell Dev. Biol. Plant* **1993,** *29*, 125–130.

Karp, A. Can Genetic Instability be controlled in Plant Tissue Cultures? *Nwsl. Intl. Assn. Plant Tissue Cult.* **1989,** *58*, 2–11.

Karp, A. On the Current Understanding of Somaclonal Variation. In *Oxford Surveys of Plant Molecular and Cell Biology* (Vol. 7); Miflin B. J., Ed.; Oxford University Press: Oxford, **1991;** pp 1–58.

Karp, A. Origins, Causes and uses of Variation in Plant Tissue Cultures. In *Plant Cell and Tissue Culture*; Vasil I. K., Thorpe T. A., Eds.; Kluwer Academic Publishers: Dordrecht, **1994;** pp 139–152.

Karp, A.; Seberg, O. L. E.; Buiatti, M. Molecular Techniques in the Assessment of Botanical Diversity. *Ann. Bot.* **1996,** *78*, 143–149.

Kawiak, A.; Lojkowska, E. Application of RAPD in the Determination of Genetic Fidelity in Micropropagated *Drosera* Plantlets. *In Vitro Cell Dev. Biol. Plant* **2004,** *40*, 592–595.

Khlestkina, E.; Roder, M.; Pshenichnikova, T.; Borner, A. Functional Diversity at the RC (Red Coleoptile) Gene in Bread Wheat. *Mol. Breed.* **2010,** *25*, 125–132.

Kole, P.; Chawla, H. Variation of *Helminthosporium* Resistance and Biochemical and Cytological Characteristics in Somaclonal Generations of Barley. *Biol. Plant.* **1993,** *35*, 81–86.

Krikorian, A. D.; Irizarry, H.; Cronauer-Mitra, S. S.; Rivera, E. Clonal Fidelity and Variation in Plantain (*Musa* AAB) Regenerated from Vegetative Stem and Floral Axis Tips in vitro. *Ann. Bot.* **1993,** *71*, 519–535.

Kunitake, H.; Koreeda, K.; Mii, M. Morphological and Cytological Characteristics of Protoplast-Derived Plants of Statice (*Limonium perezii* Hubbard). *Sci. Hortic.* **1995,** *60*, 305–312.

Larkin, P.; Skowcroft, W. Somaclonal Variation-A Novel Source of Variability from Cell Cultures for Plant Improvement. *Theor. Appl. Genet.* **1981,** *60*, 197-214.

Larkin, P.J. Introduction. In *Somaclonal variation and Induced Mutations in Crop Improvement*; Jain S. M., Brar D. S., Ahloowalia B. S., Eds.; Kluwer Academic Publishers: Dordrecht, **1998,** pp 3–13.

Lee, M.; Phillips, R. L. The Chromosomal Basis of Somaclonal Variation. *Annu. Rev. Plant Physiol. Plant Mol. Biol.* **1988,** *39*, 413–437.

Levinson, G.; Gutman, G. A. Slipped-Strand Mispairing: A Major Mechanism for DNA Sequence Evolution. *Mol. Biol. Evol.* **1987,** *4*, 203–221.

Li, Y. C.; Korol, A. B.; Fahima, T.; Beiles, A.; Nevo, E. Microsatellites: Genomic Distribution, Putative Functions and Mutational Mechanisms: A Review. *Mol. Ecol.* **2002,** *11*, 2453–2465.

Litt, M.; Luty, J. A. A Hypervariable Microsatellite Revealed by in vitro Amplification of a Dinucleotide Repeat within the Cardiac Muscle Actin Gene. *Am. J. Hum. Genet.* **1989,** *44*, 397–401.

LoSchiavo, F.; Pitto, L.; Giuliano, G.; Torti, G.; Nuti-Ronchi, V.; Marazziti, D.; Vergara, R.; Orselli, S.; Terzi, M. DNA Methylation of Embryogenic Carrot Cell Cultures and its Variations as Caused by Mutation, Differentiation, Hormones and Hypomethylating Drugs. *Theor. Appl. Genet.* **1989,** *77*, 325–331.

Mandal, A.; Maiti, A.; Chowdhury, B.; Elanchezhian, R. Isoenzyme Markers in Varietal Identification of Banana. *In Vitro Cell Dev. Biol. Plant* **2001,** *37*, 599–604.

Martin, K.; Pachathundikandi, S.; Zhang, C.; Slater, A.; Madassery, J. RAPD Analysis of a Variant of Banana (*Musa* sp.) cv. Grande Naine and its Propagation via Shoot Tip Culture. *In Vitro Cell Dev. Biol. Plant* **2006,** *42*, 188–192.

Matheka, J. M.; Magiri, E.; Rasha, A. O.; Machuka, J. In vitro Selection and Characterization of Drought Tolerant Somaclones of Tropical Maize (*Zea mays* L.). *Biotechnol.* **2008,** *7*, 641–650.

McClintock, B. The Origin and Behavior of Mutable Loci in Maize. *Proc. Natl. Acad. Sci.* U.S.A. **1950,** *36*, 344–355.

McClintock, B. The Significance of Responses of the Genome to Challenge. *Science* **1984,** *226(4676)*, 792-801.

Mohanty, S.; Panda, M.; Subudhi, E.; Nayak, S. Plant Regeneration from Callus Culture of *Curcuma aromatica* and in vitro Detection of Somaclonal Variation through Cytophotometric Analysis. *Biol Plant.* **2008,** *52*, 783–786.

Mujib, A. Colchicine Induced Morphological Variants in Pineapple. *Plant Tissue Cult. Biotechnol.* **2005,** *15*, 127–133.

Mujib, A.; Banerjee, S.; Dev Ghosh, P. Callus Induction, Somatic Embryogenesis and Chromosomal Instability in Tissue Culture Raised Hippeastrum (*Hippeastrum hybridum* cv. United Nations). *Propag. Ornam. Plants* **2007,** *7*, 169–174.

Nehra, N. S.; Kartha, K. K.; Stushnott, C.; Giles, K. L. The Influence of Plant Growth Regulator Concentrations and Callus Age on Somaclonal Variation in Callus Culture Regenerants of Strawberry. *Plant Cell Tissue Org. Cult.* **1992,** *29*, 257–268.

Oono, K. Putative Homozygous Mutations in Regenerated Plants of Rice. *Mol. Gen. Genet.* **1985,** *198*, 377–384.

Palombi, M. A.; Lombardo, B.; Caboni, E. In vitro Regeneration of Wild Pear (*Pyrus pyraster* Burgsd) Clones Tolerant to Fe-Chlorosis and Somaclonal Variation Analysis by RAPD Markers. *Plant Cell Rep.* **2007,** *26*, 489–496.

Pathak, A.; Dwivedi, M.; Laddha, N. C.; Begum, R.; Joshi, A. Detection of Somaclonal Variants using RAPD Marker in *Bacopa monnieri* and *Tylophora indica*. *J. Agri. Tech.* **2013,** *9(5)*, 1253–1260.

Peschke, V. M.; Phillips, R. L. Genetic Implications of Somaclonal Variation in Plants. *Adv. Genet.* **1992,** *30*, 41–75.

Peschke, V. M.; Phillips, R. L.; Gengenbach, B. G. Discovery of Transposable Element Activity among Progeny of Tissue Culture Derived Maize Plants. *Science* **1987,** *238*, 804–807.

Pierik, R. L. M. *In Vitro Culture of Higher Plants*; Kluwer Academic Publishers: Dordrecht, **1987,** pp 231–238.

Pietsch, G.; Anderson, N. Epigenetic Variation in Tissue Cultured *Gaura lindheimeri*. *Plant Cell Tissue Org. Cult.* **2007,** *89*, 91–103.

Piola, F.; Rohr, R.; Heizmann, P. Rapid Detection of Genetic Variation within and among in vitro Propagated Cedar (*Cedrus libani* Loudon) Clones. *Plant Sci.* **1999,** 141, 159–163.

Popescu, A. N.; Isac, V. S.; Coman M. S. R. Somaclonal Variation in Plants Regenerated by Organogenesis from Callus Cultures of Strawberry (*Fragaria × ananassa*). *Acta Hortic.* **1997,** *439*, 89–96.

Radhakrishnan, R.; Ranjitha Kumari, B. D. Morphological and Agronomic Evaluation of Tissue Culture Derived Indian Soybean Plants. *Acta Agric. Slov.* **2008,** *91*, 391–396.

Rahman, M. H.; Rajora, O. P. Microsatellite DNA Somaclonal Variation in Micropropagated Trembling Aspen (*Populus tremuloides*). *Plant Cell Rep.* **2001,** *20*, 531–536.

Raimondi, J. P.; Masuelli, R. W.; Camadro, E. L. Assessment of Somaclonal Variation in Asparagus by RAPD Fingerprinting and Cytogenetic Analyses. *Sci. Hortic.* **2001,** *90*, 19–29.

Rani, V.; Raina, S. Genetic Fidelity of Organized Meristem Derived Micropropagated Plants: A Critical Reappraisal. *In Vitro Cell Dev. Biol. Plant* **2000.** *36*, 319–330.

Rietveld, R. C.; Bressan, R. A.; Hasegawa, P. M. Somaclonal Variation in Tuber Disc-Derived Populations of Potato. II. Differential Effect of Genotype. *Theor. Appl. Genet.* **1993,** *87*, 305–313.

Roels, S.; Escalona, M.; Cejas, I.; Noceda, C.; Rodriguez, R.; Canal, M. J.; Sandoval, J.; Debergh, P. Optimization of Plantain (*Musa* AAB) Micropropagation by Temporary Immersion System. *Plant Cell Tissue Organ Cult.* **2005,** *82*, 57–66.

Sahijram, L.; Soneji, J.; Bollamma, K. Analyzing Somaclonal Variation in Micropropagated Bananas (*Musa* spp.). *In Vitro Cell Dev. Biol. Plant* **2003,** *39*, 551–556.

Sanchez-Teyer, L. F.; Quiroz-Figueroa, F.; Loyola-Vargas, V.; Infante, D. Culture-Induced Variation in Plants of *Coffea arabica* cv. Caturra rojo, Regenerated by Direct and Indirect Somatic Embryogenesis. *Mol. Biotechnol.* **2003,** *23*, 107–115.

Sandoval, J.; Kerbellec, F.; Cote, F.; Doumas, P. Distribution of Endogenous Gibberellins in Dwarf and Giant Off-Types Banana (*Musa* AAA cv. 'Grand nain') Plants from in vitro Propagation. *Plant Growth Regul.* **1995,** *17*, 219–224.

Saradhi, P. P.; Alia. Production and Selection of Somaclonal Variants of *Leucaena leucocephala* with High Carbon Dioxide Assimilating Potential. *Energ. Convers. Manage.* **1995,** *36*, 759–762.

Sengar, A.; Thind, K.; Kumar, B.; Pallavi, M.; Gosal, S. In vitro Selection at Cellular Level for Red Rot Resistance in Sugarcane (*Saccharum* sp.). *Plant Growth Regul.* **2009,** *58*, 201–209.

Shah, S. H.; Wainwright, S. J.; Merret, M. J. Regeneration and Somaclonal Variation in *Medicago sativa* and *Medicago media*. *Pak. J. Biol. Sci.* **2003,** *6*, 816–820.

Sharma, S.; Bryan, G.; Winfield, M.; Millam, S. Stability of Potato (*Solanum tuberosum* L.) Plants Regenerated via Somatic Embryos, Axillary Bud Proliferated Shoots, Microtubers and True Potato Seeds: A Comparative Phenotypic, Cytogenetic and Molecular Assessment. *Planta* **2007,** *226*, 1449–1458.

Sheidai, M.; Aminpoor, H.; Noormohammadi, Z.; Farahani, F. RAPD Analysis of Somaclonal Variation in Banana (*Musa acuminate* L.) cultivar Valery. *Acta Biol. Szeged.* **2008,** *52*, 307–311.

Shepard, J. F.; Bidney, D.; Shahin, E. Potato Protoplasts in Crop Improvement. *Science* **1980,** *208*, 17–24.

Shepherd, K.; Dos Santos, J. A. Mitotic Instability in Banana Varieties. I. Plants from Callus and Shoot Tip Cultures. *Fruits* **1996,** *51*, 5–11.

Sibi, M. La notion de programme genétique chez les vegetaux superieurs II. Aspect experimental: Obtention de variants par culture de tissus in vitro sur *Lactuca sativa* L. Apparition de vigueur chez les croisements. *Ann. Amelior. Plantes* **1976,** *26*, 523-547.

Singh, A.; Lai, M.; Singh, M.; Lai, K.; Singh, S. Variations for Red Rot Resistance in Somaclones of Sugarcane. *Sugar Tech.* **2000,** *2*, 56–58.

Singh, G.; Sandhu, S.; Meeta, M.; Singh, K.; Gill, R.; Gosal, S. In vitro Induction and Characterization of Somaclonal Variation for Red Rot and other Agronomic Traits in Sugarcane. *Euphytica* **2008,** *160*, 35–47.

Sivanesan, I. Shoot Regeneration and Somaclonal Variation from Leaf Callus Cultures of *Plumbago zeylanica* Linn. *Asian J. Plant Sci.* **2007,** *6*, 83–86.

Skirvin, R. M.; McPheeters, K. D.; Norton, M. Sources and Frequency of Somaclonal Variation. *Hort. Sci.* **1994,** *29*, 1232–1237.

Skirvinm, R. M.; Janick, J. Tissue Culture-Induced Variation in Scented *Pelargonium* spp. *J. Amer. Soc. Hort. Sci.* **1976,** *101*, 281–290.

Squirrell, J.; Hollingsworth, P. M.; Woodhead, M.; Russell, J.; Lowe, A. J.; Gibby, M.; Powell, W. How Much Effort is required to Isolate Nuclear Microsatellites from Plants? *Mol. Ecol.* **2003,** *12*, 1339–1348.

Srivastava, S.; Gupta, P. S.; Srivastava, B. L. Genetic Relationship and Clustering of Some Sugarcane Genotypes Based on Esterase, Peroxidase and Amylase Isozyme Polymorphism. *Cytologia* **2005,** *70*, 355–363.

Stover, R. H. Somaclonal Variation in Grande Naine and Saba Bananas in the Nursery and in the Field. In *ACIAR Proceeding No. 21,* Persley G.J., De Langhe E. Eds.; Canberra, **1987**.

Sultana, R.; Tahira, F.; Tayyab, H.; Khurram, B.; Shiekh, R. RAPD Characterization of Somaclonal Variation in Indica Basmati Rice. *Pak. J. Bot.* **2005,** *37*, 249–262.

Swartz, H. J. Post Culture Behaviour, Genetic and Epigenetic Effects and Related Problems. In *Micropropagation: Technology and Application*; Debergh P. C., Zimmerman R. H., Eds.; Kluwer Academic Publishers: Dodrecht, **1991;** pp 95–122.

Thomas, J.; Raj Kumar, R.; Mandal, A. K. A. Metabolite Profiling and Characterization of Somaclonal Variants in Tea (*Camellia* spp.) for Identifying Productive and Quality Accession. *Phytochem.* **2006,** *67*, 1136–1142.

Venkatachalam, L.; Sreedhar, R. V.; Bhagyalakshmi, N. Molecular Analysis of Genetic Stability in Long-Term Micropropagated shoots of Bananas using RAPD And ISSR Markers. *Electron. J. Biotechnol.* **2007,** *10*, 1–8.

Vidal, M. D. C.; De Garcia, E. Analysis of a *Musa* spp. Somaclonal Variant Resistant to Yellow Sigatoka. *Plant Mol. Biol. Rep.* **2000,** *18*, 23–31.

Wang, Y.; Wang, F.; Zhai, H.; Liu, Q. Production of a Useful Mutant by Chronic Irradiation in Sweetpotato. *Sci. Hortic.* **2007,** *111*, 173–178.

Weising, K.; Nybom, H.; Wolff, K.; Kahl, G. *DNA Fingerprinting in Plants: Principles, Methods, and Applications*; CRC Press: New York, **2005,** p 472.

Williams, J. G. K.; Kubelik, A. R.; Livak, K. J.; Rafalski, J. A.; Tingey, S. V. DNA Polymorphisms Amplified by Arbitrary Primers are Useful as Genetic Markers. *Nucleic Acids Res.* **1990,** *18*, 6531–6535.

Yadav, P.; Suprasanna, P.; Gopalrao, K.; Anant, B. Molecular Profiling using RAPD Technique of Salt and Drought Tolerant Regenerants of Sugarcane. *Sugar Tech.* **2006,** *8*, 63–68.

Zhang, M.; Wang, H.; Dong, Z.; Qi, B.; Xu, K.; Liu, B. Tissue Culture Induced Variation at Simple Sequence Repeats in Sorghum (*Sorghum bicolor* L.) is Genotype-Dependent and Associated with Down-Regulated Expression of a Mismatch Repair Gene, MLH3. *Plant Cell Rep.* **2010,** *29*, 51–59.

Zhong, Z.; Smith, H. G.; Thomas, T. H. In vitro Culture of Petioles and Intact Leaves of Sugar Beet (*Beta vulgaris*). Plant Growth Regul. **1993,** *12*, 59–66.

Zietkiewicz, E.; Rafalski, A.; Labuda, D. Genome Fingerprinting by Simple Sequence Repeat (SSR)-Anchored Polymerase Chain Reaction Amplification. *Genomics* **1994,** *20*, 176–183.

CHAPTER 10

SOMACLONAL VARIATION: A TISSUE CULTURE APPROACH TO CROP IMPROVEMENT

KUMARI RAJANI[1], RAVI RANJAN KUMAR[2*], TUSHAR RANJAN[3], GANESH PATIL[4], ANAND KUMAR[5] and JITESH KUMAR[2]

[1]*Department of Seed Science and Technology, Bihar Agricultural University, Sabour, Bhagalpur, Bihar, India*

[2]*Department of Molecular Biology and Genetic Engineering, Bihar Agricultural University, Sabour, Bhagalpur, Bihar, India*

[3]*Department of Basic Science and Humanities Genetics, Bihar Agricultural University, Sabour, Bhagalpur, Bihar, India*

[4]*Vidya Pratishthan's College of Agricultural Biotechnology, Vidya Nagari, Baramati, India*

[5]*Department of Plant Breeding and Genetics, Bihar Agricultural University, Sabour, Bhagalpur, Bihar, India*

[*]*Corresponding author. E-mail: ravi1709@gmail.com*

CONTENTS

ABSTRACT

Plant breeders have successfully recombined the desired genes from cultivated crop germplasm and related wild species through hybridization program, and have been able to develop new cultivars with superior agronomic traits, such as high yield, biotic and abiotic resistance etc. Even though conventional breeding is the main traditional method, so far, these methods have managed to feed the world's ever-growing population. Tissue culture provides an excellent opportunity to create the range of genetic variability in crops that can be used in breeding programs. In the past, tissue culture cycle was offered as a method for cloning a specific genotype and today this is a common method for propagating plants with commercial importance. Somaclonal variation has widely been used for improvement of crops for superior agronomic traits. Somaclonal variation has been most successful in crops with limited genetic systems and/or narrow genetic bases. The somaclonal variation has not only been used in field crops but also in important horticultural and ornamental crops for genetic improvement. This chapter elaborates the use of this technique and its scope of genetic improvement in important crop species in a very simple and holistic manner.

10.1 INTRODUCTION

Genetic variation is an essential component of any conventional crop improvement program. The typical crop improvement cycle takes 10–15 years to complete and includes germplasm manipulations, genotype selection and stabilization, variety testing, variety increase, proprietary protection, and crop production stages. Plant tissue culture is an enabling technology from which many novel tools have been developed to assist plant breeders (Krishna et al., 2016). Somaclonal variation is akin to variations induced with chemical and physical mutagenic agents (Mohan-Jain, 2001) and offers an excellent opportunity to uncover natural variability for its potential exploitation in crop improvement.

Somaclonal variation is the variation seen in plants that have been produced by plant tissue culture. Genetic variation arising from tissue culture of plants has been termed somaclonal variation (Larkin and Scowcroft, 1981). Somaclonal variation is defined as genetic and phenotypic variation among clonally propagated plants of a single donor clone (Olhoft and Phillips, 1999). Variation has been observed in a wide range of species from plants derived from a variety of explants, using different cultural methods

(D'Arnato, 1977). Chromosomal rearrangements are an important source of this variation. The term "somaclonal variation" is a phenomenon of broad taxonomic occurrence, reported for species of different ploidy levels, and for outcrossing and inbreeding, vegetatively and seed propagated, and cultivated and noncultivated plants (Miguel and Marum, 2011). Characters affected include both qualitative and quantitative traits. Somaclonal variation is not restricted to, but is particularly common in, plants regenerated from callus. The variations can be genotypic or phenotypic, which in the latter case can be either genetic or epigenetic in origin. Typical genetic alterations are: changes in chromosome numbers (polyploidy and aneuploidy), chromosome structure (translocations, deletions, insertions, and duplications), and DNA sequence (base mutations). A typical epigenetics-related event would be gene methylation (Jaligot et al., 2011).

Somaclonal variation results from both preexisting genetic variation within the explants and the variation induced during the tissue culture phase (Evans et al., 1984). There are two types of somaclonal variation: heritable (genetic) and epigenetic. Heritable variation is stable through the sexual cycle or repeated asexual propagation; epigenetic variation may be unstable even when asexually propagated. Epigenetic variation is also known as developmental variation, and includes persistent changes in phenotype that involve the expression of particular genes (Hartmann and Kester, 1983). The best known example of epigenetic variation is the loss of auxin, cytokinin, or vitamin requirements by callus. Other epigenetic changes include extreme vigor ex vitro associated with either the reversion to juvenility (Swartz et al, 1981) or virus elimination (Abo El-Nil and Hilderbradt, 1971). Transient dwarfism is probably epigenetic also, and may be due to a carryover effect of growth regulators from the tissue culture medium (Nwauzoma et al., 2013).

Based on genetic changes, there are four types of somaclonal variation:

1. Point mutations (e.g., Adh mutants in wheat)
2. Cytoplasmic (maternal inheritance)
3. Gene amplification (e.g., increase gene copy no.)
4. Activation of transposable element

On the basis of genetic changes, the following types of somaclonal variants are available:

1. Cytogenetic (changes to genome structure)
2. Aneuploidy: gain or loss of one or more chromosomes
3. Polyploidy: gain or loss of an entire genome

4. Translocation: arms of chromosomes switched
5. Inversion: piece of chromosome inverted

Epigenetic

a. Change in phenotype that is not stable during sexual propagation
b. May or may not be stable during asexual propagation
c. Usually undesirable in a breeding program, not always undesirable in propagation
d. Habituation (most studied epigenetic change)

The continual endeavors of man to exploit the natural variation present in base population of crop plant to obtain improved varieties were achieved initially by multiplication of best available material. This was followed by selective cropping and cross hybridization to obtain hybrid crop which eventually led to the perpetuation of desirable germplasm. However, this also led to inbreeding depression. Breeding for a "plant ideotype" is a postulation in agriculture. Plant improvement is a multidisciplinary activity concerned with the optimization of genetic attributes within the constraints of the environment, and of environmental factors within the constraints of the genetic material (Byth et al., 1980). Conventional breeding exploits the natural variation existing in plant populations to recover elite crops. However, the available genetic variability in gene pools is one of the limits to crop improvement. The best available germplasm may be subjected to a tissue culture cycle with or without selection pressure and regenerants may be selected for superiority for one or more traits while retaining all the original characters. Such incremental improvement in desirable traits could therefore lead to the formation of new alleles spontaneously generated in vitro. Tissue culture methods leading to somaclonal variation could be capitalized upon to accelerate progress in conventional breeding.

10.2 INDUCTION AND MECHANISM OF SOMACLONAL VARIATION

Various types of somaclonal mutations including point mutations, gene duplication, chromosome rearrangement, chromosome number variations have been reported. These changes might be naturally accumulated in somatic cells of plant and tissue culture provides opportunity for these mutations so that they emerge in plants derived from tissue culture (Larkin, 2004).

The mechanism of somaclonal variation includes extensive genetic drift, and epigenetic factors are also involved in complicating it. Somaclonal variation depends on plant growth regulators, variability of variety, age of variety in culture, ploidy level, explants source, genotype, and other culture conditions (Karp, 1995; Rasheed et al., 2005). The presence of some chemical material such as 2, 4-D also causes to increase this variation ratio (Rasheed et al., 2005). The experiments performed with numerous regenerated plants derived from protoplasts separated from a single leaf have demonstrated that somaclonal variation occurs in some cases during the culture cycle (Larkin, 2004). Somatic mutation often does not transfer to the next generation and the primary regenerated plants become final product while the stable gametic variation (meiotic) is transferred to progeny. So, assaying varieties from tissue culture by using molecular markers to utilize this variety in plant breeding are more important (Rasheed et al., 2005).

10.2.1 SOURCES OF VARIATIONS

Although somaclonal variation has been studied extensively, the mechanisms by which it occurs remain largely either unknown or at the level of theoretical speculation in perennial fruit crops (Skirvin et al., 1993; Skirvin et al., 1994). Tissue culture is an efficient method of clonal propagation; however, the resulting regenerants often have a number of somaclonal variations (Larkin and Scowcroft, 1981). These somaclonal variations are mainly caused by newly generated mutations arising from tissue culture process (Sato et al., 2011b).

The triggers of mutations in tissue culture had been attributed to numerous stress factors, including wounding, exposure to sterilants during sterilization, tissue being incomplete (protoplasts as an extreme example), imbalances of media components such as high concentration of plant growth regulators (auxin and cytokinins), sugar from the nutrient medium as a replacement of photosynthesis in the leaves, lighting conditions, the disturbed relationship between high humidity and transpiration (Smulders and de Klerk, 2011). Much of the variability expressed in micropropagated plants may be the result of, or related to, oxidative stress damage inflicted upon plant tissues during in vitro culture (Tanurdzic et al., 2008; Nivas and Dsouza, 2014). Oxidative stress results in elevated levels of pro-oxidants or reactive oxygen species (ROS) such as superoxide, hydrogen peroxide, hydroxyl, peroxyl, and alkoxyl radicals. These ROS may involve in altered hyper- and hypomethylation of DNA (Wacksman, 1997), changes in chromosome number from polyploidy to aneuploidy, chromosome strand breakage, chromosome rearrangements, and DNA base deletions and substitutions (Czene and Harms-Ringdahl, 1995), which in turn may lead to mutations in plant cells in vitro. Somaclonal variation shows a similar spectrum of genetic variation to induced mutation as both of them result in qualitatively analogous gamut of DNA changes (Cassells et al., 1998). Different factors affect the frequency of development of somaclones under *in vitro* conditions.

10.2.1.1 EXPLANT SOURCE

Differences in both the frequency and nature of somaclonal variation may occur when regeneration is achieved from different tissue sources. Highly differentiated tissues such as roots, leaves, and stems generally produce more variations than explants with preexisting meristems, such as axillary buds and shoot tips (Krishna et al., 2016). Somaclonal variation can also arise from somatic mutations already present in the donor plant, i.e., presence of chimera in explants.

10.2.1.2 MODE OF REGENERATION

Both culture initiation and subsequent subculture expose explants to oxidative stress, which may result in mutations. It seems evident that "extreme" procedures such as protoplast culture and also callus formation impose stress (Smulders and de Klerk, 2011). Investigations indicate more chromosome variability in the callus phase than in adventitious shoots (Saravanan et al., 2011), indicating a loss of competence in the more seriously disturbed genomes.

10.2.1.3 EFFECT OF LENGTH OF CULTURE PERIOD AND NUMBER OF SUBCULTURE CYCLE

The longer a culture is maintained in vitro, the greater the somaclonal variation is (Sun et al., 2013). The rapid multiplication of a tissue, during micropropagation, may affect its genetic stability. Khan et al. (2011) reported that after the eighth subculture, the number of somaclonal variants increased with a simultaneous decrease in the multiplication rate of propagules in banana.

10.2.1.4 CULTURE ENVIRONMENT

External factors like growth regulators, temperature, light, osmolarity, and agitation rate of the culture medium are known to influence the cell cycle in vivo in plants, considerably, which indicates that inadequate control of cell cycle in vitro is one of the causes of somaclonal variation. Normal cell cycle controls, which prevent cell division before the completion of DNA replication, are presumed to be disrupted by tissue culture, resulting in chromosomal breakage (Nwauzoma and Jaja 2013).

10.2.2 MECHANISM OF SOMACLONAL VARIATION

The somaclonal variation may be attributed to (1) preexisting variation in the somatic cells of the explant or (2) variation generated during tissue culture (epigenetic). Often both factors may contribute.

The original ploidy level of the plant or plant organ from which the explant is taken may play an important role in somaclonal variation. Meristematic explants, such as apical meristem derived from either shoot apex or axillary bud, have a lesser degree of genetic variability as compared with plants regenerated from nonmeristematic explants which generally produce genetic variability (Krishna et al., 2016). Cells of meristematic explant divide by normal mitosis and cells are maintained at a uniform diploid level. However, the cells in nonmeristematic explants are derivation of the meristematic part of the plant and during their subsequent differentiation, do not divide by normal mitosis, but undergo DNA duplication and endoreduplication. Endoreduplication leads to the formation of chromosomes with four chromatids, chromosomes with eight chromatids, and polytene condition (Cubas et al., 1999).

When the cells of various genomic constitutions of the initial explants are induced to divide in cultures, the cells may exhibit changes in chromosome number such as aneuploids and polyploids. Organogenesis and or embryogenesis occur mostly from diploid cells.

Therefore, preexisting variation in explant tissue always rules out the somaclonal variation in the culture. The presence of several chromosomal aberrations such as reciprocal translocation, deletion, inversion, chromosome reunion, multicentric, acentric fragments, heteromorphic pairing, etc. were found among the somaclones of barley, ryegrass, garlic, and oat. Besides these changes, there are examples of phenotypic variation which can be observed in plants regenerated from cultured cells or protoplasts where no apparent chromosomal abnormalities are seen (Bairu et al., 2011). The schematic diagram of mechanism of somaclonal variation is shown in Figure 10.1.

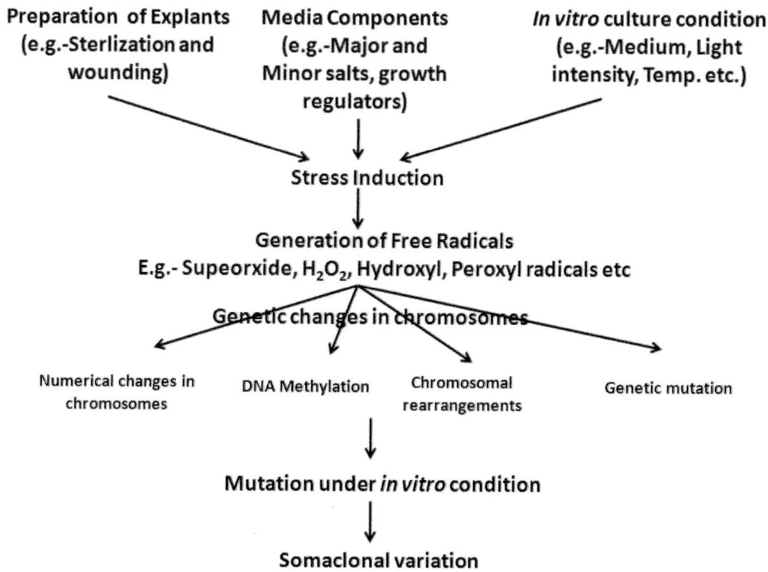

FIGURE 10.1 Mechanism of somaclonal variation in micropropagated plants as a result of oxidative burst upon in vitro culture.

10.2.3 REDUCING SOMACLONAL VARIATION

Different steps can be used to reduce somaclonal variation. It is well known that increasing numbers of subculture increases the likelihood of somaclonal variation, so the number of subcultures in micropropagation protocols should be kept to a minimum. Regular reinitiation of clones from new explants might reduce variability over time. Another way of reducing somaclonal variation is to avoid 2,4-D in the culture medium, as this hormone is known to introduce variation. Vitrification, commonly referred to as hyperhydricity in the tissue culture world, may be a problem in some species. Hyperhydricity is a physiological malformation that results in excessive hydration, low lignification, impaired stomatal function, and reduced mechanical strength of tissue culture-generated plants. In case of forest trees, mature elite trees can be identified and rapidly cloned by this technique. High production cost has limited the application of this technique to more valuable ornamental crops and some fruit trees.

10.3 GENETIC AND MOLECULAR BASIS OF SOMACLONAL VARIATION

Genetic basis of somaclonal variation was not determined in early reports of this phenomenon; more recent work has included genetic analysis of the variation. The origin of somaclonal variation is both the inherent variation in tissue which is placed in culture and changes which are associated with the passage through tissue culture. Variation in plant phenotype is determined by genetic and epigenetic factors. Phenotypic and DNA variation among putative plant clones is termed somaclonal variation (Kaeppler et al., 2000).

Several mechanisms for somaclonal variation have been proposed, which include changes in chromosome number (Mujib et al., 2007; Leva et al., 2012), point mutations (D'Amato, 1985; Ngezahayo et al., 2007), somatic crossing over and sister chromatid exchange (Duncan 1997; Bairu et al., 2011), chromosome breakage and rearrangement (Czene and Harms-Ringdahl, 1995; Alvarez et al., 2010), somatic gene rearrangement, DNA amplification (Karp, 1995; Tiwari et al., 2013), changes in organelle DNA (Cassells and Curry, 2001; Bartoszewski et al., 2007), DNA methylation (Guo et al., 2007; Linacero et al., 2011), epigenetic variation (Kaeppler et al., 2000; Guo et al., 2006; Smulders and de Klerk, 2011), histone modifications and RNA interference (Miguel and Marum, 2011), segregation of preexisting chimeral tissue (Brar and Jain, 1998; Va´zquez, 2001; Ravindra et al., 2012; Nwauzoma and Jaja, 2013), and insertion or excision of transposable elements (Gupta, 1998; Sato et al., 2011b). In particular, transposable elements are one of the causes of genetic rearrangements in in vitro culture (Hirochika et al., 1996; Sato et al., 2011a).

Somaclonal variation caused by the process of tissue culture is also called tissue culture-induced variation to more specifically define the inducing environment. Somaclonal variation can be manifested as either somatically or meiotically stable events. Somatically stable variation includes phenotypes such as habituation of cultures and physiologically induced variation observed among primary regenerants. This type of variation is often not transmitted to subsequent generations and is of most impact in situations where the primary regenerant is the end product such as the amplification of ornamental plants or trees for direct use. Meiotically heritable variation also occurs and is important in situations where the end product of the tissue culture is propagated and sold as seed. Mechanisms producing both somatically and meiotically heritable variations also contribute to the decline in vigor and regenerability of cultures over time. The loss of culture health with time is a major detriment to the efficiency of transgenic plant production and much effort has been devoted to avoiding this problem. Epigenetic control of gene expression can be defined as a somatically or meiotically heritable alteration in gene expression that is potentially reversible and is not due to sequence modification. Epigenetic aspects of somaclonal variation would therefore involve mechanisms of gene silencing or gene activation that were not due to chromosomal aberrations or sequence change. These changes might be unstable or reversible somatically or through meiosis, although certain epigenetic systems outside of tissue culture are quite stable for many generations (Patterson et al., 1993; Cubas et al., 1999). Therefore, epigenetic changes induced by tissue culture could be manifested as the activation of quiescent loci or as epimutation of loci sensitive to chromatin-level control of expression.

10.4 APPLICATION OF SOMACLONAL VARIATION FOR CROP IMPROVEMENT

Tissue culture provides opportunity to show the range of genetic variability in plants that can be used in plant breeding programs. In the past, tissue culture cycle was offered as a method for cloning a specific genotype and today this is a common method for propagating plants with commercial importance (Larkin and Scowcroft, 1981; Rasheed, 2005). Previously, it was expected that all of regenerated plants from cell or tissue culture process have the same genetic structure with original mother plant, therefore, it was accepted as a rule that plantlet derived from tissue culture should exactly resemble parental plant (Larkin and Scowcroft, 1981). But phenotypic variability was observed with high frequency among regenerated plants (Rasheed et al., 2005). Various references have introduced somaclonal variation as a novel and useful source in plant breeding (Kang-le, 1989; Zong-xiu, 1983). The successful use of somaclonal variation much more depends on its genetic stability in the next generations (Mohan-Jain, 2001). Somaclonal variation leads to the creation of additional genetic variability. Characteristics for which somaclonal mutants can be enriched during in vitro culture includes resistance to disease pathotoxins, herbicides, and tolerance to environmental or chemical stress, as well as for increased production of secondary metabolites (Jaligot et al, 2011). The somaclonal variation is utilized to develop several horticultural and cereal crops in order to develop improved crop species with novel traits. A few examples of somaclonal variation in different crop species are presented in Table 10.1.

TABLE 10.1 *In vitro* Selection of Desirable Traits in Plants through Somaclonal Variation.

Sl. No.	Plant	Trait improved	Reference
1.	Apple (*Malus domestica* Borkh.)	Resistance to *Erwinia amylovora*	Chevreau et al. (1998)
2.	Brinjal (*Solanum melongena* L.)	Stress-tolerant somaclone selection	Ferdausi et al. (2009)
3.	Banana (*Musa acuminate* L.)	Semi-dwarf and resistant to Fusarium wilt TC1-229	Tang et al. (2000)
		Var. CUDBT-B1, reduced height and early flowering	Martin et al. (2006)
		Var. Tai-Chiao No. 5, superior horticultural traits and resistance to Fusarium wilt	Lee et al. (2011)
4	Capsicum (*Capsicum annuum* L.)	Yellow fruited var. Bell sweet	Morrison et al. (1989)
5.	Carrot (*Daucus carota* L.)	Resistance to leaf spot (*Alternaria dauci*)	Dugdale et al. (2000)
6.	Garlic (*Allium sativum* L.)	Resistance against the pathogenic fungus *Sclerotium cepivorum*	Zhang et al. (2012
7.	Ginger (*Zingiber officinale* Rosc.)	Tolerant to wilt pathogen (*Fusarium oxysporum* f. sp. *zingiberi* Trujillo)	Bhardwaj et al. (2012)
8.	Pea (*Pisum sativum* L.)	Resistance to *Fusarium solani*	Horacek et al. (2013)
9.	Potato (*Solanum tuberosum* L.)	Somaclones for heat tolerance	Das et al. (2000)
		High-yielding genotype SVP-53	Hoque and Morshad (2014)
		Increased phytonutrient and antioxidant components over cv. "Russet Burbank"	Nassar et al. (2014)
10.	Rice (*Oryza sativa*)	Submerge tolerance	Joshi and Rao (2009)
		Enhanced salinity stress	Kumari et al. (2015)
11.	Strawberry (*Fragaria* sp.)	Resistant to *Verticillium dahlia* Kleb	Zebrowska (2010)

TABLE 10.1 *(Continued)*

Sl. No.	Plant	Trait improved	Reference
		"Serenity," a paler skin-colored, late season, resistant to powdery mildew and Verticillium wilt somaclonal variant of the short-day cv. "Florence"	Whitehouse et al. (2014)
12.	Sweet orange (*Citrus sinensis* (L.) Osb.)	Somaclone of OLL (Orie Lee Late) sweet orange; late maturing; suitable for fresh market or processing, exceptional juice quality and flavor	Grosser et al. (2015)
13.	Turmeric (*Curcuma longa* L.)	High essential oil-yielding somaclones	Kar et al. (2014)
		Turmeric somaclone resistant to *Fusarium oxysporum* f. sp. *zingiberi*	Kuanar et al. (2014)
14.	Wheat (*Triticum aestivum*)	Resistance to *Bipolaris sorokiniana*, *Magnaporthe grisea* or *Xanthomonas campestris* pv. *undulosa*	Mehta and Angra (2000)
		Enhanced salinity tolerance	Benabdelhafid et al. (2015)

10.5 ADVANTAGE AND LIMITATION OF SOMACLONAL VARIATION IN CROP IMPROVEMENT

Spontaneous heritable variation has been known to plant breeders before the science of genetics was established and the art of plant breeding practiced. Occurrence of "sports" spontaneous mutations, "bolters," "off-type," and "freaks" in the vegetatively propagated crop plants has been observed by farmers in sugarcane, potato, banana, and floricultural plants since plants were domesticated.

10.5.1 ADVANTAGES

Somaclonal variation has been most successful in crops with limited genetic systems (e.g., apomicts, vegetative reproducers) and/or narrow genetic bases. In ornamental plants, for instance, the exploitation of in vitro-generated variability has become part of the routine breeding practice of many commercial enterprises.

Somaclonal variations occur in high frequencies.

- Some changes can be novel and may not be achieved by conventional breeding.
- In vitro screening reduces the time for isolation of a somaclone with desirable trait.
- Sometimes new desirable characters may occur which are not available in the germplasm.
- It is cheaper than other genetic engineering methods.

10.5.2 LIMITATIONS

Somaclonal variations can become a part of plant breeding provided they are heritable and genetically stable. Only a limited number of promising varieties so far had been released using somaclonal variations. This is perhaps due to the lack of interaction between plant breeders and tissue culture scientists, and non predictability of somaclones. Further, though the new varieties have been produced by somaclonal variation, improved variants have not been selected in a large number of cases as:

- These variations are not stable after selfing or crossing.

- The variations are unpredictable in nature and uncontrollable.
- Selected cell lines often reduce their regeneration potential.
- Many selected clones show undesirable features like reduced fertility, growth, and even overall performance.

10.5.3 STRATEGIES TO OVERCOME THE CONSTRAINTS

- The breeding objective should be simple and improve one character at a time. If we require more than one trait, stepwise improvement must be possible.
- An easy and efficient screening technique should be needed to select a desired trait in somaclonal variants.
- Molecular markers and in vitro selection techniques for various diseases are very helpful in the identification of valuable variants.
- Comparative study of plants produced through somaclonal variation and conventionally propagated plants involving field trials before cultivation.

KEYWORDS

- **genetic improvement**
- **somaclones**
- **genetic variation**
- **explant**
- **mutation**
- **culture environment**

REFERENCES

Abo El-Nil, M. M.; Hildebrandt, A. C. Differentiation of Virus-symptomless Geranium Plants from Anther Callus. *Plant Dis. Reptr.* **1971,** *55*, 1017–1020.

Bairu, M. W.; Aremu, A. O.; Staden, J. V. Somaclonal Variation in Plants: Causes and Detection Methods. *Plant Growth Regul.* **2011,** *63*, 147–173.

Alvarez, M. E.; Nota, F.; Cambiagno, D. A. Epigenetic Control of Plant Immunity. *Mol. Plant Pathol.* **2010,** *11*, 563–576.

Bartoszewski, G.; Havey, M. J.; Zio΄kowska, A.; D'ugosz, M.; Malepszy, S. The Selection of Mosaic (MSC) Phenotype after Passage of Cucumber (*Cucumis sativus* L.) through Cell Culture—A Method To Obtain Plant Mitochondrial Mutants. *J. Appl. Genet.* **2007,** *48*, 1–9.

Benabdelhafid, Z.; Bouldjadj, R.; Ykhlef, N.; Djekoun, A. Selection for Salinity Tolerance and Molecular Genetic Markers in Durum Wheat (*Triticum durum* Desf.). *Int. J. Advanced Res.* **2015,** *3* (10), 397–406.

Bhardwaj, S. V.; Thakur, T.; Sharma, R.; Sharma, P. In Vitro Selection of Resistant Mutants of Ginger (*Zingiber officinale* Rosc.) against Wilt Pathogen (*Fusarium oxysporum* f. sp. Zingiberi Trujillo). *Plant Dis. Res.* **2012,** *27*, 194–199.

Brar, D. S.; Jain, S. M. Somaclonal Variation: Mechanism and Applications in Crop Improvement. In *Somaclonal Variation and Induced Mutations in Crop Improvement;* Jain, S. M., Brar, D. S., Ahloowalia, B. S., Eds.; Kluwer Academic Publishers: Dordrecht, **1998;** pp 15–37.

Byth, D. E.; Wallis, E. S.; Saxena, K. B. Adaptation and Breeding Strategies for Pigeon Pea. In *ICRISAT, Proceedings of International Workshop on Pigeon Pea*, Patancheru, Andhra Pradesh, India, 1980, Vol. 1, **1980;** pp 450-465.

Cassells, A. C.; Curry, R. F. Oxidative Stress and Physiological, Epigenetic and Genetic Variability in Plant Tissue Culture: Implications for Micropropagators and Genetic Engineers. *Plant Cell Tissue Organ Cult.* **2001,** *64*, 145–157.

Cassells, A. C.; Deadman, M. L.; Brown, C. A.; Griffin, E. Field Resistance to Late Blight *(Pytophtora infectans* (Mont.) De Bary in Potato (*Solanum tuberosum* L.) Somaclones Associated with Instability and Pleiotropic Effects. *Euphytica* **1998,** *57*, 157–167.

Chevreau, E.; Brisset, M. N.; Paulin, J. P.; James, D. J. Fire Blight Resistance and Genetic Trueness-to-Type of Four Somaclonal Variants from the Apple Cultivar Green Sleeves. *Euphytica* **1998,** *104*, 199–205.

Cubas, P.; Vincent, C.; Coen, E. An Epigenetic Mutation Responsible for Natural Variation in Floral Symmetry. *Nature* **1998,** *401*, 157–161.

Czene, M.; Harms-Ringdahl, M. Detection of Single-strand Breaks and Form Amidoprymidine-DNA Glycosylase-sensitive sites in DNA of Cultured Human Fibroblasts. *Mutat. Res.* **1995,** *336*, 235–242.

D'Arnato, F. Cytogenetics of Differentiation in Tissue and Cell Cultures. In *Plant Cell, Tissue and Organ Culture,* Reinert, J., Bajaj, Y. P. S., Eds., Springer: New York, **1977;** pp 362–347.

Das, A.; Gosal, S. S.; Sidhu, J. S.; Dhaliwal, H. S. Induction of Mutations for Heat Tolerance in Potato by Using *In Vitro* Culture and Radiation. *Euphytica* **2000,** *114*, 205–209.

Dugdale, L. J.; Mortimer, A. M.; Isaac, S.; Collin, H. A. Disease Response of Carrot and Carrot Somaclones to *Alternaria dauci*. *Plant Pathol.* **2000,** *49*, 57–67.

Duncan, R. R. Tissue Culture-induced Variation and Crop Improvement. *Adv. Agron.* **1997,** *58*, 201–240.

Evans, D. A.; Sharp, W. R.; Medina-Filho, H. P. Somaclonal and Gametoclonal Variation. *Am. J. Bot.* **1984,** *71* (6), 759–774.

Ferdausi, A.; Nath, U. K.; Das, B. L.; Alam, M. S. *In Vitro* Regeneration System in Brinjal (*Solanum melongena* L.) for Stress Tolerant Somaclone Selection. *J. Bangladesh Agric. Univ.* **2009,** *7* (2), 253–258.

Grosser, J. W.; Gmitter, F. G., Jr.; Dutt, M.; Calovic, M.; Ling, P.; Castle, B. Highlights of the University of Florida, Citrus Research and Education Center's comprehensive Citrus Breeding and Genetics Program. *Acta Hortic.* **2015,** *1065*, 405–413.

Guo, W.; Gong, L.; Ding, Z.; Li, Y.; Li, F.; Zhao, S.; Liu, B. Genomic Instability in Phenotypically Normal Regenerants of Medicinal Plant *Codonopsis lanceolata* Benth. Et Hook. f., as Revealed by ISSR and RAPD Markers. *Plant Cell Rep.* **2006,** *25*, 896–906.

Guo, W.; Wu, R.; Zhang, Y.; Liu, X.; Wang, H.; Gong, L.; Zhang, Z.; Liu, B. Tissue Culture Induced Locus-specific Alteration in DNA Methylation and Its Correlation with Genetic Variation in *Codonopsis lanceolata* Benth. et Hook. f. *Plant Cell Rep.* **2007,** *26*, 1297–1307.

Hartmann, H. T.; Kester, D. E. *Plant Propagation*; Prentice-Hall: Englewood Cliffs, New Jersey, **1983**.

Hoque, M. E.; Morshad, M. N. Somaclonal Variation in Potato (*Solanum tuberosum* L.) Using Chemical Mutagens. *Agriculturists* **2014,** *12* (1), 15–25

Horacek, J.; Svabova, L.; Sarhanova, P.; Lebeda, A. Variability for Resistance to *Fusarium solani* Culture Filtrate and Fusaric Acid among Somaclones in Pea. *Biol. Plant.* **2013,** *57* (1), 133–138.

Jaligot, E.; Adler, S.; Debladis, É.; Beulé, T.; Richaud, F.; Ilbert, P.; Finnegan, E. J.; Rival, A. Epigenetic Imbalance and the Floral Developmental Abnormality of the In Vitro Regenerated Oil Palm *Elaeis guineensis*. *Ann. Bot.* **2011,** *108* (8), 1453–1462.

Joshi, R. K.; Rao, G. J. N. Somaclonal Variation in Submergence Tolerant Rice Cultivars and Induced Diversity Evaluation by PCR Markers. *Int. J. Genet. Mol. Biol.* **2009,** *1* (5), 80–88.

Kaeppler, S. M.; Kaeppler, H. F.; Rhee, Y. Epigenetic Aspects of Somaclonal Variation in Plants. *Plant Mol. Biol.* **2000,** *43*, 179–188.

Kang-Le, Z.; Zong-Ming, Z.; Guo-Liang, W.; Yu-Kun, L.; Zhen-Min, X. Somatic Cell Culture of Rice Cultivars with Different Grain Types: Somaclonal Variation in Some Grain and Quality Characters. *Plant Cell Tissue Organ Cult.* **1989,** *18*, 201–208.

Kar, B.; Kuanar, A.; Singh, S.; Mohanty, S.; Joshi, R. K.; Subudhi, E.; Nayak, S. In Vitro Induction, Screening and Detection of High Essential Oil Yielding Somaclones in Turmeric (*Curcuma longa* L.). *Plant Growth Regul.* **2014,** *72* (1), 59–66.

Karp, A. Somaclonal Variation as a Tool for Crop Improvement. *Euphytica* **1995,** *85*, 295–302.

Khan, S.; Saeed, B.; Kauser, N. Establishment of Genetic Fidelity of In Vitro Raised Banana Plantlets. *Pak. J. Bot.* **2011,** *43*, 233–242.

Krishna, H.; Alizadeh, M.; Singh, D.; Singh, U.; Chauhan, N.; Eftekhari, M.; Sadh, R. K. Somaclonal Variations and Their Applications in Horticultural Crops Improvement. *Biotechnology* **2016,** *6* (1), 1–18.

Kuanar, A.; Nayak, P. K.; Subudhi, E.; Nayak, S. In Vitro Selection of Turmeric Somaclone Resistant to *Fusarium oxysporum* f. sp. Zingiberi. *Proc. Natl. Acad. Sci. India Sect. B Biol. Sci.* **2014,** *84*, 1077–1082.

Kumari, R.; Sharma, V. K.; Kumar, H. Seed Culture of Rice Cultivars under Salt Stress, *Int. J. Pure. App. Biosci.* **2015,** *3* (1), 191–202.

Larkin, P. Somaclonal Variation: Origins and Causes. *Encycl. Plant Crop Sci.* **2004,** *1*, 1158–1161.

Larkin, P. J.; Scowcroft, W. R. Somaclonal Variation—A Novel Source of Variability from Cell Culture for Plant Improvement. *Theor. Appl. Genet.* **1981,** *60*, 167–214.

Lee, S. Y.; Su, Y. U.; Chou, C. S.; Liu, C. C.; Chen, C. C.; Chao, C. P. Selection of a New Somaclone Cultivar 'Tai-Chiao No. 5' (AAA, Cavendish) with Resistance to Fusarium Wilt of Banana in Chinese Taipei. *Acta Hortic.* **2011,** *897*, 391–397.

Leva, A. R.; Petruccelli, R.; Rinaldi, L. M. R. Somaclonal Variation in Tissue Culture: A Case Study with Olive. In *Recent Advances in Plant In Vitro Culture;* Leva, A. R., Rinaldi, L. M. R. Eds., INTECH Open Access Publisher: Croatia, **2012;** pp 123–150.

Linacero, R.; Rueda, J.; Esquivel, E.; Bellido, A.; Domingo, A.; Va´zquez, A. M. Genetic and Epigenetic Relationship in Rye, *Secale cereale* L., Somaclonal Variation within Somatic Embryo-derived Plants. *In Vitro Cell Dev. Biol. Plant*. **2011,** *47*, 618–628.

Martin, K.; Pachathundikandi, S.; Zhang, C.; Slater, A.; Madassery, J. RAPD Analysis of a Variant of Banana (*Musa* sp.) cv. Grande Naine and Its Propagation via Shoot Tip Culture. *In Vitro Cell Dev. Biol. Plant* **2006,** *42*, 188–192.

Mehta, Y. R.; Angra, D. C. Somaclonal Variation for Disease Resistance in Wheat and Production of Dihaploids through Wheat x Maize Hybrids. *Genet. Mol. Biol.* **2000,** *23* (3), 617–622.

Miguel, C.; Marum, L. An Epigenetic View of Plant Cells Cultured In Vitro: Somaclonal Variation and Beyond. *J. Exp. Bot.* **2011,** *62* (11), 3713–3725.

Mohan-Jain, S. Tissue Culture-derived Variation in Crop Improvement. *Euphytica* **2001,** *118*, 153–166.

Morrison, R. A.; Loh, W. H. T.; Green, S. K.; Griggs, T. D.; McLean, B. T. Tissue Culture of Tomato and Pepper: New Tools for Plant Breeding. Tomato and Pepper Production in the Tropics. In *Proceedings of International Symposium on Integrated Management Practices*, Tainen, Shanhua, Taiwan, AVRDC, **1989,** pp 44–50.

Mujib, A.; Banerjee, S.; Dev, G. P. Callus Induction, Somatic Embryogenesis and Chromosomal Instability in Tissue Culture Raised Hippeastrum (*Hippeastrum hybridum* cv. United Nations). *Propag. Ornam. Plants* **2007,** *7*, 169–174.

Ngezahayo, F.; Dong, Y.; Liu, B. Somaclonal Variation at the Nucleotide Sequence Level in Rice (*Oryza sativa* L.) as Revealed by RAPD and ISSR Markers, and by Pairwise Sequence Analysis. *J. Appl. Genet.* **2007,** *48*, 329–336.

Nivas, S. K.; D'Souza, L. Genetic Fidelity in Micropropagated Plantlets of *Anacardium occidentale* L. (Cashew) an Important Fruit Tree. *Int. J. Sci. Res.* **2014,** *3*, 2142–2146.

Nwauzoma, A. B.; Jaja, E. T. A. Review of Somaclonal Variation in Plantain (*Musa* spp.): Mechanisms and Applications. *J. Appl. Biosci.* **2013,** *67*, 5252–5260.

Olhoft, P. M.; Phillips, R. L. Genetic and Epigenetic Instability in Tissue Culture and Regenerated Progenies. In *Plant Responses to Environmental Stresses: From Phytohormones to Genome Reorganization;* Lerner, H. R., Ed., Marcel Dekker: New York, **1999,** pp 111–148.

Patterson, G. I.; Thorpe, C. J.; Chandler, V. L. Paramutation, an Allelic Interaction, Is Associated with a Stable and Heritable Reduction of Transcription of the Maize *b* Regulatory Gene. *Genetics* **1993,** *135*, 881–894.

Rasheed, S.; Fatima, T.; Husnain, T.; Bashir, K.; Riazuddin, S. RAPD Characterization of Somaclonal Variation in Indica Basmati Rice. *Pak. J. Bot.* **2005,** *37*, 249–262.

Ravindra, N. S.; Ramesh, S. I.; Gupta, M. K.; Jhang, T.; Shukla, A. K.; Darokar, M. P.; Kulkarni, R. N. Evaluation of Somaclonal Variation for Genetic Improvement of Patchouli (*Pogostemon patchouli*), an Exclusively Vegetatively Propagated Aromatic Plant. *J. Crop Sci. Biotechnol.* **2012,** *15*, 33–39.

Saravanan, S.; Sarvesan, R.; Vinod, M. S. Identification of DNA Elements Involved in Somaclonal Variants of *Rauvolfia serpentina* (L.) Arising from Indirect Organogenesis as Evaluated by ISSR Analysis. *Indian J. Sci. Technol.* **2011,** *4*, 1241–1245.

Sato, M.; Kawabe, T.; Hosokawa, M.; Tatsuzawam, F.; Doi, M. Tissue Culture Induced Flower-color Changes in Saintpaulia Caused by Excision of the Transposon Inserted in the Flavonoid 39, 59 Hydroxylase (F3959H) Promoter. *Plant Cell Rep.* **2011a,** *30*, 929–939.

Sato, M.; Hosokawa, M.; Doi, M. Somaclonal Variation Is Induced De Novo via the Tissue Culture Process: A Study Quantifying Mutated Cells in Saintpaulia. *PLoS ONE* **2011b,** *6*, 235–241.

Skirvin, R. M.; Norton, M.; McPheeters, K. D. Somaclonal Variation: Has It Proved Useful for Plant Improvement? *Acta Hortic.* **1993,** *336*, 333–340.

Skirvin, R. M.; McPheeters, K. D.; Norton, M. Sources and Frequency of Somaclonal Variation. *Hort. Sci.* **1994,** *29*, 1232–1237.

Smulders, M.; de Klerk, G. Epigenetics in Plant Tissue Culture. *Plant Growth Regul.* **2011,** *63*, 137–146.

Sun, S.; Zhong, J.; Li, S.; Wang, X. Tissue Culture-induced Somaclonal Variation of Decreased Pollen Viability in Torenia (*Torenia fournieri* Lind.). *Bot. Stud.* **2013,** *54* (1), 36.

Swartz, H. J.; Galletta, G. H.; Zimmerman, R. H. Field Performance and Phenotypic Stability of Tissue Culture-propagated Strawberries. *J. Am.. Soc. Hort. Sci.* **1981,** *106*, 667–673.

Takahashi, H.; Matsumoto, T.; Takai, T. Somaclonal Variants from Strawberry cv. Morioka-16, M16-AR 1, 2 and 3 Resistant to Several Isolates of *Alternaria alternate* Strawberry Pathotype Occurring in Tohoku and Hokkaido [Japan]. *J. Jpn. Soc. Hortic. Sci.* **1993,** *61*, 821–826.

Tang, C. Y.; Liu, C. C.; Hwang, S. C. Improvement of the Horticultural Traits of Cavendish Banana (Musa spp., AAA group I). Selection and Evaluation of a Semi-dwarf Clone Resistant to Fusarium Wilt. *J. Chin. Soc. Hortic. Sci.* **2000,** *46*, 173–182.

Tiwari, J. K.; Chandel, P.; Gupta, S.; Gopal, J.; Singh, B. P.; Bhardwaj, V. Analysis of Genetic Stability of In Vitro Propagated Potato Microtubers Using DNA Markers. *Physiol. Mol. Biol. Plants* **2013,** *19*, 587–595.

Va´zquez, A. M. Insight into Somaclonal Variation. *Plant Biosyst.* **2001,** *135*, 57–62.

Wacksman, J. T. DNA Methylation and the Association between Genetic and Epigenetic Changes: Relation to Carcinogenesis. *Mutat. Res.* **1997,** *375*, 1–8.

Whitehouse, A. B.; Johnson, A. W.; Passey, A. J.; McLeary, K. J.; Simpson, D. W. Serenity: A Paler Skin-coloured Somaclonal Variant of the Short-day Cultivar Florence. *Acta Hortic.* **2014,** *1049*, 819–821.

Zebrowska, J. I. *In Vitro* Selection in Resistance Breeding of Strawberry (Fragaria x Ananassa Duch.). *Commun. Agric. Appl. Biol. Sci.* **2010,** *75*, 699–704.

Zong-xiu, S.; Chang-zhang, Z.; Kang-le, Z.; Xiu-fang, Q.; Ya-ping, F. Somaclonal Genetics of Rice (*Oryza sativa* L.). *Theor. Appl. Genet.* **1983,** *67*, 67–73.

PART III

Techniques in Molecular Biology

CHAPTER 11

RESTRICTION ENDONUCLEASES

SHIV SHANKAR[1*], IMRAN UDDIN[2], and
SEYEDEH FATEMEH AFZALI[3]

[1]*Department of Food Engineering and Bionanocomposite Research Institute, Mokpo National University, 61 Dorimri, Chungkyemyon, Muangun 534729, Jeonnam, Republic of Korea*

[2]*Nanotechnology Innovation Centre, Department of Chemistry, Rhodes University, PO Box 94, Grahamstown, South Africa*

[3]*Department of Biological Science, Faculty of Science, Universiti Tunku Abdul Rahman, Malaysia*

[*]*Corresponding author. E-mail: shivbiotech@gmail.com*

CONTENTS

ABSTRACT

Restriction endonucleases are an integral part of genetic engineering. The birth of genetic engineering and the advancement in the molecular techniques in modern research were possible due to the discovery of the restriction endonucleases. Various types of restriction endonucleases have been discovered and named according to the recognition and cleavage position sites in the DNA sequences. This chapter has focused on types of restriction endonuclease, their mechanism of action, and the interaction with DNA. At the end of this chapter, recent developments of restriction endonucleases and restriction mapping have been discussed

11.1 GENERAL INTRODUCTION

The study of genetic materials (genetic engineering) has contributed significant advancement in many areas of modern research and development. The birth of genetic engineering was possible due to the discovery of special enzymes that cut DNA. Many endeavors of molecular-level engineering rely on biological material such as nucleic acids and restriction enzymes. The field of recombinant DNA and genetic engineering depend on enzymes and techniques that permit the precise cutting, splicing, and sequencing DNA molecules; recognition of recombinant products; and the introduction of recombinant molecules into the cells of any organism. The study of gene themselves became possible with the advent of endonuclease enzymes in bacteria. Endonucleases are enzymes that cleave the phosphodiester bond within a polynucleotide chain. Some endonucleases, such as deoxyribonuclease I cut the DNA relatively nonspecifically (without regard to sequence), while many others, typically called restriction endonucleases or restriction enzymes, cleave only at very specific nucleotide sequences (Cox et al., 2005). Restriction enzymes are endonucleases that are found in eubacteria and archaea and recognize a specific DNA sequence (Stephen et al., 2011). The nucleotide sequence recognized by the restriction enzymes for cleavage is called the restriction site. Generally, the restriction site are a palindromic sequence of about four to six nucleotides in length. Most restriction endonucleases cut the DNA strand unevenly, leaving complementary single-stranded ends. These ends can reconnect through hybridization and are called as "sticky ends," which can be joined through the phosphodiester bonds by the DNA ligase. The hundreds of restriction endonucleases are well-known that are specific for unique restriction sites. The DNA fragments from different origin that are cut by the same endonuclease can be joined to make recombinant DNA. Recombinant DNA is formed by the joining of two or more genes into new combinations (Cox et al., 2005). Restriction enzymes are usually classified into three types that are different in structure and whether they cut DNA at their recognition site or if their cleavage and recognition sites are separate from one another. To cleave DNA, all restriction enzymes make at least two incisions through each sugar–phosphate backbone of the DNA double helix.

Restriction enzymes are found in archaea and bacteria that provide a defense mechanism against invading viruses (Albert and Linn, 1969; Kruger and Bickle, 1983). The restriction enzymes selectively cut foreign DNA inside a prokaryote in a process called restriction. However, the DNA of host organism is protected by a modification by an enzyme, methyltransferase blocks cleavage. These two processes establish the restriction modification system (Kobayashi, 2001). More than 4000 restriction enzymes have been studied in detail, and more than 600 of these are available commercially (Roberts et al., 2007). These enzymes have been used routinely for DNA modification by researchers and are a valuable tool in molecular cloning

11.2 BACKGROUND OF RESTRICTION ENDONUCLEASE

The name restriction enzyme has originated from the studies of phage λ and the phenomenon of host-controlled restriction and modification of a bacterial virus (Winnacker, 1987). The process was first recognized in the work done in the laboratories of Salvador Luria and Giuseppe Bertani in early 1950s (Luria and Human, 1952; Bertani and Weigle, 1953). It was found that a bacteriophage λ which can grow well in one strain of bacteria, such as *Escherichia coli* K, when allowed to grown in another strain, such as *E. coli* C, its yields can drop significantly. The host cell, *E. coli* C, is called as the restricting host and have the capability to decrease the phage activity. If a phage λ grown in one strain, the ability of that phage to grow in the other strains also becomes restricted. In the 1960s, Werner Arber and Matthew Meselson showed that the restriction was instigated by an enzymatic breakdown of the phage λ DNA. The enzyme involved in the breakdown of phage DNA was coined as a restriction enzyme (Meselson and Yuan, 1968; Dussoix and Arber, 1962; Lederberg and Meselson, 1964).

The restriction endonuclease studied by Arber and Meselson were type I restriction enzymes, which cleaves DNA randomly away from the recognition site. The isolation and characterization of the first type II restriction enzyme, *Hin*dII, from the bacterium *Haemophilus influenzae* was carried out by Hamilton O. Smith, Thomas Kelly, and Kent Wilcox in 1970. The type II restriction enzymes are more useful for laboratory use, as they cut the DNA within their recognition sequence. Later, Daniel Nathans and Kathleen Danna showed that the cleavage of simian virus 40 (SV40) DNA by restriction enzymes produce particular fragments which can be separated by polyacrylamide gel electrophoresis. This result showed that the restriction enzymes can also be useful in the mapping of the DNA (Danna and Nathans, 1971). For this work, Werner Arber, Daniel Nathans, and Hamilton O. Smith was awarded the 1978 Nobel Prize in Physiology or Medicine. The innovation of restriction enzymes paved the way of DNA manipulation, resulting in the development of recombinant DNA technology, which has various applications such as the large scale production of proteins, such as human insulin used by diabetics. The discovery of restriction endonucleases was an important discovery for predicting the DNA structure and function that further became a backbone for molecular biology studies (Szybalski et al., 1991).

11.3 RECOGNITION SITES OF RESTRICTION ENDONUCLEASE

Restriction enzymes identify a specific sequence of nucleotides and make a double-stranded cut in the DNA. The recognised DNA sequences can be classified by the total number of bases in its recognition site, usually between 4 and 8 bases. Also, the number of bases in the sequence that determines how often the site will appear in any given genome. For example, a 4-bp sequence would theoretically occur once every $(4)^4$ or 256 bp, 6 bases at every $(4)^6$ or 4096 bp, and 8 bases at every $(4)^8$ or 65,536 bp. Most of the sequences recognized by restriction enzymes are palindromic sequences. The base sequence that reads the same forward and backward is called as a palindromic sequence. Theoretically, there are two types of palindromic sequences possible in DNA. First, the mirror-like palindrome that is similar to those found in the ordinary text, in which a sequence reads in the same manner forward and backward on a single strand of DNA strand, e.g., GTAATG. The second is inverted repeat palindrome that reads the sequence same forward and backward; however, the forward and backward sequences are present in complementary DNA strands (i.e., of double-stranded DNA), as in GTATAC (GTATAC being complementary to CATATG). The inverted palindromes are more common than mirror-like palindromes.

*Eco*RI digestion produces "sticky ends," GAATTC, whereas *Sma*I restriction enzyme cleavage produces "blunt ends," CCC/GGG. The recognition sequences in DNA differ for each restriction enzyme, producing DNA of different length and sequence, as well as they differ in their strand orientation (5′ end or the 3′ end). The cut end can be a sticky end "overhang" or blunt end for an enzyme restriction. The restriction enzymes that recognize the same DNA sequence are known as neoschizomers. These often cleave in different locations of the sequence. However, different enzymes that have recognition and cleavage sequence in the same location are known as isoschizomers.

It is known that chromosomes are huge biomolecules that have many genes, and to locate a specific gene physically or manipulate them was impossible before the invention of restriction endonucleases. Previously, scientist isolated and purified the bacterial chromosomes that contain many genes. They used to break the chromosome into smaller segments using physical force that resulted in a random break in the chromosomes and cloned these fragments randomly. So, for many years, physical manipulation of DNA was virtually impossible. It was initially known due to their ability to breakdown/restrict foreign DNA. Restriction enzymes appear to be made exclusively by prokaryotes. It can detect the foreign DNA very easily, such as infecting bacteriophage DNA, and protect the cell from invasion by cleavage of foreign DNA into small pieces making them nonfunctional. There are multiple functions performed by the restriction enzymes, which cut the DNA/RNA of foreign viruses invading bacteria DNA or DNA/RNA of any of the types of organism. This make them as important and useful tools for molecular genetics. It is generally accepted that restriction enzymes are remarkable tools for the biologists for their investigations in gene organizations, function, and expression. Beside the wide applications of restriction enzymes, the structures and catalytic dynamics and mechanism are a hot topic of research for future development (Bourniquel and Bickle, 2002; Mark et al., 1996; Roberts et al., 2003; Titheradge et al., 2001).

11.4 DISCOVERY OF RESTRICTION ENZYMES OR RESTRICTION ENDONUCLEASES

Restriction enzymes were discovered in 1970, and Werner Arber, Hamilton Smith, and Daniel Nathans received the 1978 Nobel Prize for the discovery (Dussoix and Arber, 1962; Linn and Arber, 1968; Loenen et al., 2014). Restriction enzymes cleave DNA at a specific recognition site and have many uses in molecular biology, genetics, and biotechnology. More than 4000 restriction enzymes are known today, of which more than 621 are commercially available (Avery et al., 1944). The first restriction enzyme isolated was *Hind* II, but many other restriction enzymes were discovered and characterized later (Kelly and Smith, 1970; Smith and Wilcox, 1970). Restriction enzyme for the first time originated from the studies of phage λ. The discovery of the restriction endonucleases permits researchers to cleave DNA at specific sites, which is a great benefit over chemical or physical cleavage that results in random fragmentation of DNA. P. Berg developed a revolutionary idea to create recombinant DNA for the first time in 1972. Restriction endonucleases are mostly present in bacteria. However, their presence has been confirmed in archaebacteria, viruses, and even in eukaryotes. The discovery of restriction enzymes paved the way for scientists to cut the DNA into specific pieces. Every time a given piece of DNA was cut with a given enzyme, the same fragments were produced. These defined pieces could be put back together in new ways. So, in conclusion, cutting DNA molecules in a particular region and reproducible order opened new gate of experimental possibilities.

11.5 TYPES OF RESTRICTION ENDONUCLEASES

The naturally occurring restriction endonucleases are divided into three main groups (types I, II, and III), depending on their enzyme cofactor requirements, composition, nature of their target sequence, and the position of their DNA cut-site relative to the target sequence. However, type IV and type V are also reported (Bickle and Krüger, 1993; Boyer, 1971; Yuan, 1981). All types of restriction endonucleases recognize specific DNA sequences and carry out the endonucleolytic cleavage of DNA to give specific fragments with terminal 5′-phosphates. On the detailed biochemical characterization of purified restriction enzyme, it became apparent that restriction endonucleases are different in their basic enzymology. In particular to their sub-unit composition, cofactor requirement, and mode of cleavage, they have been divided into different groups. Type I restriction endonucleases cleave at sites remote from recognition site, requiring both ATP and S-adenosyl-L-methionine to function and are a multifunctional protein with both restriction and methylase activities. Type II restriction endonucleases cleave within or at short, specific distances from recognition site; most of these restriction endonucleases require magnesium, and the single function (restriction) enzymes are independent of methylase. Type III restriction endonucleases cleave DNA at sites that are at a short distance from recognition site and require ATP (but do not hydrolyse it). The S-adenosyl-L-methionine stimulates reaction, but is not required, and exists as part of a complex with a modification methylase. Type IV restriction endonucleases target modified DNA, e.g., methylated, hydroxymethylated and glucosyl-hydroxymethylated DNA. Type III restriction endonucleases cut the DNA at recognition site and then dissociate from the substrate. However, type I enzyme binds to the recognition sequence but cleave at random sites, when the DNA loops back to the bound enzyme (Eskin and Linn, 1972). Neither type I nor type III restriction enzymes are widely used in molecular cloning. Type I and II enzymes are mostly used in research and development. All of them need a divalent metal cofactor (Mg^{2+}) for their function and activity. All three types of restriction enzymes, their structure, and mode of action summarized in Table 11.1.

TABLE 11.1 Type of Restriction Endonuclease, Recognition, and Cleavage Sites.

Restriction Endonuclease	Structure	Recognition Site	Restriction and Methylation	Cleavage Site
Type I	Bifunctional enzyme (3 subunits)	Bipartite and asymmetric	Naturally exclusive	Nonspecific >1000 bp from recognition site
Type II	Separate endonuclease and methylase	4–6 bp sequence, often palindromic	Separate reaction	Same as or close to recognition site
Type III	Bifunctional enzyme (2 subuints)	5–7 bp asymmetric sequence	Simultaneous	24–26 bp Downstream of recognition site

11.5.1 TYPE I RESTRICTION ENDONUCLEASES

The first restriction enzymes identified were Type I restriction enzymes in two different strains *E. coli* K-12 and *E. coli* B (Murray, 2000). These enzymes cleave the DNA at a site that differs and at a random distance of around 1000-bp far from their recognition site. The cleavage of DNA at these random sites follows a process of DNA translocation that confirms that these restriction enzymes are also molecular motors. The recognition site is asymmetrical and is composed of two specific portions: first containing 3–4 nucleotides and second containing 4–5 nucleotides and separated by about 6–8 nucleotides long nonspecific spacer. These enzymes are multifunctional and possess both restriction and modification activities, which depend on upon the target DNA methylation status. The S-adenosylmethionine (AdoMet) is a cofactor that hydrolyzes ATP and requires magnesium (Mg^{2+})

ions for their full activity. Type I restriction enzymes have three subunits called HsdR, HsdM, and HsdS. HsdR is required for restriction; HsdM is for adding methyl groups to host DNA; and HsdS is for the specificity of the recognition site in addition to both restriction (DNA cleavage) and modification (DNA methyltransferase) activity.

11.5.2 TYPE II RESTRICTION ENDONUCLEASES

In general, type II restriction endonucleases differ from type I restriction endonucleases in various ways. They form homodimers, with recognition sites that are usually palindromic and 4–8 nucleotides long. They recognize the sequence and cleave DNA at the same site, and they do not use ATP or AdoMet for their activity but usually require only Mg^{2+} as a cofactor (Pingoud and Jeltsch, 2001). The type II restriction endonucleases are the most commonly used restriction enzymes. In the 1990s and early 2000s, various new type II restriction endonucleases were discovered which did not follow all the essential criteria of this enzyme class. Therefore, the nomenclature for new subfamily was developed to divide this family into subcategories depending on the deviations from typical characters of type II enzymes. These subgroups are defined using a letter suffix.

Type IIB restriction endonucleases, such as *Bpl*I and *Bcg*I are multimers that contain more than one subunit. They cut DNA on both sides of their recognition to cut out the recognition site. They require both Mg^{2+} cofactors and AdoMet. Type IIE restriction endonucleases, such as *Nae*I, cut DNA following the interaction with two copies of their recognition sequence (Pingoud and Jeltsch, 2001). The first recognition site acts as the target for cleavage; however, the other acts as an allosteric effector, which speeds up or improves the efficiency of enzyme cleavage. Similar to type IIE restriction endonucleases, type IIF restriction endonucleases, such as *Ngo*MIV, interact with two copies of their recognition sequence, but it cleaves both sequences at the same time. Type IIG restriction endonucleases (*Eco*57I) have a single subunit, like classical Type II restriction enzymes, but it requires the cofactor AdoMet to be active. Type IIM restriction endonucleases, such as *Dpn*I, can recognize and cut methylated DNA. Type IIS restriction endonucleases (e.g., *Fok*I) cleave DNA at a defined distance from their non-palindromic asymmetric recognition sites, and these enzymes are widely used to perform in vitro cloning techniques such as Golden Gate cloning. These enzymes may function as dimers. Similarly, type IIT restriction endonucleases (*Bpu*10I and *Bsl*I) are composed of two different subunits. Some recognize asymmetric sequences, while others have palindromic recognition sites.

11.5.3 TYPE III RESTRICTION ENDONUCLEASE

Type III restriction endonucleases, such as *Eco*P15, identify two separate non-palindromic sequences that are inversely oriented. They cut the DNA about 20–30 bp away from the recognition site. These enzymes consist of more than one subunit and require AdoMet and ATP cofactors for their roles in DNA methylation and restriction, respectively (Meisel et al., 1992). They are part of prokaryotic DNA restriction-modification mechanisms, which protect the organism against the invading foreign DNA. The type III enzymes are hetero-oligomeric, multifunctional proteins that are composed of two subunits called Res and Mod. Mod subunit recognizes the DNA sequence specific for the system and is a methyltransferase, and it is functionally similar to the M and S subunits of type I restriction endonuclease. The Res is required for restriction, even though it has no enzymatic activity of its own. Type III restriction enzymes recognize 5–6-bp long asymmetric DNA sequences and cleave the DNA 25–27 bp downstream to recognition site and generate single-stranded 5′ protrusions. They require the presence of two inversely oriented unmethylated recognition sites for the restriction to happen. These enzymes methylate the only one strand of DNA, at the N6 position of adenosyl residues, so the newly replicated DNA contains only one strand methylated that is sufficient to protect against restriction. Type III restriction enzymes belong to the beta-subfamily of N6 adenine methyltransferases that contain the nine motifs, which characterize this family, including the AdoMet binding pocket (FXGXG), motif I, and motif IV.

11.5.4 TYPE IV RESTRICTION ENDONUCLEASES

Type IV restriction endonucleases recognize methylated DNA and are exemplified by the McrBC and Mrr systems of *E. coli*.

11.5.5 TYPE V RESTRICTION ENDONUCLEASES

Type V restriction endonucleases, such as cas9–gRNA complex from CRISPRs, utilize guide RNAs to target specific non-palindromic sequences found on invading organisms. They can cut DNA of variable length and suitable guide RNA is provided. The flexibility and ease of use of these enzymes make them promising for future genetic engineering applications.

11.6 NOMENCLATURE

Since their discovery in the 1970s, a large number of restriction enzymes have been identified. More than 3500 different Type II restriction enzymes have been characterized (Pingoud, 2004). Each restriction enzyme is named after the bacterium from which it was isolated, using a naming system based on the bacterial genus, species, and strain. For example, the name of the *Eco*RI restriction enzyme was derived as E from the first letter of genus *Escherichia,* co from first two letters of genus *coli,* R from the strain RY13, and I for first identified (order of identification in bacteria).

Restriction endonucleases are present in the bacteria, presumably to destroy the DNA from foreign sources (e.g., infecting bacteriophage) by cutting the foreign DNA at the specific restriction sites. The host bacteria DNA is, however, protected from cleavage because specific recognition sites are modified by methylation at one of the bases on the site, making the site as no longer a substrate for the restriction endonuclease cleavage. The host bacteria used to propagate cloned DNA in the laboratory are usually a mutant in the host restriction genes. Therefore, their intracellular enzyme activities do not destroy the foreign recombinant sequences. The specific cleavage sites for various restriction endonucleases have been defined. They cut the DNA within or near to their particular recognition sequences that typically are four to six nucleotides long with a twofold axis of symmetry. For example, the restriction endonuclease *Eco*RI, used for cloning of the gene, requires that the 6 bp occur in the following specific order:

5′-----GAATTC------3′

3′------CTTAAG------5′

*Eco*RI recognizes this sequence and cleaves the sequence in a unique fashion, resulting in two terminals with protruding 5′ ends:

5′-----G AATTC------3′

3′-----CTTAA G-------5′

These cut ends are complementary (sticky) and can be enzymatically reattached to any other *Eco*RI generated termini by T4 DNA ligase. Many restriction enzymes, like *Eco*RI, generate fragments with protruding 5′ ends. However, others (e.g., *Pst*I) generate fragments with 3′ protruding, cohesive termini, whereas still others (e.g., *Bal*I) cleave at the axis of symmetry to produce blunt-ended fragments. Each restriction endonuclease has a specific sequence and number of nucleotides required to create the recognition site. Some restriction endonucleases do not require a particular nucleotide in every position of the recognition site.

These enzymes allow cloning and purification as well as sequence determination. During evolution, different alleles can require mutations in the sequences next to the hybridization position that may result in a different length of a particular restriction fragment. This phenomenon is called a restriction fragment length polymorphism (RFLP). Such RFLP are useful for the identification of genetic diseases because the gene defect linked to the RFLP can be identified in the absence of a phenotypic abnormality. RFLP can also be used for forensic purposes. The restriction analysis of amplified internal transcribed spacer (ITS) of ribosomal DNA digested with several endonucleases and the sequence analysis of the ITS have been applied to differentiate the closely related species. For example, *E. australis* and *E. fawcettii* are distinguished by the endonuclease restriction analysis of the amplified ITSs of ribosomal DNA. *E. fawcettii* isolated from Florida and Australia could be separated by random amplified DNA polymorphism (RAPD) analysis (Tan et al., 1996).

11.7 MECHANISM OF ACTION OF RESTRICTION ENDONUCLEASE

The mode of action of restriction endonuclease enzymes is well-known and well-studied. The detailed mechanism and its mode of action are well-documented in Figure 11.1. The restriction endonuclease binds to the recognition site and checks for the presence of methyl group on the DNA at a particular nucleotide. If there is a methylation in the recognition sequence, then it does not cut the DNA. If only one strand of the DNA is methylated in the recognition sequence and the other strand is not methylated, then the only type I and type III restriction endonucleases will methylate the other strand at the required position. The restriction endonuclease takes the methyl group from AdoMet using modification site present in the restriction enzymes. However, type II restriction endonuclease takes the help of another enzyme called methylase to methylate the DNA. If there is no methyl group on both the strands of DNA, then restriction endonuclease cut the DNA. Owing to the methylation mechanism, restriction endonuclease, although present in bacteria, does not cleave the bacterial DNA but cleaves the foreign DNA. However, there are some restriction endonucleases that functions exactly in reverse mode. They cut the DNA if they are methylated.

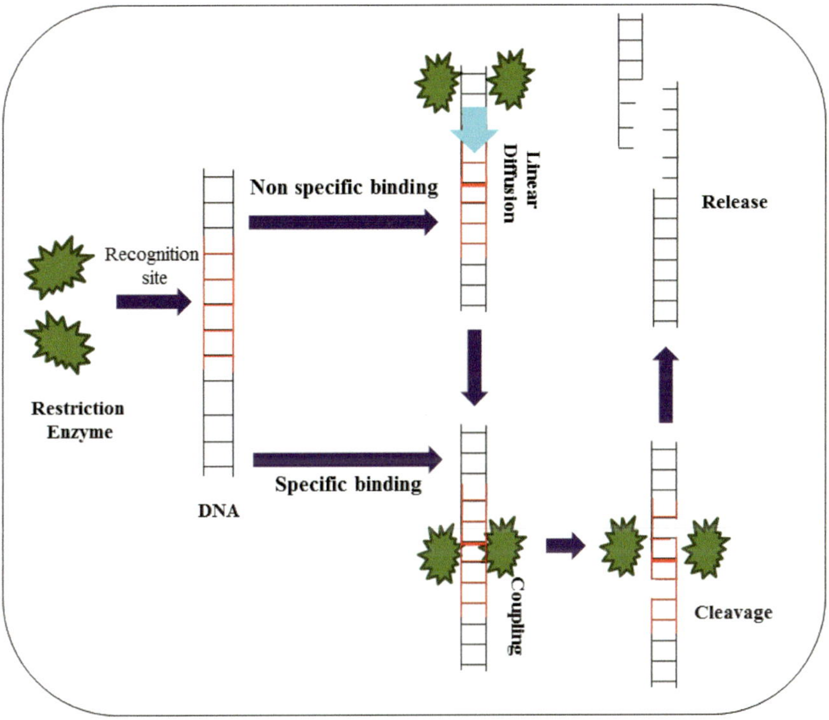

FIGURE 11.1 Mechanism of action of type II restriction endonucleases.

Restriction endonuclease recognizes the binding site to the respective DNA, which cleaved the respective DNA molecule and restriction endonuclease released (Pingoud and Jeltsch, 2001). Restriction enzymes bind recognition site by two modes: specific binding and nearby nonspecific binding. In the nonspecific binding, distance between recognition sites is far away, and the restriction enzyme travel along the DNA strand till it captures the recognition site. Further, restriction enzymes detect recognition site, hydrolyze the sugar phosphate bonds of the DNA, and enzyme released after that cleave DNA molecules. To begin, all restriction endonucleases will bind DNA specifically and, with much less strength, nonspecifically (Mark et al., 1996).

This is a characteristic of many proteins that interact with DNA. It is probable that even nonspecific DNA binding will induce a conformational change in the restriction enzyme dimer that will result in the protein adapting to the surface of the DNA strands. These changes are not the same as those that occur when the dimer binds to the recognition site though. As the dimer slides along the DNA strands, it searches for recognition elements and, when these are encountered, an interaction between the protein and the DNA ensues in which the nonspecific complex is converted into a specific complex. This requires significant conformational changes in both the protein and the DNA as well as the expulsion of water molecules from the protein/DNA interface so that more intimate contacts can be established. In general, intimate contact is held by 15–20 hydrogen bonds that form between the protein and the DNA in the recognition site. These bonds are shown to be facilitated through specific amino acids, primarily ASP, and GLU held in a proper three-dimensional configuration (Wright et al., 1999).

11.8 INTERACTION OF RESTRICTION ENDONUCLEASE WITH THE DNA

The restriction endonucleases interact with the DNA in a complex mode. Owing to the large size of a normal DNA substrate, the reaction of a the restriction enzyme with the DNA cannot be simply formulated as a sequence of two or three steps. The reaction cycle starts with a nonspecific binding to the macromolecular DNA, which is followed by a random diffusional walk of restriction endonuclease on the DNA. If a recognition site is not away from the initial site of contact, it will most likely be located within one binding event. However, at the recognition site, conformational changes take place that constitutes the recognition process and leads to the activation of the catalytic centers. The product is released after phosphodiester bond cleavage in both strands, either by a transfer of enzyme to nonspecific sites on the same the DNA molecule or by direct dissociation of the enzyme–product complex. Often this step is rate limiting for the DNA cleavage by restriction enzymes under multiple turnover conditions.

11.9 ISOSCHIZOMERS AND NEOSCHIZOMERS

Isoschizomers are a pair of restriction endonuclease with same recognition sequence and cutting pattern, but isolated from a different strain of bacteria. For example, *Sph*I from *Streptomyces phaeochromogenes* (CGTAC/G) and *Bbu*I from *Bacillus sp.* (CGTAC/G) are isoschizomers of each other. Among isoschizomers, the first enzyme discovered which recognizes a given sequence is known as the prototype, and all subsequently identified enzymes that recognize that sequence is isoschizomers. Isoschizomers are usually isolated from different strains of bacteria and, therefore, may require different reaction conditions.

Neoschizomers or heteroisoschizomers are a pair of restriction endonuclease with identical recognition sequence but a different cutting pattern. For example, *Sma*I from *Serratia marcescens* (CCC/GGG) and *Xma*I from *Xanthomonas malvacearum* (C/CCGGG) are neoschizomers of each other.

In a few cases, only one enzyme out of a pair of isoschizomers can recognize both, the methylated and unmethylated forms of restriction sites. However, the other restriction enzyme can recognize only the unmethylated form of the restriction site. This property of some of the isoschizomers allows the identification of methylation state of the restriction site while isolating them from a bacterial strain. For example, the restriction enzymes *Msp*I and *Hpa*II are isoschizomers, as they both recognize the sequence 5′-CCGG-3′ when it is unmethylated. However, the second C of the sequence is methylated, only *Msp*I can recognize it while *Hpa*II can not.

11.10 COMMONLY USED RESTRICTION ENDONUCLEASES

Restriction endonucleases in combination with DNA ligases facilitated a robust "cut and paste" workflow to move defined DNA fragment from one organism to another. This technique helps to incorporate exogenous DNA into natural plasmids to create the vehicle for cloning-plasmid vectors that self-propagate in *E. coli* (Cohen et al., 1973). These became the backbone of many recent vectors and enabled the cloning of DNA for more research in the production of recombinant proteins. Another application of restriction enzymes is post cloning, which is a confirmatory tool to ensure that the insertions have taken place correctly. The traditional cloning work along with DNA amplification technologies (PCR and RT-PCR) has become a mainstream application for restriction endonuclease and is a pathway to study molecular mechanisms. Type II restriction enzymes are most commonly used in molecular biology applications, as they recognize stereotypical sequences and produce a predictable cleavage pattern (Cohen et al., 1967). Table 11.2 shows commonly used restriction enzymes and their recognition site.

TABLE 11.2 Commonly Used Restriction Enzymes and Their Recognition Site.

Enzyme	Recognition Sequence
AatII	GACGT/C
*Acl*I	AA/CGTT
*Age*I	A/CCGGT
*Avr*II	C/CTAGG
*Bam*HI	G/GATCC
*Bmt*I	GCTAG/C
*Bsa*I	GGTCTC
*BspH*I	T/CATGA
*Bvb*I	GCAGC
*Cla*I	AT/CGAT
*EcoR*I	G/AATTC
*EcoR*II	CCWGG/GGWCC
*EcoR*V	GAT/ATC
*Fse*I	GGCCGG CC
*Hpa*I	GTT/AAC
*Kas*I	G GCGCC
*Kpn*I	GGTAC/C
*Mbo*I	GATCGATC
*Mfe*I	C/AATTG
*MspA1*I	CMG/CKG
*Nde*I	CA/TATG
*Nde*II	GATCGATC
NgoMIV	G/CCGGC
*Not*I	GC/GGCCGC
*Nsi*I	ATGCA/T
*Pme*I	GTTT/AAAC
*Pst*I	CTGCA/G
*Pvu*I	CGAT/CG
*Pvu*II	CAG/CTG
*Sau*I	CCTNAGG
*Sma*I	CCC/GGG
*Sst*I	GAGCTC
*Stu*I	AGG/CCT
XbaI	T/CTAGA
*Xma*I	C/CCGGG

11.11 RECENT DEVELOPMENT OF RESTRICTION ENDONUCLEASES

It seems improbable that today's Biomedical Sciences and the Biotechnology industry would have developed without restriction enzymes. The development of restriction enzyme technology occurred through the appropriation of known tools and procedures in novel ways that had broad applications for analyzing and modifying the gene structure and organization of complex genomes (Berg and Mertz, 2010). During the past three decades, more than 3500 restriction enzymes have been identified, but only about 500 have been commercialized. Some enzyme may have undesirable qualities include a tendency toward star activity, low yield upon purification, and poor stability. Star activity is the alteration or relaxation of the specificity of restriction enzyme-mediated cleavage of DNA that can occur under low ionic strength, high pH, and high (>5% v/v) glycerol concentrations that differ significantly from those optimal for the enzymes. The glycerol concentration is of particular interest, since commercial restriction enzymes are commonly supplied in a buffer that contains a substantial amount of glycerol (50% v/v is typical). The insufficient dilution of the enzyme solution can cause star activity; this problem most often arises during double or multiple digests.

Isoschizomer and neoschizomer have more suitable traits than the classic enzyme and, thus, may be a more attractive alternative for advanced molecular biology studies. New isoschizomers are easily purified, have cheaper costs with a better value. The neoschizomers are sometimes developed for the same reasons as isoschizomers, to remove undesirable properties and increase yield. The neoschizomers may offer alternative ends upon cleavage, which are advantageous in one reaction or another (Roberts, 1990). Novel technologies can expand the scientific ability to create new DNA molecules. For example, the potential to generate new recognition specificity in the *Mme*I family restriction endonuclease, the engineering of more restriction endonuclease may create new tools for DNA manipulation and epigenome analysis. Development of new applications of these enzymes will take the role of restriction endonuclease beyond molecular cloning by the development of biotechnology and introduce us a new area of research. A human gene with potent antiretroviral activity has recently been found to encode a new member of the family of cytidine deaminases, which is involved in mRNA editing, immunoglobulin gene class switching, and hypermutation. This enzyme attacks viral DNA and acts as a new soldier in the battle between host and virus. Therefore, this enzyme can be added to a list of antiviral host defense mechanisms (Goff, 2003).

Restriction site-associated DNA (RAD) tags are a genome-wide representation of every site of a particular restriction enzyme by short DNA tags. Rapid and cost-effective polymorphism identification and genotyping by RAD markers was done by Miller et al. (2007). Their results demonstrated that these markers can be identified and typed on pre-existing microarray formats and the method to develop RAD marker microarray resources, which allow high-throughput genotyping with high resolution in both model and non-model systems.

Terminal restriction fragment length polymorphism (T-RFLP) is a high-throughput fingerprinting technique used to monitor changes in microbial communities, which offers a compromise between the information gained and labor intensity. This is a new approach, where the progress in T-RFLP analysis of 16S rRNA and genes allows researchers to make experimental and statistical choices appropriate for the hypotheses of their studies (Pingoud et al., 2014).

The application of restriction enzymes together with PCR has revolutionized molecular cloning; however, it is restricted by the manipulated DNA sequences content. Uracil excision-based cloning is ligase and a sequence independent technique, which allows sequencing in simple one-tube reactions with higher accuracy (Nørholm, 2010). In Nørholm's study, a different uracil excision-based molecular tools developed in an open-source fashion that is simple, cheap and comprehensive toolkit with different applications in molecular cloning.

In the past few years, a huge effort has been made for constructing recombinant DNA molecules, which is traditionally constructed using type II restriction enzymes and ligase (Szybalski et al., 1991). In particular, this approach is slow, monotonous, and limited to the creation of small size constructs with only few genes. So, it seems to be difficult for large constructs that will be cut many times by available restriction enzymes. Recently, a different approach including PCR-based assembly, ligation-independent cloning, recombinase-based cloning, and homologous recombination-based cloning have been developed to overcome these limitations and dispel the problems coming from the multiple occurrences of restriction sites in large constructs (Gibson et al., 2010).

11.12 FAST DIGEST RESTRICTION ENDONUCLEASES

Restriction digestion (or hydrolysis) is the process of cutting DNA molecules into smaller pieces with special enzymes called restriction endonucleases.

Their biochemical activity is the hydrolysis of the phosphodiester backbone at specific sites in a DNA sequence that may be used for analysis or other processing. The resulting digested DNA is usually amplified using PCR, to make it suitable for analytical techniques such as molecular cloning, chromatography, agarose gel electrophoresis, and genetic fingerprinting. It is also named DNA fragmentation, but this term is used for other procedures in molecular biology as well. Digestion technique may be used for cleaving DNA fragments at specific sites with a specific size (Sanniyasi et al., 2013). For example, to digest DNA, *Bam*HI enzyme finds the sequence GGATCC on each strand and nicks the phosphodiester backbone between the G nucleotides, which breaks the hydrogen bonds and the two fragments move away from each other (Fig. 11.2).

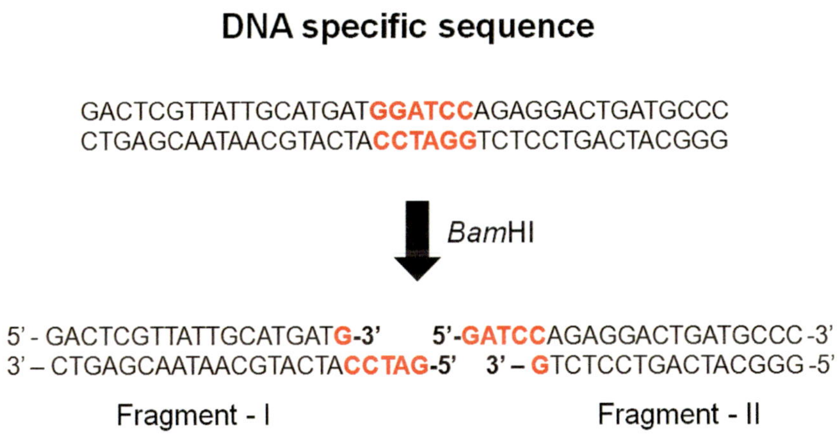

FIGURE 11.2 Digestion of DNA sequence with the *Bam*HI enzyme at GGATCC site.

Several factors need to be considered when setting up a restriction endonuclease digestion, including the amounts of DNA, enzyme and buffer components, and correct reaction volume which allows you to achieve optimal digestion. Generally, 1 unit of restriction enzyme will completely digest 1 μg of substrate DNA in a 50 μl reaction in 1 h. Adjusting the enzyme and DNA reaction volume ratio can help to achieve maximal success in the restriction endonuclease reactions. A 30-min incubation at 37°C is the optimum digestion condition or samples can be digested by placing tubes at 37°C water bath, allowing them to incubate overnight as the water cools to room temperature (King et al., 1997).

Partial digestion happens where the DNA is exposed to the restriction enzyme for only a short time; then not all sites are cleaved due to a limitation in enzyme activity, which results in fragments ranging in size from the smallest to the longest (some or no sites are cut) (Church and Gilbert, 1984). Another common task named double digest is digesting DNA with two restriction enzymes which require different buffers. Under nonstandard conditions, the restriction enzymes cleave at sites that are similar, but not identical, to its normal recognition sequence causes altered cutting which is called "star" or "relaxed" activity. A common example of an enzyme that exhibits star activity resulting in almost completely nonspecific digestion of DNA is *Eco*R I. While the normal specific recognition site of *Eco*R I is G↓AATTC, it may change to N↓AATTN under nonstandard conditions or changes to Pu↓PuATPyPy with a much higher frequency (Strong et al., 1997). Fast digest enzymes are an advanced line of restriction enzymes that create a new standard in DNA digestion and are commercially available. They usually save time and effort, increase the output, does not require overnight digestions, and the star activity is eliminated due to short reaction times. With high-quality, efficient restriction enzymes, the incubation time can be significantly shortened, allowing for rapid screening of clones or preparation of digested DNA for downstream applications. The fast digest restriction enzyme applications, their advantages, and disadvantages need to be studied profoundly to contribute to the molecular biology research in future.

11.13 RESTRICTION MAPPING

Restriction mapping is a technique used to map the unknown segments of DNA by breaking it into pieces and then identifying the locations of the breakpoints. This method depends on the use of proteins called restriction enzymes, which can cut or digest the DNA at specific sequences called restriction sites. After a DNA segment has been digested using a restriction enzyme, the resulting fragments can be examined using gel electrophoresis, which is used to separate pieces of DNA according to their size. The common method for constructing a restriction map involves digesting the unknown DNA sample in three ways. The two portions of the DNA sample are individually digested with different restriction enzymes, and a third portion of the DNA sample should be double-digested with both restriction enzymes at the same time. Then, each digestion sample should be separated using gel electrophoresis and the sizes of the DNA fragments are recorded. The total length of the fragments in each digestion will be equal. However, because the length of each individual DNA fragment depends upon the positions of its restriction sites, each restriction site can be mapped according to the lengths of the fragments. The information from the double-digestion is particularly useful for correctly mapping the restriction sites. The final drawing of the DNA segment that shows the positions of the restriction sites is called a restriction map.

11.14 CONCLUSIONS AND FUTURE PROSPECT

Restriction endonuclease has made major contributions to many aspects of nucleic acid analysis and its applications. Many of the developments reached maturity already and are now used in different contexts. We have shown many concepts that can now be translated for research and application. However, there are many aspects that do not yet have a final solution. Extending the length of DNA sequences is critical. The more we refine the tools for genome analysis, the more we find that structures of genomes are more diverse than we thought. We are discovering that the concept of a single reference for a species is not entirely correct. Technologies to analyze long DNA molecules at high resolution need further development. In general, technologies to carry out nucleic acid analyses at very low concentration still require further development. A further opportunity that would be very beneficial and would advance the integration of these technologies into a routine is the development of fully integrated cartridge-based systems in which procedures can be done far more reliably, reproducibly, and with a dramatically reduced risk of contamination. Care needs to be taken that good standards of data analysis are installed in common practice. An eye needs to be kept on computational efficiency of analysis and storage as large-scale projects are initiated.

KEYWORDS

- **restriction endonucleases**
- **recognition sequences**
- **isoschizomers**
- **neoshizomers**

REFERENCES

Arber, W.; Linn, S. DNA Modification and Restriction. *Annu. Rev. Biochem.* **1969,** *38*, 467–500.

Avery, O. C.; MacLeod, M.; McCarty, M. Studies on the Chemical Nature of the Substance Inducing Transformation of Pneumococcal Types Induction of Transformation by a Desoxyribonucleic Acid Fraction Isolated from Pneumococcus Type III. *J. Exp. Med.* **1944,** *79*, 137–158.

Berg, P.; Mertz, J. E. Personal Reflections on the Origins and Emergence of Recombinant DNA Technology. *Genetics*. **2010,** *184*, 9–17.

Bertani, G.; Weigle, J. J. Host Controlled Variation in Bacterial Viruses. *J. Bacteriol.* **1953,** *65*, 113–121.

Bickle, T. A.; Krüger, D. H. Biology of DNA Restriction. *Microbiol. Rev.*, **1993,** *57*, 434–450.

Bourniquel, A. A.; Bickle, T. A. Complex Restriction Enzymes: NTP-driven Molecular Motors. *Biochimie*. **2002,** *84*, 1047–1059.

Boyer, H. W. DNA Restriction and Modification Mechanisms in Bacteria. *Annu. Rev. Microbiol.* **1971,** *25*, 153–176.

Church, G. M.; Gilbert, W. Genomic Sequencing. *Proc. Natl. Acad. Sci. USA.* **1984,** *81*, 1991–1995.

Cohen, S. N.; Chang, A. C.; Boyer, H. W.; Helling, R. B. Construction of Biologically Functional Bacterial Plasmids In vitro. *Proc. Natl. Acad. Sci. USA.* **1973,** *70*, 3240–3244.

Cohen, S. N.; Maitra, U.; Hurwitz, J. Role of DNA in RNA Synthesis. XI. Selective Transcription of Gamma DNA Segments In vitro by RNA Polymerase of *Escherichia coli*. *J. Mol. Biol.* **1967,** *26*, 19–38.

Cox, M.; Nelson, D. R.; Lehninger, A. L. *Lehninger: Principles of Biochemistry*; W.H. Freeman: San Francisco, 2005; p 952. ISBN 0-7167-4339-6.

Danna, K.; Nathans, D. Specific Cleavage of Simian Virus 40 DNA by Restriction Endonuclease of *Hemophilus influenzae*. *Proc. Natl. Acad. Sci. USA.* **1971,** *68*, 2913–2917.

Dussoix, D.; Arber, W. Host Specificity of DNA Produced by *Escherichia coli*. II. Control Over Acceptance of DNA from Infecting Phage Lambda. *J. Mol. Biol.* **1962,** *5*, 37–49.

Eskin, B.; Linn, S. The Deoxyribonucleic Acid Modification and Restriction Enzymes of *Escherichia coli* B II. Purification, Subunit Structure, and Catalytic Properties of the Restriction Endonuclease. *J. Biol. Chem.* **1972,** *247*, 6183–6191.

Gibson, D. G.; Glass, J. I.; Lartigue, C.; Noskov, V. N.; Chuang, R. Y.; Algire, M. A.; Venter, J. C. Creation of a Bacterial Cell Controlled by a Chemically Synthesized Genome. *Science*, **2010,** *329*, 52–56.

Goff, S. P. Death by Deamination: A Novel Host Restriction System for HIV-1. *Cell*, **2003,** *114*, 281–283.

Kelly, T. J.; Smith, H. O. A Restriction Enzyme from *Hemophilus influenzae*: II. Base Sequence of the Recognition Site. *J. Mol. Biol.* **1970,** *51*, 393–409.

King, D. P.; Zhao, Y.; Sangoram, A. M.; Wilsbacher, L. D.; Tanaka, M.; Antoch, M. P.; Turek, F. W. Positional Cloning of the Mouse Circadian Clockgene. *Cell*, **1997,** *89*, 641–653.

Kobayashi I. Behavior of Restriction-modification Systems as Selfish Mobile Elements and Their Impact On Genome Evolution. *Nuc. Acids Res.* **2001,** *29*, 3742–3756.

Krüger, D. H.; Bickle, T. A. Bacteriophage Survival: Multiple Mechanisms for Avoiding the Deoxyribonucleic Acid Restriction Systems of Their Hosts. *Microbiol. Rev.* **1983,** *47*, 345–360.

Lederberg, S.; Meselson, M. Degradation of Non-replicating Bacteriophage DNA in Non-accepting Cells. *J. Mol. Biol.* **1964,** *8*, 623–628.

Linn, S.; Arber, W. Host Specificity of DNA Produced by *Escherichia coli*, X. In vitro Restriction of Phage fd Replicative Form. *Proc. Natl. Acad. Sci. USA.* **1968,** *59*, 1300–1306.

Loenen, W. A. M.; Dryden, D. T. F.; Raleigh, E. A.; Wilson, G. G.; Murrayy, N. E. Highlights of the DNA Cutters: A Short History of the Restriction Enzymes. *Nuc. Acids Res.* **2014,** *42*, 3–19.

Luria, S. E.; Human, M. L. A Non Hereditary, Host-induced Variation of Bacterial Viruses. *J. Bacteriol.* **1952,** *64*, 557–569.

Mark, V. B.; Freyer, G. A.; Micklos, D. A. *Laboratory DNA Science: An Introduction to Recombinant DNA Techniques and Methods of Genome Analysis;* The Benjamin/Cummings Publishing Company: Menlo Park, **1996**.

Meisel, A.; Bickle, T. A.; Krüger, D. H.; Schroeder, C. Type III Restriction Enzymes Need Two Inversely Oriented Recognition Sites for DNA Cleavage. *Nature*, **1992,** *355*, 467–469.

Meselson, M.; Yuan, R. DNA Restriction Enzyme from *E. coli*. *Nature*, **2007,** *217*, 1110–1114.

Miller, M. R.; Dunham, J. P.; Amores, A.; Cresko, W. A.; Johnson, E. A. Rapid and Cost-effective Polymorphism Identification and Genotyping Using Restriction Site Associated DNA (RAD) Markers. *Genome Res.* **2007,** *17*, 240–248.

Murray, N. E. Type I Restriction Systems: Sophisticated Molecular Machines (A Legacy of Bertani and Weigle). *Microbiol. Mol. Biol. Rev.* **2000,** *64*, 412–434.

Nørholm, M. H. A Mutant Pfu DNA Polymerase Designed for Advanced Uracil-excision DNA Engineering. *BMC Biotechnol.* **2010,** *10*, 21.

Pingoud, A.; Jeltsch, A. Structure and Function of Type II Restriction Endonucleases. *Nuc. Acids Res.* **2001,** *29*, 3705–3727.

Pingoud, A. Restriction Endonucleases. In *Nucleic Acids and Molecular Biology*; Springer, **2004;** p 3.

Pingoud, A.; Wilson, G. G.; Wende, W. Type II Restriction Endonucleases—A Historical Perspective and More. *Nuc. Acids Res.* **2014,** *42*, 7489–7527.

Roberts, R. J. Restriction Enzymes and Their Isoschizomers. *Nucleic Acids Res.* **1990,** *18*, 2331–2365.

Roberts, R. J.; Vincze, T.; Posfai, J.; Macelis, D. REBASE—Enzymes and Genes for DNA Restriction and Modification. *Nuc. Acids Res.* **2007,** *35*, D269–270.

Roberts, R. J.; Vincze, T.; Posfai, J.; Macelis, D. REBASE: Restriction Enzymes and Methyltransferases. *Nuc. Acids Res.* **2003,** 31: 418–420.

Sanniyasi, E.; Selvarajan, R.; Velu, P.; Mylsamy, P. Construction of Plant Expression Vector of Synthetic *bt-cry1Ac* Gene for Genetic Transformation. *Adv. Biores.* **2013,** *53*, 11736–11740.

Smith, H. O.; Wilcox, K. W. A Restriction Enzyme from *Hemophilus influenzae*: I. Purification and General Properties. *J. Mol. Biol.* **1970,** *51*, 379–391.

Stephen, T. K.; Jocelyn, E. K.; Lewin, B.; Goldstein, E. *Lewin's genes X;* Jones and Bartlett: Boston, **2011**.

Strong, S. J.; Ohta, Y.; Litman, G. W.; Amemiya, C. T. Marked Improvement of PAC and BAC Cloning is Achieved Using Electroelution of Pulsed-field Gel-separated Partial Digests of Genomic DNA. *Nuc. Acids Res.* **1997,** *25*, 3959–3961.

Szybalski, W.; Kim, S. C.; Hasan, N.; Podhajska, A. J. Class-IIS Restriction Enzymes—A Review. *Gene*, **1991,** *100*, 13–26.

Tan, M. K.; Timmer, L. W.; Broadbent, P.; Priest, M, C. Differentiation by Molecular Analysis of *Elsinoë spp.* Causing Scab Diseases of Citrus and Its Epidemiological Implications. *Phytopathology*, **1996,** *86*, 1039–1044.

Titheradge, A. J.; King, J.; Ryu, J.; Murray, N. E. Families of Restriction Enzymes: An Analysis Prompted by Molecular and Genetic Data for Type ID Restriction and Modification Systems. *Nuc. Acids Res.* **2001,** *29*, 4195–4205.

Winnacker, E. L. Isolation, Identification, and Characterisation of DNA fragments (Chapter 2). In *Genes to Clones;* Verlagsgesellschaft, **1987**.

Wright, J. D.; Jack, W. E.; Modrich, P. The Kinetic Mechanism of EcoRI Endonuclease. *J Biol. Chem.* **1999,** *5*, 31896–31902.

Yuan, R. Structure and Mechanism of Multifunctional Restriction Endonucleases. *Annu. Rev. Biochem.* **1981,** *50*, 285–319.

CHAPTER 12

LIGATION OF INSERT DNA INTO CLONING VECTOR

TUSHAR RANJAN[1], PANKAJ KUMAR[2], BISHUN DEO PRASAD[2*], SANGITA SAHNI[3], VAISHALI SHARMA[4], SONAM KUMARI[2], RAVI RANJAN KUMAR[2], MAHESH KUMAR[2], VIJAY KUMAR JHA[5], and PRASANT KUMAR[6]

[1]*Department of Basic Science and Humanities Genetics, Bihar Agricultural University, Sabour, Bhagalpur, Bihar, India*

[2]*Department of Molecular Biology and Genetic Engineering, Bihar Agricultural University, Sabour, Bhagalpur, Bihar, India*

[3]*Department of Plant Pathology, T.C.A., Dholi, Muzaffarpur, Bihar, India*

[4]*Department of Botany, Patna University, Patna, Bihar, India*

[5]*DOS in Biotechnology, University of Mysore, Mysore, Karnataka, India*

[6]*Department of Fundamental and Applied Science, C. G. Bhakta Institute of Biotechnology, Uka Tarsadia University, Bardoli, Surat, Gujarat, India*

[*]*Corresponding author. E-mail: dev.bishnu@gmail.com*

CONTENTS

ABSTRACT

This single cell harboring recombinant DNA can then be grown exponentially to generate a large number of bacteria (descendants), each of these contain copies of the original recombinant molecule. Thus, both the resulting bacterial population and the recombinant DNA molecule are commonly referred to as "clones". In all living organisms, the basic chemical structure of DNA is fundamentally same and molecular cloning takes advantage of this fact. Hence, if a segment of DNA from the species *A* is inserted into a DNA segment containing the information required for DNA replication, and the resulting recombinant DNA is introduced into the species *B* from which the replication sequences have been derived, then the foreign DNA will be replicated along with the host cell's genome (DNA) in the transgenic organism (host). In a fascinating way, we can conclude that molecular cloning is very much similar to polymerase chain reaction (PCR) where it permits the multiplication of DNA sequence. The basic difference between these two popular methods is that molecular cloning involves replication of the DNA in a living host (microorganism) while PCR replicates DNA in an in vitro solution, i.e., free of living cells.

12.1 INTRODUCTION AND HISTORICAL BACKGROUND OF MOLECULAR CLONING

Earlier days, prior to the 1970s, our understanding of genetics and core molecular biology was severely hindered because of our inability to isolate and study the individual gene from organisms. But our perception has been changed dramatically with the advent of molecular cloning techniques and methods. Microbiologists have made successful attempts toward understanding the molecular mechanisms through which bacteria restrict the growth of bacteriophage and isolated many restriction endonucleases or enzymes that could cleave DNA molecules only when specific DNA sequences are encountered. Study showed that restriction enzymes have potential to cleave chromosome-length DNA molecules at specific locations, and those specific fragments of the larger DNA molecule could be separated and purified by size fractionation. Using another set of an enzyme, known as DNA ligase discovered in 1967, cleaved fragments generated by restriction enzymes could be joined in new combinations, termed recombinant DNA molecule (Nathans and Smith, 1975). By joining a DNA fragment of interest with the vector DNA (e.g., plasmids), which naturally multiplicates or replicates inside the bacteria, huge population of purified recombinant DNA molecules could be produced in bacterial cultures. The recombinant DNA molecules or chimeric DNA was first time generated and studied in 1972 (Fig. 12.1). Figure 12.1 represents the milestones of molecular cloning till date. Molecular cloning is a combination of experimental methods in the field of molecular biology that are basically employed to assemble recombinant DNA molecules and direct their multiplication (replication) within suitable host organisms. The word *cloning* basically refers to the replication of one molecule to produce a large population of cells with identical DNA molecules. Molecular cloning basically involves DNA sequences from two different organisms: One is the species that is the source of the DNA to be cloned, and the other species serves as the living host or the reactor for replication of the recombinant DNA construct (Patton et al., 2009; Watson, 2007). Molecular cloning methods are central to many concurrent areas of biotechnology and medicine where this technology provides biologically active molecules at low cost. In a traditional molecular cloning experiment, the DNA of interest to be cloned (obtained from an organism) is treated with the restriction enzymes in the test tube to generate smaller DNA fragments. Later on, these fragments are joined with vector DNA with the help of ligase enzyme to generate recombinant DNA molecules. The recombinant DNA or chimeric DNA molecules are then introduced into a host organism (*Escherichia coli*) via the process called transformation. This will generate a similar kind of population of organisms in which recombinant DNA molecules are replicated along with the host's own DNA molecule. Since this host harbors foreign DNA fragments (inter- or intraspecies), these hosts are called transgenic or genetically modified microorganisms (GMO). This overall process considers advantage of the fact that a single bacterial cell at a time can be induced to take up and replicate a single recombinant DNA molecule (Brown, 2006).

This single cell harboring recombinant DNA can then be grown exponentially to generate a large number of bacteria (descendants), each of these contain copies of the original recombinant molecule. Thus, both the resulting bacterial population and the recombinant DNA molecule are commonly referred to as "clones" (Brown, 2006). In all living organisms, the basic chemical structure of DNA is fundamentally same and molecular cloning takes advantage of this fact. Hence, if a segment of DNA from the species *A* is inserted into a DNA segment containing the information required for DNA replication, and the resulting recombinant DNA is introduced into the species *B* from which the replication sequences have been derived, then the foreign DNA will be replicated along with the host cell's genome (DNA) in

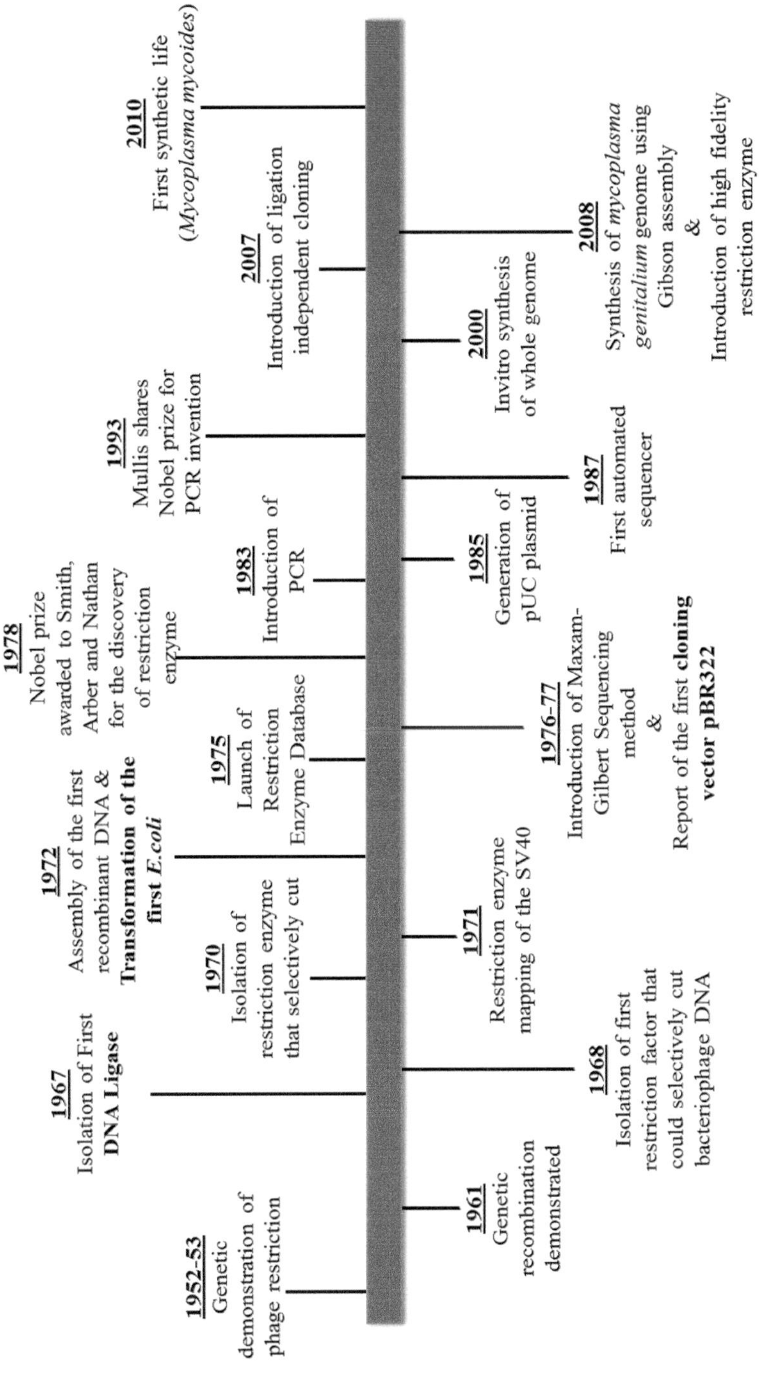

FIGURE 12.1 A brief history or milestones of molecular cloning.

the transgenic organism (host). In a fascinating way, we can conclude that molecular cloning is very much similar to polymerase chain reaction (PCR) where it permits the multiplication of DNA sequence. The basic difference between these two popular methods is that molecular cloning involves replication of the DNA in a living host (microorganism) while PCR replicates DNA in an in vitro solution, i.e., free of living cells (Cohen et al., 1975).

12.2 STEPS IN MOLECULAR CLONING

In an ideal molecular cloning experiment, the cloning of any desired DNA segment essentially involves seven critical steps: (1) selection of suitable host organism and cloning vector, (2) construction of appropriate vector DNA, (3) preparation of DNA (or insert) to be cloned, (4) creation of recombinant DNA, i.e., ligation of insert to the appropriate vector backbone, (5) transformation of recombinant DNA (insert + vector) into host organism, (6) selection of true recombinants, i.e., organisms containing recombinant DNA molecules, (7) screening for true clones with desired DNA inserts by exploring their biological properties (Fig. 12.2). Figure 12.2 represents the steps involved in the molecular cloning and Table 12.1 indicates the critical enzymes involved in the recombinant DNA technology.

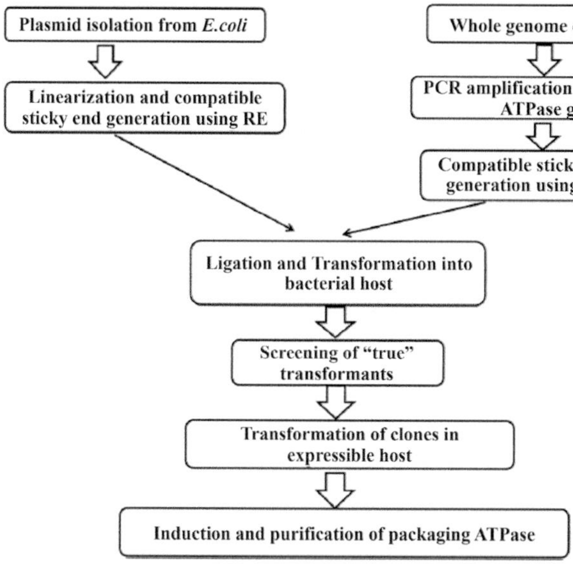

FIGURE 12.2 Critical steps involved in molecular cloning or genetic engineering or recombinant DNA technology.

TABLE 12.1 Enzymes Used in the Molecular Cloning.

Enzymes	Function
Type II restriction endonuclease	Cleaves DNA at specific sequence
DNA ligase	Joins two DNA fragments
DNA polymerase I	Fill gaps in duplexes by stepwise addition of nucleotides to the 3′ end
Reverse transcriptase	Make a DNA copy of RNA molecule
Polynucleotide kinase	Add phosphate to 5′-OH end of a polynucleotide to permit ligation
Terminal transferase	Add homopolymer tails to 3′-OH group
Exonuclease III	Remove nucleotides from 3′ end of DNA strand
Lambda exonuclease	Remove nucleotides from 5′ end of DNA strand to expose single stranded 3′end
Alkaline phosphatase	Remove terminal phosphate from either the 5′ or 3′ end of DNA

However, a large number of host organisms and molecular cloning vectors are available these days, but the great concern of molecular cloning experiment begins with the selection of appropriate strain of *E. coli* and a cloning vector or plasmid. In molecular experiment, *E. coli* and plasmid vectors are in common use because of their availability, versatility, and most importantly recombinant organisms could be grown rapidly with the minimal equipment. If the desired DNA to be cloned is exceptionally huge (>100 kb), a bacterial artificial chromosome (BAC) or yeast artificial chromosome (YAC) vector is often used (Brown, 2006). "Ori C" or origin of DNA replication is basically needed not only for the vector but also for the linked recombinant sequences (insert) to replicate inside the host organism. Generally, multiple cloning sites (MCS) with one or more unique restriction endonuclease recognition sites serve as the place where foreign DNA molecule may be incorporated. A plasmid does encode selectable genetic marker gene which basically enables the survival of host cells that have taken up vector sequences (i.e., true transformants). Another reporter or tag gene is also possessed by the vector that can be used for screening of true recombinant cells containing the foreign DNA from false positive colonies. If we want to purify and characterize a particular protein from the recombinant organisms, then an expression vector must be chosen that contains suitable signals for transcription and translation of desired genes. If transfer of DNA from bacteria to plants is desired, then a multiple host range vector or shuttle vector may be chosen.

12.3 PREPARATION OF CLONING VECTOR AND DNA INSERT

One should take care while choosing restriction enzymes to generate a configuration at the cleavage site in vector that is compatible with the ends of the foreign insert DNA. This is achieved by cleaving the vector DNA and foreign DNA with the same set of restriction enzymes. If necessary, short double-stranded segments of DNA (known as linkers) containing desired restriction sites may be added to create end structures that are compatible with the vector. Highly advanced vectors contain a MCS that are unique within the vector DNA backbone sequence. This helps in cleaving of a vector only at a single site and the MCS are mainly located within a gene (generally, beta-galactosidase reporter gene) whose inactivation helps in distinguishing of recombinant from nonrecombinant organisms while screening. To increase the ratio of recombinant to nonrecombinant clones, the cleaved vector is usually treated with an alkaline phosphatase enzyme that dephosphorylates the vector ends. Now, replication can only be restored if foreign DNA is integrated into the cleavage site because vector molecules with dephosphorylated ends are unable to replicate by host replicating enzymes (Russell and Sambrook, 2001). While cloning of genomic DNA, PCR methods are often used for the amplification of specific DNA or RNA (RT-PCR) sequences prior to molecular cloning (Fig. 12.3). The amplified DNA is then treated with a restriction enzyme to generate fragments with ends capable of being linked to the vector (Russell and Sambrook, 2001).

12.4 DNA LIGASE CREATED CHIMERIC DNA OR RECOMBINANT DNA

DNA ligase lies at the heart of molecular cloning and generation of recombinant DNA molecule is not possible without this. Both vector and foreign DNA are simply mixed together at appropriate concentrations in the presence of an enzyme DNA ligase that covalently links the ends together. Intriguingly, DNA ligase is capable of recognizing and acts on the ends of linear DNA molecules. This joining reaction between vector and insert DNA is often termed ligation. The ligated DNA mixture containing randomly joined ends is now ready for transformation into the host organism. The resulting ligated mixture contains the desired products (vector covalently linked to insert), but other unwanted products (such as insert-insert ligation, vector-vector ligation) are also usually present. This complex mixture is sorted out

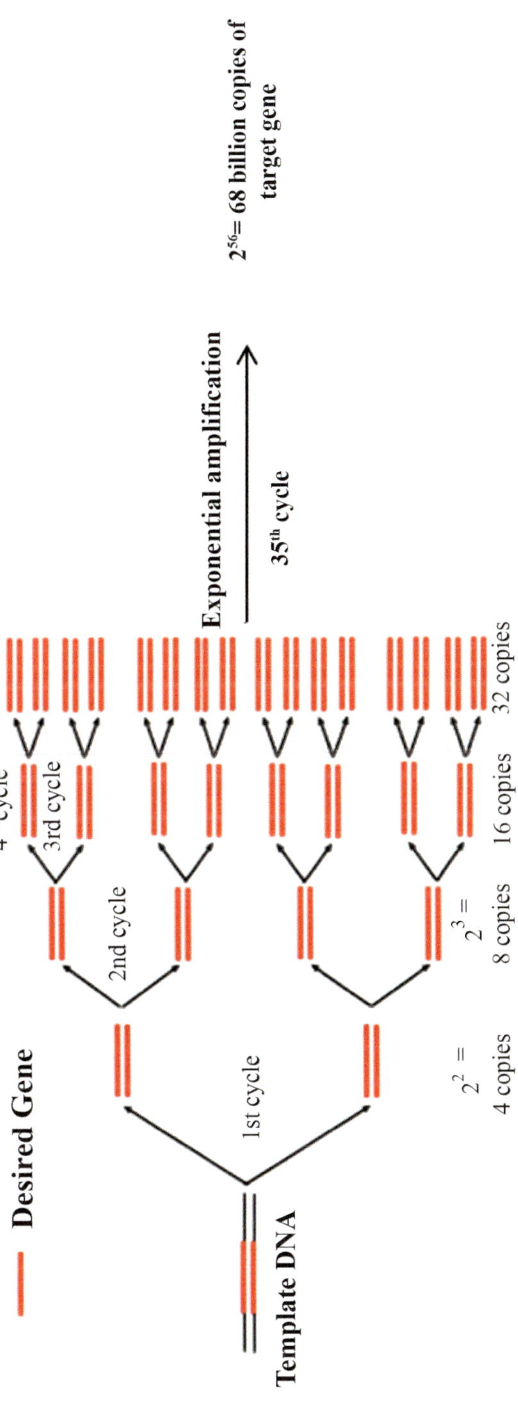

FIGURE 12.3 The principle and overview of amplification of desired insert (DNA) by PCR. (Reprinted with permission from Vierstraete, Andy, Principle of the PCR, 1999. https://users.ugent.be/~avierstr/principles/pcr.html.)

during the next step of cloning process, called as transformation, while the DNA mixture is introduced into cells (Brown, 2006; Russell and Sambrook, 2001).

Ligation refers to the joining of two or more than two DNA fragments through the formation of a phosphodiester bond. In the cell, ligases repair single- and double-strand breaks that occur during DNA replication. In the laboratory, DNA ligase is used during molecular cloning to join DNA fragments of inserts with vectors which will replicate target fragments in host organisms. All cells have DNA ligases, but two ligases are particularly important: *E. coli* DNA ligase and bacteriophage T4 DNA ligase. *E. coli* and T4 DNA ligase require NAD^+ and ATP, respectively, as a cofactor. T4 DNA ligase joins cohesive and blunt ends effectively, whereas *E. coli* DNA ligase joins only cohesive ends. T4 DNA ligase is the enzyme of choice in the laboratory for joining DNA molecules together.

12.4.1 DNA LIGASE: A GLIMPSE INTO THE STRUCTURE

In the early 1960s, two groups discovered that genetic recombination could occur through the breakage and ligation of DNA molecules closely followed by the observation that linear bacteriophage DNA is rapidly converted to covalently closed circles after infection of the host. Just 2 years later, five groups independently isolated DNA ligases and demonstrated their ability to assemble two pieces of DNA. Not long after the discovery of restriction enzymes and DNA ligases, the first recombinant DNA molecule was made. In 1972, Berg separately cut and ligated a piece of lambda bacteriophage DNA or the *E. coli* galactose operon with SV40 DNA to create the first recombinant DNA molecules. These studies pioneered the concept that, because of the universal nature of DNA, DNA from any species could be joined together. In 1980, Paul Berg shared the Nobel Prize in Chemistry with Walter Gilbert and Frederick Sanger (the developers of DNA sequencing), for his fundamental studies of the biochemistry of nucleic acids, with particular regard to recombinant DNA (Jackson et al., 1972).

In the area of molecular biology, DNA ligase is a specific type of enzyme, a ligase, (EC 6.5.1.1) that facilitates the joining of DNA strands together by catalyzing the formation of a phosphodiester bond via the mechanism of nucleophilic attack. It plays a role in repairing single-strand breaks in duplex DNA in living organisms, but some forms (such as DNA ligase IV) may specifically repair double-strand breaks (i.e., a break in both complementary strands of DNA). Single-strand breaks are repaired by DNA ligase using the complementary strand of the double helix as a template, with DNA ligase creating the final phosphodiester bond to fully repair the DNA. Thus, DNA ligase is an enzyme that has a very critical role in the process of DNA replication and DNA repair. The strands of DNA are antiparallel, meaning that they are parallel to each other but in the opposite direction. DNA helicase opens the DNA so that both strands are exposed. During the process of replication, this double helix is unwound, and nucleotide bases are added to both strands going in the 5′ to 3′ direction. For one strand, a new continuous strand of DNA is synthesized. For the other strand, small fragments of DNA are synthesized as more and more of the DNA is opened and replicated. These fragments are called Okazaki fragments. Since the DNA is not one continuous piece at this point, something has to be done in order to make it into one continuous piece. This is where DNA ligase comes in. DNA ligase makes the DNA continuous by using the template strand to fill in the complementary bases (Bode and Kaiser, 1965; Cozzarelli et al., 1967; Meselson and Weigle, 1961).

Although DNA ligases have been discovered many years ago, we did not have much idea about their catalytic mechanism. However, in the recent years, crystal structures of many DNA ligases have been solved and given insights into the detailed molecular architecture of a DNA ligase, mainly the ATP-dependent ligase of T7 phage. Determination of these crystal structures has given us valuable molecular insights into the enzyme catalytic mechanism of nucleotidyl transferases. Recently, elucidation of the half-length structures, for example, adenylation domain of *Bacillus stearothermophilus* (*Bst*) ligase and the full-length *Thermus filiformis* (*Tfi*) ligase has greatly deepened our understanding of the multidomain NAD^+-dependent ligases. Here, we revisit the current knowledge of the structures aspect of DNA ligases and discuss the detailed structural and mechanistic implications arising from these recent study (Engler and Richardson, 1982; Singleton et al., 1999). The discovery of crystal structures of the ATP-dependent and NAD^+-dependent ligase has revealed that DNA ligases have a remarkably highly modular architecture with a unique arrangement of two or more discrete domains (Fig. 12.4). Recent study indicated that five classes of motifs [such as nucleotide-binding domain, oligomer-binding (OB) fold, zinc finger, helix–hairpin–helix (HhH) motif, and BRCA1 C-terminus (BRCT) domain] have been found in both sequence and at the structural level (Fig. 12.4). The detailed structure of bacteriophage T7 ligase reported that the enzyme consists of two distinct domains (Fig. 12.4), a larger N-terminal domain (green) designated as domain 1 and a small C-terminal domain (red) designated as domain 2 (Fig. 12.4). The domain 1 spreading over residues

1–240 is made up of three antiparallel β-sheets flanked by six α-helices. This domain basically possesses the ATP-binding and hydrolysis pocket situated just beneath the β-sheets. Ubiquitously, adenylation domain 1 of most of the DNA ligases is connected to a highly conserved domain 2 (Fig. 12.4, in red). High resolution crystal structures of T7 and *Tfi* DNA ligase revealed that OB fold is basically derivative of a Greek key motif, commonly observed in the structures of many DNA binding proteins. This finding nicely explains how ligase interacts with the DNA molecule and facilitates nucleophilic attack. Apart from that four cysteine residues are also found to be conserved in the C-terminal region of NAD^+-dependent ligases and indicate their role in zinc binding while interacting with the DNA. Recent atomic emission spectroscopy data has strengthened this theory and confirmed that *Tfi* ligase binds zinc ions. In the *Tfi* ligase structure, a zinc ion is tetrahedrally coordinated by the four conserved cysteine residues (Cys406, Cys409, Cys422, and Cys427). This single zinc finger forms a subdomain (3a) of the larger domain 3 of *Tfi* ligase (Fig. 12.4). It is assumed that all the NAD^+-dependent ligases have a similar kind of zinc finger motifs, since the four cysteine residues which coordinate the zinc ion are strictly conserved throughout eubacterial DNA ligases. As we know zinc fingers motifs act as DNA recognition modules, often recognizing specific DNA sequences (Fig. 12.4) (Kodama et al., 1991; Subramanya et al., 1996).

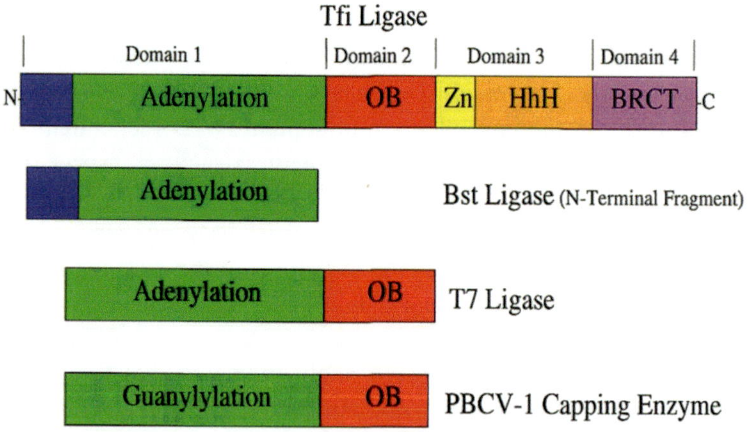

FIGURE 12.4 Domain structure of DNA ligases. Schematic representation of the domain architecture of the known DNA ligase structure. The domains are color coded: subdomain 1a, blue; subdomain 1b (adenylation), green; domain 2 (Oligo binding), red; subdomain 3a (zinc finger), yellow; subdomain 3b (helix-hairpin-helix), orange; domain 4 (BRCA1 C-terminus), pink. (Reprinted from Doherty. A. J.; Suh, S. W. Structural and Mechanistic Conservation in DNA Ligases. Nucleic Acids Res. 2000, 28(21), 4051–4058. https://www.ncbi.nlm.nih.gov/pmc/articles/PMC113121/ © 2000 Oxford University Press.)

12.4.2 TYPES OF DNA LIGASE

There are basically three types of DNA ligase on the basis of their origin or source.

1. *E. coli* DNA ligase

The *E. coli* DNA ligase is encoded by the *lig* gene. Prokaryotic DNA ligase utilizes energy by cleaving nicotinamide adenine dinucleotide (NAD) for phosphodiester bond formation. It does not ligate blunt-ended DNA except under conditions of molecular crowding with polyethylene glycol (PEG),and also cannot join RNA to DNA efficiently (Foster and Slonczewski, 2010).

2. T4 DNA ligase

The DNA ligase from bacteriophage T4 is the most commonly used ligase in laboratory research. It can quickly ligate sticky ends of DNA, as well as RNA and RNA-DNA hybrids, but not single-stranded nucleic acids. It can also ligate blunt-ended DNA with much greater efficiency than prokaryotic DNA ligase. Unlike *E. coli* DNA ligase, T4 DNA ligase cannot utilize NAD as an energy source and critically requires ATP as a cofactor. Interestingly, some engineering has been done to improve the in vitro activity of T4 DNA ligase. One successful approach, for example, tested T4 DNA ligase were fused to several alternative DNA binding proteins and found that enzymes are more active in blunt-end ligations than wild-type T4 DNA ligase (Wilson et al., 2013).

3. Mammalian ligases

In mammals, there are four specific types of ligase. DNA ligase from eukaryotes and some microbes uses adenosine triphosphate (ATP) rather than NAD.

a. **DNA ligase I:** Ligates the nascent DNA of the lagging strand after the Ribonuclease H has removed the RNA primer from the Okazaki fragments.
b. **DNA ligase III:** Complexes with DNA repair protein XRCC1 to aid in sealing DNA during the process of nucleotide excision repair and recombinant fragments. Of the all known mammalian DNA ligases, only Lig III has been found to be present in mitochondria.

c. **DNA ligase IV**: Complexes with XRCC4 catalyze the final step in the nonhomologous end joining DNA double-strand break repair pathway. It is also required for V(D)J recombination, the process that generates diversity in immunoglobulin and T-cell receptor loci during immune system development (Russell and Sambrook, 2001).

12.4.3 MECHANISM OF THE DNA LIGASE REACTION

The mechanism of DNA ligase is to form two covalent phosphodiester bonds between 3′ hydroxyl ends of one nucleotide, ("acceptor") and 5′ phosphate end of another ("donor") (Fig. 12.5). ATP is required for the ligase reaction which proceeds in three steps:

1. Adenylation (addition of AMP) of a lysine residue in the active center of the enzyme, pyrophosphate is released;
2. Transfer of the AMP to the 5′ phosphate of the so-called donor, formation of a pyrophosphate bond;
3. Formation of a phosphodiester bond between the 5′ phosphate of the donor and the 3′ hydroxyl of the acceptor.

Despite their occurrence in all organisms, DNA ligases show a wide diversity of amino acid sequences, molecular sizes, and properties. The enzymes fall into two groups based on their cofactor specificity, those requiring NAD^+ for activity and ATP. The eukaryotic, viral, and archael bacteria encoded enzymes all require ATP. NAD^+-requiring DNA ligases have only been found in prokaryotic organisms (Fig. 12.5).

A direct physical interaction between domains of DNA ligase has been demonstrated by several techniques such as gel filtration. Two crystal structures of PBCV-1 mRNA capping enzyme have provided conclusive evidence for such a conformational change during the guanylylation reaction in the capping enzymes. This involves a 13 Å movement of the C-terminal OB domain 2 toward domain 1. This reaction is equivalent to the adenylation reaction catalyzed by DNA ligases. During this conformational change by PBCV-1 capping enzyme, conserved residues in motifs V and VI, located in the OB domain, are positioned in the active site and form specific interactions with the nucleotide. Motif VI encompasses the last strand (in blue) of the "OB fold" domain 2 (Fig. 12.4). Two residues (R295 and K298) from the corresponding motif in the capping enzyme bind and position the

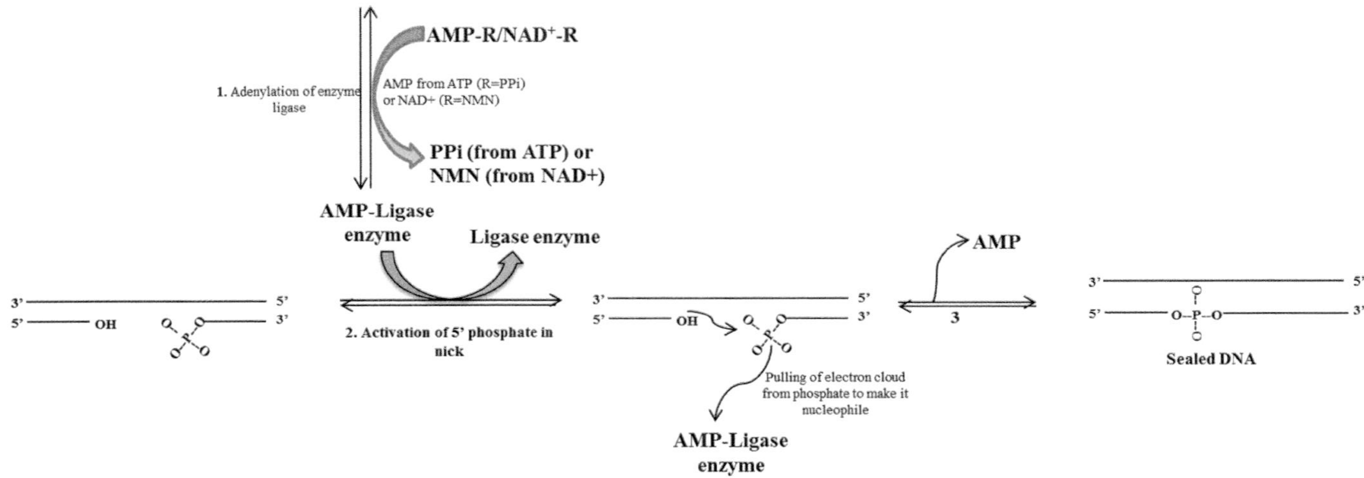

FIGURE 12.5 Mechanism of action of ligation. Steps (1) and (2) lead to the activation of 5′ phosphate in the nick. An AMP group is first transferred to lysine residue on the enzyme and then to 5′ phosphate in the nick. In Step (3), 3′-OH group attacks this phosphate and displaces AMP, producing a phosphodiester bond to seal the nick. The source of AMP is NAD+ and ATP in prokaryotes and eukaryotes, respectively. Viral ligase also uses ATP as a source of AMP. (Modified from Recombinant DNA Technology, http://sivabio.50webs.com/dnacloning.htm)

triphosphate tail of GTP for a direct in-line attack by the active-site lysine. It is likely that this motif also plays a similar role in the adenylation reaction of ligases and this is supported by mutagenesis studies on the PBCV-1 DNA ligase (Dhoerty and Wigley, 1999; Dohetry and Suh, 2000; Singleton et al., 1997).

The term recombinant DNA encapsulates the concept of recombining fragments of DNA from different sources into a new and hopefully useful DNA molecule. Joining linear DNA fragments together with covalent bonds is called ligation. More specifically, DNA ligation involves creating a phosphodiester bond between the 3′ hydroxyl of one nucleotide and the 5′ phosphate of another. The enzyme used to ligate DNA fragments is T4 DNA ligase which originates from the T4 bacteriophage. This enzyme will ligate DNA fragments having overhanging, cohesive that are annealed together. T4 DNA ligase will also ligate fragments with blunt ends although higher concentrations of the enzyme are usually recommended for this purpose (Fig. 12.6). A ligation reaction requires three ingredients in addition to water: (1) two or more fragments of DNA that have either blunt or compatible cohesive ("sticky") ends; (2) a buffer which contains ATP. The buffer is usually provided or prepared as a 10X concentrate which, after dilution, yields an ATP concentration of roughly 0.25–1 mM. Most restriction enzyme buffers will work if supplemented with ATP; (3) T4 DNA ligase; a typical reaction for inserting a fragment into a plasmid vector (sub cloning) would utilize about 0.01 (sticky ends) to 1 (blunt ends) units of ligase. The optimal incubation temperature for T4 DNA ligase is 16°C and when very high efficiency ligation is desired (e.g., making libraries) this temperature is recommended. However, ligase is active at a broad range of temperatures, and for routine purposes such as sub cloning, convenience often dictates incubation time and temperature—ligations performed at 4°C overnight or at room temperature for 30 min to a couple of hours usually work well (Gefter et al., 1967; Russell and Sambrook, 2001).

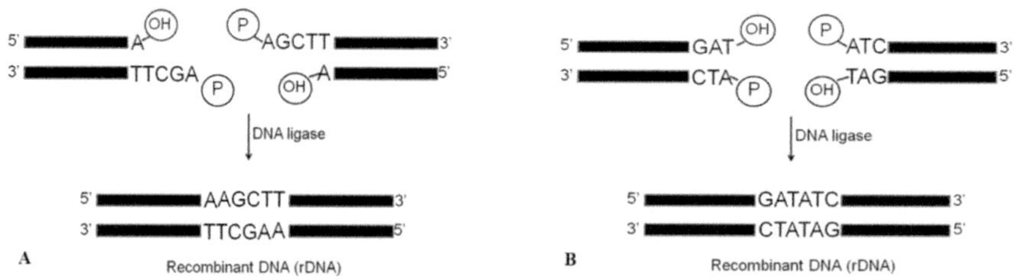

FIGURE 12.6 Ligation of sticky (A) and blunt (B) end of DNA fragment.

12.4.4 STEPS REQUIRED FOR LIGATION

A. Preparation of vector DNA

1. Digest the vector DNA using restriction digestion
   a. Use blunt-end cutter RE (e.g., *Eco* RV, *Sma* I, *Pvu* II, etc.) for blunt-end ligation
   b. Use sticky-end cutter RE (e.g., *Hind* III, *Eco* RI, *Bam* HI, etc.) for sticky-end ligation
2. Check for complete digestion on agarose gel
3. Elution of digested linear vector DNA by running in LMP agarose
4. Check the concentration of DNA by running a gel and spectrophotometer analysis

B. Preparation of Insert DNA

1. For blunt-end ligation

a. PCR product (using proofreading DNA polymerase enzyme like *Pfu*, Phusion, *XT*, etc.) can be used as insert
b. Digest plasmid DNA from which insert is going to be prepared using blunt-end cutter RE
c. Digest genomic DNA from which insert is going to be prepared using blunt-end cutter RE

2. For sticky-end ligation

a. Digest plasmid DNA from which insert is going to be prepared using sticky end cutter RE
b. Digest genomic DNA from which insert is going to be prepared using sticky-end cutter RE
c. PCR product using Taq. DNA polymerase (Taq. DNA polymerase has a tendency to add an extra A)
d. PCR product (amplified using specific primer having enzyme sites at 5′ end) digested using sticky-end cutter RE

12.4.5 CALCULATE THE AMOUNT OF INSERT AND VECTOR NEEDED FOR THEIR LIGATION

For sticky-end ligations, an average inserts to vector molar ratio of about 3:1. A 5:1 to 10:1 ratio is suggested for blunt-end ligations. Please note that this is a molar ratio: You need to consider the size of the DNA, not just its concentration. The amount of insert required in ligation reaction can be calculated by using following given formula:

$$\text{Amount of insert required }(\text{ng}) = \frac{\text{vector (ng)} \times \text{size of insert (kb)}}{\text{size of vector (kb)}} \times \text{molar ratio of} \frac{\text{insert}}{\text{vector}}$$

Example:

How much 0.5 kb insert DNA needs to be added to 100 ng of 3.0 kb vector in a ligation reaction for a desired vector:insert ratio of 1:3?

$$\text{Amount of insert required }(\text{ng}) = \frac{100\text{ ng vector} \times 0.5\text{ kb insert}}{3.0\text{ kb}} \times \frac{3}{1} = 50\text{ ng.}$$

Ligation reaction setup:

Vector DNA: 1 μL (50–100 ng)
Insert DNA:— μL
Ligation buffer (5X): 2 μL
T4 DNA ligase: 1 μL (200 U for sticky end and 400 U for blunt-end ligation)
Sterile RO water: — μL
Total volume: 10 μL

Mix gently by flicking with your finger and spin for 2 s in a microcentrifuge to get the liquid down to the bottom of the tube. Incubate at 16°C overnight. Ligation reaction can also be incubated from 4°C to 20°C for 4–24 h.

For ligation reaction following controls needs to be included:

	Ligation 1	Ligation 2	Ligation 3	Ligation 4
Vector	√	√	–	√
Insert	√	–	√	√
Buffer	√	√	√	√
DNA ligase	√	√	√	–

12.5 MEASUREMENT OF LIGASE ACTIVITY

There are at least three different units used to measure the activity of DNA ligase:

1. **Weiss unit**: The amount of ligase that catalyzes the exchange of 1 nmole of ^{32}P from inorganic pyrophosphate to ATP in 20 min at 37°C. This is the one of the most commonly used unit.
2. **Modrich–Lehman unit**: This is rarely used, and one unit is defined as the amount of enzyme required to convert 100 nmoles of d $(A\text{-}T)_n$ to an Exonuclease-III resistant form in 30 min under standard conditions.
3. Many commercial suppliers of ligases use an arbitrary unit based on the ability of ligase to ligate cohesive ends. These units are often more subjective than quantitative and lack precision (Russell and Sambrook, 2001).

12.6 APPLICATION OF DNA LIGASE IN MOLECULAR BIOLOGY RESEARCH

DNA ligases have become indispensable tools in modern molecular biology research for generating recombinant DNA sequences. For example, DNA ligases are used with restriction enzymes to insert DNA fragments, often genes, into plasmids. Controlling the optimal temperature is a vital aspect of performing efficient recombination experiments involving the ligation of cohesive-ended fragments. Most experiments use T4 DNA ligase (isolated from bacteriophage T4) which is most active at 37°C. However, for optimal ligation efficiency with cohesive-ended fragments or sticky ends, the optimal enzyme temperature needs to be balanced with the melting temperature T_m of the sticky ends being ligated. The homologous pairing of the sticky ends will not be stable because the high temperature disrupts hydrogen bonding. A ligation reaction is most efficient when the sticky ends are already stably annealed, and disruption of the annealing ends would therefore result in low ligation efficiency. The shorter the overhang, lower the T_m.

Since blunt-ended DNA fragments have no cohesive ends to anneal, the melting temperature is not a factor to be considered within the normal temperature range of the ligation reaction. However, the higher the temperature, the lower the chance that the ends to be joined will be aligned to allow for ligation (molecules move around the solution more at higher temperatures). The limiting factor in blunt-end ligation is not the activity of the ligase but rather the number of alignments between DNA fragment ends that occur. The most efficient ligation temperature for blunt-ended DNA would therefore be the temperature at which the greatest number of alignments can occur. The majority of blunt-ended ligations are carried out at 14°C–25°C overnight. The absence of stably annealed ends also means that the ligation efficiency is lowered, requiring a higher ligase concentration to be used (Baneyx et al., 1993; Tabor, 2001).

12.7 INTRODUCTION OF CHIMERIC DNA INTO HOST ORGANISM

When microorganisms are able to take up and replicate DNA from their local environment, the process is termed transformation, and cells that are in a physiological state such that they can take up DNA are said to be competent. Generally, heat shock and $CaCl_2$ method is more reliable and frequently used for bacterial transformation. In this method recombinant DNA molecule enters inside the cell via the transitory hole in the plasma membrane induced by heat shock. $CaCl_2$ basically helps in binding of DNA molecules and their docking to the plasma membrane surface. In mammalian cell culture, the analogous process of introducing DNA into cells is commonly termed transfection. Both transformation and transfection usually require preparation of the cells through a special growth regime and chemical treatment process that will vary with the specific species and cell types that are used. Electroporation uses high voltage electrical pulses to translocate DNA across the cell membrane (and cell wall, if present). In contrast, transduction involves the packaging of DNA into virus-derived particles, and using these virus-like particles to introduce the encapsulated DNA into the cell through a process resembling viral infection. Although electroporation and transduction are highly specialized methods, they may be the most efficient methods to move DNA into cells (Lederberg, 1994; Wirth et al., 1994).

12.8 SELECTION OF ORGANISMS CONTAINING VECTOR SEQUENCES

The introduction of recombinant DNA into the chosen host organism is usually a low-efficiency process and only a small fraction of the cells actually takes up DNA. We deal with this issue through a step of genetic selection, in which cells that have not taken up DNA are selectively killed, and only those cells that can actively replicate DNA containing the selectable marker gene encoded by the vector are able to survive. When bacterial cells are used as host organisms, the selectable marker is usually a gene that confers resistance to an antibiotic that would otherwise kill the cells. Cells harboring the plasmid will survive when exposed to the antibiotic while those that have failed to take up plasmid sequences will die (Brown, 2006; Russell and Sambrook, 2001).

12.8.1 SCREENING FOR CLONES WITH DESIRED DNA INSERTS AND BIOLOGICAL PROPERTIES

Modern bacterial cloning vectors (such as pGEM) use the blue-white screening system to distinguish colonies of transgenic cells from those that contain the parental vector (i.e., vector DNA with no recombinant sequence inserted). In these vectors, foreign DNA is inserted into a sequence that encodes an essential part of beta-galactosidase, an enzyme whose activity results in formation of a blue-colored colony on the culture medium. Insertion of the foreign DNA into the beta-galactosidase coding sequence disables the function of the enzyme so that colonies containing transformed DNA remain white. Therefore, we can easily identify and conduct further studies on transgenic bacterial clones while ignoring those that do not contain recombinant DNA (blue). The total population of individual clones obtained in a molecular cloning experiment is often termed a DNA library. Further screening may be accomplished through a very wide range of methods, including nucleic acid hybridizations, PCR, restriction fragment analysis, and at last DNA sequencing (Brown, 2006; Russell and Sambrook, 2001).

12.9 CONCLUSION

Molecular cloning has paved a way directly to the elucidation of the complete DNA sequence of the genomes of a very large number of species and to an exploration of genetic diversity within individual species. DNA ligases have become vital tools in modern molecular biology research for generating recombinant DNA molecules (Brown, 2006). The DNA segment, which may be a gene, can be isolated and purified from a prokaryotes or eukaryotes. Following isolation of the gene of interest (insert), both the vector and insert must be cut with restriction enzymes and then purified. At the level of individual genes, molecular clones are used to generate probes that are being used for examining how genes are expressed, and how that expression is related to other processes in biology, including the metabolic environment, extracellular signals, development, learning, senescence, and cell death. Cloned genes can also provide tools to examine the biological function and importance of individual genes by allowing investigators to inactivate the genes, or make more subtle mutations using regional mutagenesis or site-directed mutagenesis (Russell and Sambrook, 2001). Obtaining the molecular clone of a gene can lead to the development of organisms that produce the protein product of the cloned genes, termed a recombinant protein.

KEYWORDS

- **molecular cloning**
- **T4 DNA ligase**
- **vector**
- **polymerase chain reaction (PCR)**
- **restriction enzyme**
- **ligation**
- **transformation**

REFERENCES

Baneyx, F.; Lucotte, G. *Introduction to Molecular Cloning Techniques;* John Wiley & Sons: Chichester, **1993.** p. 156.

Bode, V. C.; Kaiser, A. D. . Changes in the Structure and Activity of λ DNA in a Superinfected Immune Bacterium. *J. Mol. Biol.* **1965,** *14*(2),399–417.

Brown, T. *Gene Cloning and DNA Analysis: An Introduction.* Blackwell: Cambridge, MA, **2006**.

Cohen, S. N.; Chang, A. C.; Boyer, H. W.; Helling, R. B. Construction of Biologically Functional Bacterial Plasmids In Vitro. *Proc. Natl. Acad. Sci. U. S. A.* **1973,** *70*(11), 3240–3240.

Cozzarellig, N. R.; Melechen, N. E.; Jovin, T. M.; Kornberg, A. Polynucleotide Cellulose as a Substrate for a Polynucleotide Ligase Induced by Phage T4. *Biochem. Biophys. Res. Commun.* **1967,** *28*(4), 578–586.

Doherty. A. J.; Suh, S. W. Structural and Mechanistic Conservation in DNA Ligases. *Nucleic Acids Res.* **2000,** *28*(21), 4051–4058.

Dhoerty, A. J.; Wigley, D. B. Functional Domains of an ATP-dependent DNA Ligase. *J. Mol. Biol.* **1999,** *285*(1), 63–71.

Engler, M. J.; Richardson, C. C. In *The Enzymes;* Boyer, P. D., Ed.; Academic Press: New York, NY, **1982,** Vol. 15, pp. 3–29.

Foster, J. B.; Slonczewski, J. *Microbiology: An Evolving Science* (2nd Edn.); W. W. Norton & Company: New York, **2010**.

Gefter, M. L; Becker, A.; Hurwitz, J. The Enzymatic Repair of DNA. I. Formation of Circular Lambda-DNA. *Proc. Natl. Acad. Sci. U. S. A.* **1967,** *58*(1), 240–247.

Subramanya, H. S.; Doherty, A. J.; Ashford, S. R.; Wigley, D. B. Crystal Structure of an ATP-dependent DNA Ligase from Bacteriophage T7. *Cell* **1996,** *85*(4), 607–615.

Jackson, D. A.; Symons, R. H.; Berg, P. Biochemical Method for Inserting New Genetic Information into DNA of Simian Virus 40: Circular SV40 DNA Molecules Containing Lambda Phage Genes and the Galactose Operon of *Escherichia coli*. *Proc. Natl. Acad. Sci. U. S. A*. **1972,** *69*(10), 2904–2909.

Kodama, K.; Barnes, D. E.; Lindahl, T. In vitro Mutagenesis and Functional Expression in *Escherichia coli* of a cDNA Encoding the Catalytic Domain of Human DNA Ligase I. *Nucleic. Acids Res.* **1991,** *19*(22), 6093–6099.

Lederberg, J. The Transformation of Genetics by DNA: An Anniversary Celebration of Avery, MacLeod, and McCarty. *Genet.* **1994,** *136*(2): 423–6.

Lodish, H.; Berk, A.; Zipursky, S. L. *Molecular Cell Biology* (4th Edn.); W. H. Freeman: New York, **2000;** Section 7.1, DNA Cloning with Plasmid Vectors.

Meselson, M.; Weigle, J. J. Chromosome Breakage Accompanying Genetic Recombination in Bacteriophage. *Proc. Natl. Acad. Sci. U. S. A.* **1961,** *47*(6), 857–868.

Nathans, D.; Smith, H. O. Restriction Endonucleases in the Analysis and Restructuring of DNA Molecules. *Annu. Rev. Biochem.* **1975,** *44*, 273–293.

Lehninger, A. L; Nelson, D. L.; Cox, M. M. *Principle of Biochemistry* (5th Edn.); **2008**.

Patten, C. L.; Glick, B. R.; Pasternak, J. *Molecular Biotechnology: Principles and Applications of Recombinant DNA;* ASM Press: Washington, D.C., **2009**.

Russell, D. W.; Sambrook, J. *Molecular Cloning: A Laboratory Manual;*Cold Spring Harbor Laboratory: Cold Spring Harbor, NY, **2001**.

Singleton, M. R.; Håkansson, K.; Timson, D. J.; Wigley, D. B. Structure of the Adenylation Domain of an NAD^+-dependent DNA Ligase. *Struct. Fold. Des.* **1999,** *7*(1), 35–42.

Tabor, S. DNA Ligases. In *Current Protocols in Molecular Biology;* Chapter 3: Unit 3.14. doi:10.1002/0471142727.mb0314s08.

Watson, J. D. *Recombinant DNA: Genes and Genomes: A Short Course;* W.H. Freeman: San Francisco, **2007**.

Wilson, R. H.; Morton, S. K.; Deiderick, H.; Gerth, M. L.; Paul, H. A.; Gerber, I. Patel, A.; Ellington, A. D.; Hunicke-Smith, S. P.; Patrick, W. M. Engineered DNA Ligases with Improved Activities In Vitro". *Protein Eng. Des. Sel.* **2013,** *26*(7), 471–478.

Wirth, R.; Friesenegger, A.; Fiedler, S. Transformation of Various Species of Gram-negative Bacteria Belonging to 11 Different Genera by Electroporation. *Mol. Gen. Genet.* **1989,** *216*(1), 175–177.

CHAPTER 13

BLOTTING TECHNIQUES

PRASANT KUMAR[1*], MITESH DWIVEDI[1], CHANDRA PRAKASH[2], SANGITA SAHNI[3], and BISHUN DEO PRASAD[4]

[1]*C. G. Bhakta Institute of Biotechnology, Uka Tarsadia University, Tarsadi, Surat, Gujarat 394350, India*

[2]*Department of Microbiology and Biotechnology Centre, Faculty of Science, Maharaja Sayajirao University of Baroda, Vadodara, Gujarat 390002, India*

[3]*Department of Plant Pathology, Tirhut College of Agriculture, Dholi, RAU, Pusa, Bihar, India*

[4]*Department of Plant Breeding and Genetics, Bihar Agricultural University, Sabour, Bihar 813210, India*

[*]*Corresponding author. E-mail: prasantmmbl@gmail.com*

CONTENTS

ABSTRACT

During the past few years the development of nucleic acid and protein blotting techniques has changed to such a extent that it is important to incorporate the changes and modifications that have been proved to be necessary for the improvement of the technique. Blotting of nucleic acid and protein on membrane is one of the most important techniques in the molecular biology. Various types of blotting techniques have been developed according to the erratic nature of the transferring material like DNA, RNA and protein. This chapter has focused on the principle of blotting technique, its applications and different methodologies have been used for past several years. Various modifications have also been introduced in the blotting techniques and their advantages and disadvantages are discussed. At the end of this chapter, the blotting of combination of transferring material like DNA-protein on the PVDF has been discussed.

13.1 INTRODUCTION

Biomolecules like deoxyribonucleic acids (DNA), ribonucleic acids (RNA), and proteins are basic, essential, and primary molecules for all molecular biology-related research. DNA is an almost unique molecule made of four different nitrogenous bases called adenine, thymine, guanine, and cytosine, which are abbreviated as A, T, G, and C, deoxyribose sugar and phosphate. Each base pair with complementary strand base and have one of four compositions, that is, AT, TA, CG, or GC. For each position on the strand, it is possible to control which base pair is present. A always pairs with T and not G or C, G pair with C and not to A or T. Nitrogenous base pair with complementary base through hydrogen bond. Due to these limitations, the only possible base pairs are AT and GC or TA and CG. A molecule of DNA is organized in the form of two complementary strands of nucleotides wound in a double helix. One of these strands is called the sense strand, and other the antisense strand or similarly coding and noncoding strand. The sequences present in the strand are read in a very sophisticated way to synthesize mRNA and translate the information in the form of a protein.

Extraction of DNA, RNA, and protein are the first and the most basic methods used in molecular biology (Tan and Yiap, 2009). These biomolecules can be isolated from any biological material for subsequent downstream processes, analytical, or preparative purposes. Blotting techniques are used to identify the unique nucleic acid such as DNA and RNA or proteins sequences. All the blotting techniques are based on hybridization, in these techniques the immobilization of samples (nucleic acid or protein) onto a solid support, generally nylon or nitrocellulose membranes (Primrose and Twyman, 2006). The blotted samples (nucleic acid or protein) are then used as "targets" in subsequent hybridization experiments. They have been developed to be highly specific and sensitive and have become one of the important tools used in both molecular biology and genetics. A blot in molecular biology and genetics refers to methods of transferring DNA, RNA, or protein onto the carriers.

13.2 PRINCIPLE OF BLOTTING TECHNIQUES

Blotting technique is used for transferring biomolecules such as DNA, RNA, or protein from gel to a solid support such as nitrocellulose filter or nylon membrane, that is, blot. To detect a specific fragment of DNA or RNA, or a specific protein, a "probe" is applied to the blot, which will hybridize or bind specifically to the target molecule of interest. For DNA or RNA, the probe is a complementary single-strand sequence and for protein, the probe is an antibody to the protein of interest.

The four basic steps to be followed in all the blotting techniques are: (1) electrophoretic separation of nucleic acid fragments or of protein in the sample; (2) transfer of the gel and immobilization on the membrane support; (3) binding of analytical probe to the target molecule on the membrane; and (4) visualization of bound probe (Hayes et al., 1989).

13.3 TYPES OF BLOTTING

Blotting techniques are of various types based on the biomolecules used to identify in the samples.

1. *Southern Blotting*: Used to detect specific DNA sequences from a mixture of DNA molecules.
2. *Northern Blotting*: Used to detect specific messenger RNA (mRNA) sequences from a mixture of RNA molecules.
3. *Western Blotting*: Used to detect specific protein molecules from the mixture of protein samples.

4. *Eastern Blotting*: Used to analyze posttranslational modifications of proteins (often carbohydrate epitopes in glycosylation).
6. *Southwestern Blotting*: Used to investigate DNA–protein interactions (Siu et al., 2008).

13.4 SOUTHERN BLOTTING

Southern blotting technique was invented in 1975 by Edward M. Southern. It was a simple but one of the best idea at that time that led Edward M. Southern to invent a method that revolutionized the study of DNA.

In the early 1960s, Julius Marmur and Paul Doty published a study that accurately described the conditions for the optimal renaturation of DNA complementary strands, upon denaturation by high temperatures (Francesca, 2007). They proposed that high temperature was required to block the formation of weak bonds between noncomplementary strands and to guarantee the proper pairing of complementary nitrogen bases.

At the same time, other scientists were busy in the development of electrophoresis. The electrophoresis technique began with the pioneering work of the Swedish biochemist, Arne Tiselius, who published his first paper on electrophoresis in 1937. By the end of the year 1960s, sophisticated gel electrophoresis techniques made it possible to separate biological molecules based on minute physical and chemical differences. In the 1970s, the powerful tool of DNA gel electrophoresis was developed. This process uses electricity to separate DNA fragments by size as they migrate through a gel matrix.

In year 1970, discovery of restriction endonucleases took place by Werner Arber, Hamilton Smith, and Daniel Nathans. Smith and his colleagues showed that restriction endonuclease from *Haemophilus influenzae* was capable of double-strand break at specific sequences of DNA. Later on, many restriction enzymes were discovered and characterized which have similar properties. The restriction enzymes isolated from *H. influenzae* are abbreviated as Hind II and it was the first restriction enzyme.

These above discoveries and Edwin Southern idea of transferring the DNA samples from gel into nitrocellulose membrane led a remarkable breakthrough in the field of molecular biology by discovering the technique called Southern blotting. Southern blotting technique finally made it possible to detect a specific DNA sequence from a smear without having to purify it away from the rest of the genome (Southern, 2006).

13.4.1 PRINCIPLE

The key to this method is hybridization where a double-stranded DNA molecule is formed between a single-stranded DNA probe and a single-stranded target DNA (Fig. 13.1). These hybridization reactions are specific, that is, the probes will only bind to targets with a complementary sequence. The probe can find one molecule of target in a mixture of millions of related but noncomplementary molecules.

13.4.2 STEPS IN SOUTHERN BLOTTING

The whole process of Southern techniques involves a number of steps (Fig. 13.1):

Step 1: Extraction and purification of DNA to be analyzed.

Step 2: Digestion of DNA with the restriction enzymes.

Step 3: Loading the digested DNA samples on the well made in agarose gel and run the gel electrophoresis. DNA are negatively charged molecules hence migrate toward positive terminal. The distance a specific fragment migrates is inversely proportional to the fragment size.

Step 4: Soaking the agarose gel into alkali; alkaline condition help to denature the double-stranded DNA fragments.

Step 5: Transferring the gel to membrane by blotting.

Step 6: Designing and labelling of probe.

Step 7: Hybridizing labelled short single-strand probes with the complementary denatured DNA strand.

Step 8: Autoradiography helps to detect the labelled probe hybridized with the complementary denatured DNA strand.

13.4.3 MODIFICATION IN SOUTHERN BLOTTING

The method used currently is basically the same with some modifications. The modifications principally for blotting material to enhance the effectiveness of gel transfer to the membrane. Initially when the blotting techniques started, scientists used nitrocellulose sheet which was later on replaced by nylon membranes (Twomey and Krawetz, 1990; Brown, 2001).

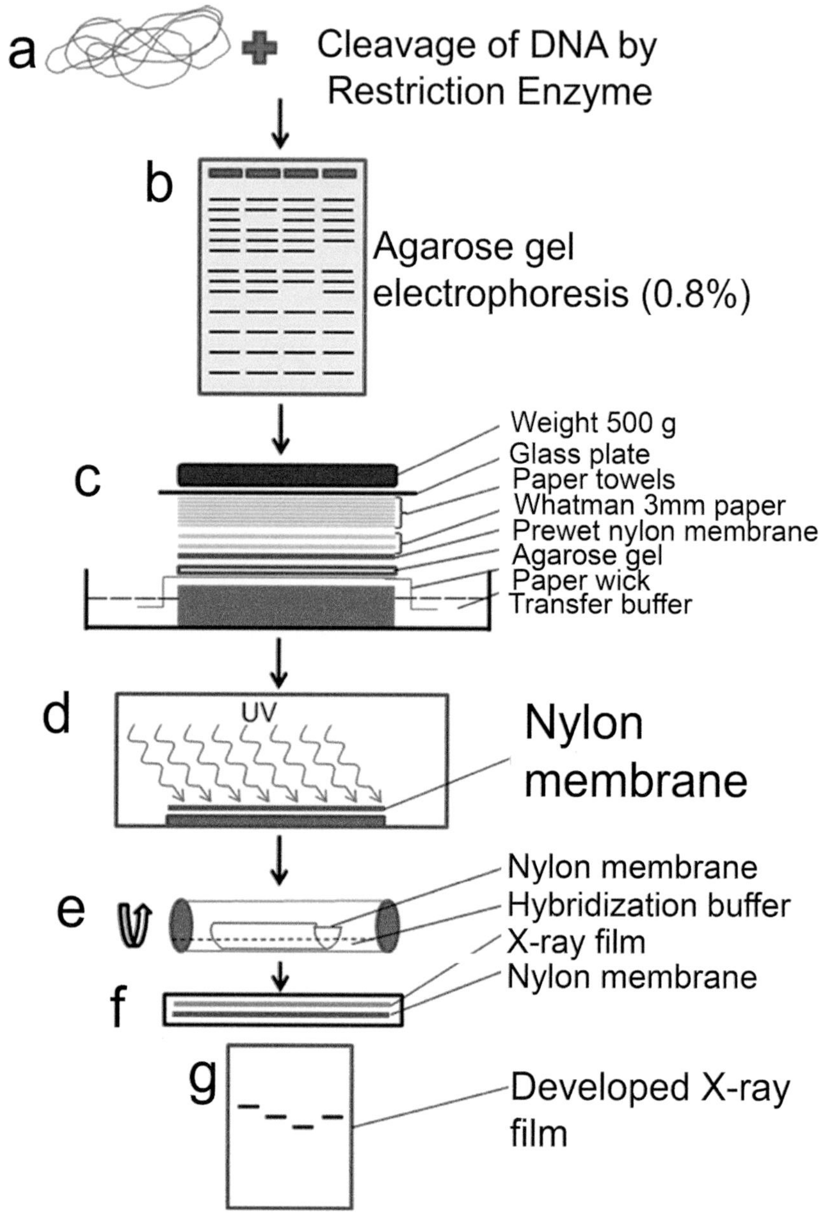

FIGURE 13.1 Southern blotting technique. (a) Genomic DNA preparation and restriction digestion with appropriate restriction endonuclease. (b) Digested genomic DNA separation on Agarose gel (0.8%) by electrophoresis. (c) Blotting method by upward capillary transfer. (d) Cross linking of DNA with nitrocellulose or nylon membrane by UV irradiation. (e) DNA-DNA hybridization in hybridization tube at stringent temperature. (f) Labelled blot exposed to X-ray film in dark X-ray film cassette. (g) Detection of bands by developing and fixing X-ray film.

Nylon membrane has three major advantages over nitrocellulose sheet.

1. Nylon membranes are less brittle and easy to handle compared to nitrocellulose sheets; hence, it is user friendly. A single blot can be rehybridized for more than 10 times.
2. Nylon membranes are much more sensitive as compared with nitrocellulose membranes. Because of the nature of membrane under certain conditions (a positively charged membrane and an alkaline transfer buffer) the transferred DNA becomes covalently bound to the membrane during the transfer process. This is not the case with a nitrocellulose membrane, which initially binds DNA in a semipermanent manner; immobilization occurs only when the membrane is baked at 80°C. Transfer onto a positively charged nylon membrane can therefore reduce the possible loss of DNA that might occur by leaching through the membrane during the blotting process; it is also quicker, the transfer time being reduced from 18 h to 2 h.
3. Nylon membranes efficiently bind DNA fragments down to 50 bp in length, whereas nitrocellulose membranes are effective only with molecules length more than 500 bp.

However, nitrocellulose cannot be completely outdated because it has one significant advantage compared with nylon membranes: The background noise during the hybridization is quite low especially with probes that have been labelled with nonradioactive markers.

Capillary method of transfer is replaced by noncapillary method such as electroblotting. Vacuum blotting is more popular in case of agarose gels, a vacuum pressure is generated to draw the buffer through the gel and membrane more rapidly occurs by simple capillary action, thus enabling the transfer time to be reduced to as little as 30 min.

13.4.4 ADVANTAGES

1. Detection of foreign DNA integration and its correct position into the plant genome.
2. Detection of multiple copies of the transgene in a genome.

13.4.5 DISADVANTAGES

1. Labor intensive.
2. Costly due to requirement of expensive chemical and restriction enzymes.
3. High amount of pure genomic DNA is required.

13.4.6 APPLICATIONS

1. Isolation of specific DNA in a DNA sample.
2. Identification of clone (Brown, 2001) and to isolate desired DNA for construction of rDNA. This technique is not only used to verify the transgene but also used to identify the number of copies inserted into genome.
3. It is used in genetic mapping mainly in restriction fragment length polymorphism (RFLP) analysis.
4. Detection of minute quantities of DNA in DNA fingerprinting.
5. Identification of mutations such as insertion, deletion, and gene rearrangements.
6. Used in phylogenetic analysis.

13.5 NORTHERN BLOTTING

Northern blotting also known as northern hybridization is a technique used for detection and quantification of specific RNA sequences mainly mRNA (Streit et al., 2008). This technique was developed in 1977 by James Alwine, David Kemp, and George Stark at Stanford University, who named it on the basis of its analogy to Southern blotting. It is usual to separate the mRNA transcript by gel electrophoresis under denaturing conditions since this improves the resolution and allows a more accurate estimation of the size of the transcripts (Wilson and walker, 2010). RNA–DNA hybridization occurs in northern blotting. The technique of Southern blotting used for the DNA transfer from gel to membrane cannot be replicated in the same manner for the blot transfer of mRNA (Gupta, 2006). Northern blotting does not require mRNA digestion with restriction endonucleases. In this case, formaldehyde is used for mRNAs denaturation and hydrogen bond break.

The mRNA separated by gel electrophoresis was in normal condition not to bind nitrocellulose. Instead mRNA bands from the gel were blots transferred onto a chemically reactive nitrocellulose sheet prepared by diazotization of aminobenzoyloxymethyl filter paper. Later it was found that RNA bands can indeed be blotted onto nitrocellulose membranes under appropriate conditions and suitable nylon membranes have been developed. Once covalently bound, the RNA is available for hybridization with radiolabelled DNA probes.

13.5.1 PRINCIPLE

Northern blotting principle is that RNA are separated by size and detected on a membrane using a hybridization probe with a base sequence complementary to all, or a part, of the sequence of the target mRNA (Trayhurn, 1996).

13.5.2 STEPS IN NORTHERN BLOTTING

The northern blotting involves the following steps (Fig. 13.2):

Step 1: Extract total RNA from a homogenized tissue sample or cells. Further eukaryotic mRNA can then be isolated by using oligo (dT) cellulose chromatography to isolate only mRNAs as the mRNA synthesis generate poly-A tail.

Step 2: The isolated mRNA is then separated by gel electrophoresis. The distance a specific fragment migrates is inversely proportional to the fragment size.

Step 3: The RNA samples separated on the basis of size are transferred to a nylon membrane employing a capillary or vacuum-based system for blotting.

Step 4: Similar to Southern blotting, the membrane filter is revealed to a labelled DNA probe that is complementary to the mRNA sequences.

Step 5: The labelled membrane is then subjected to autoradiography to detect the hybridization between probe and mRNA sequences.

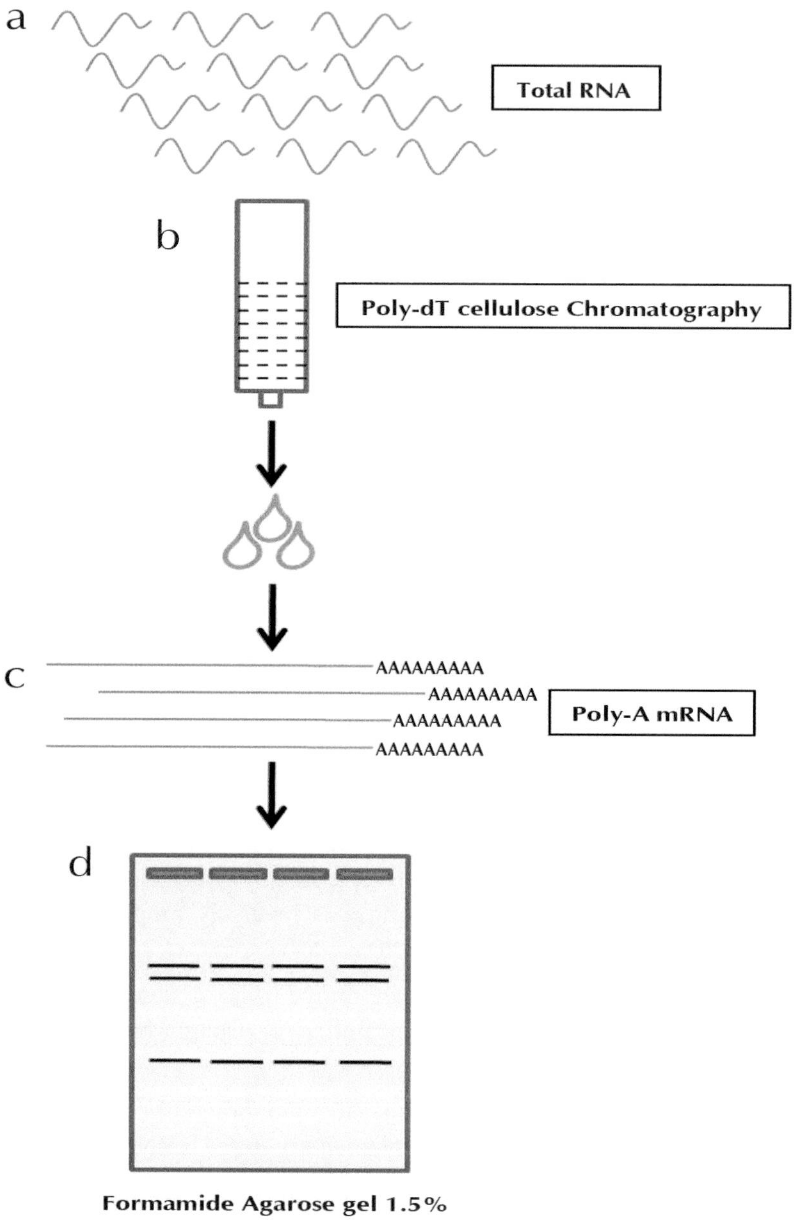

FIGURE 13.2 Northern Blotting. RNA preparation and separation on gel. (a) Total RNA isolation in RNAs free environment. (b) Purification of mRNA by mRNA trapping oligo (dT) cellulose chromatography. (c) Total mRNA separated containing poly-A tail. (d) Poly-A tail containing mRNA separated on denaturing formamide or formaldehyde containing Agarose gel (1.5%) by electrophoresis.

13.5.3 ADVANTAGES

Following are the advantages of northern blotting (Streit et al., 2008):

1. Sequences with even partial homology can be used as hybridization probes.
2. Specificity is relatively high.
3. mRNA transcript size can be detected.
4. RNA splicing is visible because alternatively spliced transcripts can be detected.
5. The cost of running many gels is low once the equipment is set up.
6. Blots can be stored for several years and reprobed if necessary.
7. The strength of this method is its simplicity.

13.5.4 DISADVANTAGES

Following are the disadvantages of northern blotting (Streit et al., 2008):

1. Risk of mRNA degradation during electrophoresis.
2. Quality and quantification of expression are negatively affected.
3. High doses of radioactivity and formaldehyde are a risk for workers and the environment.
4. The sensitivity of northern blotting is relatively low in comparison with that of RT-PCR.
5. Detection with multiple probes is difficult.
6. Use of ethidium bromide, DEPC, and UV light needs special training and attention.

13.5.5 APPLICATIONS

1. Transcripts of specific genes can be studied and measured.
2. The size of mRNA transcripts can be determined.

13.6 WESTERN BLOTTING

Western blotting is an important technique used in cell and molecular biology. In the late 1970s, Towbin et al. (1979) enabled proteins to be electrophoretically separated using polyacrylamide–urea gels and transferred onto a nitrocellulose membrane. Burnette (1981) later employed the more widely used sodium dodecyl sulfate–polyacrylamide gels (SDS–PAGE), which eventually led to this method being termed "western blotting." It is also called protein blotting or immunoblotting and has rapidly become a powerful tool for studying proteins (Jensen, 2012). Western blotting technique is used to identify a specific protein within a complex mixture of heterologous proteins, which was named for its similarity to Southern blotting, which detects DNA fragments, and northern blotting, which detects mRNAs.

For sample preparation, cells and tissues should be rapidly frozen with liquid nitrogen to avoid protease degradation of proteins or collected and lysed as quickly as possible (Jensen, 2012). Solid tissue is mechanically broken down, usually using a homogenizer or by sonication in a lysis buffer. The prepared sample is then assayed for protein content so that a consistent amount of protein can be taken from each different sample. In western blotting, a protein mixture is electrophoretically separated on an SDS–PAGE, a slab gel infused with SDS, a dissociating agent (Fig. 13.3).

The protein bands are transferred to a nylon or polyvinylidene difluoride (PVDF) or nitrocellulose membrane by electrophoresis and the individual protein bands are identified by flooding the membrane with radiolabelled or enzyme-linked polyclonal or monoclonal antibody specific for the protein of interest (Laddha et al., 2013; Supporting data). In particular, the membrane is probed with a primary antibody directed against the specific protein. Depending on the source of the antibody (mouse, rabbit, other), the membrane is probed with a secondary antibody that specifically binds to the primary antibody (e.g., antimouse IgG antibody). The Antigen–antibody complexes that form on the band containing the protein can be recognized and visualized in a variety of ways. However, the most generally used detection procedures employ enzyme-linked antibodies against the protein. After binding of the enzyme antibody conjugate, addition of a chromogenic substrate that produces a highly colored and insoluble product causes the appearance of a colored band at the site of the target antigen. The site of the protein of interest can be determined with much higher sensitivity if a chemiluminescent compound along with suitable enhancing agents is used to produce light at the antigen site. For example, the secondary antibody can be tagged with horseradish peroxidase. The result is an antibody "sandwich" where the primary antibody binds the target protein and the secondary antibody binds the primary antibody. Adding hydrogen peroxide and luminol sets off a luminescent reaction wherever an antibody sandwich has assembled. The light emitted can then be detected by autoradiography or phosphorimaging. Only protein bands that bind to the primary antibody will be detected.

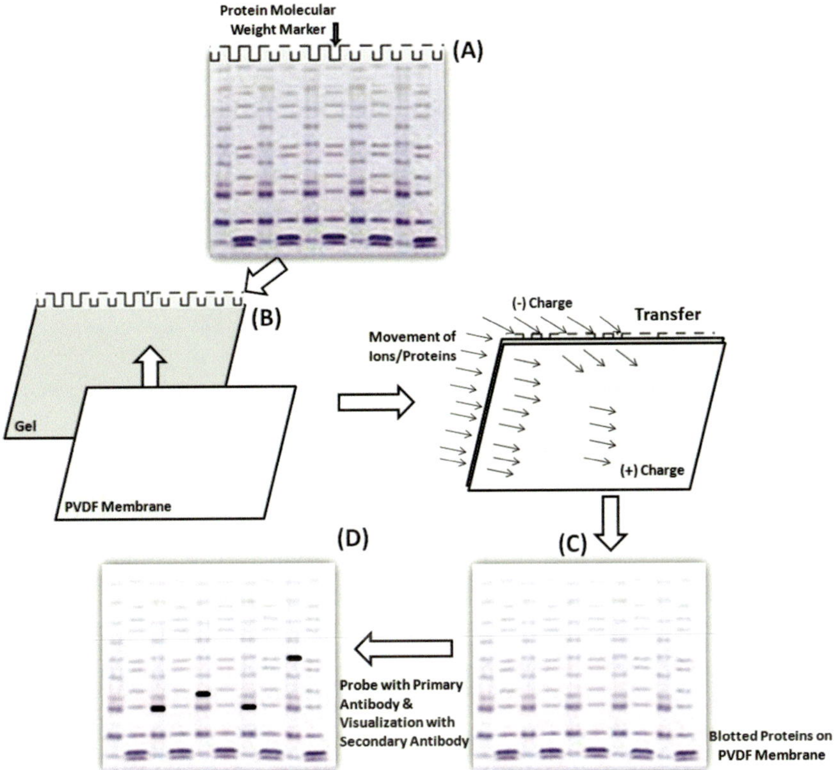

FIGURE 13.3 Western blotting. (a) Mixtures of proteins are loaded onto separate lanes of the SDS polyacrylamide gel. After electrophoresis, this gel can be stained with Coomassie blue or fluorescent dyes to give the pattern shown. (b) To identify which band in each lane is the protein of interest, the unstained protein bands are electrophoretically transferred to a PVDF membrane as shown. (c) The membrane is then probed with primary antibody directed against the protein of interest. Bands to which these antibodies have bound are then visualized by adding a secondary antibody directed against the first antibody. The secondary antibody is tagged with the enzyme horseradish peroxidase (HRP). (d) After binding of the secondary antibody, hydrogen peroxide and luminol are added to the membrane. HRP, in the presence of H_2O_2, oxidizes luminol, which enters an excited state that decays and emits light. Thus, areas where the primary antibody bind to the gel (detecting the target protein) will emit light that can be detected with autoradiography (bands corresponding to the target protein are in bold).

If the protein of interest is bound by a radioactive antibody, its position on the blot can be determined by exposing the membrane to a sheet of X-ray film, a procedure called "autoradiography." Thus, the technique is used to determine if, and at what levels, specific proteins are present in a cell extract. It can also be used to detect the processing of a protein into smaller fragments (Fig. 13.3).

In addition, western blotting can also identify a specific antibody in a mixture. In this case, known antigens of well-defined molecular weight are separated by SDS–PAGE and blotted onto nitrocellulose. The separated bands of known antigens are then probed with the sample suspected of containing antibody specific for one or more of these antigens. Reaction of an antibody with a band is detected by using either radiolabelled or enzyme-linked secondary antibody that is specific for the species of the antibodies present in the test sample.

13.6.1 PRINCIPLE

Western blotting is an analytical technique where the protein sample is run on an SDS–PAGE and transferred to a solid membrane (Fig. 13.3). The transferred protein is detected by a specific primary and secondary labelled antibody. These antibodies bind to a specific sequence of amino acids. As the amino acid sequences differ from protein to protein, these can help target specific proteins in a complex mixture. First, proteins are separated on the basis of size. Second, antibodies are used to identify the target protein. Lastly, a substrate reacting with the enzyme is used for the protein/antibody complex detection.

13.6.2 STEPS IN WESTERN BLOTTING

The western blotting involves the following steps (Fig. 13.3) (Blancher and Jones, 2001):

Step 1: Extraction and quantification of protein samples.

Step 2: Resolution of the protein sample in sodium dodecyl sulfate polyacrylamide denaturing gel (SDS–PAGE) electrophoresis.

Step 3: Transfer of the separated polypeptides to a membrane support.

Step 4: Blocking nonspecific binding sites on the membrane.

Step 5: Addition of antibodies.

Step 6: Detection.

13.6.3 ADVANTAGES

1. Protein contents of the samples can be compared.
2. It can be used to determine the size of the target protein or polypeptide.
3. Western blotting utilizes not only antigens but also antisera as a diagnostic tool.
4. Immunogenic responses from infectious agents are easy to detect by this technique.

13.6.4 DISADVANTAGES

1. Time consuming.
2. Optimization of experimental conditions is required.
3. Trained person is required to perform the western blotting.
4. A nonspecific protein has a slight chance of reacting with the secondary antibody, resulting in the labelling of an incorrect protein.

13.6.5 APPLICATIONS

1. It can identify the nature of the protein or epitope effectively. Also, it can be implicated as a tool of quantitative analysis of the micromolecule antigen in cooperation with immunoprecipitation.
2. It can be used for epitope mapping which can identify the process of the binding sites, or "epitope," of antibodies on their target antigens (proteins). The identification and characterization of the epitopes can help us to discover and develop new therapeutics, diagnostics, and vaccines.
3. It can be useful for amino acid composition and sequence analysis of extremely trace protein or peptide (10 pmol) transferred to PVDF membrane.
4. It can be useful for spots imprinting analysis, available chromatography components analysis, sucrose gradient analysis, or pulse tracking experimental analysis.
5. It is used to test the endogenous or exogenous expression phosphoprotein so as to detect the phosphorylation signal.
6. It can be useful for protein resilience in the function experiment.
7. It is useful for structure domain analysis.
8. It is used to analyze protein expression level, for example, analysis of regulation protein expressed in the cell cycle, etc.

13.7 SOUTHWESTERN BLOTTING

Southwestern blotting, first described by Bowen and colleagues, is a powerful technique for identifying and characterizing DNA-binding proteins by their ability to bind to specific oligonucleotide probes (Handen and Rosenberg, 1997). This method combines the features of Southern and western blotting techniques. In this technique, the proteins are first separated based on size on a SDS–PAGE, then renatured by removing SDS in the presence of urea and transferred by electroblotting to a nylon or nitrocellulose membrane support, and detected by their ability to bind radiolabelled DNA probe (Siu et al., 2998; Labbé et al., 2009). The interaction of the probe with the protein(s) is later visualized by autoradiography. Autoradiography reveals if a specific protein band binds the labelled DNA. Later on the protein can be excised and identified.

13.7.1 ADVANTAGES

1. It is mainly used for identification of sequence-specific DNA-binding proteins, for example, histone proteins.
2. It provides information regarding the molecular weight of unknown protein.

13.8 EASTERN BLOTTING

Every living organism present on Earth is placed into two groups, that is, prokaryote and eukaryote based on the presence of nucleus. In prokaryotes, no special modification of mRNA is required and translation of the message starts even before the transcription is complete. In eukaryotes, however, mRNA is further processed to remove the introns (a process known as "Splicing"). Similarly, after translation the eukaryotic protein or polypeptides are modified in various ways to complete their structure, or regulate their activity within the cell; such modifications in protein are called post-translational modifications. These posttranslational modifications are various additions or alterations to the chemical structure of eukaryotic protein. Post-translational modifications modulate the activity of most eukaryote proteins (Matthias and Jensen, 2003). Analysis of these modifications presents formidable challenges but their determination generates indispensable insight into biological function.

Strategies developed to characterize individual proteins are now systematically applied to protein populations. Eastern blotting technique is used to analyze posttranslational modifications of proteins or peptides, such as addition of lipids, phosphomoieties, and glycoconjugates. It is most commonly used to identify carbohydrate epitopes in glycosylation. In this technique, proteins or lipids are blotted from SDS–PAGE gel onto a nitrocellulose or PVDF membrane. These proteins are then analyzed for posttranslational modifications using a specific probe that is capable to identify lipids and carbohydrates.

13.8.1 ADVANTAGES

1. It is used to identify the glycosides from natural products, for example, solasodine glycosides and ginsenosides, etc. (Tanaka et al., 2012).
2. It is used in visual screening of toxic compounds, aristolochic acid I, and aristolochic acid II in plant extracts or tissues of *Aristolochia* and *Asarum* species.
3. It is useful in easy identification of flavone glycoside, baicalin in the extracts of crude drugs, and Kampo medicines (Fukuda et al., 2006).
4. It is used in fingerprinting of natural products.

KEYWORDS

- **Southern blotting**
- **northern blotting**
- **western blotting**
- **southwestern blotting**
- **eastern blotting**

REFERENCES

Alwine, J. C.; Kemp, D. J.; Stark, G. R. Method for Detection of Specific RNAs in Agarose Gels by Transfer to Diazobenzyloxymethyl-paper and Hybridization with DNA Probes. *Proc. Natl. Acad. Sci. U.S.A.* **1977,** *74*(12), 5350–5354.

Blancher, C.; Jones, A. SDS-PAGE and Western Blotting Techniques. *Methods Mol. Med.* **2001,** *57*, 145–162.

Brown, T. A. Southern Blotting and Related DNA detection techniques. Encyclopedia of Life Sciences, Nature Publishing Group, **2001**. www.els.net. DOI:10.1038/nrg2244.

Francesca, P. *Southern Migration. Nature Milestones DNA Technologies*, 2007, 1.

Gupta, P. K. *Biotechnology and Genomics*, 5th Edn. Rastogi Publication, **2006,** 122–128.

Handen, J. S; Rosenberg, H. F. An Improved Method for Southwestern Blotting. *Front. Biosci.* **1997,** *2*, c9–11.

Hayes, P. C., Wolf, C. R.; Hayes, J. D. Blotting Techniques for the Study of DNA, RNA, and Proteins. *Br. Med. J.* **1989,** *299*, 965–968.

Jensen, E. C. The Basics of Western Blotting. *Anat. Rec.* **2012,** *295*, 369–371.

Labbé, S., Harrisson, J. F.; Séguin, C. Identification of Sequence-specific DNA-binding Proteins by Southwestern Blotting. *Methods Mol. Biol.* **2009,** *543*, 151–161.

Laddha, N. C.; Dwivedi, M.; Gani, A. R.; Shajil, E. M.; Begum, R. Involvement of Superoxide Dismutase Isoenzymes and Their Genetic Variants in Progression of and Higher Susceptibility to Vitiligo. *Free Radic. Biol. Med.* **2013,** *65*, 1110–1125.

Prakash, C.; Manjrekar, J.; Chattoo, B. B. Skp1, a Component of E3 Ubiquitin Ligase is Necessary for Growth, Sporulation, Development and Pathogenicity in Rice Blast Fungus (*Magnaporthe oryzae*) *Mol. Plant. Pathol.* **2015.** DOI: 10.1111/mpp.12336.

Primrose, S. B.; Twyman, R. *Principles of Gene Manipulation*, 7th Edn. 2006, 8–25.

Siu, F. K.; Lee, L.T.; Chow, B. K. Southwestern Blotting in Investigating Transcriptional Regulation. *Nat Protoc.* **2008,** *3*(1), 51–58.

Southern. E. Southern Blotting. *Nat. Protoc.* **2006,** *1*(2), 518–525.

Streit, S.; Michalski, C. W.; Erkan, M.; Kleeff, J.; Friess, H. Northern Blot Analysis for Detection and Quantification of RNA in Pancreatic Cancer Cells and Tissues. *Nat. Protoc.* **2008,** *4*(1), 37–43.

Tan, S. C.; Yiap, B. C. DNA, RNA, and Protein Extraction: The Past and The Present. *J. Biomed. Biotechnol.* **2009,** *574398*, 1–10.

Twomey, T. A.; Krawetz, S. A. Parameters Affecting Hybridization of Nucleic Acids Blotted onto Nylon or Nitrocellulose Membranes. *Biotechniques* **1990,** *8*(5), 478–482.

Wilson, K; Walker, J. *Principles and Techniques* of biochemistry and Molecular biology 7th Edn, Cambridge University Press, **2010,** 172–173.

Tanaka, H.; Putalun, W.; Shoyama, Y. Fingerprinting of Natural Product by Eastern Blotting Using Monoclonal Antibodies. *Chromatogr. Res. Int.* **2012,** *130732*, 1–7.

Trayhurn, P. Northern blotting. *Proc. Nutr. Soc.* **1996,** *55*(1B), 583–589.

Matthias, M.; Jensen, O. N. Proteomic Analysis of Post-translational Modifications. *Nat. Biotechnol.* **2003,** *21*, 255–261.

Fukuda, N.; Shan, S.; Tanaka, H.; Shoyama, Y. New Staining Methodology: Eastern Blotting for Glycosides in the Field of Kampo Medicines, *J. Nat. Med.* **2006,** *60*(1), 21–27.

CHAPTER 14

ADVANCES IN PCR TECHNOLOGY AND RNA INTERFERENCE

SUHAIL MUZAFFAR*

National Centre for Biological Sciences, GKVK Campus, Bellary Road, Bangalore 560065, India

**Corresponding author. E-mail: suhail.bt@gmail.com*

CONTENTS

ABSTRACT

Molecular biology has become a crucial tool for identifying new genes with importance in medicine, agriculture, environment, and health. One of the essential molecular techniques with multiple advancements and applications is polymerase chain reaction (PCR). This technique was originally developed by Kary Mullis in 1983 and since then it has become an essential tool for the amplification of nucleic acids. PCR is a fairly simple technique used to detect and amplify a nucleic acid sequence. In the recent years, modifications have been introduced in this technique to develop different variants of the basic PCR method. Another advanced technique developed recently is RNA interference. RNA interference is a gene regulatory mechanism that downregulates the expression of genes by either suppressing the transcription or by activating a sequence-specific mRNA degradation. The natural functions of RNAi are to regulate the developmental programs of eukaryotic organisms and protection of genome against viruses and transposons. In eukaryotic organisms, two major categories of small RNAs including short interfering RNAs (siRNAs) and microRNAs (miRNAs) regulate endogenous gene expression and defend the genome from invasive nucleic acids. RNAi provides a rapid means of silencing a specific gene by introducing double-stranded DNA sequence into the cell. Pioneering work on RNAi was reported in plants, and later on such mechanism was also described in nematodes, insects, and human cells.

14.1 ADVANCES IN PCR TECHNOLOGY

14.1.1 OVERVIEW

The discovery of polymerase chain reaction (PCR) has dramatically altered how molecular studies are conducted to answer different scientific questions and has simplified the complex molecular biology techniques. PCR-based strategies have led to various scientific endeavours such as the Human Genome Project. It is widely used for identification of novel genes, diagnosis of diseases, quantitative gene expression, and whole-genome studies in a very sensitive and efficient manner. PCR-based assay is widely used in forensic medicine to identify criminals and paternity testing.

The PCR is a molecular biology technique used for amplification of a single copy or a few copies of a specific portion of DNA to millions of identical copies. The method basically relies on thermal cycling, which consists of repeated cycles of heating and cooling of the DNA in a reaction system for the enzymatic replication of the DNA. The first step consists of DNA melting, which involves separation of the two strands of the DNA double helix at a high temperature. The second step involves lowering of the temperature and binding of primers to the complementary DNA (cDNA) sequence on the template strand. DNA polymerase has the ability to add a nucleotide only onto a pre-existing 3′-OH group; therefore, it needs a primer sequence to which it can add the first nucleotide. After binding of primers to template DNA, amplification of the DNA starts between the specific primers.

PCR was developed by Kary Mullis in 1983, for which he received a Nobel Prize in 1993 (Mullis and Faloona, 1987; Mullis et al., 1986). This was one of the most significant breakthroughs in the area of molecular biology, and the awareness of the PCR technique spread throughout the international scientific community. PCR is often an indispensable technique in the area of molecular biology, genetic engineering, metagenomics, and medical and agricultural research labs around the globe. One of the important developments in PCR technology was the discovery and use of a thermostable enzyme *Taq* DNA polymerase (Chien et al., 1976).

14.1.2 COMPONENTS OF A PCR REACTION

- **DNA template:** It is the DNA sample that contains a target sequence that needs to be amplified.
- **Primers:** These are short pieces (10–30 nucleotides) of single-stranded DNA that bind to the complementary target sequence on the template.
- **Nucleotides (dNTPs):** These are four basic components of the DNA, which are used as building blocks for new DNA strands.
- **DNA polymerase:** It is an enzyme that synthesizes new DNA from the complementary target sequence. *Taq* DNA polymerase (from *Thermus aquaticus* is the first and most commonly used thermostable enzyme for PCR. *Pfu* DNA polymerase (from *Pyrococcus furiosus*) is used for its higher fidelity.
- **Buffer solution:** It provides a suitable pH and salt concentration for optimum activity of the DNA polymerase.

14.1.3 ADVANCES IN THE PCR TECHNOLOGY

On the basis of applications of PCR, a number of variants of this technique have emerged. Some of which are optimizations of the basic method to suit specific requirements, while others completely modify the technique in order to formulate some novel and creative applications. Some of the common variations of basic PCR are listed below.

14.1.3.1 ASSEMBLY PCR

It is also known as Polymerase Cycling Assembly or PCA. This technique is used for the synthesis of long DNA sequences by performing PCR on a large number of oligonucleotides with short overlapping regions, to assemble multiple DNA fragments into a single fragment. It involves two steps, an initial PCR with overlapping primers and a second one using the products of the first reaction as the template to generate the final full-length product (Fig. 14.1). Assembly PCR is often used as the substitute for ligation-based DNA assembly (Stemmer et al., 1995). A modification of assembly PCR,

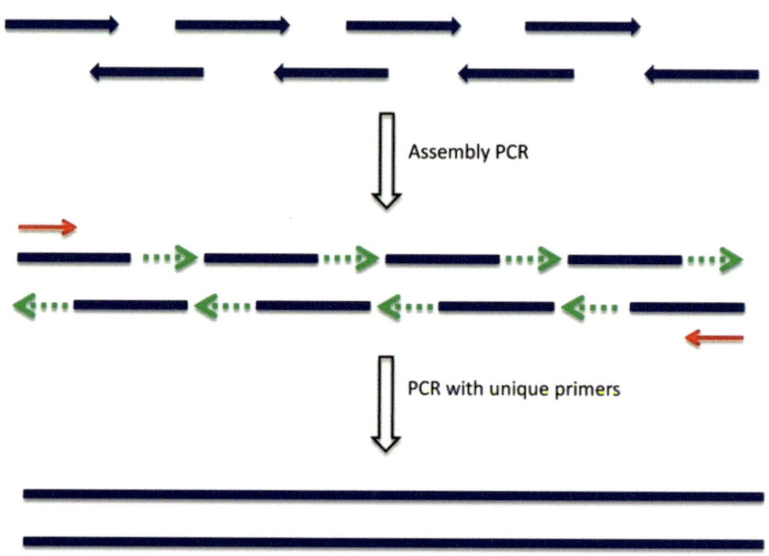

FIGURE 14.1 Assembly PCR. It involves the artificial synthesis of long DNA sequences by performing PCR on several fragments of long oligonucleotides containing small overlapping regions. Red arrows indicate the unique primers used for amplification of final full-length product in which assembly PCR product is used as the template.

"Gibson assembly" is a molecular cloning technique used to join multiple DNA fragments in a single PCR reaction. (Gibson et al., 2009). This method requires three enzyme activities, i.e., exonuclease, DNA ligase, and DNA polymerase. Based on the sequence similarity, it can simultaneously combine more than 10 DNA fragments with 20–50 nt overlapping regions. This method is comparatively cheaper and more efficient than the conventional cloning.

14.1.3.2 ASYMMETRIC PCR

This technique involves specific amplification of one of the two strands of DNA. Initially, it generates a significant amount of double-stranded DNA, which is then used to generate single-stranded DNA due to unequal concentrations of the two primers (Fig. 14.2). It is generally used for sequencing or hybridization probing, where amplification of only one complementary strand is necessary. Asymmetric PCR requires extra cycles of PCR due to the slow amplification after the limiting primer has been used up in the later stages of reaction (Innis et al., 1988).

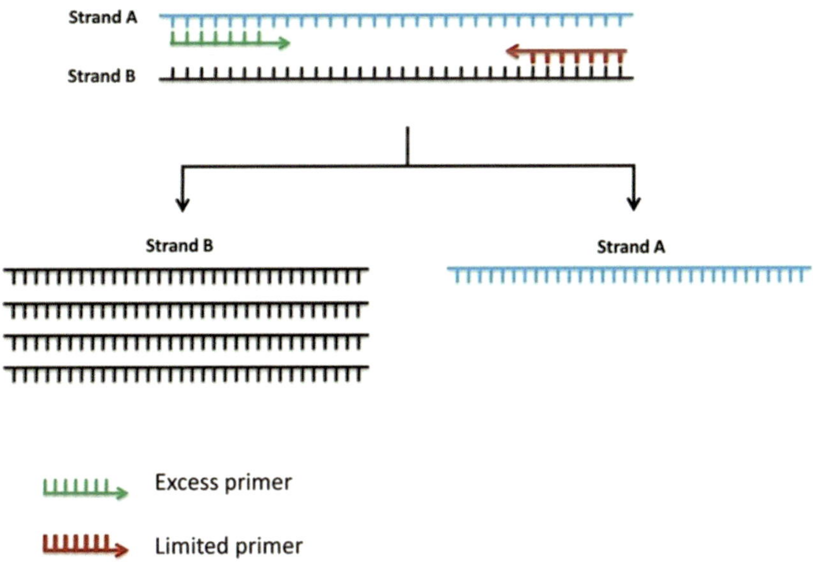

FIGURE 14.2 Asymmetric PCR. It involves a preferential amplification of one of the strands of DNA template. One of the primers (green) is added in access and another primer (red) is added at a minimal concentration in order to specifically generate strand B, whereas strand A is used up in double strand formation.

14.1.3.3 INVERSE PCR

Inverse PCR (iPCR) is also known as inverted or inside-out PCR and is commonly used to identify the flanking regions of an insert into the genome. It involves restriction digestion of mutant genomic DNA with the specific enzymes that cut outside the inserted plasmid, followed by circularization (self-ligation) with T4 DNA ligase and transformation of the ligated fragments into *Escherichia coli* (Fig. 14.3). The plasmid bearing antibiotic resistance (usually ampicillin) forms resistant colonies (Ochman et al., 1988). After the plasmid isolation, PCR is carried out using the primers complementary to the known internal regions (like left and right T-DNA boarders). The amplified product is sequenced to identify the flanking regions of the genome where the plasmid has got integrated. iPCR has numerous applications in molecular biology including the identification of viral integration and flanking transposable elements in the plant and animal genomes.

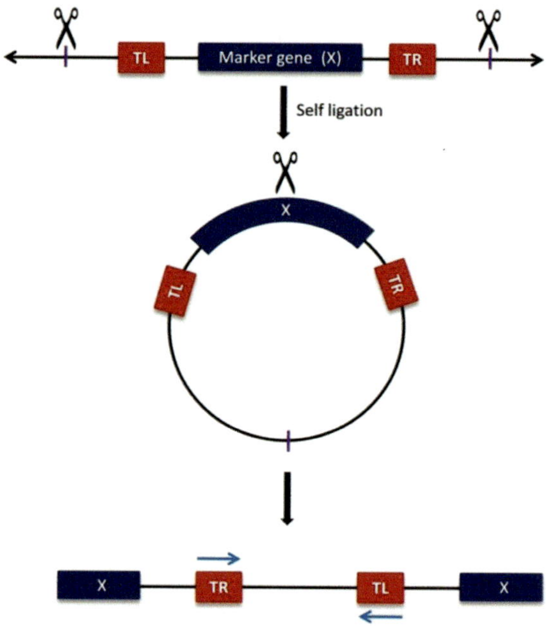

FIGURE 14.3 Inverse PCR. It is based on the standard PCR method, except that it uses primers oriented in the reverse direction. The integrated marker gene (X) is surrounded by left (TL) and right (TR) border of T-DNA of known sequences. The flanking regions outside TL and TR are digested with restriction enzymes followed by circularization of the digested fragment using self-ligation. The second restriction digestion is carried out in the marker to form a linear fragment and PCR of this fragment between TL and TR is done. The amplified product is sequenced to identify the flanking regions.

14.1.3.4 MULTIPLEX-PCR

Multiplex-PCR refers to the use of *Taq* polymerase to amplify several different targeted DNA sequences simultaneously. It consists of multiple sets of primers for the simultaneous amplification of many target sequences in a single PCR reaction (Fig. 14.4). Using this procedure, additional information can be gained from a single reaction that normally would require a lot of reagents, effort, and time to perform if the reactions are carried out for individual target amplifications. One of the primary requirements for multiplex-PCR is that annealing temperatures for each of the primer sets must be optimized to work in a single reaction (Markoulatos et al., 2002; Chamberlain et al., 1988). Also, the target amplicons size should vary significantly in order to form distinct bands when visualized after gel electrophoresis. Some of the various applications of multiplex-PCR are the identification of pathogens, forensic studies, mutation analysis, and high throughput SNP genotyping (Hayden et al., 2008).

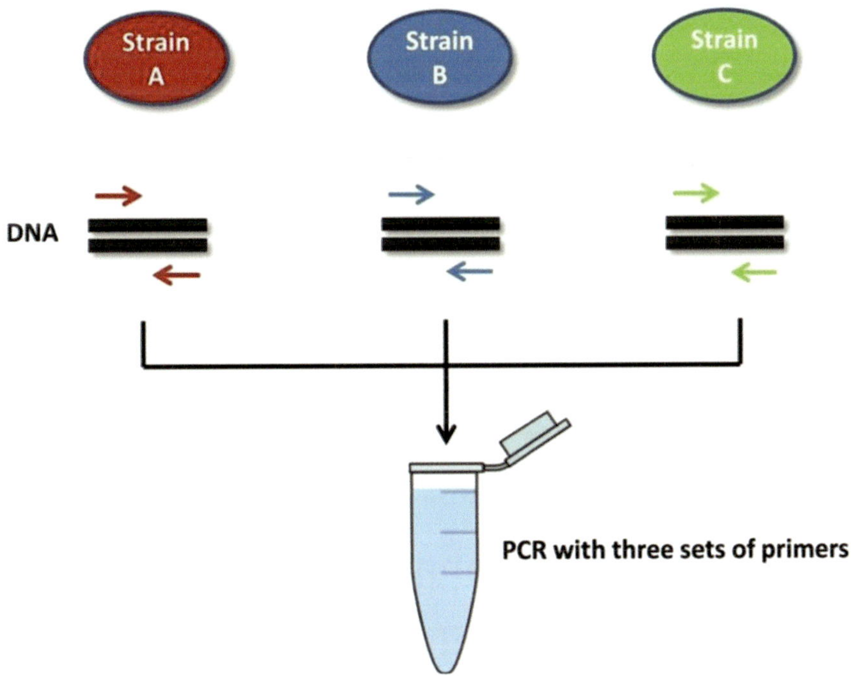

FIGURE 14.4 Multiplex-PCR. In this method, more than one target DNA sequences can be amplified using multiple primer pairs. Genomic DNA from three different bacterial strains (A, B, and C) can be isolated and used for specific amplification of genes using three different primer pairs (red, blue, and green arrows).

14.1.3.5 REVERSE TRANSCRIPTION PCR

Reverse Transcription PCR (RT-PCR) is commonly used in molecular biology to detect mRNA expression of genes or to identify the sequence of an RNA transcript (Freeman et al., 1999). In this method, the RNA template is first converted into a cDNA with reverse transcriptase enzyme. Then, the cDNA is used as a template for the exponential amplification using PCR with a specific set of primers (Fig. 14.5). RT-PCR is a qualitative study of gene expression and is a more sensitive method of RNA expression than other commonly used techniques like Northern blot analysis. Northern blotting has a lot of shortcomings like time consumption, large quantities of RNA required for detection, and inaccuracy at lower concentrations of sample RNA. However, the discovery of RT-PCR has displaced Northern blotting as the method of choice for RNA quantification due to its simplicity, time efficiency, and accuracy (Bustin et al., 2005). RT-PCR has various applications, including the detection and prognosis of cancer and to monitor response to chemotherapy. It is also commonly used to study the genomes various RNA viruses such as Influenza virus A and HIV.

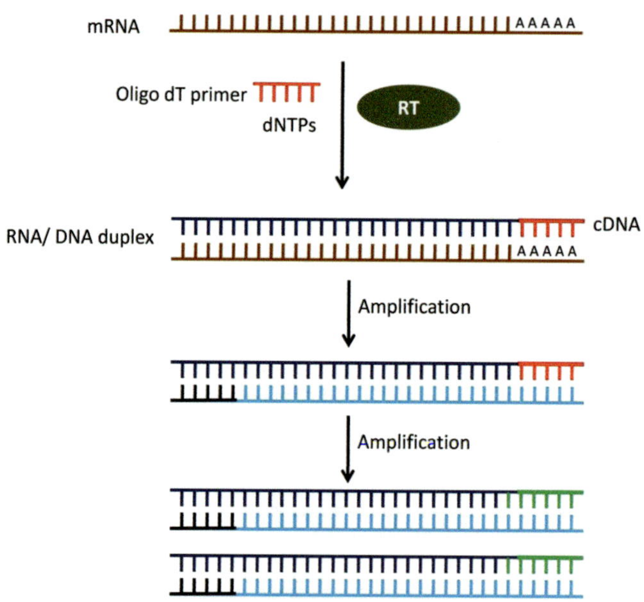

FIGURE 14.5 Reverse transcription PCR. This technique is commonly used to detect mRNA expression. Oligo dT primers (Red) bind to poly A tail of mRNA and add dNTPs to form cDNA in a reaction catalyzed by reverse transcriptase. This cDNA is then used as a template for the amplification of specific genes using gene-specific primers (Green and black).

14.1.3.6 QUANTITATIVE REAL-TIME PCR

Quantifying gene expression by traditional detection methods is inefficient and mostly not very accurate. Quantitative real-time PCR (QRT-PCR or real-time PCR) is a very precise method used to measure both relative as well as the absolute quantity of DNA, cDNA, or RNA in a given sample. The reaction is carried out in a modified thermal cycler, which has inbuilt capacity to illuminate a given nucleic acid sample with a beam of a specific wavelength and detect the emitted fluorescence from the excited fluorophore (Fig. 14.6). Real-time PCR uses fluorescent dyes, such as Sybr Green or fluorophore-containing DNA probes like TaqMan, for real-time quantification of the amplified product. Presently, there are four different fluorescent DNA probes used for the detection of PCR products by real-time PCR. These include TaqMan, SYBR Green, Molecular Beacons, and Scorpions. Real-time PCR uses two common methods for quantification of nucleic acids, i.e., relative quantification and absolute quantification (Dhanasekaran et al., 2010).The fluorescence-based real-time quantitative PCR has become the standard technology for the detection of nucleic acids in every area of microbiology, biomedical research, and in forensic applications. The various applications of real-time PCR include the expression analysis of genes for functional genomics, disease diagnostics, detection of pathogens, genotyping, and food safety tests (Jernigan et al., 2011; Filion, 2012; Yeh et al., 2004).

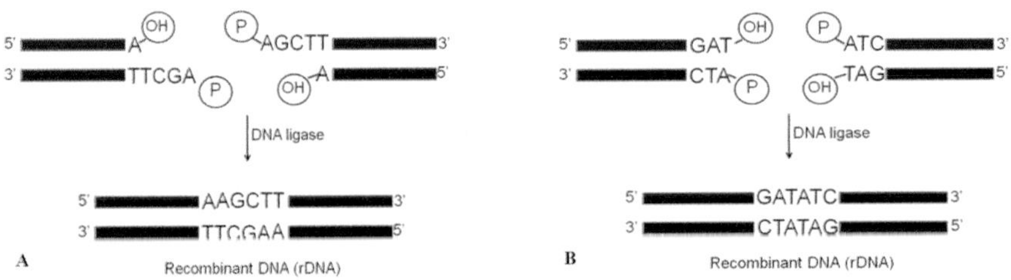

FIGURE 14.6 Quantitative real-time PCR. SYBR green is a cyanine dye which preferentially binds to double-stranded DNA and gives fluorescence only when bound to DNA. Quantitative real-time PCR technique exploits this property of SYBR green to monitor the expression levels of specific genes at various cycles of PCR.

14.1.3.7 THERMAL ASYMMETRIC INTERLACED PCR

Thermal asymmetric interlaced PCR (TAIL-PCR) is a powerful tool to identify an unknown DNA sequence flanking a known sequence. This technique was developed by Liu and Whittier in 1995 (Liu and Whittier, 1995). It is often carried out in three sets of reactions known as primary, secondary, and tertiary reactions and these sets vary in the primers and temperature profile used. TAIL-PCR makes use of "nested" or "specific" primers with different annealing temperatures and a degenerate primer is used to amplify in the other direction from the unknown sequence Fig. 14.7). This technique is an efficient tool for amplifying insert end segments from BAC and YAC clones. The availability of a large number of whole-genome sequences makes TAIL-PCR an attractive tool for the identification of insertional segments by direct sequencing of TAIL-PCR products.

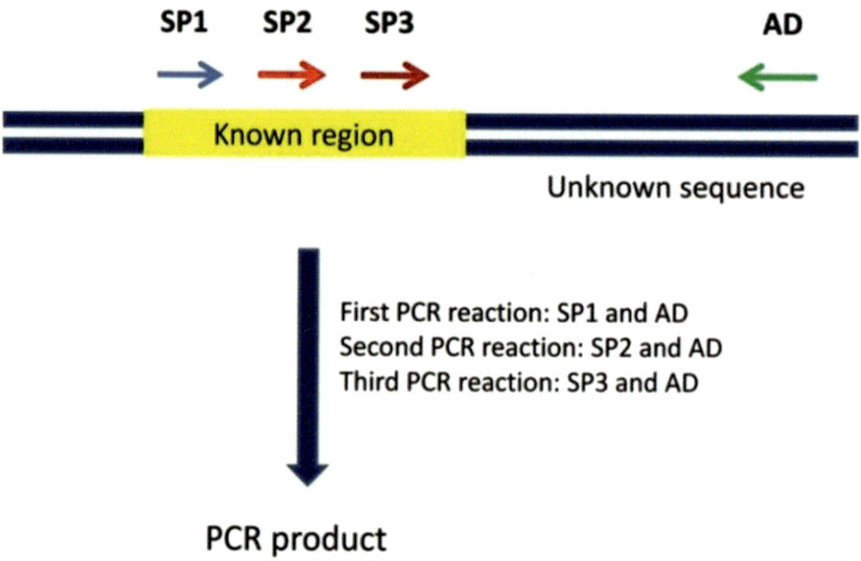

FIGURE 14.7 Schematic diagram of TAIL-PCR. This technique is an efficient tool for the identification of unknown DNA fragments adjacent to the known sequences. It involves three consecutive rounds of PCR, carried out with a set of three nested specific primers (SP_1, SP_2, and SP_3) and an arbitrary degenerate (AD) primer.

14.1.3.8 MINIPRIMER PCR

This method utilizes a thermostable S-Tbr polymerase that can extend from short primers, as short as 9 or 10 nucleotides, known as smalligos. This method is helpful for amplification of target sequences with smaller primer binding sites and has been used to amplify conserved DNA sequences, such as the 16S rRNA gene (Isenbarger et al., 2008). Miniprimer PCR is used in microbial ecology to study the diversity of microbial populations in numerous complex samples.

14.2 RNA INTERFERENCE

14.2.1 OVERVIEW

RNA interference (RNAi) is a conserved gene regulatory mechanism that suppresses the transcript level of a specific gene by activating the RNA degradation process (Agrawal et al., 2003). RNAi, mediated by small RNAs, has become an effective genetic tool in functional genomics, molecular medicine, gene therapy, and infection biology. RNAi was first discovered in the nematode *Caenorhabditis elegans* as a sequence-specific gene silencing in response to the double-stranded RNA (dsRNA) (Fire et al., 1998). It was observed that the dsRNA mixture is more the potent silencing agent than the antisense RNAs (Fire et al., 1998). RNAi has been described as phenotypically different but mechanistically similar process in plants, fungi, and animals and has been termed as cosuppression or PTGS, quelling, and RNAi respectively (Agrawal et al., 2003). Andrew Fire and Craig C. Mello were awarded the 2006 Nobel Prize in Physiology or Medicine for their work on RNAi in *C. elegans*.

14.2.2 DISCOVERY

Discovery of RNA silencing was serendipitous in plants, when scientists wanted to upregulate the expression of a gene for chalcone synthase (*chsA*), an enzyme involved in the production of anthocyanin pigments in petunia flowers. Instead of becoming violet, the flowers developed white sectors and lost both endogene as well as transgene chalcone synthase activity (Napoli et al., 1990). Subsequently, many similar reports of cosuppression were published by other laboratories (Ingelbrecht et al., 1994; Van der Krol et al., 1990). This type of RNA-based gene regulation in plants was termed as post-transcriptional gene silencing (PTGS). PTGS can be triggered by both sense and antisense transgenes, and it has been suggested that similar mechanisms operate in both the instances (Di Serio et al., 2001).

The homology-based gene silencing was also observed independently in different fungal systems, including *Neurospora crassa, and the phenomenon was termed as quelling* (Romano and Macino, 1992). An *al1* gene, which codes for an orange pigment, was expressed in a *N. crassa* strain to enhance the production of the pigment. Some of the transformants showed albino phenotypes, therefore, indicating the quelling phenomenon. It was observed that the level of unspliced *al1* mRNA was not affected; however, the native *al1* mRNA was significantly reduced. It indicates that the phenotype is because of quelling, a post-transcriptional gene regulation phenomenon.

RNAi came into the limelight with the observation that dsRNA could directly silence the genes through RNAi in the nematode *C. elegans (Fire et al., 1998)*. A dsRNA corresponding to a 742-nucleotide segment of *unc22* was injected into the body of an adult nematode. *Unc22* encodes a myofilament protein, and the decrease in *unc22* expression causes a strong twitching phenotype. The injected nematodes showed a weak twitching, whereas the progeny individuals showed the increased twitching phenotype, indicating the silencing of *Unc22* (Fig. 14.8). This discovery was remarkable for two reasons. First, RNAi can be used to specifically knockdown the expression of any gene in a cell. Second, initially scientists believed that introns were just junk DNA but now it was revealed that the introns code for RNAi elements.

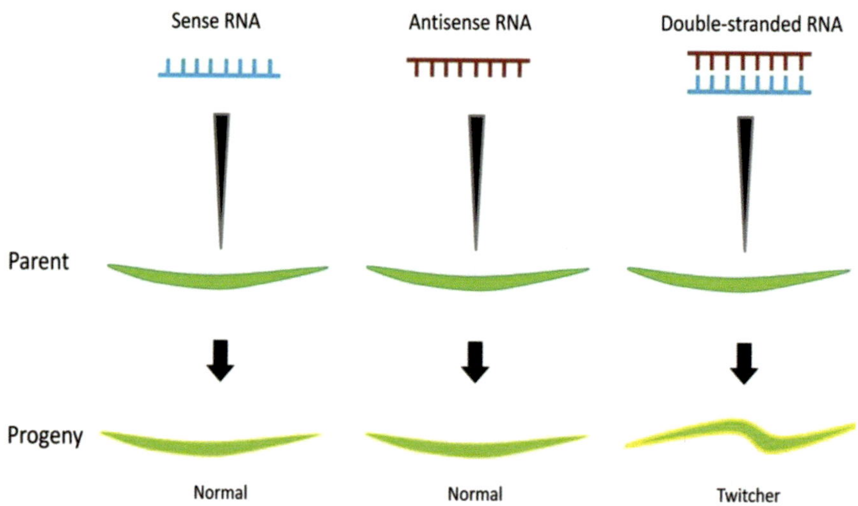

FIGURE 14.8 Discovery of RNA interference. Andrew Fire and Craig Mello injected sense and antisense RNA molecules, encoding a muscle protein, *Unc22* in *C. elegans* but did not find any changes in the phenotype of the worms. However, when both sense and antisense RNAs were injected together, it was found that the worms developed a peculiar twitching behaviour. This phenotype was caused by silencing of the native *Unc22* gene.

14.2.3 MECHANISM OF RNAI

Endogenous dsRNA initiate the process of RNAi by activating the RNase III ribonuclease protein called as Dicer. The Dicer cleaves longer dsRNA fragments into small fragments of approximately 21 base pairs with a 2-nucleotide overhang at the 3′ end (Zhang, 2004; Siomi, 2009). The Dicer contains two RNase domains and one PAZ domain and catalyzes the first step in the RNAi pathway. The formation of small interfering RNAs (siRNA) is amplified through the synthesis of secondary siRNAs, for which the Dicer-produced "primary" siRNAs are used as templates (Baulcombe, 2007). The siRNA–Dicer complex, Argonaute protein, and some additional proteins form an RNA-induced silencing complex (RISC), for the degradation of target mRNA by Argonaute (Hammond et al., 2000; Zamore et al., 2000). Out of eight Argonaute proteins, only Ago-2 possesses an active catalytic domain for cleavage activity (Meister et al., 2004). After fragmentation by Dicer, the two strands of dsRNA are separated into the guide strand and passenger strand. The guide strand remains bound to the Argonaute protein and directs RISC complex to the target mRNA and the passenger strand is degraded (Gregory et al., 2005) (Fig. 14.9).

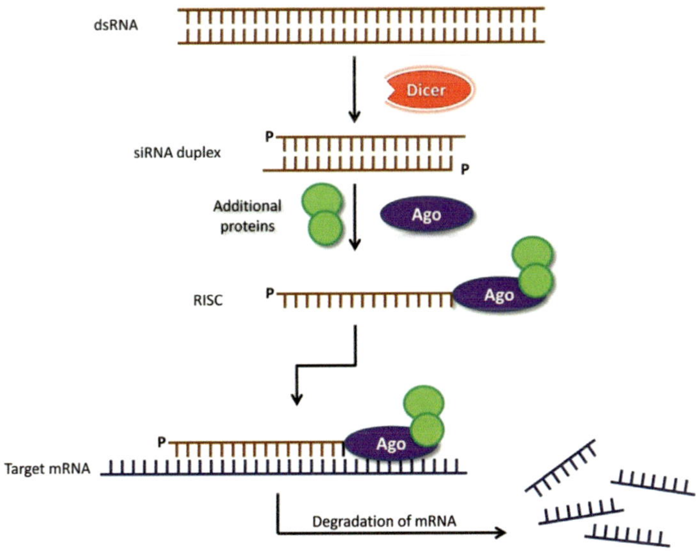

FIGURE 14.9 Mechanism of RNA interference. Long dsRNA or miRNA precursors are cleaved to siRNA/miRNA duplexes by RNase III like enzyme Dicer. These short dsRNAs are then unwound and assembled into RNA induced silencing complex upon binding to Argonaute and some additional proteins. Antisense siRNA guides RISC to target mRNA and initiates target mRNA degradation.

14.2.4 COMPONENTS OF RNAI MACHINERY

14.2.4.1 DICER

Dicer is endoribonuclease belonging to the RNase III family, which cleaves dsRNA and pre-microRNA (pre-miRNA) into double-stranded siRNA and miRNA, respectively. In humans, it is encoded by the *DICER1* gene and contains helicase as well as PAZ (Piwi/Argonaute/Zwille) domains (Matsuda et al., 2000). The PAZ domain binds to the 2-nucleotide 3′ overhang of dsRNA, while the RNase III catalytic domains initiate the cleavage of the strands (MacRae et al., 2006).

14.2.4.2 ARGONAUTE

Argonaute proteins are the active components of RISC, involved in degradation of the target mRNA strand complementary to the bound siRNA (Gregory et al., 2005). Argonaute proteins RDE-1, QDE2, and AGO1 are central for RNAi in worms, fungi, and plants, respectively (Bartel, 2004). Argonaute proteins bind to small non-coding RNAs like siRNAs, microRNAs (miRNAs), and piwi-interacting RNAs (piRNAs). These small RNAs guide Argonaute proteins to the specific target mRNAs through sequence complementarity and leads to mRNA cleavage or translation inhibition. The Argonaute (*AGO*) gene family contains four characteristic domains: N- terminal, PAZ, mid, and C-terminal PIWI domain (Hutvagner and Simard, 2008). The PAZ domain is an RNA-binding component of the protein that recognizes the 3′ end of siRNA and miRNA and can bind both double- and single-stranded RNAs (Lingel et al., 2003).

14.2.4.3 DROSHA

Drosha is a Class 2 ribonuclease III, encoded by the *DROSHA* gene in humans (Filippov et al., 2000; Wu et al., 2000). It is the core nuclease that initiates the miRNA processing in the nucleus (Fortin et al., 2002). RNA polymerase II transcribes miRNA as primary miRNAs (pri-miRNAs) which can be thousands of nucleotides long. Drosha cleaves the flanks of pri-miRNAs to produce approximately 70 nucleotide long stem-loop structures, known as precursor miRNAs (pre-miRNAs). These pre-miRNAs contain the mature miRNA in either the 5′ or 3′ end of the stem, and further cleavage by Drosha establishes either the 5′ or the 3′ end of the mature miRNA (Tomari and Zamore, 2005a).

14.2.4.4 PASHA

Pasha (also known as DGCR8), is a nuclear protein, required for microRNA processing during RNAi. Drosha is a part of microprocessor protein complex, which also contains Pasha, a dsRNA-binding protein (Denli, 2004). Pasha is crucial for Drosha activity and binds to single-stranded pri-miRNA for processing and maturation of miRNA (Han, 2006).

14.2.4.5 SIRNAS AND MIRNAS

Although several classes of small RNAs have emerged, but they have been broadly classified into three main categories: miRNAs, siRNAs, and piRNAs. These small RNAs are found in eukaryotes, although the Argonaute proteins have been found in bacterial and archaeal species (Carthew and Sontheimer, 2009). Primarily, miRNAs and siRNAs can be distinguished in two major ways. First, miRNAs are endogenous products of an organism's own genome, whereas siRNAs are mostly exogenous in origin. Second, miRNAs are processed from stem-loop precursors with incomplete double-stranded character, whereas siRNAs are excised from long and fully complementary ds-RNAs (Tomari and Zamore, 2005b). However, both siRNAs and miRNAs associate with RISCs for post-transcriptional gene regulation (Hammond et al., 2000).

siRNA are dsRNA molecules, (20–25 base pairs in length) playing a key role in the RNAi pathway by silencing the expression of specific genes with the complementary base pairing. These RNA molecules have phosphorylated 5′ ends and hydroxylated 3′ ends with overhanging nucleotides. The Dicer enzyme catalyzes the formation of siRNAs from long dsRNAs with a small hairpin RNAs (Bernstein et al., 2001). Elbashir and colleagues reported that synthetic siRNAs could induce RNAi in mammalian cells (Elbashir et al., 2001). This discovery led to a new era in the area scientific research, where scientists started harnessing RNAi for functional genomics, developmental biology, biomedical research, and drug development. Some siRNAs have also been credited as defenders of genome integrity in response to foreign or invasive nucleic acids such as viruses, transgenes, and transposons (Carthew and Sontheimer, 2009).

miRNAs are small, non-coding, endogenous RNAs (approximately 22 nt) involved in the post-transcriptional regulation of gene expression in plants, animals, and some viruses. Some of the functions attributed to the miRNA include cell division, cell death, and neuronal patterning in nematodes;

modulation of hematopoietic lineage differentiation in mammals; fat metabolism in flies; and leaf and flower development in plants (Bartel, 2004). The first miRNA, *lin-4* was discovered as a regulator of genes that control developmental timing in the larva of *C. elegans*, (Lee et al., 1993). It was found that *lin-4* does not code for a protein but produces a pair of small RNAs of 22 and 61 nt in length; the longer one was predicted to fold into a stem-loop proposed to be the precursor of the shorter one. It was found that these *lin-4* RNAs had antisense complementarity to a number of sites in the 3′ UTR of the *lin-14* gene and mediate the repression of *lin-14 (Wightman et al., 1993).* The shorter *lin-4* RNA is now considered as the founding member of regulatory RNAs called as miRNAs (Lee and Ambros, 2001). Several other miRNAs have been discovered in animals and plants and different functions have been attributed to them (Table 14.1).

TABLE 14.1 MicroRNA and Their Functions.

miRNA	Target Gene(s)	Biological Role
Nematodes (*C. elegans*)		
lin-4 RNA	lin-14 Probable transcription factor	Regulation of developmental transitions
let-7 RNA	lin-41 Probable RNA-binding protein	Regulation of developmental transitions
Insects (*Drosophila melanogaster*)		
Bantam miRNA	Hid pro-apoptotic protein	Regulation of apoptosis and growth
miR-14	Unknown	Fat metabolism and apoptosis
Mammals (*Homo sapiens*)		
miR-1	*GJA1/KCNJ2*	Heart muscle development
miR-15/16	*Bcl2*	Apoptosis
miR-155	Unknown	Adaptive immunity
miR-181	Unknown	Hematopoietic differentiation
Plants (*Arabidopsis thaliana*)		
miR-172	*AP2* and related transcription factors	Regulation of flower development
miR-JAW	*TCP4* and related transcription factors	Leaf development and embryonic patterning
miR-159	*MYB33* and related transcription factors	Regulation of flower development

KEYWORDS

- **asymmetric PCR**
- **multiplex-PCR**
- **thermal asymmetric interlaced PCR**
- **quantitative real-time PCR**
- **RNA interference**

REFERENCES

Agrawal, N.; Dasaradhi, P.; Mohmmed, A.; Malhotra, P.; Bhatnagar, R. K.; Mukherjee, S. K. RNA Interference: Biology, Mechanism, and Applications. *Microbiol. Mol. Biol. Rev.* **2003,** *67*, 657–685.

Bartel, D. P. MicroRNAs: Genomics, Biogenesis, Mechanism, and Function. *Cell*, **2004,** *116*, 281–297.

Bernstein, E.; Caudy, A. A.; Hammond, S. M.; Hannon, G. J. Role for a Bidentate Ribonuclease in the Initiation Step of RNA Interference. *Nature*, **2001,** *409*, 363–366.

Bustin, S.; Benes, V.; Nolan, T.; Pfaffl, M. Quantitative Real-time RT-PCR–A Perspective. *J. Mol. Endocrinol.* **2005,** *34*, 597–601.

Carthew, R. W.; Sontheimer, E. J. Origins and Mechanisms of miRNAs and siRNAs. *Cell* **2009,** *136*, 642–655.

Chamberlain, J. S.; Gibbs, R. A.; Rainer, J. E.; Nguyen, P. N.; Thomas, C. Deletion Screening of the Duchenne Muscular Dystrophy Locus via Multiplex DNA Amplification. *Nucleic Acids Res.* **1988,** *16*, 11141–11156.

Chien, A.; Edgar, D. B.; Trela, J. M. Deoxyribonucleic Acid Polymerase From the Extreme Thermophile *Thermus aquaticus*. *J. Bacteriol.* **1976,** *127*, 1550–1557.

Dhanasekaran, S.; Doherty, T. M.; Kenneth, J.; Group, T. T. S. Comparison of Different Standards for Real-time PCR-based Absolute Quantification. *J. Immunol. Methods*, **2010,** *354*, 34–39.

Di Serio, F.; Schöb, H.; Iglesias, A.; Tarina, C.; Bouldoires, E.; Meins, F. Sense-and Antisense-mediated Gene Silencing in Tobacco is Inhibited by the Same Viral Suppressors and is Associated with Accumulation of Small RNAs. *Proc. Natl. Acad. Sci.* **2001,** *98*, 6506–6510.

Elbashir, S. M.; Harborth, J.; Lendeckel, W.; Yalcin, A.; Weber, K.; Tuschl, T. Duplexes of 21-nucleotide RNAs Mediate RNA Interference in Cultured Mammalian Cells. *Nature*, **2001,** *411*, 494–498.

Filion, M. *Quantitative Real-time PCR in Applied Microbiology;* Horizon Scientific Press, **2012**.

Filion M (2012) Quantitative real-time PCR in Applied Microbiology. *Caister Academic Press Place, Norfolk*, **2012**, 242 p.

Filippov, V.; Solovyev, V.; Filippova, M.; Gill, S. S. A Novel Type of RNase III Family Proteins in Eukaryotes. *Gene*, **2000,** *245*, 213–221.

Fire, A.; Xu, S.; Montgomery, M. K.; Kostas, S. A.; Driver, S. E.; Mello, C. C. Potent and Specific Genetic Interference by Double-stranded RNA in *Caenorhabditis elegans*. *Nature*, **1998,** *391*, 806–811.

Fortin, K. R.; Nicholson, R. H.; Nicholson, A. W. Mouse Ribonuclease III. cDNA Structure, Expression Analysis, and Chromosomal Location. *BMC Genomics*, **2002,** *3*, 26.

Freeman, W. M.; Walker, S. J.; Vrana, K. E. Quantitative RT-PCR: Pitfalls and Potential. *Biotechniques*, **1999,** *26*, 112–125.

Gibson, D. G.; Young, L.; Chuang, R. -Y.; Venter, J. C.; Hutchison, C. A.; Smith, H. O. Enzymatic Assembly of DNA Molecules Up To Several Hundred Kilobases. *Nature Methods*, **2009,** *6*, 343–345.

Gregory, R. I.; Chendrimada, T. P.; Cooch, N.; Shiekhattar, R. Human RISC Couples MicroRNA Biogenesis and Posttranscriptional Gene Silencing. *Cell*, **2005,** *123*, 631–640.

Hammond, S. M.; Bernstein, E.; Beach, D.; Hannon, G. J. An RNA-directed Nuclease Mediates Post-transcriptional Gene Silencing in Drosophila cells. *Nature*, **2000,** *404*, 293–296.

Hayden, M. J.; Nguyen, T. M.; Waterman, A.; Chalmers, K. J. Multiplex-ready PCR: A New Method for Multiplexed SSR and SNP Genotyping. *BMC Genomics*, **2008,** *9*, 80.

Hutvagner, G.; Simard, M. J. Argonaute Proteins: Key Players in RNA Silencing. *Nat. Rev. Mol. Cell Biol.* **2008,** *9*, 22–32.

Ingelbrecht, I.; Van Houdt, H.; Van Montagu M.; Depicker, A. Posttranscriptional Silencing of Reporter Transgenes in Tobacco Correlates with DNA Methylation. *Proc. Natl. Acad. Sci.* **1994,** *91*, 10502–10506.

Innis, M. A.; Myambo, K. B.; Gelfand, D. H.; Brow, M. DNA Sequencing with *Thermus aquaticus* DNA Polymerase and Direct Sequencing of Polymerase Chain Reaction-Amplified DNA. *Proc. Natl. Acad. Sci.* **1988,** *85*, 9436–9440.

Isenbarger, T. A.; Finney, M.; Ríos-Velázquez, C.; Handelsman, J.; Ruvkun, G. Miniprimer PCR, A New Lens for Viewing the Microbial World. *Appl. Environ. Microbiol.* **2008,** *74*, 840–849.

Jernigan, D.; Lindstrom, S.; Johnson, J.; Miller, J.; Hoelscher, M.; Humes, R.; Shively, R.; Brammer, L.; Burke, S.; Villanueva, J. Detecting 2009 Pandemic Influenza A (H1N1) Virus Infection: Availability of Diagnostic Testing Led to Rapid Pandemic Response. *Clin. Infect. Dis.* **2011,** *52*, S36–S43.

Lee, R. C.; Ambros, V. An Extensive Class of Small RNAs in *Caenorhabditis elegans*. *Science*, **2001,** *294*, 862–864.

Lee, R. C., Feinbaum, R. L.; Ambros, V. The *C. elegans* Heterochronic Gene *lin-4* Encodes Small RNAs with Antisense Complementarity to *lin-14*. *Cell*, **1993,** *75*, 843–854.

Lingel, A.; Simon, B.; Izaurralde, E.; Sattler, M. Structure and Nucleic-acid Binding of the Drosophila Argonaute 2 PAZ Domain. *Nature*, **2003,** *426*, 465–469.

Liu, Y. -G.; Whittier, R. F. Thermal Asymmetric Interlaced PCR: Automatable Amplification and Sequencing of Insert End Fragments From P1 and YAC Clones for Chromosome Walking. *Genomics*, **1995,** *25*, 674–681.

MacRae, I. J.; Zhou, K.; Li, F.; Repic, A.; Brooks, A. N.; Cande, W. Z.; Adams, P. D.; Doudna, J. A. Structural Basis for Double-stranded RNA Processing by Dicer. *Science*, **2006,** *311*, 195–198.

Markoulatos, P.; Siafakas, N.; Moncany, M. Multiplex Polymerase Chain Reaction: A Practical Approach. *J. Clin. Lab. Anal.* **2002,** *16*, 47–51.

Matsuda, S.; Ichigotani, Y.; Okuda, T.; Irimura, T.; Nakatsugawa, S.; Hamaguchi, M. Molecular Cloning and Characterization of a Novel Human Gene (HERNA) Which Encodes a Putative RNA-helicase. *Biochimica et Biophysica Acta (BBA)-Gene Structure and Expression*, **2000,** *1490*, 163–169.

Meister, G.; Landthaler, M.; Patkaniowska, A.; Dorsett, Y.; Teng, G.; Tuschl, T. Human Argonaute2 Mediates RNA Cleavage Targeted by miRNAs and siRNAs. *Molecular Cell*, **2004,** *15*, 185–197.

Mullis, K.; Faloona, F.; Scharf, S.; Saiki, R.; Horn, G.; Erlich, H. Specific Enzymatic Amplification of DNA In vitro: The Polymerase Chain Reaction. *Cold Spring Harb. Symp. Quant. Biol.* **1986,** *51*, 263–273.

Mullis, K. B.; Faloona, F. A. Specific Synthesis of DNA In vitro via a Polymerase-catalyzed Chain Reaction. *Methods Enzymol.* **1987,** *155*, 335.

Napoli, C.; Lemieux, C.; Jorgensen, R. Introduction of a Chimeric Chalcone Synthase Gene Into Petunia Results In Reversible Co-suppression of Homologous Genes in Trans. *The Plant Cell*, **1990,** *2*: 279–289.

Ochman, H.; Gerber, A. S.; Hartl, D. L. Genetic Applications of An Inverse Polymerase Chain Reaction. *Genetics*, **1988,** *120*, 621–623.

Romano, N.; Macino, G. Quelling: Transient Inactivation of Gene Expression in *Neurospora crassa* by Transformation with Homologous Sequences. *Mol. Microbiol.* **1992,** *6*, 3343–3353.

Stemmer, W. P.; Crameri, A.; Ha, K. D.; Brennan, T. M.; Heyneker, H. L. Single-step Assembly of a Gene and Entire Plasmid from Large Numbers of Oligodeoxyribonucleotides. *Gene*, **1995,** *164*, 49–53.

Tomari, Y.; Zamore, P. D. MicroRNA Biogenesis: Drosha Can't Cut It Without a Partner. *Curr. Biol.* **2005a,** *15*, R61–R64.

Tomari, Y.; Zamore, P. D. Perspective: Machines for RNAi. *Genes Dev.* **2005b,** *19*, 517–529.

Van der Krol, A. R.; Mur, L. A.; Beld, M.; Mol, J.; Stuitje, A. R. Flavonoid Genes in Petunia: Addition of a Limited Number of Gene Copies May Lead to a Suppression of Gene Expression. *The Plant Cell*, **1990,** *2*, 291–299.

Wightman, B.; Ha, I.; Ruvkun, G. Posttranscriptional Regulation of the Heterochronic Gene *lin-14* by lin-4 Mediates Temporal Pattern Formation in *C. elegans*. *Cell*, **1993,** *75*, 855–862.

Wu, H.; Xu, H.; Miraglia, L. J.; Crooke, S. T. Human RNase III is a 160-kDa Protein Involved in Preribosomal RNA Processing. *J. Biol. Chem.* **2000,** *275*, 36957–36965.

Yeh, S. -H.; Tsai, C. -Y.; Kao, J. -H.; Liu, C. -J.; Kuo, T. -J.; Lin, M. -W.; Huang, W. -L.; Lu, S. -F.; Jih, J.; Chen, D. -S. Quantification and Genotyping of Hepatitis B Virus in a Single Reaction by Real-time PCR and Melting Curve Analysis. *J. Hepatol.* **2004,** *41*, 659–666.

Zamore, P. D.; Tuschl, T.; Sharp, P. A.; Bartel, D. P. RNAi: Double-stranded RNA Directs the ATP-dependent Cleavage of mRNA at 21 to 23 Nucleotide Intervals. *Cell*, **2000,** *101*, 25–3

CHAPTER 15

TILLING: GENOME POKING WITH DILIGENCES AND CONSTRAINTS

G. THAPA[1*] and J. G. HEHIR[2]

[1]*Molecular Plant Pathogen Interaction Group, Earth Institute, Science East, University College of Dublin, Belfield, Dublin 4, Ireland*

[2]*Department of Crop Science, Oak Park Crops Research Centre, Teagasc, Carlow, Ireland*

Corresponding author. E-mail: gthaps@gmail.com

CONTENTS

ABSTRACT

The rapid advancement of large-scale genome sequencing projects has opened new avenues for the application of conventional mutation techniques in not only forward but also in reverse genetics strategies. Targeting induced local lesions in genomes (TILLING), developed as an alternative to cumbersome insertional mutagenesis, takes advantage of classical mutagenesis, sequence availability and high throughput screening for nucleotide polymorphisms in a targeted sequence. Since it is readily applicable to most plants, it remains a dominant non-genetic manipulation (GM) method for obtaining desirable mutations in known genes. The application of next-generation sequencing (NGS) has been a breakthrough change, which permits multiplexing of gene targets and genomes. In this chapter, we review here the discovery and technology with more focus on the ease of applications, Ecotilling, constraints, recent developments in polyploids crops, and the future of TILLING for crop breeding.

15.1 INTRODUCTION

The best way to annotate a gene's function is to phenotype the differences that occur in the organism when mutated. In recent years, updated databases of model plant species with improvised bioinformatics tools equip the prediction of gene function based on comparative genomics, but it still requires validation. In such cases, genetic mutation becomes a persuasive tool that authenticates association between the biochemical function of a gene product and its role in vivo. Several recent molecular biology techniques have been developed to generate or identify mutations in cloned genes to determine the function of regulatory and resistance genes (R-gene) (Gilchrist and Haughn, 2005). In plants, the most commonly used reverse genetic approaches are posttranscriptional gene silencing (PTGS) (Chuang et al., 2000) and insertional mutagenesis (Alonso et al., 2003).

These techniques have led to the identification of numerous beneficial mutations; however, they have limitations. PTGS is labor intensive, can give ambiguous results, and is unsuitable for isolating mutants that have lethal or sterile phenotypes. Insertional mutagenesis has been reported with low frequency of mutations per genome, necessitating the screening of large mutagenized population to identify mutations in any given gene (Krysan et al., 1999). In addition, the mutant alleles are likely to result in a complete loss-of-function for the gene product which, if the effect is lethal or detrimental, might limit further downstream analysis.

The use of chemical mutagens represents an alternative method with mitigated drawbacks of the methods mentioned above. Ethylmethane sulfonate (EMS) has been the most commonly used mutagen in plants and induces large numbers of recessive mutations per genome (Brockman et al., 1984). Other alkylating agents such as ethylnitrosourea (ENU) have also been used effectively. EMS mainly induces canonical C-to-T substitutions resulting in C/G to T/A transitions (Krieg, 1963; Kim et al., 2006), and also generates G/C to C/G or G/C to T/A transversions through 7-ethylguanine hydrolysis or A/T to G/C transitions through 3-ethyladenine pairing errors (Krieg, 1963; Kim et al., 2006; Kodym and Afza, 2003; Greene et al., 2003). If EMS mutagenesis is facilitated with a highly sensitive detection method such as illumina sequencing and next generation sequencing, it is estimated that a population of 3000 plants would contain ~20 mutations (knockouts as well as allelic variants) in any given gene in *Arabidopsis* (Greene et al., 2003), with similar results across other diploid species. The benefits of Targeting Induced Local Lesions IN Genomes (TILLING) are that it can be applied to comparatively small populations of any plant species and does not require growth facilities for transgenic plants. One limitation of TILLING is that most of the positive pools (those containing mismatches) do not contain knockouts of the target gene, that is, most of the detected mutations are silent. A further limitation to the widespread adoption of TILLING-based methods is that the equipment required for the detection of mutations is expensive. The aim of this chapter is to provide an overview of the TILLING method and to highlight its potential applications and drawbacks in the field of plant genomic research.

15.2 TILLING AND AREAS FOR IMPROVEMENT

TILLING is a functional genomics method that can target potentially any gene by inducing chemically induced mutations in populations with wider applicability to many sexual species even in relatively small populations (Comai and Henikoff, 2006; Wang et al., 2012). The focal doctrine of TILLING is the discernment of rare genetic mutations in pooled DNA samples from large mutant populations in a first round of screening followed by deconvolution of the pool to single out the distinctive sample carrying the mutation (Fig. 15.1).

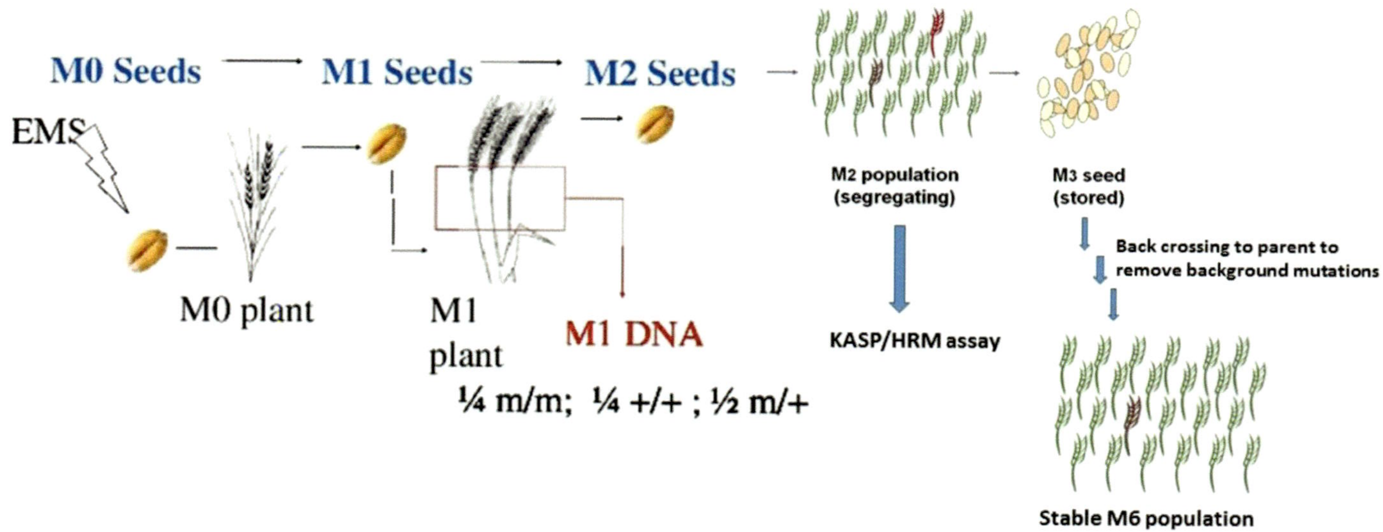

FIGURE 15.1 A scheme of EMS TILLING employed in wheat to create a mutant population. EMS treatment was applied on seeds of spring wheat Cadenza (from 0.5% to 1.2%) and 2016 M_1 plants DNA was collected for tilling assessment by direct Sanger sequencing of *Tabri1* kinase domain and M_2 population were used for KASP assay.

The majority of contemporary TILLING protocols employ two distinctive and consecutive approaches: one to screen the pools and another to pick out the individual. For example, McCallum et al. (2000) used denaturing high-performance liquid chromatography (DHPLC, the Transgenomic WAVE System; Jones et al., 1999) from pooled DNA samples to detect mutations in small PCR fragments that had been denatured by heating and subsequently cooled to promote the formation of heteroduplexes. PCR and sequencing of these fragments from individual DNA samples then identified both the individual mutant and the exact position and nature of the mutation. While DHPLC has largely been superseded by higher throughput and gel-based techniques using fluorescent PCR primers for visualization, conventional sequencing remains the most common method to confirm the putative mutations detected in the first round. TILLING allows the identification of single-base-pair (bp) allelic variation in a target gene in a high-throughput manner (Colbert et al., 2001; Till et al., 2003). It has various benefits over other methods used to detect single-bp polymorphisms. Other gel-based assays, such as single-strand conformation polymorphism (SSCP) and denaturing gradient gel electrophoresis (DGGE), do not pinpoint the location or the type of polymorphism present in the DNA fragment (DeFrancesco et al., 2001). Techniques that rely on denaturation kinetics can be observed with quantitative polymerase chain reaction (PCR) but merely work for small fragments of DNA (Gundry et al., 2003). Other techniques, such as array hybridization, are only effective in discovering ~50% of simple nucleotide polymorphisms (SNPs) (Borevitz et al., 2003). Many plant researchers consider insertion mutants as their first port of call with the availability of genome sequences of several crops; being relatively cheaper; the whole process is often an *in-silico* exercise; they produce nonfunctional alleles in most instances. In contrast, only 5% of mutant alleles recovered in TILLING screens of EMS mutagenized populations are truncations (Table 15.1) (Greene et al., 2003; Perry et al., 2009).

The data presented in Table 15.1 show the number of lines screened to be 95% confident of identifying a nonsense (truncation) mutation in a typical 1.0 kbp of coding region. On the other hand, insertional mutagenesis requires an efficient transformation system and cannot therefore be readily carried out in most plant species. Moreover, it often requires large population sizes to have a high probability of identifying an insertion in the target gene. This feature makes it best suited for species with small genomes because the development of the populations is expensive and requires complex identification strategies. Although recent rapid sequencing technologies have largely overcome the latter, insertional mutagenesis is still not a valid alternative for many crop species.

TABLE 15.1 Estimated Truncation Mutation Frequency in Mutagenized Populations of Different Plant Species.

Species	Ploidy	Mutagen	Mutation frequency (per 106 bp)	Number of lines required	Reference
Arabidopsis	Diploid	EMS	3.3	~18,000	Greene et al. (2003)
Rice	Diploid	MNU	7.4	~8000	Suzuki et al. (2008)
Barley	Diploid	EMS	1.0	~60,000	Caldwell et al. (2004)
		NaN_3	2.6	~23,000	Talamè et al. (2008)
Durum wheat	Tetraploid	EMS	25	~2400	Slade et al. (2005)
Bread wheat	Hexaploid	EMS	42	~1400	Slade et al. (2005); Uauy et al. (2009); Wang et al. (2012)

(From Parry, M. A. J.; Madgwick, P. J.; Bayon, C.; Tearall, K.; Hernandez-Lopez, A.; Baudo, M.;Rakszegi, M.; Hamada, M.; Al-Yassin. A.; Ouabbou, H.; Labhilili, M.; Phillips, A. L. Mutation Discovery for Crop Improvement. J. Exp. Bot. 2009, 60(10). Reprinted with permission from Oxford University Press.)

Many plants of agronomic importance correspond to polyploid species, including oilseed rape (Brassica *napus*), cotton (*Gossypium hirsutum*), and both bread and durum wheat (*Triticum aestivum* and *T. turgidum*) (Uauy et al., 2009; Wang et al., 2012; Tsai et al., 2013). In these species, most genes are represented by multiple *homologues* with high sequence identity. This redundancy limits the use of forward genetic screens based on phenotype as the effects of single gene knockouts are frequently masked by functional complementation with homologues present in the other genomes (Lawrence and Pikaard, 2003). Phenotypic screening for desired traits is not always easy in the case of wheat (hexaploid). To visualize a phenotypic variation in a recessive trait, all three homologues must be mutated and homozygous, an extremely rare event (Dong et al., 2009). Despite this obvious drawback, the multiple copies of each gene make polyploid species absolute fit for TILLING as they can tolerate very high mutation densities. This is well exemplified by the multiple TILLING populations, which have been published in recent years in various crops such as in tetraploids (durum wheat and tobacco) and one mutation per 32 kb in hexaploid (bread wheat and oat) (Uauy et al., 2009; Wang et al., 2012; Tsai et al., 2013). Though TILLING deals with polyploidy and mutation detection effectively, the time needed for back crossing to remove background mutation increases the time period required for phenotyping a trait. Furthermore, the specificity of mutation remains the focus for improvement and it has thus led to new site-directed mutation technology in recent years.

15.3 ADVANTAGES OF TILLING

15.3.1 SIMPLE PROCEDURE

Reverse genetics has been instrumental in validation of a gene's function with known sequence by phenotypic analysis of cells or organisms, when this gene is impaired. In plants, the most commonly used reverse-genetic approaches are insertional mutagenesis, antisense RNA, and RNA interference (RNAi), and are not similarly applicable to all organisms. For example, though T-DNA insertional mutagenesis results in gene knockout rate of 70% of *Arabidopsis* genes in silico (Alonso et al., 2003), there are over 200,000 T-DNA insertional populations in rice with only few published reports about gene knockout (An et al., 2005). Though antisense RNA and RNAi techniques are commonly used to silence expression of genes (Waterhouse et al., 1998; Yan et al., 2005), RNAi suppression generates unpredictable outcomes and the whole procedure requires time consuming techniques such as cloning, transformation, and transgenic analysis (Que et al., 1998). Consequently, reverse genetic approaches fail to meet the demands of high-throughput and large-scale detection of mutants. Therefore, the TILLING approach, which combines the traditional chemical mutagenesis and the double-dye far-red fluorescent detecting technique and requires no complicated manipulations or expensive apparatus, emerges as viable alternative. It enables screening of mutant pools easily for investigating the functions of specific genes, avoiding both the perplexing gene separation steps and tedious tissue-culture procedures involved in antisense RNA and RNAi. The series of mutant alleles corresponding to a specific gene can be produced from numerous point mutants induced by chemical mutagenesis.

15.3.2 HIGH SENSITIVITY

Original work by McCallum and colleagues (2000) in *Arabidopsis* demonstrated the high sensitivity of TILLING. Seven different PCR fragments that ranged from 345 to 970 bp in size were examined for a total of 2 Mb of DNA sequence screened by DHPLC to detect mutations in chromomethylase (CMT) genes among 835 M2 plants in *Arabidopsis*. Thirteen chromatographic alterations were detected and confirmed to be mutations by amplification and sequencing. No PCR errors were found, indicating an error rate of $<10^{-6}$ (*Arabidopsis* Genome Initiative, 2000). Combination of *CelI*, double-end fluorescent dyes labeling, and LI-COR system as an alteration to

DHPLC maintained and secured the high sensitivity of the modified high-throughput TILLING (Oleykowski et al., 1998; Till et al., 2004b). Obviously, the LI-COR double end-labeling system for detecting *CelI* cleavage products satisfies the demands of high-throughput and large-scale mutant detection as shown in Figure 15.2 (Parry et al., 2009).

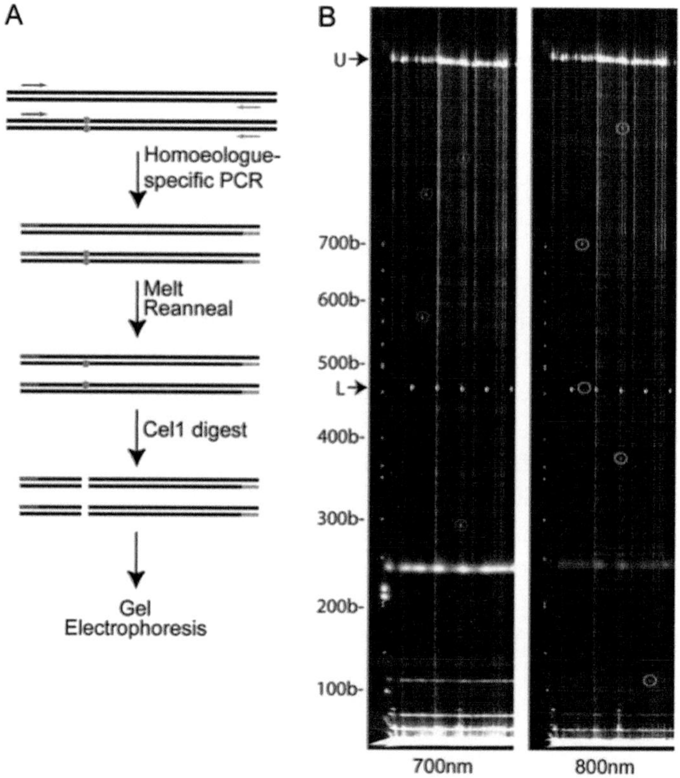

FIGURE 15.2 Overview of mutation detection in wheat by *Cel1*-based TILLING. (A) Schematic of PCR and *Cel1* digestion. (B) Typical TILLING runs in wheat. The target was 1289 bp of the bread wheat *A20ox1A* gene, amplified using homologue-specific primers labelled with the Li-Cor dyes IRD-700 and IRD-800. The DNA samples were from 80 M2 lines of an EMS-mutagenized population of bread wheat cv. Cadenza, pooled twofold. After PCR, melting/reannealing, *Cel1* digestion, and cleanup, the 40 samples were run on a Li-Cor 4300 Genetic Analyser; fragments generated by *Cel1* are ringed in the 700 nm (*red*) and 800 nm (*green*) channels. Lane (*L*) and size markers were also loaded onto the gel and the full-length (undigested) PCR product is indicated (*U*). (From Parry, M. A. J.; Madgwick, P. J.; Bayon, C.; Tearall, K.; Hernandez-Lopez, A.; Baudo, M.;Rakszegi, M.; Hamada, M.; Al-Yassin. A.; Ouabbou, H.; Labhilili, M.; Phillips, A. L. Mutation Discovery for Crop Improvement. J. Exp. Bot. 2009, 60(10). Reprinted with permission from Oxford University Press.)

The advent of high-resolution melting (HRM) (Rouleau et al., 2009), Kompetitive Allele Specific PCR (KASP) (Semagn et al., 2014) and low-cost high-throughput sequencing has added advantage to its screening sensitivity (Rigola et al., 2009; Tsai et al., 2011). Tsai et al. (2011) recently described the use of illumina sequencing and single-nucleotide polymorphism analysis in multidimensional pools as a method for efficient mutation discovery in rice and wheat. Mutations are identified by comparison with a wild-type reference sequence, thus circumventing the requirement for detection by enzymatic cleavage or PCR-based melting techniques. However, this technique still relies on PCR amplification of gene of interest and comparison.

15.3.3 HIGH EFFICIENCY

The high mutation-detecting efficiency of TILLING is attributed to its high frequencies of mutagenesis and high-throughput screening capacity (McCallum et al., 2004, Greene et al., 2003). Based on these results, the most suitable fragment is selected in a specific gene of interest. Because of the ability of chemical mutagenesis to induce high density of mutations in multiple locus, genome wide saturated mutagenesis can be achieved using a relatively small mutant population. The frequencies of mutagenesis are different in various species and receptors (Henikoff et al., 2003). If the suitable mutagenic densities satisfied the requirement, the size of mutant population will be confirmed. According to the general estimation made by the *Arabidopsis* TILLING Project, ~7 mutations per 1 Mb could be identified after screening the mutant *Arabidopsis* plant lines (Colbert et al., 2001). Based on the above estimation, a total of 10,000 mutant plants will achieve satisfied mutant densities (Henikoff et al., 2004). For the highest mutation rates of ~10,000 per genome as observed in the *Arabidopsis* TILLING Project, this corresponds to more than 20 mutations, adequate to provide a probability better than at least one knockout lesion in a typical gene, in addition to an allelic series of a dozen or more missense mutations.

The integration of automatic robotics in high-throughput gel and outstanding LICOR analyzer system supported by HRM, KASP, and low cost illumine sequencing provides TILLING with a very powerful screening capacity now that was not realized in the *Arabidopsis* TILLING Project as can be seen from Figure 15.3 (Till et al., 2003).

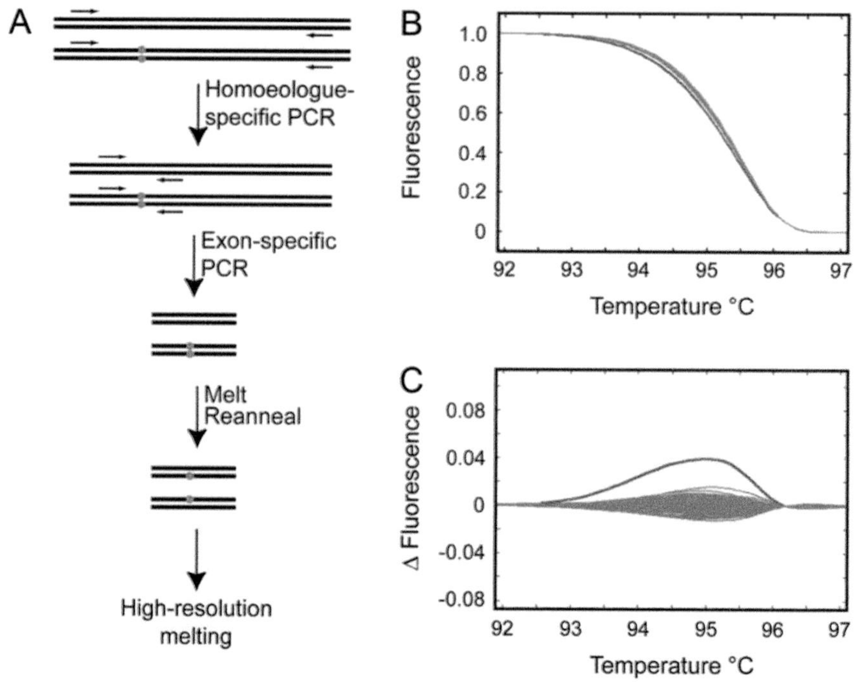

FIGURE 15.3 Mutation detection in wheat using high-resolution melt analysis (HRM). (A) Schematic of PCR and heteroduplex production. The first-round homologue-specific PCR targeted 971 bp of the bread wheat *GA20ox1D* gene; the second round amplicon was 201 bp of its first exon. LCGreen Plus dye was included in the second round PCR for melt detection and heteroduplex melt analysis was carried out on the Lightscanner (Idaho Technology). (B) Temperature-normalized fluorescence and (C) difference fluorescence plotted against temperature from melt analysis of the second-round PCR product from 192 samples (pooled twofold) from an EMS-mutagenized population of bread wheat cv. Cadenza. Wild-type samples are in gray, the single putative mutant pool identified through its different melt profile (and confirmed by sequencing of the two individual samples) is in red (From Parry, M. A. J.; Madgwick, P. J.; Bayon, C.; Tearall, K.; Hernandez-Lopez, A.; Baudo, M.;Rakszegi, M.; Hamada, M.; Al-Yassin. A.; Ouabbou, H.; Labhilili, M.; Phillips, A. L. Mutation Discovery for Crop Improvement. J. Exp. Bot. 2009, 60(10). Reprinted with permission from Oxford University Press.)

15.4 APPLICATIONS OF TILLING

15.4.1 GENE DISCOVERY

The TILLING technique was first utilized in *Arabidopsis*, a workshop to develop TILLING as a service to the *Arabidopsis* community, known as the *Arabidopsis* TILLING Project, which during the first year itself detected, sequenced, and delivered over 1000 mutations in more than 100 genes in

Arabidopsis (Till et al., 2003). With its success, Codons Optimized to Detect Deleterious Lesions (CODDLE; http://www.proweb.org/coddle/) was also developed as a general tool for polymorphism analysis and set up numerous collaborations with other groups to apply the methods to additional species. Another software developed, the conservation-based SIFT program, was able to predict, with ~75% accuracy, whether an amino-acid change results in damage to a protein (Ng et al., 2003). As such the application of TILLING and its related software and approaches were used to investigate new function of genes in other crops. Perry and colleagues scored ~45,600 M_2 progeny of 4190 EMS-mutagenized M_1 plants of *L. japonicas* and isolated mutants affecting metabolism, morphology, and the root-nodule symbiosis. This allowed for the assembly of trait-specific or theme-based TILLING populations, enriched for mutants that are affected in a particular developmental process (Perry et al., 2003).

High-throughput TILLING was also used in maize, an important crop with a large genome, wherein pools of DNA samples were screened for mutations in 1 kb segments from 11 genes and 17 independent induced mutations were obtained from a population of 750 pollen-mutagenized maize plants (Till et al., 2004a). Furthermore, the TILLING strategy also succeeded to create and identify genetic variation in wheat, thereby showing a great potential as a tool for genomic research in polyploidy plants (Slade et al., 2005). Researchers at the International Rice Research Institute (IRRI) screened pooled DNA samples from 0.8% and 1.0% EMS-mutated 2000 lines that revealed independent mutations in two genes viz., *pp2A4* encoding serine/threonine protein phosphatase catalytic subunit and *cal7* encoding callose synthase, yielding a mutation density of ~0.5 mutation per Mb. Sequencing mutated loci confirmed that they were G/C-to-A/T transition mutations (Wu et al., 2005; Till et al., 2004a; Till et al., 2007).

15.4.2 DNA POLYMORPHISM ASSESSMENT

DNA polymorphism widely exists in various species and plays an important role in biological evolution. The methods currently available for revealing DNA polymorphism encompass DNA sequencing, SSCP, hybridization, and microarray, and these methods have their own advantages and limitations. Although DNA sequencing is simple and straightforward, it is rather costly and time consuming. SSCP provides a high-throughput strategy for polymorphism detection with low efficient mutation detection ranging from 200 to 300 bp length of target DNA sequence. Microarray apart from being highly cost affair has low detecting frequency of less than 50% (Tillib et al., 2001; Borevitz et al., 2003).

Since the success of TILLING in different crop species, viz., *Arabidopsis* (McCallum et al., 2000; Till et al., 2006), rice (Sato et al., 2006; Till et al., 2007; Suzuki et al., 2008), barley (Caldwell et al., 2004), soybean (Cooper et al., 2008), maize (Till et al., 2004b), and *Triticum* (Slade et al., 2005; Uauy et al., 2009) and in nonplant species such as *Caenorhabditis elegans, Drosophila,* and zebrafish (Gilchrist et al., 2006; Winkler et al., 2005; Wienholds et al., 2003), the application of the TILLING technique has been extended to survey natural variation in genes, which has been termed as Eco-TILLING (ecotypic TILLING). It has been developed to detect DNA polymorphism present in naturally occurring variations from single nucleotide polymorphism (SNP), small fragment insertion and deletions to simple sequence repeat (SSR) (Henikoff and Comai, 2003; Comai et al., 2004). Eco-TILLIG incorporates the use of specific primers targeting two loci of the target gene of plant samples from different populations of the plant species. The nucleotide change in each of the two loci is determined by sequencing the amplified target fragments. Thus, SNP detection by Eco-TILLING is very cost-effective and requires only a small proportion of the whole cost of the conventional approach of sequencing a genetic locus in every individual.

PCR amplification of specific alleles (PASA) is another mutation detection method. This technique thus detects mutation without prior sequence information. Whereas the allele-specific PCR method (ASPCR) (Pacey-Miller et al., 2003; Schmid et al., 2003) is based on the use of available sequence information to design allele-specific primers that will anneal with the resistant biotype and not the wild type. The ASPCR technique is useful only for the screening of known mutations, whereas the Eco-TILLING technique offers the advantage of detecting new mutations as well as screening for known ones. This can be done at a fraction of the cost of SNP/haplotyping methods, which requires large-scale sequencing. Therefore, Eco-TILLING can be used, in single-nucleotide mutation detection of herbicide-R genes, as a powerful, low-cost and high-throughput reverse genetic method which cannot be practically achieved through full sequencing. The outcome of base changes is functional, synonymous, or deleterious is determined by the protein coding, phenotyping, and validation by various genetic approaches such as transgenic, silencing, and in/del (CRISPR) genome editing.

15.4.3 CROP BREEDING

Conventional mutation breeding, either by radiation or by chemical treatment, has a proven influence on production of many varieties including high-yielding rice, barley, and wheat (Ahloowalia et al., 2004). Unlike conventional mutation breeding, in which the mutation frequency is unknown or estimated only from mutations conveying a visible phenotype, TILLING provides a direct measure of induced mutations. Furthermore, TILLING allows not only the rapid, parallel screening of several genes but also a prediction of the number of alleles that will be identified on the basis of the mutation frequency and library size.

The commercial applications of TILLING to identify extensive allelic series of the *waxy* genes in both hexaploid bread wheat and tetraploid pasta wheat has been demonstrated for practical crop improvement (Slade et al., 2005). In Slade's study, they found a mutation frequency of one in 24 kb in hexaploid wheat and a frequency of one in 40 kb in tetraploid wheat. This is approximately fivefolds higher than that was observed in *Arabidopsis*. Because of this remarkable mutation frequency, Slade and colleagues identified 196 new alleles in the A and D genome *waxy* genes in only 1152 individual plants screened in hexaploid TILLING population and 50 new alleles in only 768 individuals in tetraploid wheat TILLING populations. These allelic series in both hexaploid and tetraploid wheat include multiple truncations and splice junction mutations as well as numerous missense mutations with predicted deleterious effects on the function of the waxy enzyme. These new alleles in *waxy* genes in wheat represent a useful resource for breeding a range of *waxy* and partial *waxy* wheat. Even more importantly, this work represents proof-of-concept for TILLING in other genes whose modification may be desired in wheat or other crops.

15.4.4 GENOME EDITING APPROACHES

The focal area to improve in TILLING, that is, to develop efficient and reliable methods to make precise, targeted editing to the genome of living cells, has led to new efficient approaches. Recently, the bacterial clustered regularly interspaced short palindromic repeats (CRISPR)-associated protein-9 nuclease (Cas9)-based technique from *Streptococcus pyogenes,* has had great success in humans and it has also been successfully used in plants (Cong et al., 2013). Apart from homologous recombination and RNAi, other recent approaches to targeted genome modification—zinc-finger nucleases (ZFNs) (Miller et al., 2005) and transcription-activator-like effector nucleases (TALENs) (Mussolini et al., 2011) has been used to generate permanent mutations by introducing double-stranded breaks to activate repair pathways. The requirement of high cost, expertise, and time constraints has limited there use for large-scale and high-throughput studies.

There has been recent report of preassembled complexes of purified Cas9 protein and guide RNA being used in plant protoplasts of *Arabidopsis thaliana*, tobacco, lettuce, and rice to achieve targeted mutagenesis in regenerated plants at frequencies of up to 46% (Woo et al., 2015). Other reports illustrating the first time applications of the CRISPR-Cas9/sgRNA system to plants and its success story have been demonstrated (Nekrasov et al., 2013; Shan et al., 2013; Li et al., 2013; Feng et al., 2013; Mao et al., 2013; Xie et al., 2013; Miao et al., 2013; Jiang et al., 2013).

The rapid progress in developing CRSPR-Cas9 due to its simplicity, high efficiency, and versatility of the system has led it to become one of the new plant breeding techniques (NPBTs). NPBTs are currently debated by advisory and regulatory authorities in Europe and worldwide in relation to the GMO legislation (Kuzma et al., 2011; Lusser et al., 2012; Lusser et al., 2013; Pauwels et al., 2014). Of the designer nuclease systems currently available for precision genome engineering, the CRISPR/Cas system is by far the most user friendly in terms of making possible plant genome modifications, which are indistinguishable from those introduced by conventional breeding and chemical or physical mutagenesis. As a result, crop varieties produced using CRSPR-Cas9 technologies may be classified as non-GM and will possibly bypass GMO legislation and would have progressive impact in the plant biotechnology and breeding sector.

15.5 PERSPECTIVES

The available TILLING data from different projects across the world suggest that polyploidy not genome size may be involved in conferring tolerance to high mutation density, a finding consistent with an early analysis (Stadler, 1929). There is a dearth of information on the molecular mechanisms underlying variable mutation yields in plants, which could have been harnessed for improving TILLING populations. Now, we may hypothesize that genetic redundancy may shield polyploids from the deleterious consequences of mutation or that polyploids may be physiologically more tolerant of genotoxic treatments or more susceptible to mutagenesis, due to adaptive changes affecting DNA repair and genome maintenance taking place after polyploidization. So, the use of polyploid lines would be effective for TILLING to have populations with high density of mutations, which will be ideally suited for genomic (whole genome or exome) approaches, instead of TILLING one gene at time, to collect all significant mutations.

15.6 CONCLUSIONS

TILLING is a technique that adds significantly to the arsenal of reverse genetics tools that are available to researchers wanting to capitalize on the information being provided by genome-sequencing projects. It is efficient and cost-effective, and both mutagenized and natural populations of any organism can be screened. As no reverse genetics technique developed to date is ideal for all purposes, TILLING complements other techniques and in the absence of site-directed mutagenesis in flowering plants, it is one of the few methods of detecting missense mutations in a high-throughput manner available to plant geneticists. The value of such missense mutations has long been recognized by classical genetics researchers as being essential for the elucidation of complex gene functions and gene interactions. TILLING competes easily with direct sequencing as a means of quickly identifying point mutations in genes of interest. Although improvements in pooled sequencing procedures or the development of novel mutagenesis techniques may eventually make TILLING obsolete, at present, it remains the technique of choice for medium-to-high-throughput reverse genetics in many organisms.

KEYWORDS

- **TILLING**
- **genome editing**
- **posttranscriptional gene silencing**
- ***Cell***
- **eco-TILLING**

REFERENCES

Ahloowalia, B. S.; Maluszynski, M.; Nichterlein, K. Global Impact of Mutation-derived Varieties. *Euphytica* **2004,** *135*(2), 187–204.

Alonso, J. M.; Stepanova, A. N.; Leisse, T. J.; Kim, C. J.; Chen, H.; Shinn, P.; Stevenson, D. K.; Zimmerman, J.; Barajas, P.; Cheuk, R.; Gadrinab, C.; Heller, C.; Jeske, A.; Koesema, E.; Meyers, C. C.; Parker, H.; Prednis, L.; Ansari, Y.; Choy, N.; Deen, H.; Geralt, M.; Hazari, N.; Hom, E.; Karnes, M.; Mulholland, C.; Ndubaku, R.; Schmid, I. T.; Guzman, P.; Aguilar-Henonin, L.; Schmid, M.; Weigel, D.; Carter, D. E.; Marchand, T.; Risseeuw, E.; Brogden, D.; Zeko, A.; Crosby, W. L.; Berry, C. C.; Ecker, J. R. Genome-wide Insertional Mutagenesis of *Arabidopsis* thaliana. *Science* **2003,** *301*(5633), 653–657.

An, G.; Lee, S.; Kim, S. H.; Kim, S. R. Molecular Genetics using T-DNA in Rice. *Plant Cell Physiol.* **2005,** *46*(1), 14–22.

Arabidopsis Genome Initiative Analysis of the Genome Sequence of the Flowering Plant *Arabidopsis thaliana. Nature* **2000,** *408*(6814), 796–815.

Borevitz, J. O.; Liang, D.; Plouffe, D.; Chang, H.S.; Zhu, T.; Weigel, D.; Berry, C. C.; Winzeler, E.; Chory, J. Large-scale Identification of Single-feature Polymorphisms in Complex Genomes. *Genome Res.* **2003,** *13*, 513–523.

Brockman, H. E.; de Serres, F. J.; Ong, T. M.; DeMarini, D. M.; Katz, A. J.; Griffiths, A. J.; Stafford, R.S. Mutation Tests in *Neurospora crassa*. A report of the U.S. Environmental Protection Agency Gene-Tox Program. *Mutat Res.* **1984,** *133*(2), 87–134.

Caldwell, D. G.; McCallum, N.; Shaw, P. et al. A Structured Mutant Population for Forward and Reverse Genetics in Barley (*Hordeum vulgare* L.). *Plant J.* **2004,** *40*, 143–150.

Chuang, C.F.; Meyerowitz, E. M Specific and Heritable Genetic Interference by Double-stranded RNA in *Arabidopsis thaliana*. *Proc. Natl. Acad. Sci. USA*. **2000,** *97*, 4985–4990.

Colbert, T.; Till, B. J.; Tompa, R.; Reynolds, S.; Steine, M. N.; Yeung, A. T.; McCallum, C. M.; Comai, L.; Henikoff, S. High-throughput Screening for Induced Point Mutations. *Plant Physiol.* **2001,** *126*(2), 480–484.

Comai, L.; Henikoff, S. TILLING: Practical Single-nucleotide Mutation Discovery. *Plant J.* **2006,** *45*, 684–694.

Comai, L.; Young, K.; Till, B. J.; Reynolds, S. H.; Greene, E. A.; Codomo, C. A.; Enns, L. C.; Johnson, J. E.; Burtner, C.; Odden, A. R.; Henikoff, S. Efficient Discovery of DNA Polymorphisms in Natural Populations by EcoTILLING. *Plant J.* **2004,** *37*(5), 778–786.

Cong, L.; Ran, F. A.; Cox, D.; Lin, S.; Barretto, R.; Habib, N.; Hsu, P. D.; Wu, X.; Jiang, W.; Marraffini, L. A.; Zhang, F. Multiplex Genome Engineering using CRISPR/Cas Systems. *Science* **2013,** *339*(6121), 819–823.

Cooper, J. L.; Till, B. J.; Laport, R. G.; Darlow, M. C.; Kleffner, J. M.; Jamai, A.; El-Mellouki, T.; Liu, S.; Ritchie, R.; Nielsen, N.; Bilyeu, K. D.; Meksem, K.; Comai, L.; Henikoff, S. TILLING to Detect Induced Mutations in Soybean. BMC Plant Biol. **2008,** *8*, 9.

DeFrancesco, L.; Perke, J. M. In Search of Genomic Variation: A Wealth of Technologies Exists to Find Elusive Genetic Polymorphisms. *Scientist* **2001,** *15*, 24.

Dong, C.; Dalton-Morgan, J.; Vincent, K.; Sharp, P. A modified TILLING Method for Wheat Breeding. *Plant Genome* **2009,** *2*, 39–47.

Feng, Z.; Zhang, B.; Ding, W.; Liu, X.; Yang, D. L.; Wei, P.; Cao, F.; Zhu, S.; Zhang, F.; Mao, Y.; Zhu, J. K. Efficient Genome Editing in Plants using a CRISPR/Cas System. *Cell Res.* **2013,** *23*, 1229–1232.

Gilchrist, E. J.; Haughn, G. W. TILLING without a Plough: A New Method with Applications for Reverse Genetics. *Curr. Opin. Plant Biol.* **2005,** *8*, 1–5.

Gilchrist, E.J.; O'Neil, N. J.; Rose, A. M. et al. TILLING is an Effective Reverse Genetics Technique for *Caenorhabditis elegans*. *BMC Genom.* **2006,** *7*, 262.

Greene, E. A.; Codomo, C. A.; Taylor, N. E.; Henikoff, J. G.; Till, B. J.; Reynolds, S. H.; Enns, L. C.; Burtner, C.; Johnson, J. E.; Odden, A. R.; Comai, L.; Henikoff, S. Spectrum of Chemically Induced Mutations from a Large-scale Reverse-genetic Screen in *Arabidopsis*. *Genetics* **2003,** *164*, 731–740.

Gundry, C. N.; Vandersteen, J. G.; Reed, G. H.; Pryor, R. J.; Chen, J.; Wittwer, C. T. Amplicon Melting Analysis with Labelled Primers: A Closed-Tube Method for Differentiating Homozygotes and Heterozygotes. *Clin. Chem.* **2003,** *49*, 396–406.

Henikoff, S.; Comai, L. Single-nucleotide Mutations for Plant Functional Genomics. *Ann. Rev. Plant Biol.* **2003,** *54*, 375–401.

Henikoff, S.; Till, B. J.; Comai, L. TILLING. Traditional Mutagenesis Meets Functional Genomics. *Plant Physiol.* **2004,** *135*(2), 630–636.

Jiang, W.; Zhou, H.; Bi, H.; Fromm, M.; Yang, B.; Weeks, D. P. Demonstration of CRISPR/Cas9/sgRNA-mediated Targeted Gene Modification in *Arabidopsis*, Tobacco, Sorghum and Rice. *Nucleic Acids Res.* **2013,** *41*(20), 188.

Jones, A. C.; Austin, J.; Hansen, N.; Hoogendoorn, B.; Oefner, P. J.; Cheadle, J. P.; O'Donovan, M. C. Optimal Temperature Selection for Mutation Detection by Denaturing HPLC and Comparison to Single-stranded Conformation Polymorphism and Heteroduplex Analysis. *Clin. Chem.* **1999,** *45*, 1133–1140.

Kim, Y.; Schumaker, K. S.; Zhu, J. K. EMS Mutagenesis of *Arabidopsis*. *Methods Mol. Biol.* **2006,** *323*, 101–103.

Kodym, A.; Afza, R. Physical and Chemical Mutagenesis. *Methods Mol Biol.* **2003,** *236*, 189–204.

Krieg, D. R. Ethyl Methanesulfonate-induced Reversion of Bacteriophage T4rII Mutants. *Genetics* **1963,** *48*, 561–580.

Krysan, P. J.; Young, J. C.; Sussman, M. R. T-DNA as an Insertional Mutagen in *Arabidopsis*. *Plant Cell* **1999,** *11*, 2283–2290.

Kuzma, J.; Kokotovich, A. Renegotiating GM Crop Regulation. Targeted Gene-modification Technology Raises New Issues for the Oversight of Genetically Modified Crops. *EMBO Rep.* **2011,** *12*, 883–888.

Lawrence, R. J.; Pikaard, C. S. Transgene-induced RNA Interference: A Strategy for Overcoming Gene Redundancy in Polyploids to Generate Loss of-function Mutations. *Plant J.* **2003,** *36*, 114–121.

Li, J. F.; Norville, J. E.; Aach, J.; McCormack, M.; Zhang, D.; Bush, J.; Church, G. M.; Sheen, J. Multiplex and Homologous Recombination-mediated Genome Editing in *Arabidopsis* and *Nicotiana benthamiana* using Guide RNA and Cas9. *Nat Biotechnol.* **2013,** *31*, 688–691.

Lusser, M.; Davies, H. V. Comparative Regulatory Approaches for Groups of New Plant Breeding Techniques. *N Biotechnol.* **2013,** *30*, 437–446.

Lusser, M.; Parisi, C.; Plan, D.; Rodriguez-Cerezo, E. Deployment of New Biotechnologies in Plant Breeding. *Nat Biotechnol.* **2012,** *30*, 231–239.

Mao, Y.; Zhang, H.; Xu, N.; Zhang, B.; Gao, F.; Zhu. J. K. Application of the CRISPR-Cas System for Efficient Genome Engineering in Plants. *Mol. Plant* **2013,** *6*(6), 2008–2011.

McCallum, C. M.; Comai, L.; Greene, E. A.; Henikoff, S. Targeted Screening for Induced Mutations. *Nat. Biotechnol.* **2000,** *18*, 455–457.

Miao, J.; Guo, D.; Zhang, J.; Huang, Q.; Qin, G.; Zhang, X.; Wan, J.; Gu, H.; Qu, L. J. Targeted Mutagenesis in Rice using CRISPR-Cas System. *Cell Res.* **2013,** *23*, 1233–1236.

Miller, J. C.; Urnov, F. D.; Lee, Y. L.; Rock, J.; Sorba, T.; Patterson, S.; Gregory, P. D.; Holmes, M. C.; Rebar, E. J. Development of Zinc Finger Nucleases for Therapeutic Gene Correction of Sickle Cell Anemia. *Mol. Ther.* **2005,** *11*, S35–S35.

Mussolino, C.; Morbitzer, R.; Lütge, F.; Dannemann, N.; Lahaye, T.; Cathomen, T. A Novel TALE Nuclease Scaffold Enables High Genome Editing Activity in Combination with Low Toxicity. *Nucleic Acids Res.* **2011,** *39*(21), 9283–9293.

Nekrasov, V.; Staskawicz, B.; Weige, D.; Jones, J. D.; Kamoun, S. Targeted Mutagenesis in the Model Plant *Nicotiana benthamiana* using Cas9 RNA-guided Endonuclease. *Nat. Biotechnol.* **2011,** *31*, 691–693.

Oleykowski, C. A.; Bronson Mullins. C. R.; Godwin, A. K.; Yeung, A. T. Mutation Detection using a Novel Plant Endonuclease. Nucleic Acids Res. **1998,** *26*(20), 4597–4602.

Pacey-Miller, T.; Henry, R. Single-nucleotide Polymorphism Detection in Plants using a Single-stranded Pyrosequencing Protocol with a Universal Biotinylated Primer. *Anal. Biochem.* **2003,** *317*, 166–170.

Parry, M. A. J.; Madgwick, P. J.; Bayon, C.; Tearall, K.; Hernandez-Lopez, A.; Baudo, M.; Rakszegi, M.; Hamada, M.; Al-Yassin. A.; Ouabbou, H.; Labhilili, M.; Phillips, A. L. Mutation Discovery for Crop Improvement. *J. Exp. Bot.* **2009,** *60*(10), 2817–2825.

Pauwels, K.; Podevin, N.; Breyer, D.; Carroll, D.; Herman, P. Engineering Nucleases for Gene Targeting: Safety and Regulatory Considerations. *N. Biotechnol.* **2014,** *31*(1), 18–27.

Perry, J.; Brachmann, A.; Welham, T.; Binder, A.; Charpentier, M.; Groth, M.; Haage, K.; Markmann, K.; Wang, T. L.; Parniske, M. TILLING in *Lotus japonicus* Identified Large Allelic Series for Symbiosis Genes and Revealed a Bias in Functionally Defective Ethyl Methanesulfonate Alleles Toward Glycine Replacements. *Plant Physiol.* **2009,** *151*, 1281–1291.

Perry, J.A.; Wang, T. L.; Welham, T. J.; Gardner, S.; Pike, J. M.; Yoshida, S.; Parniske, M. A TILLING Reverse Genetics Tool and a Web-accessible Collection of Mutants of the Legume *Lotus japonicus*. *Plant Physiol.* **2003,** *131*, 866–871.

Que, Q.; Jorgensen, R. A. Homology-based Control of Gene Expression Patterns in Transgenic Petunia Flowers. *Dev. Genet.* **1998,** *22*(1), 100–109.

Rigola, D.; van Oeveren, J.; Janssen, A.; Bonné, A.; Schneiders, H.; van der Poel. H. J.; van Orsouw, N. J.; Hogers, R. C.; de Both, M. T.; van Eijk. M. J. High Throughput Detection of Induced Mutations and Natural Variation using Key Point Technology. *PLoS One.* **2009,** *4*, e4761.

Rouleau, E.; Lefo, C.; Bourdon, V.; Coulet, F.; Noguchi, T.; Soubrier, F.; Bièche, I.; Olschwang, S.; Sobol, H.; Lidereau, R. Quantitative PCR High-resolution Melting (qPCR-HRM) Curve Analysis, a New Approach to Simultaneously Screen Point Mutations and Large Rearrangements: Application to MLH1 Germline Mutations in Lynch Syndrome. *Hum. Mutat.* **2009,** *30*(6), 867–875.

Sato, Y.; Shirasawa, K.; Takahashi, Y.; Nishimura, M.; Nishio, T. Mutant Selection from Progeny of Gamma-ray-irradiated Rice by DNA Heteroduplex Cleavage using Brassica Petiole Extract. *Breed Sci.* **2006,** *56*, 179–183.

Schmid, K. J.; Sorensen, T. R.; Stracke, R.; Torjek, O.; Altmann, T.; Mitchell-Olds, T.; Weisshaar, B. Large-scale Identification and Analysis of Genome-wide Single-nucleotide Polymorphisms for Mapping in *Arabidopsis thaliana*. Genome Res. **2003,** *13*, 1250–1257.

Semagn, K.; Babu, R.; Hearne, S.; Olsen, M. Single Nucleotide Polymorphism Genotyping using Kompetitive Allele Specific PCR (KASP): Overview of the Technology and Its Application in Crop Improvement. *Mol. Breed.* **2014,** *33*, 1–14.

Shan, Q.; Wang, Y.; Li, J.; Zhang, Y.; Chen, K.; Liang, Z.; Zhang, K.; Liu, J.; Xi, J. J.; Qiu, J. L., Gao, C. Targeted Genome Modification of Crop Plants using a CRISPR-Cas System. *Nat. Biotechnol.* **2013,** *31*, 686–688.

Slade, A. J.; Fuerstenberg, S. I.; Loeffler, D.; Steine, M. N.; Facciotti, D. A Reverse Genetic, Nontransgenic Approach to Wheat Crop Improvement by TILLING. *Nat. Biotechnol.* **2005,** *23*(1), 75–81.

Stadler, L. J. Chromosome Number and the Mutation Rate in *Avena* and *Triticum*. *Proc. Natl. Acad. Sci. USA.* **1929,** *15*, 876–881.

Steine, M. N.; Comai, L.; Henikoff, S. High-throughput TILLING for *Arabidopsis*. In *Methods in Molecular Biology*; Salinas, J. S. J. J., Ed.; Humana Press Inc.: Clifton, **2006**; pp 127–135.

Suzuki, T.; Eiguchi, M.; Kumamaru, T.; Satoh, H.; Matsusaka, H.; Moriguchi, K; Nagato, Y.; Kurata, N. MNU-induced Mutant Pools and High Performance TILLING Enable Finding of Any Gene Mutation in Rice. *Mol. Genet. Genom.* **2008,** *279*, 213–223.

Till, B. J.; Burtner, C.; Comai, L.; Henikoff, S. Mismatch Cleavage by Single-strand Specific Nucleases. *Nucleic Acids Res.* **2004b,** *32*(8), 2632–2641.

Till, B. J.; Cooper, J.; Tai, T. H.; Colowit, P.; Greene, E. A.; Henikoff, S.; Comai, L. Discovery of Chemically Induced Mutations in Rice by TILLING. *BMC Plant Biol.* **2007,** *7*, 19.

Till, B. J.; Reynolds, S. H.; Weil, C. et al. Discovery of Induced Point Mutations in Maize Genes by TILLING. *BMC Plant Biol.* **2004a,** *4*, 12.

Till, B. J.; Reynolds, S. H.; Greene, E. A.; Codomo, C. A.; Enns, L. C.; Johnson, J. E.; Burtner, C.; Odden, A. R.; Young, K.; Taylor, N. E.; Henikoff, J. G.; Comai, L.; Henikoff, S. Large-scale Discovery of Induced Point Mutations with High-throughput TILLING. *Genome Res.* **2003,** *13*(3), 524–530.

Till, B. J.; Colbert, T.; Codomo, C.; Enns, L.; Johnson, J.; Reynolds, S. H.; Henikoff, J. G.; Greene, E. A.; Steine, M.N.; Comai, L.; Henikoff, S. High-throughput TILLING for *Arabidopsis*. *Methods Mol Biol.* **2006**, *323*, 127–135.

Tillib, S. V.; Mirzabekov, A. D. Advances in the Analysis of DNA Sequence Variations using Oligonucleotide Microchip Technology. *Curr. Opin. Biotechnol.* **2001,** *12*(1), 53–58.

Tsai, H.; Howell, T.; Nitcher, R.; Missirian, V.; Watson, B.; Ngo, K. J.; Lieberman, M.; Fass, J.; Uauy, C.; Tran, R. K. et al. Discovery of Rare Mutations in Populations: TILLING by Sequencing. *Plant Physiol.* **2011,** *156*, 1257–1268.

Tsai, H.; Missirian, V.; Ngo, K. J.; Tran, R. K.; Chan, S. R.; Sundaresan, V.; Comai, L. Production of a High-efficiency TILLING Population through Polyploidization. *Plant Physiol.* **2013,** *161*, 1604–1614.

Uauy, C.; Paraiso, F.; Colasuonno, P.; Tran, R.; Tsai, H.; Berardi, S.; Comai, L.; Dubcovsky, J. A modified TILLING Approach to Detect Induced Mutations in Tetraploid and Hexaploid Wheat. *BMC Plant Biol.* **2009,** *9*, 115.

Wang, T. L.; Uauy, C.; Robson, F.; Till, B. TILLING in Extremis. Plant *Biotechnol. J.* **2012,** *10*, 761–772.

Waterhouse, P. M.; Graham, M. W.; Wang, M. B. Virus Resistance and Gene Silencing in Plants can be Induced by Simultaneous Expression of Sense and Antisense RNA. *Proc. Natl. Acad. Sci.* USA. **1998,** *95*(23), 13959–13964.

Wienholds, E.; van Eeden, F.; Kosters, M. et al. Efficient Target-selected Mutagenesis in Zebrafish. *Genome Res.* **2003,** *13*, 2700–2707.

Winkler, S.; Schwabedissen, A.; Backasch, D. et al. Target-selected Mutant Screen by TILLING in *Drosophila. Genome Res.* **2005,** *15*, 718–723.

Woo, J.W.; Kim, J.; Kwon, S.; Corvalán, C.; Cho, S. W.; Kim, H.; Kim, S. G.; Kim, S. T.; Choe, S.; Kim, J. S. DNA-free Genome Editing in Plants with Preassembled CRISPR-Cas9 Ribonucleoproteins. *Nat. Biotechnol.* **2015,** *33*, 1162–1164.

Wu, J. L.; Wu, C.; Lei, C.; Baraoidan, M.; Bordeos, A.; Madamba, M. R.; Ramos-Pamplona, M.; Mauleon, R.; Portugal, A.; Ulat, V. J.; Bruskiewich, R.; Wang, G.; Leach, J.; Khush, G.; Leung, H. Chemical- and Irradiation-induced Mutants of *indica* Rice IR64 for Forward and Reverse Genetics. *Plant Mol. Biol.* **2005,** *59*(1), 85–97.

Xie, K.; Yang, Y. RNA-guided Genome Editing in Plants using a CRISPR-Cas System. *Mol. Plant* **2013,** *6*(6), 1975–1983.

Yan, F.; Cheng, Z. M. Progress of RNA Interference Mechanism. *Hereditas* (Beijing) **2005,** *27*(1), 167–172.

CHAPTER 16

ADVANCES IN MOLECULAR TECHNIQUES TO STUDY DIVERSITY

PRASANT KUMAR[1*], MITESH DWIVEDI[1], MITESH B. PATEL[1], CHANDRA PRAKASH[2], and BISHUN DEO PRASAD[3]

[1]*C. G. Bhakta Institute of Biotechnology, Uka Tarsadia University, Tarsadi 394350, Surat, Gujarat, India*

[2]*Department of Microbiology and Biotechnology Centre, Faculty of Science, Genome Research Centre, Maharaja Sayajirao University of Baroda, Vadodara 390002, Gujarat, India*

[3]*Department of Molecular Biology and Biotechnology, Bihar Agricultural College, Sabour, Bhagalpur, Bihar, India*

[*]*Corresponding author. E-mail: prasantmmbl@gmail.com*

CONTENTS

ABSTRACT

The existing genetic diversity in plant population is of fundamental interest for basic science and applied aspects like the efficient management of crop genetic resources. The processes dealing with improvement of crop genetic resources require an assessment of diversity at some level, to select resistant, highly productive varieties. With recent advances in genomics and sequencing technologies, molecular techniques have initiated a new era of molecular diversity in plants. Molecular techniques have had critical roles in studies of phylogeny and species evolution as well, and have been applied to increase our understanding of the distribution and extent of genetic variation within and between the species. This chapter summarizes recent progress in the area of advancement of molecular techniques with an emphasis on novel techniques and approaches that offer new insights into the molecular diversity in plants and plant-associated microbes present in the rhizosphere. The advantages and pitfalls of commonly used molecular methods to investigate molecular diversity in plants are also discussed. In addition, the potential applications of such molecular techniques and how they can be combined for a greater comprehensive assessment of molecular diversity in plants have been discussed here.

16.1 INTRODUCTION

Traditional approaches contribute significantly in understanding the diversity but are limited by lack of precision, low level of reproducibility, and being labor intensive, and thus limiting the effectiveness for analyzing the large number of individuals. Genetic markers are mainly used to identify specific genes located in the chromosomes. Arrays of genetic marker are reported and fall into one of the three broad categories: (1) morphological and agronomic traits, (2) biochemical markers, (3) DNA-based markers. The DNA-based markers or molecular markers are revolutionary on the entire concept of marker and from last 2 decades, other approaches are least observed in many laboratories. This is mainly due to simplicity, low cost, less laborious, and very accurate as compared to other techniques. These molecular techniques are initially used to investigate ecological and microbial communities and slowly these techniques have been applied to the higher organisms such as plant and animals. In present condition, biotechnologists use vast array of techniques to investigate molecular diversity in plants and plant-associated microbes present in the rhizosphere. The rhizosphere is the narrow region of soil that is directly influenced by root secretions and soil-associated microbes. DNA-based markers are used for the study of plant growth-promoting bacteria, mycorrhiza association, and pathogen interaction to the plants. Various molecular approaches are available to analyze the plant and plant microbial diversity yet suffer from experimental bias and selectivity in their own way (Vaughan et al., 2000; Keeley et al., 2015) (Table 16.1). In the last part of this chapter, we will cover DNA barcoding as it is necessary in present and future to deal with diversity data. Nowadays DNA barcoding has become a justifiable tool for the assessment of global biodiversity patterns and it can allow diagnosis of known species to nontaxonomists.

TABLE 16.1 Application and Limitations of Various Molecular Techniques Used to Study Diversity.

Molecular Techniques	Applications	Limitations
16S and 18S rDNA sequencing	Identification of bacteria and fungi	Whole genome sequencing and large-scale cloning are laborious and costly process
SNP	Detection of DNA polymorphisms, cost-effective, and identifies a large number of SNPs in short time in any plant species	Abundance of polymorphisms is very high in the organism and low-level information obtained as compared to microsatellite; polyploid genomes are more difficult to analyze
SSCP	Rapid comparative analysis; simple, accurate, and relatively inexpensive method	Semi-quantitative; identification only possible with clone library. Skills required for handling gel and staining
DGGE/TGGE	Analyses of communities without any prior knowledge of the species. Rapid comparative analysis; identification by individual band extraction for detection of specific groups or species	Semi-quantitative; time consuming; works only for short fragment of DNA.
FISH	Detection; enumeration; comparative analysis possible with automation	Requires probe design; laborious without automation
Dot-blot hybridization	Detection; estimates relative abundance of fungal pathogens and virus infection in plants	Requires probe design; laborious method
DNA barcoding	Solves problems of specific classification in a wide range of organisms	Suboptimal analytical method, incorrectly interpreting the barcoding gap

16.2 SINGLE NUCLEOTIDE POLYMORPHISMS

16.2.1 INTRODUCTION

A single nucleotide polymorphism (SNP) refers to a single base change in DNA. These SNPs occur when there are two or more possible nucleotides seen at a specific mapped location in the genome, where the least frequent allele has an abundance of 1% or more (Brookes et al., 1999). In simple words, SNP is the polymorphism occurring between DNA samples with respect to single base. SNPs comprise the most abundant molecular markers in the genome. An International Single Nucleotide Polymorphism Consortium (ISNPC) has currently identified over 6 million SNPs, approximately one at every 1–2 kbp. SNPs may occur in noncoding regions as well as in coding regions. Some missense polymorphisms are more conservative than others, e.g., a change in the codon CUU (leucine) to AUU (isoleucine) would have minimal structural impact, whereas modification of CAU (histidine) to CCU (proline) would be expected to have dramatic structural and/or functional influence on the protein.

In plants, SNPs are found to be present in high density across the genome. In maize genome, one SNP per 70 bp and in wheat one SNP per 20 bp has been observed in some regions (Bhattramakki et al., 2000).

The SNPs have become the markers of choice. Due to their abundance in genome, they are extremely useful for creating high-density genetic map. SNPs are less mutable as compared to other markers, particularly microsatellites. The low rates of recurrent mutation make them evolutionarily stable. They are excellent markers for studying complex genetic traits and understanding the genomic evolution. This also makes them suitable and easier to follow in population studies.

SNPs act as potential useful markers for the gene mapping studies, particularly for identifying genes involved in complex diseases (Chakravarti et al., 2001). But the knowledge of frequency and distribution of these SNPs across ethnically diverse populations is essential in order to know their usefulness as markers for gene mapping studies. Additionally, the density of SNPs needed for mapping complex diseases will likely vary across populations with distinct demographic histories (Tishkoff and Verrelli, 2003).

16.2.2 SIGNIFICANCE OF SNPS IN PLANTS

SNPs have not been in regular use in plant genotyping. A large amount of SNP data is available in humans but very limited data is available on SNPs in plants. This is mainly due to the enormous cost involved in developing SNPs, but since human geneticists have developed a number of SNP genotyping assays, plant biologists can take advantage and use the already well-developed assays in human studies. SNPs have tremendous potential for germplasm fingerprinting and marker assisted selection (MAS). In several crop plants, markers associated with phenotypic traits have been used in selection for desirable traits in plant breeding programs. Polymorphisms also permit the rapid identification of genetic diversity among cultivars and genomic locations of heritable traits.

However, practical use of MAS has been very limited. The example where MAS has been actually used in plant breeding program is in the development of soybean cultivars resistant to soybean cyst nematode (SCN) (Young, 1999). SNPs found in coding sequences may result in phenotypic polymorphism or show 100% association with a particular trait. Therefore, they can be used efficiently in MAS. SNPs are present in close proximity to the coding sequences and show <100% association can also be used in MAS.

16.2.3 SNP ANALYSIS

SNP analysis techniques fall into two distinct classes:

1. SNP Identification: Detection of novel polymorphisms
2. SNP Genotyping: Identifying specific allele in a known population

16.2.3.1 SNP IDENTIFICATION METHODS

The identification and characterization of large numbers of SNPs are necessary before their use as genetic tools. The following four methods are commonly used for SNP detection (Gray et al., 2000).

16.2.3.1.1 SSCP Detection

For single-strand conformation polymorphism (SSCP) detection, the DNA fragment spanning the putative SNP is PCR amplified, denatured, and run on denaturing polyacrylamide gel. During the gel run, the single-stranded fragments adopt secondary structures according to their sequences. Fragments bearing SNPs are identified by their aberrant migration pattern and further confirmed by sequencing. Although SSCP is a widely used and relatively simple technique, it gives a variable success rate for SNP detection, typically ranging from 70% to 95%. It is labor intensive and has relatively low throughput although higher capacity methods are under development using capillary rather than gel-based detection (Orita et al., 1989).

16.2.3.1.2 Heteroduplex Analysis

This relies on the detection of a heteroduplex formed during reannealing of the denatured strands of a PCR product derived from an individual heterozygous for the SNP. The heteroduplex can be detected as a band shift on a gel or by differential retention on a HPLC column. HPLC has rapidly become a popular method for heteroduplex-based SNP detection due to simplicity, low cost, and high rate of detection, i.e., 95–100% (Lichten and Fox, 1983).

16.2.3.1.3 Direct DNA Sequencing

The favored high-throughput method for SNP detection is direct DNA sequencing. SNPs may be detected *in silico* at the DNA sequence level. The wealth of redundant sequence data deposited in public databases in recent years, in particular expressed sequence tag (EST) sequences, allows SNPs to be detected by comparing multiple versions of the same sequence from different sources.

16.2.3.1.4 Variant Detector Arrays

Variant detector arrays (VDA) technology is a relatively recent addition to the high-throughput tools available for SNP detection. This technique allows the identification of SNPs by hybridization of a PCR product to oligonucleotides arrayed on a glass chip and measuring the difference in hybridization strength between matched and mismatched oligonucleotides. The VDA detection allows rapid scanning of large amounts of DNA sequences (Wang et al., 1998).

16.2.3.1.5 High Resolution Melting

High resolution melting (HRM) is a novel, homogeneous, close-tube, post-PCR method, enabling analysis of genetic variations (SNPs, mutations, methylations) in PCR amplicons. HRM characterizes nucleic acid samples based on their disassociation (melting) behavior. Samples can be discriminated according to their sequence, length, GC content, or strand complementarity. Even single base changes such as SNPs can be readily identified (Reed et al., 2004).

16.2.3.2 SNP Genotyping Methods

SNP genotyping involves two components (Chen and Sullivan, 2003), i.e., a method for discrimination between alternate alleles and reporting the presence of the allele or alleles in the given DNA sample.

A typical genotyping protocol consists of the following steps:

1. Target fragment amplification by PCR;
2. Allelic discrimination reaction can be carried out by either of the following methods: primer extension, pyrosequencing, hybridization, and sequence specific cleavage;
3. Allele specific product identification can be done by either of the following ways: fluorescence resonance energy transfer (FRET), electrophoresis, microarray, and mass spectroscopy;
4. TaqMan assay for SNP genotyping: The TaqMan genotyping assay combines hybridization and 5′ nuclease activity of polymerase coupled with fluorescence detection. It allows screening, association, candidate region, candidate gene, and finemapping studies.

16.2.4 APPLICATIONS OF SNPS IN CROP GENETICS

DNA markers have been used for the detection of polymorphism. Among them, SSRs are usually preferred as they are highly informative but less suitable for association studies because they exhibit homoplasy (occurrence of SSR allele of identical size but of different evolutionary origin and/or conversely SSRs of different size embedded in identical haplotypes) (Rafalski, 2002). However, this can be overcome by using SNPs as they are highly informative and their assays do not require DNA separation by size. They are easier to locate in most single copy regions of genome than SSRs. Edward et al. (2001) have identified abundant SNPs in the flanking sequence of maize microsatellites.

Nasu et al. (2002) established 213 SNP markers distributed throughout the rice genome and thereby illustrating the immense potential of SNPs as molecular markers for genome research as well as molecular breeding of rice. SNPs in hexaploid wheat, which can be converted into CAPS markers, have been identified (Blake, 2004).

Use of SNP markers for linkage analysis has some of the following advantages:

1. Analysis could be performed in early growth stage of plants, requiring only a small quantity of DNA.
2. Time and labor could be saved as no electrophoresis is required in most of the SNP genotyping assays.

16.2.4.1 SNP MARKERS ASSOCIATED WITH GENES OF ECONOMIC VALUE

SNP markers associated with economic traits in plants have the potential to be used in MAS and may enable extrapolation into transgenes. For example, in *Setaria viridis* (L.) Beauv. utility of SNP present in *ACCase* (Acetyle-CoA Carboxylase) gene is herbicide resistance management. Herbicides block fatty acid synthesis by inhibiting ACCase (Délye et al., 2002). In rice (*Oryza sativa*) utility of SNP in *Wx* (waxy gene) is for development of new cultivars. *Waxy* gene controls amylose synthesis as it encodes starch synthase enzyme (Ayres et al., 1997).

16.2.4.2 EVALUATION OF DIVERSITY AND CULTIVAR IDENTIFICATION

Molecular markers have proven to be powerful tools for assessing genetic variation within and between populations of plants. SNPs have become popular tools for identifying genetic loci that contribute phenotypic variations. In addition, there are several useful SNP markers in tree species and crop plants which are of great interest.

16.3 SINGLE-STRANDED CONFORMATION POLYMORPHISM

Single-stranded conformation polymorphism (SSCP) offers a simple, inexpensive, and sensitive method for detecting whether or not DNA fragments are identical in sequence, and so it can greatly reduce the amount of sequencing necessary (Kanazawa et al., 1986; Hayashi and Yandell, 1993). It is based on the observation that the mobility of single-stranded DNA in nondenaturing polyacrylamide gels is very sensitive to primary sequence, probably because slight sequence changes have major effects on conformation. In the absence of a robust theory for predicting mobility based on conformation and conformation based on sequence, SSCP remains largely empirical.

SSCP has been extensively applied in biomedical research and there are scores of published adaptations and refinements. Despite some successful applications in population biology (Antolin et al., 1996; Bagley and Gall, 1998; Gasser et al., 1999; Steel et al., 2000), molecular ecologists are yet to fully embrace SSCP technology.

16.3.1 GENERAL METHOD (PCR–SSCP)

SSCP entails electrophoresis of single-stranded (ss) DNA fragments of suitable size through a nondenaturing polyacrylamide gel, followed by visualization. Under appropriate conditions (notably low temperature and nondenaturing conditions), DNA strands fold into structures that migrate according to their shape. DNA strands of different sequence generally do not assume the same shape and so have distinct gel mobilities. Recent evidence suggests that these mobility differences are based primarily on tertiary rather than secondary structure of the DNA molecules (Liu et al., 1999). The sensitivity of the technique is generally inversely proportional to the size of fragment [e.g., single base pair differences resolved 99% of the time for 100–300 bp fragments, >80% for 400 bp ones; references in Girman (1996)]. Fragments at least as large as 775 bp may be analyzed successfully (Ortí et al., 1997a).

Sensitivity of detecting sequence variation and the appearance of the banding patterns associated with a given sequence may alter with experimental details. When it is essential that a given variant or all variants must be revealed, one can attempt to achieve this by altering temperature, gel and buffer compositions, and running conditions (Hayashi et al., 1999; Nataraj et al., 1999). This degree of optimization will not be necessary for most applications involving screening genetic variation in molecular ecology.

16.3.2 PROTOCOL OF SSCP

The following is an example of an SSCP protocol that is used for detection of mutations in HIV-1 DNA (Fujita and Silver, 1994). PCR conditions may need to be changed depending on target and primer sequences.

16.3.2.1 AMPLIFICATION AND LABELING OF DNA BY PCR

A typical (10-~1) reaction mixture contains 1 ~g of DNA, 5 pmoles of each primer, 0.5 units of Taq polymerase, each deoxynucleoside triphosphate at 0.2 mM, 10 mM Tris-HCl (pH 8.3), 50 mM KC1, 1.5 mM MgC12, and 0.01% gelatin. DNA is labeled by adding 0.5 ~Ci of [32p]dCTP (3000 Ci/mmole) to the PCR mixture or by using ~2P-end-labeled primer(s). PCR conditions are 30 cycles of 94~ for 30 s, 55~ for 30 s, and 72~ for 30 s.

16.3.2.2 DENATURATION OF PCR PRODUCT

The amplified fragments are mixed with an equal volume of sample buffer [95% formamide, 20 mM EDTA (pH 8.0), 0.05% xylene cyanol, and 0.05% bromophenol blue], denatured at 85~ for 5 min, and cooled on ice. The artifacts owing to the presence of free primer, PCR-amplified product should be diluted 1:100 with TE buffer or passed over a Sephacryl S-300 column before denaturation.

16.3.2.3 GEL CONDITIONS

One hundred milliliters of a 5% nondenaturing gel containing 17 mL of 30% polyacrylamide-0.8% bisacrylamide mixture, 30 μL of TEMED, and 0.8 mL of 10% ammonium persulfate in lX TBE (90 mM Tris-borate at pH 8.3, 4 mM EDTA) is prepared and cast in standard sequencing gel plates using 0.4 mm spacers and a sharkstooth comb. The cast gel is cooled in a cold room for at least 30 min prior to sample loading. Denatured DNA is loaded (2 μL/well) and electrophoresed at 4μL at a constant voltage of 500 V for 12–15 h or 1100 V for 3–5 h in lx TBE electrophoresis buffer. After electrophoresis, the gel is transferred onto a sheet of Whatmann 3MM paper and vacuum-dried before autoradiography.

16.3.2.3.1 New Detection Methods

Silver staining is nearly as sensitive as radiolabeling and provides a permanent record (Ainsworth et al., 1991; Oto et al., 1993). Fluorescent dyes attached to oligonucleotides, in conjunction with sensitive fluorescence detectors, offer advantages in terms of automated data acquisition. Also, by running a marker DNA labeled with a different fluorophore in all lanes, corrections can be made for lane-to-lane variation in mobility (Makino et al., 1992; Ellison et al., 1993).

16.3.3 APPLICATIONS OF SSCP

SSCP has been used most extensively to screen for inherited mutations or detect somatic mutations in cancer cells (Jacquemier et al., 1994). Because of its sensitivity to single-base changes, SSCP has been used to search for polymorphisms in cloned or amplified DNA that can then be used as genetic markers. Extensive genetic maps using SSCP markers have been constructed in the mouse (Beier 1993; Hunter et al., 1993). SSCP can be used to purify different alleles amplified from a heterozygous individual to facilitate sequencing (Suzuki et al., 1991). SSCP can also be used to quantitative input DNA in a PCR, by adding known amounts of a sequence variant that is presumed to amplify with equal efficiency as the PCR target but migrates differently in SSCP (competitive PCR) (Yap and McGee, 1992). In microbiology, SSCP has been used to classify virus strains (Lin et al., 1993). SSCP can aid in the identification of mutations that are selected for various bacteriological or viral systems (Morohoshi et al., 1991; Fujita et al., 1992). When SSCP is used to identify sequence changes associated with a selected phenotype, it is very important to have a confirmatory test of the biological significance of identified sequence changes. Several completely in vitro systems have been described for "molecular evolution" under selective pressure (Robertson and Joyce, 1990; Bartel et al., 1991). SSCP analysis might be useful in these systems to identify strongly selected molecular variants. SSCP is a powerful method for identifying sequence variation in amplified DNA.

16.3.3.1 CONFIRMATION OF INTRASPECIFIC VARIATION REPRESENTED IN PHYLOGENETIC SEQUENCING PROJECTS

Phylogenetic approaches are used for a plethora of questions from biodiversity prioritization to phylogeography (Crozier, 1997; Moritz and Faith,

1998). For these applications, intraspecific as well as interspecific variation should be quantified. However, intraspecific variation is often undersampled owing to time and expense. Particularly at lower taxonomic levels, this carries the risk that intrataxonomic variation swamps intertaxonomic variation. Application of SSCP presents a rapid and inexpensive method to ensure adequate representation of intrataxonomic sequence variation. Example: *Cytochrome* b *variation in* different species.

16.3.3.2 SCREENING LARGE POPULATION SAMPLES FOR MITOCHONDRIAL DNA (MTDNA) SEQUENCE VARIATION

Uncovering and exploring unexpected patterns of sequence variation: mitochondrial DNA (mtDNA) cytochrome oxidase I (COI) in the different species can be analyzed.

16.3.3.3 DEVELOPING AND SCREENING SEQUENCE-VARIABLE MARKERS

Example: Nuclear intron variation: *EF1a intron variation in Sitobion* aphids. *Sitobion* aphids are useful models for studying the evolution of sex and parthenogenesis (Sunnucks et al., 1996; Simon et al., 1999). Sequence-variable markers are desirable for these projects. Conserved primers amplifying variable single copy nuclear (scn) DNA provide a source of potential nuclear marker regions (Friesen et al., 1997; Villablanca et al., 1998). However, each candidate must be assessed for sequence variation with high sensitivity, complications owing to paralogous copies must be excluded, and techniques for rapid screening are highly desirable. SSCP offers much in both marker development and screening.

SSCP has been effective in revealing variation in approximately 220 bp PCR product amplified by a pair of primers (EF1 and EF2; Palumbi, 1996) for an intron in the elongation factor gene 1a (EF1a) of aphids. For example, members of a functionally parthenogenetic set of *Sitobion* aphid genotypes can be ascribed to one of two species (Wilson et al., 1999). One species differs at only one site from the other in this EF1a region. This difference cannot be assayed by any known restriction enzyme. Three representatives of one SSCP phenotype and seven of the other have been sequenced, revealing no additional variation. Different phenotypes are seen in other *Sitobion* aphids and these always have different sequences.

16.3.3.4 GETTING THE MOST OUT OF MICROSATELLITES

Example: *length homoplasy and sequence variation:* allele length homoplasy (alleles with different evolutionary histories but having the same length; Ortí et al., 1997) is a challenge to application of microsatellites in molecular phylogeography and systematics. SSCP offers an inexpensive and rapid approach to this issue, in that it reveals same length alleles with different sequences.

16.3.3.5 INVESTIGATING COMPLEX MIXTURES OF SIMILAR LENGTH SEQUENCES

Pseudogenes and multigene families.

16.4 DENATURING GRADIENT GEL ELECTROPHORESIS (DGGE)

16.4.1 INTRODUCTION

Metagenomics has been defined as the science of biological diversity; it combines the use of molecular biology and genetics to identify and characterize genetic material from complex microbial environments. Metagenomics has great promise as a methodology to address questions in plant ecology and genetics. This approach helps in understanding the plant-microbial interaction. A full metagenomic approach is a comprehensive study of nucleotide sequence, structure, regulation, and function providing a picture of the dynamics of complex microbial communities residing in the specific ecosystem (Tringe et al., 2005). Metagenomic approach can identify the diversity, but not the relative numbers of each species residing in that particular environment. Metagenomic approaches will provide a wealth of sequence information from single ecosystems, but they are not suitable for high-throughput monitoring. Denaturing gradient gel electrophoresis (DGGE) and related fingerprinting techniques have proven their power in comparing and monitoring ecosystems at the 16S rRNA gene or 18S rRNA gene or functional gene level.

16.4.2 BASIC CONCEPT OF DGGE

DGGE and temperature gradient gel electrophoresis (TGGE) are gel-electrophoretic separation procedures for double stranded DNAs of equal size but with different base-pair composition or sequence (Muyzer and Smalla, 1998). PCR-DGGE of ribosomal DNA was introduced into microbial ecology by Muyzer and his coworkers in 1993 (Ercolini, 2004). In principle, the methods are sensitive enough to separate DNAs on the basis of single point mutations (Sheffield et al., 1989). This method allows separation of DNA fragments of the same length but with different base-pair sequences. The technique had gained the popularity in microbial ecology for analyzing the diversity of total bacterial communities. In brief, the specific genes are amplified using the appropriate primer pair, one of which has a G + C "clamp" attached to the 5′ end that prevents the two DNA strands from completely dissociating even under strong denaturing conditions. During electrophoresis through a polyacrylamide gel containing denaturants, migration of the molecule is essentially arrested once a domain in a PCR product reaches its melting temperature. Following staining of the DNA, a banding pattern emerges that represents the diversity of the rRNA gene sequences present in the sample. The intensity of an individual band is a semiquantitative measure for the relative abundance of this sequence in the population.

16.4.3 STEPS OF DGGE

The steps involved in DGGE are briefly explained here and are outlined in Figure 16.1.

1. DNA extraction: total DNA is extracted from the environmental samples.
2. PCR with specific primers: amplification of DNA using specific primers containing GC clamps targeting the gene of interest that are present in all members of the community.
3. PCR amplicon separation by DGGE: PCR amplicon separation to be performed with the help of D-code universal mutation detection system (Bio-Rad Laboratories, Richmond, Calif.) used for a DGGE analysis. The denaturing conditions are provided by urea and formamide (100% of denaturant solution consists of 7 M urea and 40% formamide) (Duarte et al., 2012). During denaturation, the two strands of a DNA molecule separate at a specific denaturant concentration, and the DNA sequence stops its migration in the gel. The gels are subjected to a constant 80 V for 12 h at 60°C, and after electrophoresis, the gels are stained with specific dye.

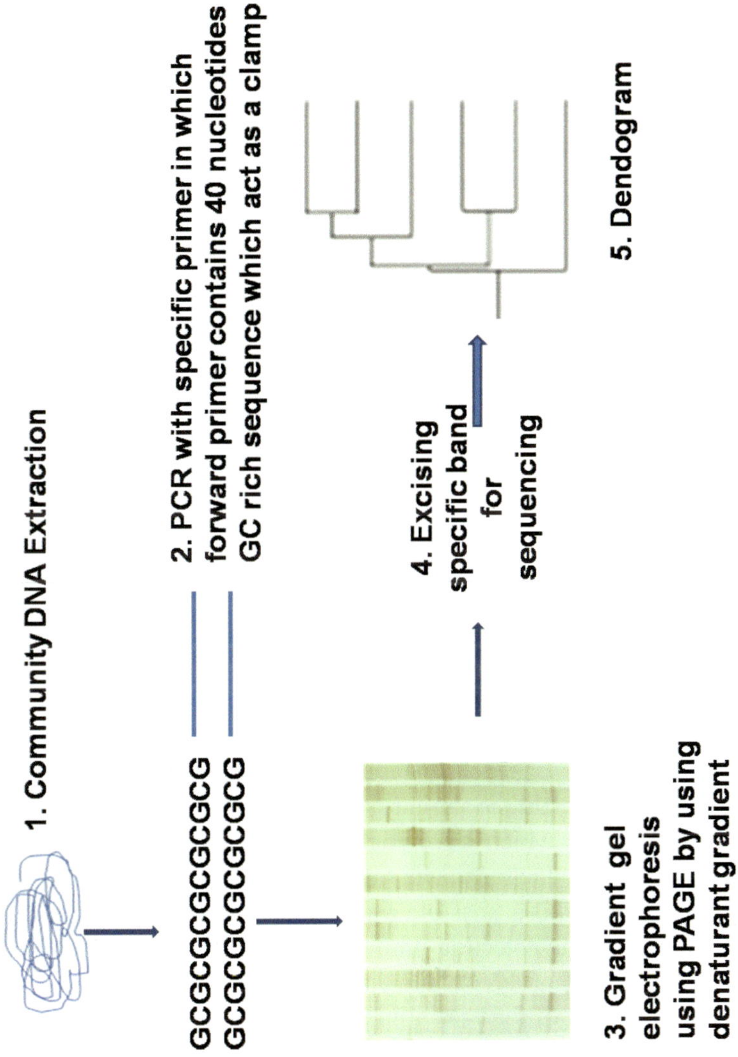

FIGURE 16.1 Steps for DGGE analysis.

4. Excision of specific band from the stain gel: The band from the gel is excise and re-PCR with specific primer without GC clamp to get the specific sequence and do further sequencing.
5. DGGE profiles: Based on the sequencing results, dendrogram are prepared to analyze the DGGE profiles.

16.4.4 ADVANTAGES AND DISADVANTAGES OF DGGE

16.4.4.1 ADVANTAGES

1. This technique helps to analyze the complex communities;
2. Distinguish homozygous versus heterozygous genotypes from a particular DNA fragments;
3. Excision of band from the DGGE and sequencing leads to the representative of species in the complex community;
4. This technique is very simple and does not require any toxic or radioactive materials.

16.4.4.2 DISADVANTAGES

1. Small fragments of less than 500 bp can be separated by this technique.
2. Ideally, one species give one band on denaturing gel but it is not true all the time, as some time more than two bands are observed on gel of single species.
3. Different DNA sequences may give band at the same place because the mobility of same is similar due to identical GC contents.
4. The community which is present in very less may not be representing in DGGE band due to lack of sufficient PCR amplicon compared to other species. This is possible when the amount of minor populations can be below the detection limit.
5. DGGE instruments are required to perform the experiment.
6. Additional cost of forward primer due to addition of GC clamp.
7. Time taken to run the gel for better separation of band is too long. It takes almost 12–16 h to run the gel for better separation of band.

16.5 DNA BARCODING

16.5.1 INTRODUCTION

DNA barcoding is a taxonomic method that uses a short genetic marker in an organism's DNA to identify it as belonging to a particular species. This method is useful for identifying living organisms to species. It makes use of a short (<1000 bp) region of the genome called barcode that evolves fast enough to differ between closely related species. When a barcode sequence has been retrieved from an unknown sample, an algorithm is used to compare it to a reference database containing barcodes from identified museum specimens, thus enabling it to be identified. In other words, DNA barcodes function as molecular identifiers for each species, in the same way as the machine-readable black-and-white barcodes are used in the retail industry to identify commercial products.

In the present conditions, DNA barcoding has become a justifiable tool for the assessment of global biodiversity patterns and it can allow diagnosis of known species to nontaxonomists. DNA barcoding is a fast, accurate, and standardized method for species-level identification by using short DNA sequences. In 2003, Hebert from the University of Guelph, Ontario, Canada, proposed a new technique called DNA barcoding. Hebert and his associates published a paper entitled "Biological identifications through DNA barcodes." They proposed a new system of species identification, i.e., the discovery of species by using a short segment of DNA from a standardized region of the genome. That DNA sequence can be used to identify different species. DNA barcoding is the standardized research that facilitates biodiversity studies like species identification and discovery. This technique helps researchers to understand genetic and evolutionary relationships by assembling molecular, morphological, and distributional data. Species-level identification by DNA barcoding is usually adapted by the recovery of a short DNA sequence from a standard part of the genome. The sequence of barcode from each unknown specimen was then compared with a library of reference barcode sequences obtained from individuals of recognized identity. DNA barcoding is an obligatory tool for species detection and specimen identification. Using standardized identification method is very advantageous for mapping of all the species on Earth, especially when DNA sequencing technology is inexpensively obtainable. The term "DNA barcode" suggests that the standardized DNA sequences can identify taxa in the same way as the 11-digit Universal Product Code identifies retail products in market. Lambert et al. (2005) scrutinized the opportunity of using DNA barcoding to measure the past diversity of the Earth's biota. The barcode of life data system (BOLD) is an informatics workbench assisting the possession, analysis, storage, and publication of DNA barcode records. It links a traditional bioinformatics opening by collecting morphological, molecular, and distributional data. BOLD is freely accessible to any researcher with awareness in DNA barcoding. It helps the assembly of records that meet the standards required to gain barcode designation in the global sequence databases by affording specialized services. BOLD could serve as the universal starting point for identification of species, which would convey users to refer to specialized databases, for example, pathogenic strains, endangered species, and disease vector species. DNA barcoding study has been assisted by the BOLD, an online resource available to the scientific community (http://www.boldsystems.org). This resource presents tools that let researchers carry out neighbor-joining clustering to identify taxa using an updated sequence library, among other things, and to store information on the different groups studied. The quarantine barcoding of life (QBOL) aspires to acquire DNA barcode data of important species of bacteria and other organisms to build an analytical tool for quarantine. Species quantification by using the total DNA barcode determines the composition of an insect's bacterial symbionts and how they alter in time, inspecting novel bacterial pathogens of insect pests and assessing hidden biodiversity of soil samples. The consortium for the barcode of life (CBOL) was launched in 2004 and now it includes more than 170 member organizations from 50 countries to promote DNA barcoding as the global standard for identification of biological specimens.

16.5.2 STEPS OF DNA BARCODING

Various steps of DNA barcoding are:

1. Collect, document and identify the specimen.
2. Isolate DNA from the samples or specimen.
3. Amplify DNA by PCR with specific primers.
4. Electrophoresis of PCR product and purify the PCR product.
5. Sequence the PCR product and analyze the results.

16.5.3 POTENTIAL BARCODING LOCI USED FOR DNA BARCODING OF PLANTS

ITS

Due to its popularity mainly in phylogenetic studies, ITS is by far the most widely sequenced locus for angiosperms (99,123 accessions in GenBank/EBI as of Sept. 2009, compared to 30,325 entries for *rbcL*, which is the most frequently sequenced plastid gene; see Chase et al., 2005). Arguably, this should make it the most suitable barcoding region if "quick and dirty" identifications are desired (e.g., in angiosperm-wide studies), as the chances of finding a high-scoring BLAST-hit are maximized.

The main reason to discredit ITS as an official barcode is its documented nonlinear pattern of evolution in some groups of plants, whereby in extreme cases multiple divergent copies may occur within the same individual. On the other hand, the presence of highly universal primers for the ITS region and its high evolutionary divergence rate (Kress et al., 2005) suggest that its use as a barcode should perhaps not be discredited entirely, at least for those groups in which "problematic" evolutionary patterns have not been observed.

matK

Although attempts were made to amplify this region using several pairs of primers claimed by some to be universal in angiosperms (Lahaye et al., 2008), it was not possible to retrieve sequences for more than about 30% of attempted reference specimens. Furthermore, a certain primer pair did not always yield a PCR product in all members of a group of seemingly closely related taxa, indicating that the primer regions themselves are not conserved. The presence of conserved flanking regions and universal primers is such a key asset for a molecular barcode (especially when dealing with material of completely unknown affinity). Similarly low success rates have been reached by others (Kress and Erickson, 2007; Fazekas et al., 2008). Surprisingly, an official proposal to adopt *matK* as a standard barcode for plants (in conjunction with *rbcL*) has recently been put forward by a leading group of plant barcoding researchers (CBOL Plant Working Group, 2009).

rpoC1

The main advantage of this chloroplast region is its very high amplification success rate, as confirmed here (94% of all reference samples were successfully sequenced) and in many other studies this locus typically scores the highest in this aspect (Sass et al., 2007). On the other hand, *rpoC1* exhibits a slower rate of evolution compared to noncoding plastid regions and some plastid genes (*matK*; Newmaster et al., 2008). Evidently, interspecific variation for this locus is too low in some plant groups to allow for it to be used as a single barcode.

16.5.4 ADVANTAGES OF DNA BARCODING

DNA barcoding have many advantages compared to traditional identification system.

1. Availability of molecular data;
2. Identification of damaged organisms or fragment of sample;
3. Immature specimen can also be easily identified;
4. Specimen cannot be identified by morphological character, the DNA barcoding play crucial role in identification;
5. Specimen can be easily identified when it has polymorphic life cycle.

16.5.5 LIMITATIONS OF DNA BARCODING

1. Barcoding has triggered a passionate debate between proponents who aim to tag the diversity of life and detractors who highlight the pitfalls of the single-gene approach and are concerned about competition for funds with traditional taxonomy.
2. However, it has been claimed that mtDNA barcode achieves accuracy close to 100% in delimiting species in some groups although in some cases it has proved less successful.
3. Limitations of existing databases—DNA barcodes cannot be a useful identification tool without a comprehensive and reliable references database.

KEYWORDS

- **molecular techniques**
- **diversity**
- **single nucleotide polymorphisms (SNPs)**

- **single-stranded conformation polymorphism (SSCP)**
- **genotyping**
- **DNA barcoding**
- **denaturing gradient gel electrophoresis (DGGE)**

REFERENCES

Ainsworth, P. J.; Surh, L. C.; Coulter-Mackie, M. B. Diagnostic Single Strand Conformational Polymorphism, (SSCP): A Simplified Non-radioisotopic Method as Applied to a Tay–Sachs B1 variant. *Nucleic Acids Res.* **1991,** *19*, 405–406.

Amanda, C. P.; Luciane, M. DNA Barcoding and Traditional Taxonomy Unified Through Integrative Taxonomy: A View that Challenges the Debate Questioning Both Methodologies. *Biota Neotrop.* **2010,** *10*(2), 339–346.

Antolin, M. F.; Bosio, C. F.; Cotton, J.; et al. Intensive Linkage Mapping in a Wasp (*Bracon hebetor*) and a Mosquito (*Aedes aegypti*) with Single-strand Conformation Polymorphism Analysis of Random Amplified Polymorphic DNA Markers. *Genetics*, **1996,** *143*, 1727–1738.

Ayres, N.M., McClung, A.M., Larkin, P.D., Bligh, H.F.J., Jones, C.A., Park, W.D. Microsatellites and a single-nucleotide polymorphism differentiate apparent amylose classes in an extended pedigree of US rice germplasm. *Theor Appl Genet*, **1997,** 94:773–781.

Bagley, M. J.; Gall, G. A. E. Mitochondrial and Nuclear Sequence Variability among Populations of Rainbow Trout (*Oncorhynchus mykiss*). *Mol. Ecol.* **1998,** *7*, 945–962.

Ball, S. L.; Armstrong, K. F. DNA Barcodes for Insect Pest Identification: A Test Case with Tussock Moths (Lepidoptera: Lymantriidae). *Can. J. Vet. Res.* **2006,** *36*(2), 337–350.

Bartel, D. P.; Zapp, M. L.; Green, M. R.; Szostak, J. W. HIV-1 Rev Regulation Involves Recognition of Non-Watson–Crick Base Pairs in Viral RNA. *Cell* **1991,** *67*, 529–536.

Beier, D. R. Single-strand Conformation Polymorphism (SSCP) Analysis as a Tool for Genetic Mapping. *Mamm. Genome* **1993,** *4*, 627–631.

Bergsten, J.; Bilton, D. T.; Fujisawa, T.; et al. The Effect of Geographical Scale of Sampling on DNA Barcoding. *Syst. Biol.* **2012,** *61*(5), 851–869.

Bhattramakki, D.; Ching, A.; Morgante, M.; Dolan, M.; Register, J.; et al. *Conserved Single Nucleotide Polymorphism (SNP) Haplotypes in Maize,.* In Book of Abstracts of Plant & Animal Genomes VIII Conference, January 7-12, 2000 Town & Country Hotel San Diego, CA.

Blake, N. K.; Sherman, J. D.; Dvorck, J.; Talbert, L. E. Genome Specific Primer Sets for Starch Biosynthesis Genes in Wheat. *Theor. Appl. Genet.* **2004,** 109(6), 1295-1302.

Bonants, P.; Groenewald, E.; Rasplus, J. Y.; et al. QBOL: A New EU Project Focusing on DNA Barcoding of Quarantine Organisms. *EPPO Bulletin* **2010,** *40*(1), 30–33.

Brookes, A. J. The Essence of SNPs. *Gene.* **1999,** *234*,177–186.

Chakravarti, A. To a Future of Genetic Medicine. *Nature* **2001,** *15*, 822–823.

Chase, M. W.; Salamin, N.; Wilkinson, M.; Dunwell, J. M.; Kesanakurthi, R. P. Land Plants and DNA Barcodes: Short-term and Long-term Goals. *Philos. Trans. R. Soc.* **2005,** *360*, 1889–1895.

Chen, X.; Sullivan, P. F. Single Nucleotide Polymorphism Genotyping: Biochemistry, Protocol, Cost, and Throughput. *Pharmacogenomics* **2003,** *3*, 77–96.

Crozier, R. H. Preserving the Information Content of Species-genetic Diversity, Phylogeny, and Conservation Worth. *Annu. Rev. Ecol. Syst.* **1997,** *28*, 243–268.

Délye, C.; Calmes, E.; Matejcek, A. SNP Markers for Black Grass (*Alopecurus myosuroides* Huds.) Genotypes Resistant to Acetyl CoA-carboxylase Inhibiting Herbicides. *Theor. Appl. Genet.* **2002,** *104*, 1114–1120.

Duarte, S; Cássio, F.; Pascoal, C. Denaturing Gradient Gel Electrophoresis (DGGE) in Microbial Ecology—Insights from Freshwaters. In *Gel Electrophoresis—Principles and Basics*; Magdeldin S., (Ed.); InTech (ISBN 978-953-51-0458-2), Croatia, **2012,** 173–196.

Edward, K. J.; Mogg, R. Plant Genotyping by Analysis of Single Nucleotide Polymorphisms, In *Plant Genotyping: The DNA Fingerprinting of Plants*; Henry, R. J., (Ed.); CAB International (ISBN 0851995152): United Kingdom, **2001,** 1–13.

Ellison, J.; Dean, M.; Goldman, D. Efficacy of Fluorescence-based PCR-SSCP for Detection of Point Mutations. *BioTechniques* **1993,** *15*, 684–691.

Ercolini, D. PCR-DGGE Fingerprinting: Novel Strategies for Detection of Microbes in Food. *J. Microbiol. Meth.* **2004,** *56*, 297–314.

Fazekas, A. J.; Burgess, K. S.; Kesanakurti, P. R.; Graham, S. W.; Newmaster, S. G.; Husband, B. C.; Percy, D. M.; Hajibabaei, M.; Barrett, S. C. H. Multiple Multilocus DNA Barcodes from the Plastid Genome Discriminate Plant Species Equally Well. *PLoS ONE,* **2008,** 3(7), e2802.

Friesen, V. L.; Congdon, B. C.; Walsh, H. E.; Birt, T. P. Intron Variation in Marbled Murrelets Detected Using Analysis of Single-stranded Conformational Polymorphism. *Mol. Ecol.* **1997,** *6*, 1047–1058.

Fujita, K.; Silver, J. Single-strand Conformational Polymorphism. *Genome Res.* **1994,** *4*, S137–S140.

Fujita, K.; Silver, J.; Peden, K. Changes in Both gp120 and gp41 Can Account for Increased Growth Potential and Expanded Host Range of Human Immunodeficiency Virus Type 1. *J. Virol.* **1992,** *66*, 4445–4451.

Gasser, R. B.; Woods, W, G.; Blotkamp, C.; et al. Screening for Nucleotide Variations in Ribosomal DNA Arrays of *Oesophagostomum bifurcum* by Polymerase Chain Reactioncoupled Single-strand Conformation Polymorphism. *Electrophoresis* **1999,** *20*, 1486–1491.

Girman, D. . The Use of PCR-based Single-stranded Conformation Polymorphism Analysis (SSCP-PCR) in Conservation Genetics. In *Molecular Genetic Approaches in Conservation;* Smith T. B., Wayne R. K., Eds.; Oxford University Press: Oxford, **1996;** pp 167–182.

Gray, I. C.; Campbell, D. A.; Spurr, N. K (.Single Nucleotide Polymorphisms as Tools in Human Genetics. *Hum. Mol. Genet.* **2000,** *9*, 2403–2408.

Hajibabaei, M.; Singer, G. A. C.; Hebert, P. D. N.; Hickey, D. A. DNA Barcoding: How It Complements Taxonomy, Molecular Phylogenetics, and Population Genetics. *Trends Genet.***2007,** *23*(4), 167–172.

Hayashi, K.; Yandell, D. W. How Sensitive is PCR-SSCP? *Hum. Mutat.* **1993,** *2*, 338–346.

Hebert, P. D. N.; Cywinska, A.; Ball, S. L.; DeWaard, J. R. Biological Identifications Through DNA Barcodes. *Proc. Biol. Sci.* **2003,** *270*(1512), 313–321.

Highsmith, W. E.; Nataraj, A. J.; Jin, Q.; et al. Use of DNA Toolbox for the Characterization of Mutation Scanning Methods. II: Evaluation of Single-strand Conformation Polymorphism Analysis. *Electrophoresis* **1999,** *20*, 1195–1203.

Hugo, E. A.; Chown, S. L.; McGeoch, M. A. The Microarthropods of Sub-Antarctic Prince Edward Island: A Quantitative Assessment. *Polar Biol.* **2006,** *30*(1), 109–119.

Hunter, K. W.; Watson, M. L.; Rochelle, J.; Ontiveros, S.; Munroe, D.; Seldin, M. F.; Housman, D. E. Single-strand Conformational Polymorphism (SSCP) Mapping of the Mouse Genome: Integration of the SSCP, Microsatellite, and Gene Maps of Mouse Chromosome 1. *Genomics* **1993,** *18*, 510–519.

Jacquemier, J.; Moles, J.; Penault-Llorca, F.; Adelaide, J.; Torrente, M.; Viens, P.; Birnbaum, D.; Theillet. C. . p53 Immunohistochemical Analysis in Breast Cancer with Four Monoclonal Antibodies: Comparison of Staining and PCR-SSCP Results. *Br. J. Cancer* **1994,** *69*, 846–852.

Kanazawa, H.; Nouni, T.; Futai, M. . Analysis of *Escherichia coli* Mutants of the H+-transporting ATPase: Determination of Altered Sites of the Structural Gene. *Method. Enzymol.* **1986,** *126*; 595–603.

Keeley, A. A.; Barcelona, M. J.; Duncan, K.; Suflita. J. M. Environmental Research Brief. *Antonie Van Leeuwenhoek* **2009,** *9*, 1–18.

Kress, J. W.; Erickson, D. L. A Two-locus Global DNA Barcode for Land Plants: The Coding *rbcL* Gene Complements the Non-coding *trnH-psbA* Spacer Region. *PLoS ONE* **2007**, 2(6), e508.

Kress, J. W.; Wurdack, K. J.; Zimmer, E. A.; Weigt, L. A.; Janzen, D. H. Use of DNA Barcodes to Identify Flowering Plants. *Proc. Natl. Acad. Sci. U. S. A.* **2005,** *102*, 8369–8374.

Lahaye, R.; van der Bank, M.; Bogarin, D.; Warner, J.; Pupulin, F.; Gigot, G.; Maurin, O.; Duthoit, S.; Barraclough, T. G.; Savolainen, V. DNA Barcoding the Floras of Biodiversity Hotspots. *Proc. Natl. Acad. Sci. U. S. A.* **2008,** *105*, 2923–2928.

Lambert, D. M.; Baker, A.; Huynen, L.; Haddrath, O.; Hebert, P. D. N.; Millar, C. D. Is a Large-scale DNA-based Inventory of Ancient Life Possible? *J. Hered.* **2005,** *96*(3), 279–284.

Lichten, M. J.; Fox, M. S. Detection of Non-homology-containing Heteroduplex Molecules. *Nucleic Acids Res.* **1983,** *11*, 3959–3971.

Lin, J. C.; De, B. K.; Lin, S. C. Rapid and Sensitive Genotyping of Epstein–Barr Virus Using Single-strand Conformation Polymorphism Analysis of Polymerase Chain Reaction Products. *J. Virol. Method.* **1993,** *43*, 233–246.

Liu, Q.; Feng, J.; Buzin, C.; et al. Detection of virtually all Mutations-SSCP (DOVAM-S): A Rapid Method for Mutation Scanning with Virtually 100% Sensitivity. *Biotechniques*, **1999,** *26*, 932–942.

Makino, R.; Yazyu, H.; Kishimoto, Y.; Sekiya, T.; Hayashi. K. F-SSCP: Fluorescence-based Polymerase Chain Reaction-single-strand Conformation Polymorphism (PCR-SSCP) Analysis. *PCR Methods Appl.* **1992,** *2*, 10–13.

Miller, S. E.. "Proposed standards for BARCODE records in INSDC (BRIs)," in Request document for continuation of support by the Alfred P. Sloan Foundation submitted by the Smithsonian Institution on behalf of Consortium for the barcode of Life: 22 January **2006** Robert H (2005), pp. 36–38.

Moritz, C.; Faith, D. P. Comparative Phylogeography and the Identification of Genetically Divergent Areas for Conservation. *Mol. Ecol.* **1998,** *7*, 419–429.

Morohoshi, F.; Hayashi, K.; Munakata, N. Molecular Analysis of Bacillus Subtilis Ada Mutants Deficient in the Adaptive Response to Simple Alkylating Agents. *J. Bacteriol.* **1991,** *173*, 7834–7840.

Muyzer, G.; Smalla, K. Application of Denaturing Gradient Gel Electrophoresis (DGGE) and Temperature Gradient Gel Electrophoresis (TGGE) in Microbial Ecology. *Antonie Van Leeuwenhoek* **1998,** *73*, 127–141.

Nasu, S.; Suzuki, J.; Ohta, K.; Hasegawa, K.; Yui, I.; et al. Search and Analysis of Single Nucleotide Polymorphisms (SNP) in Rice and Establishment of SNP Markers. *DNA Res.* **2002,** *9*, 163–171.

Nataraj, A. J.; Olivos-Glander, I.; Kusukawa, N.; Highsmith, W. E. Single-strand Conformation Polymorphism and Heteroduplex Analysis for Gel-based Mutation Detection. *Electrophoresis* **1999,** *20*, 1177–1185.

Neigel, J.; Domingo, A.; Stake, J. DNA Barcoding as a Tool for Coral Reef Conservation. *Coral Reefs* **2007,** *26*(3), 487–499.

Newmaster, S. G.; Fazekas, A. J.; Steeves, R. A. D.; Janovec, J. Testing Candidate Plant Barcode Regions in the Myristicaceae. *Mol. Ecol. Resour.* **2008,** *8*, 480–490.

Orita, M.; Iwahana, H.; Kanazawa, H.; Hayashi, K.; Sekiya, T. Detection of Polymorphisms of Human DNA by Gel Electrophoresis as Single-strand Conformation Polymorphisms. *Proc. Natl. Acad. Sci. U. S. A.* **1989,** *86*; 2766–2770.

Ortí, G.; Hare, M. P.; Avise, J. C. Detection and Isolation of Nuclear Haplotypes by PCR-SSCP. *Mol. Ecol.* **1997a,** *6*, 575–580.

Ortí, G.; Pearse, D. E.; Avise, J. C. Phylogenetic Assessment of Length Variation at a Microsatellite Locus. *Proc. Natl. Acad. Sci. U. S. A.* **1997b,** *94*, 10745–10749.

Oto, M.; Miyake, S.; Yuasa. Y. Optimization of Nonradioisotopic Single Strand Conformation Polymorphism Analysis with a Conventional Mini Slab Gel Electrophoresis Apparatus. *Anal. Biochem.* **1993,** *213*, 19–22.

Rafalski, A. Applications of Single Nucleotide Polymorphism in Crop Genetics. *Curr. Opin. Plant Biol.* **2002,** *5*; 94–100.

Ratnasingham, S.; Hebert, P. D. N. BOLD: The Barcode of Life Data System. *Mol. Ecol. Notes* **2007,** *7*(3), 355–364.

Reed, G. H.; Wittwer, C. T. Sensitivity and Specificity of Single-nucleotide Polymorphism Scanning by High-resolution Melting Analysis. *Clin. Chem.* **2004,** *50*(10), 1748–1754.

Robertson, D. L.; Joyce, G. F. Selection In Vitro of an RNA Enzyme That Specifically Cleaves Single-stranded DNA. *Nature* **1990,** *344*, 467–468.

Sass, C.; Little, D. P.; Stevenson, D. W.; Specht, C. D. DNA Barcoding in the Cycadales: Testing the Potential of Proposed Barcoding Markers for Species Identification of Cycads. *PLoS ONE* **2007**, 2(11), e1154.

Savolainen, V.; Cowan, R. S.; Vogler, A. P.; Roderick, G. K.; Lane, R. Towards Writing the Encyclopedia of Life: An Introduction to DNA Barcoding. *Proc. Biol. Sci.* **2005,** *360*(1462), 1805–1811.

Sheffield, V. A. L. C.; Coxt, D. R.; Lerman, L. S.; Mayers, R. M. Attachment of a 40-base-pair G + C-rich Sequence (GC-clamp) to Genomic DNA Fragments by the Polymerase Chain Reaction Results in Improved Detection of Single-base Changes. *Proc. Natl. Acad. Sci. U. S. A.* **1989,** *86*, 232–236.

Siddall, M. E.; Fontanella, F. M.; Watson, S. C.; Kvist, S.; Erseus, C. Barcoding Bamboozled by Bacteria: Convergence to Metazoan Mitochondrial Primer Targets by Marine Microbes. *Syst. Biol.* **2009,** *58*, 445–451.

Simon, J. C.; Baumann, S.; Sunnucks, P.; et al. Reproductive Mode and Population Genetic Structure of the Cereal Aphid *Sitobion avenae* Studied Using Phenotypic and Microsatellite Markers. *Mol. Ecol.* **1999,** *8,* 531–545.

Steel, D. J.; Trewick, S. A.; Wallis, G. P. Heteroplasmy of Mitochondrial DNA in the Ophiuroid *Astrobranchion contrictum. J. Hered.* **2000,** *91*, 146–149.

Sunnucks, P.; England, P. E.; Taylor, A. C.; Hales, D. F. Microsatellite and Chromosome Evolution of Parthenogenetic *Sitobion* Aphids in Australia. *Genetics* **1996,** *144*, 747–756.

Suzuki, Y.; Sekiya, T.; Hayashi, K. Allele-specific Polymerase Chain Reaction: A Method for Amplification and Sequence Determination of a Single Component Among a Mixture of Sequence Variants. *Anal. Biochem.* **1991,** *192*, 82–84.

Tishkoff, S. A.; Verrelli, B. C. Patterns of Human Genetic Diversity: Implications for Human Evolutionary History and Disease. *Annu. Rev. Genomics Hum. Genet.* **2003,** *4*, 293–340.

Tringe, S. G.; von Mering, C.; Kobayashi, A.; Salamov, A. A.; Chen, K.; Chang, H. W.; Podar, M.; Short, J. M.; Mathur, E. J.; Detter, J. C.; Bork, P.; Hugenholtz, P.; Rubin, E. M. Comparative Metagenomics of Microbial Communities. *Science* **2005,** *308*, 554–557.

Vaughan, E. E.; Schut, F.; Heilig, G. H.; Zoetendal, E. G.; de Vos, W. M.; Akkermans, A. D. L. A Molecular View of the Intestinal Ecosystem. *Curr. Issues Intest. Microbiol.* **2000,** *1*, 1–12.

Villablanca, F. X.; Roderick, G. K.; Palumbi, S. R. Invasion Genetics of the Mediterreanean Fruit Fly: Variation in Multiple Nuclear Introns. *Mol. Ecol.* **1998,** *7*, 547–560.

Wang, D. G.; Fan, J. B.; Siao, C. J.; Berno, A.; Young, P.; Sapolsky, R.; Ghandour, G.; Perkins, N.; Winchester, E.; Spencer J.; et al. Large-scale Identification, Mapping, and Genotyping of Single-nucleotide Polymorphisms in the Human Genome. *Science* **1998,** *280*, 1077–1082.

Wilson, A. C. C., Sunnucks, P.; Hales, D. F. Microevolution, Low Clonal Diversity and Genetic Affinities of Parthenogenetic *Sitobion* Aphids in New Zealand. *Mol. Ecol.* **1999,** 8, 1655–1666.

Yap, E. P. H.; McGee. J. O'D. Nonisotopic SSCP and Competitive PCR for DNA Quantification: p53 in Breast Cancer Cells. *Nucleic Acids Res.* **1992,** *20*, 145.

Young, N. D. A Cautiously Optimistic Vision for Marker Assisted Breeding. *Mol. Breed.* **1999,** *5*, 505–510.

CHAPTER 17

PROTEIN PURIFICATION: SCIENCE AND TECHNOLOGY

GANESH PATIL[1], RAVI RANJAN KUMAR[2*], TUSHAR RANJAN[3], KUMARI RAJANI[4], and JITESH KUMAR[2]

[1]*Vidya Pratisthan's College of Agriculture Biotechnology, Vidyanagari, Baramati 413133, India*

[2]*Department of Molecular Biology and Genetic Engineering, Bihar Agricultural University, Sabour, Bhagalpur 813210, Bihar, India*

[3]*Department of Basic Science and Humanities Genetics, Bihar Agricultural University, Sabour, Bhagalpur 813210, Bihar, India*

[4]*Department of Seed Science and Technology, Bihar Agricultural University, Sabour, Bhagalpur 813210, Bihar, India*

[*]*Corresponding author. E-mail: ravi1709@gmail.com*

CONTENTS

ABSTRACT

Protein purification is a technique by which a single protein type is isolated from a complex mixture such as a cell lysate or plant tissue. It can refer to purification of a native protein from a biological sample or of a recombinant protein using one of a variety of chemical, physical, or biological methods. Protein purification is vital for the characterization of the function, structure, and interactions of proteins. The various steps in the purification process may include cell lysis, separating the soluble protein components from cell debris, and finally separating the protein of interest from product- and process-related impurities. Separation of the protein of interest, in its desired form, from all impurities, is typically the most challenging aspect of protein purification. Extraction and purification are vital components of almost any protein-specific research effort. The methods used during these processes depend on the nature of both the protein and the solution. Sometimes the specific protein is caught in a matrix of other protein molecules, and sometimes it's surrounded by non-protein biological elements. In either case, a small sample of the protein may be need for research and analytical purposes, or a large quantity of the purified protein may be necessary for industrial or commercial reasons. In this chapter, the different methods of protein isolation and purification are discussed in depth which would be useful for the readers to understand the science and technology behind protein purification.

17.1 INTRODUCTION

Protein purification involves a series of steps intended to isolate one or a few proteins from a complex mixture of proteins, which is generally obtained from cells, tissues, or whole organisms. The purification process aims at removal of other unwanted proteins and nonprotein parts of the mixture, and finally separates the protein of interest from all other proteins (Yang et al., 2013). Separation of one protein from all others is not a straightforward task and represents one of the major bottlenecks for further exploitation of biotechnology. Separation steps are generally based on select properties of proteins such as physicochemical properties, protein size, binding affinity, and biological activity. Pure form of protein is vital for the structure and functional studies and elucidation of exact biological role and applications of proteins. The methods used in protein purification can roughly be divided into analytical and preparative methods. The difference is not exact, but the deciding factor is the amount/scale of protein to be purified by that method. Preparative methods aim to produce large quantities of the protein, whereas analytical methods aim to detect and identify a protein in a mixture. In general, the preparative methods can be used in analytical applications, but not the other way around (Berg et al., 2002).

17.2 PURPOSE OF PROTEIN PURIFICATION

Pure form of protein is needed for large variety of experiments and applications. Despite lot of advancement in science and technology, till date obtaining pure form of protein is not a simple and straightforward task as it depends on several factors. Before starting protein purification project, the following considerations need to be taken into account:

- Define objectives for purity, activity, and quantity
- Define properties of target protein and critical impurities
- Develop analytical assays for your protein (and impurities)
- Minimize sample handling at every stage
- Minimize use of additives; they will have to be removed
- Remove harmful contaminants early
- Use a different technique at each step (multiple barriers)
- Minimize number of steps (within reason)
- Depending on the objectives, one has to select appropriate techniques, methods, and tools.

17.3 PRELIMINARY STEPS IN PROTEIN PURIFICATION

17.3.1 EXTRACTION

In case of recombinant protein production, *Escherichia coli* cells harboring vectors containing genes responsible for the synthesis of desired heterologous protein are first grown in fermentation media in conical flasks or fermenters (Fig. 17.1). If the protein of interest is not secreted by the organism into the medium supernatant, the first step of each purification process is the disruption of the cells containing the protein. Depending on how fragile the protein is and how stable the cells are, one could, for instance, use one of the following methods: (1) repeated freezing and thawing, (2) sonication, (3) homogenization by high pressure (French press), (4) homogenization by grinding (bead mill), and (5) permeabilization by detergents (e.g., Triton X-100) and/or enzymes (e.g., lysozyme). Finally, the cell debris can be separated by centrifugation step so that the proteins and other soluble compounds are released in the supernatant (Scopes, 2000; Yang et al., 2013).

FIGURE 17.1 Recombinant bacteria can be grown in a flask containing growth media.

Host cell proteases are released during cell lysis, which eventually start degrading the proteins in the solution, including protein of interest. If the protein of interest is sensitive to proteolysis, it is recommended to proceed quickly, and to keep the extract cooled, to slow down the digestion. Generally, ethylene di-amine tetra-acetic acid (EDTA) is added to the lysis and extraction buffer. EDTA chelates all the metal ions present in the buffering environment of cell lysate, which are important precursors for majority of serine protease family proteases. Alternatively, protease inhibitor cocktail can be added to the lysis buffer before cell disruption. Additionally, it is recommended to add DNAse in order to reduce the viscosity of the cell lysate caused by a high DNA content (Scopes, 2000). However, it should be noted that protease inhibitor and DNAses are pretty expensive reagents and therefore it adds to the process economics. Various steps involved in recovery of recombinant proteins are represented in Figure 17.2.

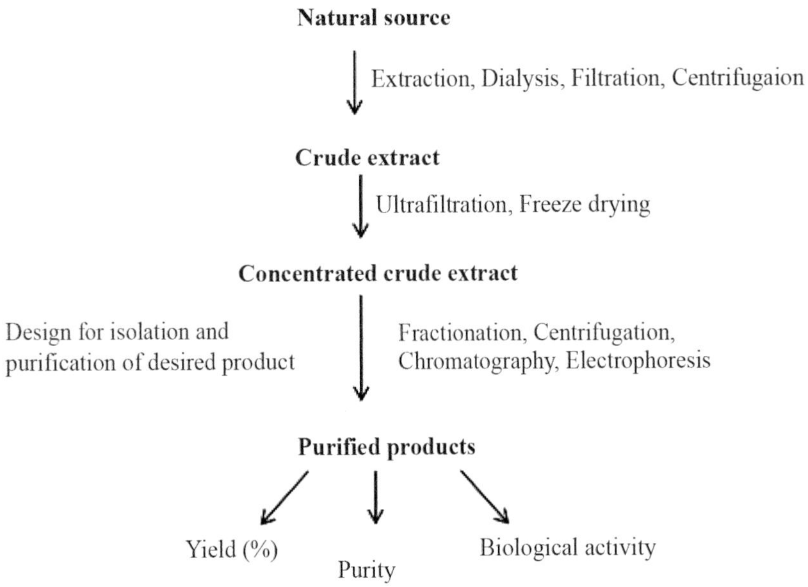

FIGURE 17.2 Overview of general steps involved in protein recovery processes.

17.3.2 PRECIPITATION AND DIFFERENTIAL SOLUBILIZATION

In bulk protein purification, a common first step to isolate proteins is precipitation with ammonium sulfate ($(NH_4)_2SO_4$). This is performed by adding increasing amounts of ammonium sulfate and collecting the different fractions of precipitate protein. Ammonium sulfate can be removed by dialysis. The hydrophobic groups on the proteins get exposed to the atmosphere, attract other protein hydrophobic groups and get aggregated. Protein precipitated will be large enough to be visible (Fig. 17.3). One advantage of this method is that it can be performed inexpensively with very large volumes. Apart from ammonium sulfate, organic solvents like ethanol and acetone can also be used as alternatives to precipitate proteins from large volumes. The first proteins to be purified are water-soluble proteins.

Purification of integral membrane proteins requires disruption of the cell membrane to isolate any one particular protein from others that are in the same membrane compartment. Sometimes, a particular membrane fraction can be isolated first, such as isolating mitochondria from cells before purifying a protein located in a mitochondrial membrane. A detergent such as sodium dodecyl sulfate (SDS) can be used to dissolve cell membranes and keep membrane proteins in solution during purification; however, because SDS causes denaturation, milder detergents such as Triton X-100 or CHAPS can be used to retain the protein's native conformation during complete purification (Ryan, 2011).

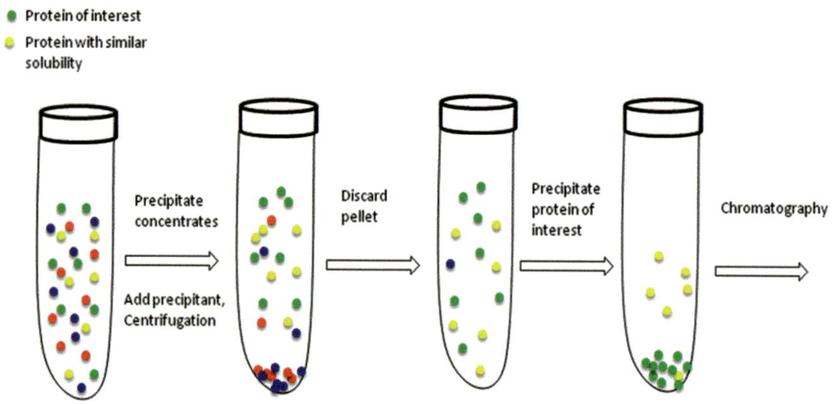

FIGURE 17.3 Fractional precipitation of proteins.

17.3.3 ULTRACENTRIFUGATION

Centrifugation applies centrifugal force to separate mixtures of particles of varying masses or densities suspended in a liquid. It is useful for separation of bacterial cells, viscous cell membranes, and cellular DNA parts when rotated at high speeds. Whereas, mixture of proteins extracted in buffer remains suspended into the solution. When suspensions of particles are spun in a centrifuge, a "pellet" may form at the bottom of the vessel that is enriched for the most massive particles with low drag in the liquid (Fig. 17.4). Noncompacted particles remain mostly in the liquid called "supernatant" and can be removed from the vessel thereby separating the supernatant from the pellet (Cole et al., 2008). Thus, centrifugation plays very important role in recovery of proteins from large complex mixtures. In inclusion bodies isolation, centrifuge helps to precipitate out dense insoluble proteinaceous material.

17.3.4 SUCROSE GRADIENT CENTRIFUGATION

In sucrose gradient centrifugation, a linear concentration gradient of sugar (typically sucrose, glycerol, or a potassium bromide, polyethylene glycol

(PEG)-based density gradient media) is generated in a tube such that the highest concentration is at the bottom and lowest at the top. A protein sample is then layered on top of the gradient and spun at high speeds such as 50,000 rpm or even higher in an ultracentrifuge. This causes heavy macromolecules to migrate toward the bottom of the tube faster than lighter material (Fig. 17.5).

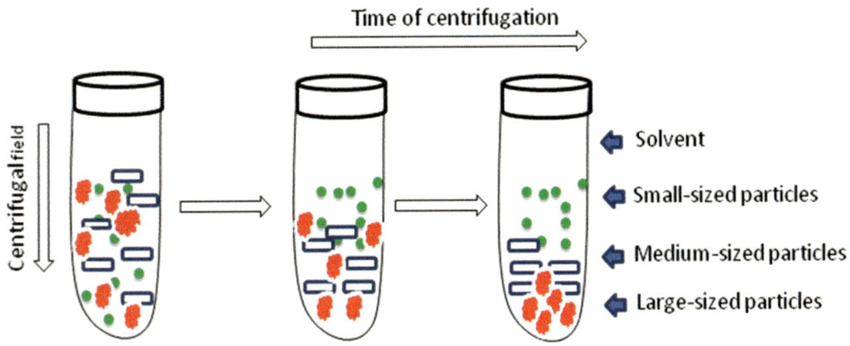

FIGURE 17.4 Fractional precipitation of proteins based on differences in the sedimentation rate of different size, shape, and density.

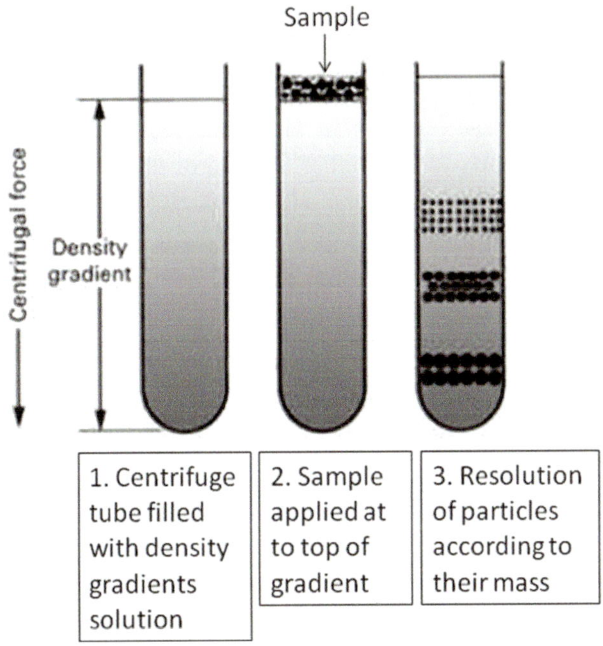

FIGURE 17.5 Sucrose density gradient centrifugation.

During centrifugation in the absence of sucrose, as particles move farther and farther from the center of rotation, they experience more and more centrifugal force (the farther they move, the faster they move). Main problem with this is that the useful separation range of within the vessel is restricted to a small observable window. Samples separated by these gradients are referred to as "rate zonal" centrifugations. After separating the protein/particles, the gradient is then fractionated and collected (Roger, 1996). It is generally used for extraction and purification of viruses from blood plasma. However, due its cumbersomeness, density gradient centrifugation is being replaced by membrane or column chromatography-based alternative methods.

17.4 PURIFICATION STRATEGIES

Selection of appropriate starting material is key to the design of an efficient purification process. In a plant or animal, a particular protein usually is not distributed homogeneously throughout the body; different organs or tissues have higher or lower concentrations of the protein. Use of only the tissues or organs with the highest concentration decreases the volumes needed to produce a given amount of purified protein. If the protein is present in low abundance, or if it has a high value, scientists may use recombinant DNA technology to develop cells that will produce large quantities of the desired protein (this is known as an expression system). Recombinant expression allows the protein to be tagged, e.g., by a His-tag, to facilitate purification, which means that the purification can be done in fewer steps. In addition, recombinant expression usually starts with a higher fraction of the desired protein than is present in a natural source (Crowe et al., 1994). Before starting protein purification activity, certain properties of proteins have to be taken into account (Table 17.1), and then one may decide on protein purification methods.

TABLE 17.1 Common Methods for Purification of Proteins Based on Their Characteristics.

Property	Methods
Solubility	Precipitation ammonium sulfate (salting out) or organic solvent such as ethanol of acetone
Size or shape	Size-exclusion chromatography
Charge or isoelectric point	Ion exchange chromatography (cation and anion exchange)
Binding or affinity	Affinity chromatography

Usually a protein purification protocol contains one or more chromatographic steps. The basic procedure in chromatography is to flow the solution containing the protein through a column packed with solid phase chromatographic resins. Chromatographic matrices are different biopolymeric beads, such as Sepharose, Agarose, charged with different chemical groups and Sephacryl, Superdax or Superose porous beads. Different proteins interact differently with the column material, and can thus be separated by the time required to pass the column, or the conditions required to elute the protein from the column. For monitoring the movement of proteins from the column, a UV detector is placed at the downstream of the column, which records their absorbance at 280 nm. Besides UV, at this point, other detectors such as conductivity and pH probe can also be fitted. These entire sets of tools required for separation of biomolecules are fitted in compact assembly of equipment called liquid chromatography system (Fig. 17.6), called UPLC (ultrapressure liquid chromatography) or FPLC (fast performance liquid chromatography). Based on how proteins interact, different chromatographic methods are classified below.

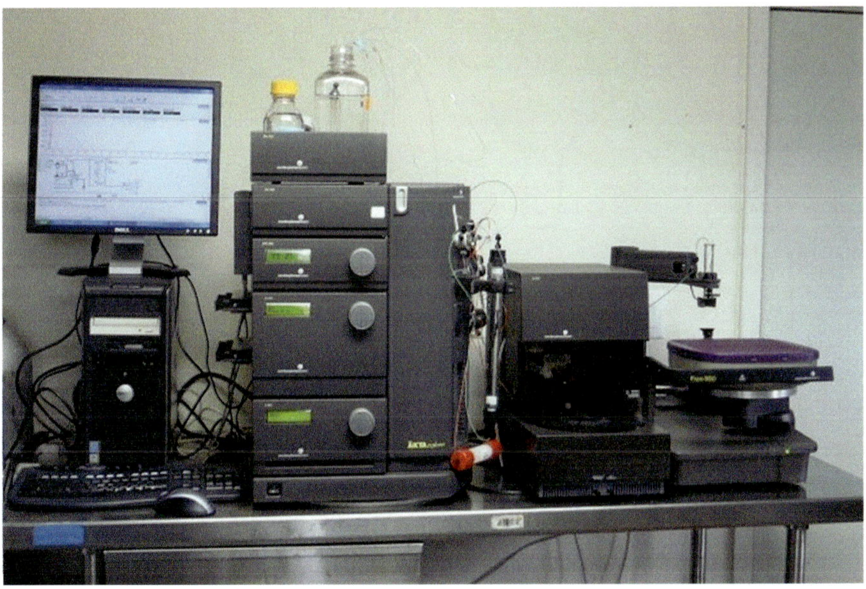

FIGURE 17.6 FPLC or ÄKTA process chromatographic system from GE Healthcare Ltd.

17.4.1 SIZE-EXCLUSION CHROMATOGRAPHY

Size-exclusion chromatography (SEC), also known as molecular sieve chromatography, is a chromatographic method in which molecules in solution are separated by their size, and in some cases molecular weight (Fig. 17.7). This technique is referred by different names such as gel-filtration chromatography or gel permeation chromatography. It is usually applied to large molecules or macromolecular complexes such as proteins and industrial polymers. SEC is a widely used polymer characterization method because of its ability to provide good molar mass distribution (Mw) results for proteins. This technique is widely used for separation of protein dimers or aggregates. Chromatography can be used to separate protein in solution under native conditions. This method can also be used for determination of molecular weight of a given protein (Kennedy, 1990). However, hydrodynamic size of proteins might differ from their theoretical mass leading to confusion.

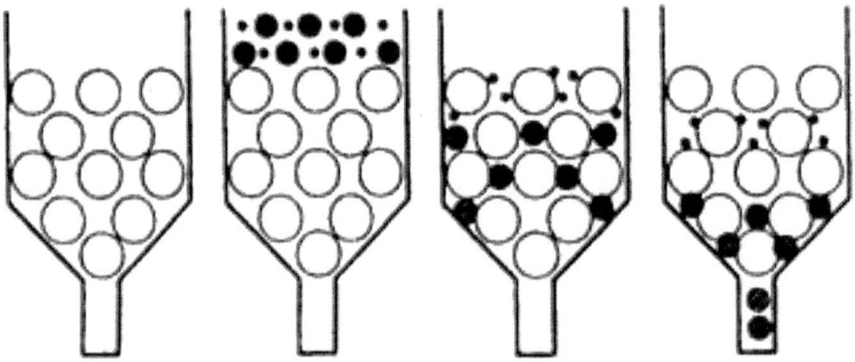

FIGURE 17.7 Schematic presentation of size-exclusion chromatography.

Drawback of SEC is limitation of sample application volume, where sample volume has to be less than 2–5% total column volume. Additionally, poor resolution limit of the resins, especially closely related molecular weight proteins, limit the widespread application of this technique.

17.4.2 ION EXCHANGE CHROMATOGRAPHY

Ion exchange chromatography separates proteins according to the degree of their ionic charge. The column to be used is selected according to its type and strength of charge. Anion exchange resins have a positive charge and are used to retain and separate negatively charged compounds, while cation exchange resins have a negative charge and are used to separate positively charged molecules (Fig. 17.8). However, irrespective proteins' ionic charge, its desired charge can be brought temporarily by adjusting pH of buffer environment (Jackson et al., 1990).

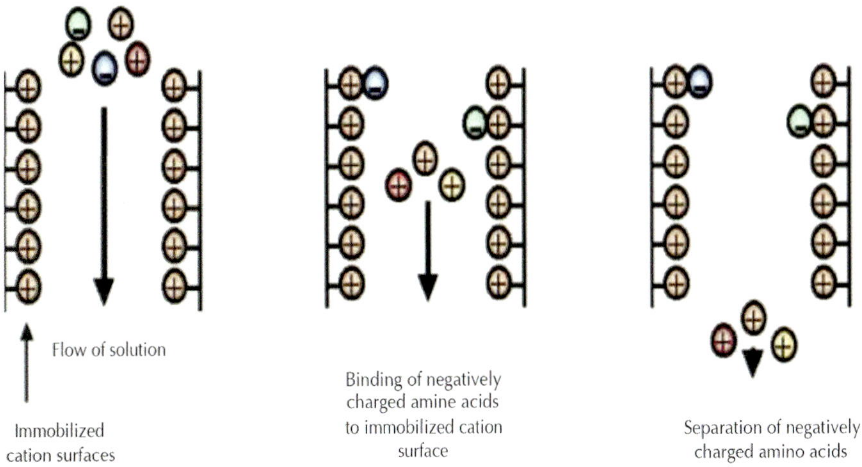

FIGURE 17.8 Schematic presentation of ion exchange chromatography.

Before the separation begins, a buffer is pumped through the column to equilibrate the opposing charged ions. After sample injection, solute molecules will exchange with the buffer ions as each competes for the binding sites on the resin. To perform protein separation, it is essential to know proteins theoretical pI. pH condition for performing the experiment is selected around two points above its pI value, e.g., if pI of a given protein is 7.0, one should select pH 7.7–8.0 for experiment using anion exchange chromatography. The length of retention for each solute depends upon the strength of its charge. The protein bound to the column can be eluted by salt gradient with increasing concentration. The most weakly charged compounds will elute first, followed by those with successively stronger charges. Because of the nature of the separating mechanism, pH, buffer type, buffer concentration, and temperature all play important roles in controlling the separation (Tatjana and Joachim, 2005).

Ion exchange chromatography is a very powerful tool for use in protein purification and is frequently used in both analytical and preparative separations. Also ion exchange resins are cheapest in class and most stable materials in comparison to other chromatographic resins.

17.4.3 AFFINITY CHROMATOGRAPHY

Affinity chromatography is a method of separation of biomolecules based on a well-defined binding affinity with specific chromatographic resins charged with ligands. Affinity chromatography is a separation technique based upon molecular conformation, which frequently utilizes application-specific resins (Fig. 17.9). These resins have ligands attached to their surfaces which are specific for the compounds to be separated. Most frequently, these ligands function in a fashion similar to that of antibody–antigen interactions. This natural binding tendency between the ligand and its target compound makes it highly specific, frequently generating a single peak, while all else in the sample is not retained (Uhlen, 2008).

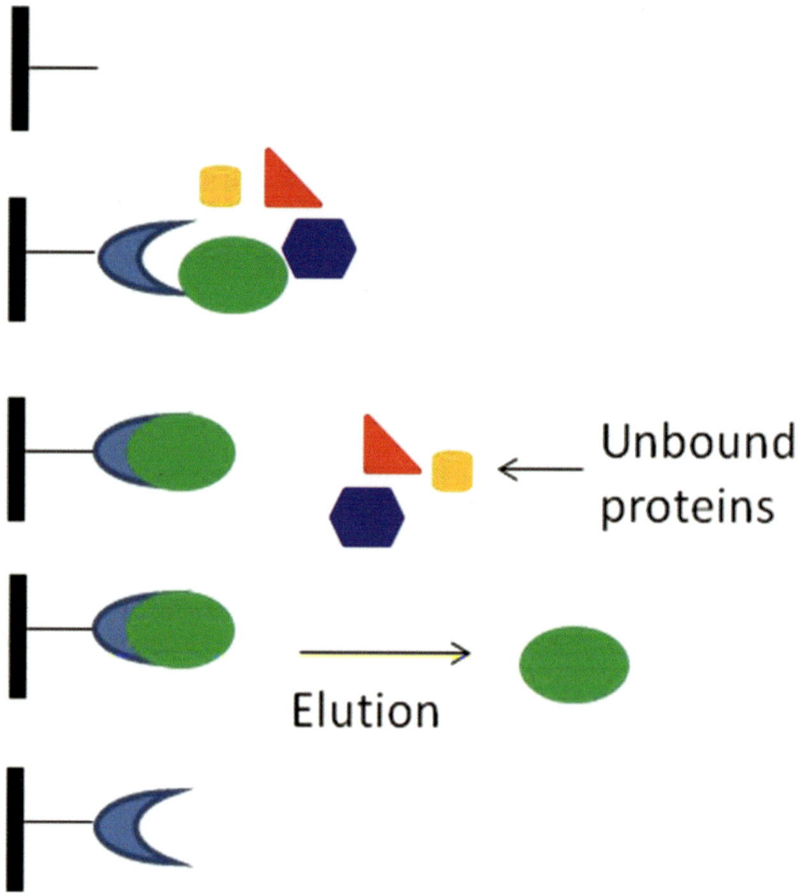

FIGURE 17.9 Schematic presentation of affinity chromatography.

Affinity chromatography is frequently used for purification of antibodies using Protein A or Protein G chromatographic resins. These resins are prepared by using purified Protein A or Protein G ligands covalently bound to Sepharose- or Agarose-based ligands. Protein A or Protein G is recombinant protein originally obtained from *Staphylococcus aureus*, having very strong affinity for Fc portion of antibodies. Among these two protein-based resins, Protein A chromatographic matrices are more frequently used. Once antibodies are bound to the resin, after washing step, bound antibodies are eluted by passing through citrate/acetate buffer, pH 3.0. However, precaution must be taken to minimize antibodies exposure to strong acidic conditions. Similarly, many membrane proteins are glycoproteins and can be purified by lectin affinity chromatography—concavelin A. Detergent-solubilized proteins can be allowed to bind to a chromatography resin that has been modified to have a covalently attached lectin. Proteins that do not bind to the lectin are washed away and then specifically bound glycoproteins can be eluted by adding a high concentration of a sugar that competes with the bound glycoproteins at the lectin binding site (Urh et al., 2009).

17.4.3.1 IMMOBILIZED METAL AFFINITY CHROMATOGRAPHY

Immobilized metal affinity chromatography (IMAC) is a widely used technique involving engineering a sequence of hexa (6) histidines—purification tag into the N- or C-terminal of the protein. The poly-histidine binds strongly to divalent metal ions such as nickel and cobalt. The protein can be passed through a column containing immobilized nickel ions, which binds the poly-histidine tag (Fig. 17.10). All untagged proteins pass through the column. The protein can be eluted with imidazole, which competes with the poly-histidine tag for binding to the column, which decreases the affinity of the tag for the resin. While this procedure is generally used for the purification of recombinant proteins with an engineered affinity tag, it can also be used for natural proteins with an inherent affinity for divalent cations. Schematic diagram (Figure 17.10) showing the steps involved in a metal binding strategy for protein purification. The use of nickel immobilized with nitrilo-tri-acetic acid (NTA) is also shown in Figure 17.10 (Porath, 1992; Crowe et al., 1994).

IMAC is perhaps the most widely used method for protein separation today. However, major limitation of this method is that proteins require removal of tag after purification with the help of proteolytic cleavage.

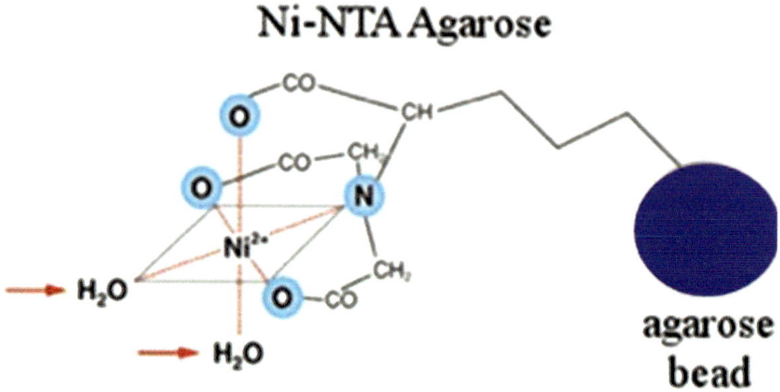

FIGURE 17.10 Nickel-affinity matrix.

17.4.3.2 IMMUNOAFFINITY CHROMATOGRAPHY

Immunoaffinity chromatography uses the specific binding of an antibody to the target protein to selectively purify the protein. The procedure involves immobilizing an antibody to a column material, which then selectively binds the protein, while everything else flows through (Fig. 17.11). The protein can be eluted by changing the pH or the salinity. This method does not involve engineering a tag, it can be used for proteins from natural sources (Ehle and Horn, 1990).

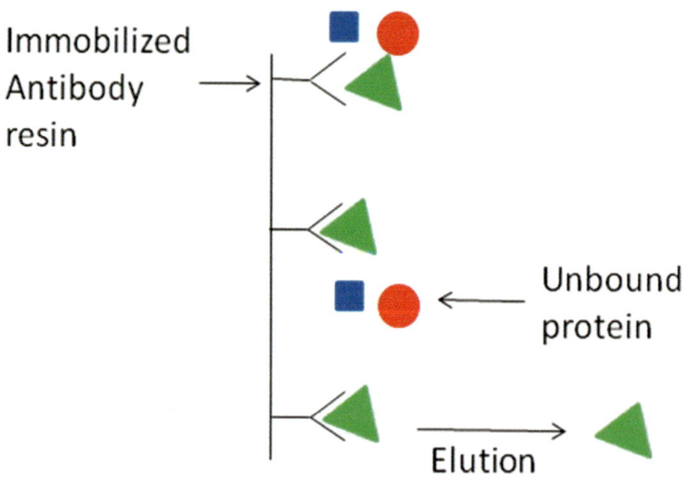

FIGURE 17.11 Antibody affinity chromatography.

17.4.4 HYDROPHOBIC INTERACTION CHROMATOGRAPHY

Hydrophobic interaction chromatography (HIC) media is amphiphilic, with both hydrophobic and hydrophilic regions, allowing for separation of proteins based on their surface hydrophobicity. In pure water, the interactions between the resin and the hydrophobic regions of protein would be very weak, but this interaction is enhanced by applying a protein sample to HIC resin in high ionic strength buffer. Here, generally protein is mixed with ammonium sulfate salt. The ionic strength of the buffer is then reduced to elute proteins in order of decreasing hydrophobicity (Smyth et al., 1978).

17.4.5 HIGH PERFORMANCE LIQUID CHROMATOGRAPHY

High performance liquid chromatography (HPLC) or high pressure liquid chromatography is a form of chromatography applying high pressure to drive the solutes through the column faster. This means that the diffusion is limited and the resolution is improved. The most common form is "reversed phase" HPLC, where the column material is hydrophobic, mostly silica-based resins are used (Fig. 17.12). The columns are prepared according to carbon load attached to silica surface such as C4, C8, and C18 columns, with pore size 300 Å. The proteins are eluted by a gradient of increasing amounts of an organic solvent, such as acetonitrile. The proteins elute according to their hydrophobicity. After purification by HPLC, the protein is in a solution that only contains volatile compounds, and can easily be lyophilized. HPLC purification frequently results in denaturation of the purified proteins and is thus not applicable to proteins that do not spontaneously refold. Also, this technique is more suitable for separation of peptides. This instrument has very sensitive detectors in comparison to other liquid chromatography formats like FPLC or UPLC. This type of chromatography is more frequently used as analytical purpose rather than preparative one, although reverse phase chromatography could be used in case of separation of a few hydrophobic proteins (Regnier, 1983).

Various steps involved in separation of proteins could be represented in a single graphical form which gives reader an idea about complexity of downstream processing. Figure 17.13 gives us overview of various steps involved in typical protein purification project.

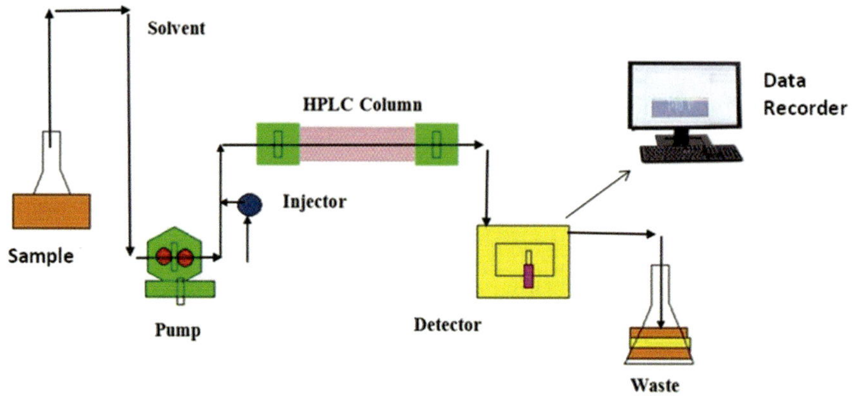

FIGURE 17.12 Schematic presentation of high performance liquid chromatography (HPLC) system.

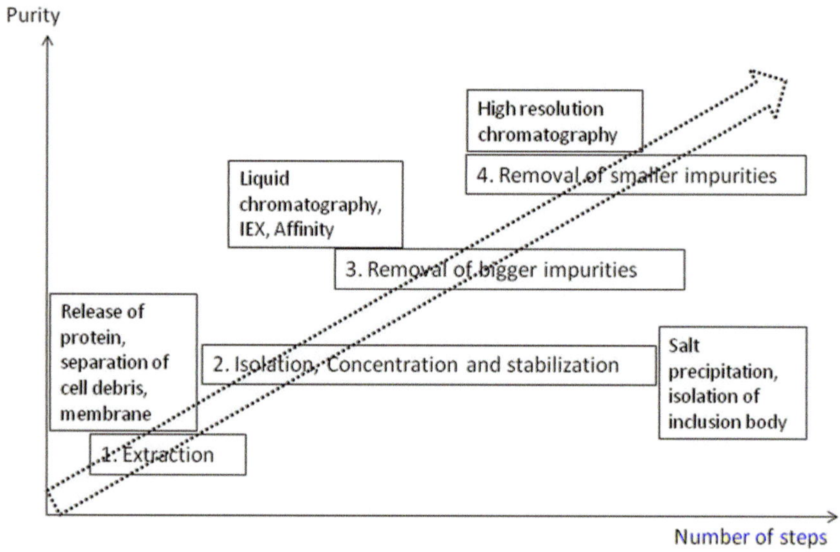

FIGURE 17.13 Protein isolation, concentration, and stabilization.

17.5 CONCENTRATION OF THE PURIFIED PROTEIN

At the end of protein purification, the protein often has to be concentrated. Generally, for this purpose a selectively permeable membrane can be mounted in a centrifuge tube on ultrafiltration device. The buffer is forced through the membrane by centrifugation, leaving the protein in the upper chamber.

17.5.1 LYOPHILIZATION

If the solution does not contain any other soluble component than the protein in question, the protein can be lyophilized (dried). This is commonly done after an HPLC run. This simply removes all volatile components, leaving the proteins behind (Mohapatra, et al, 1973).

17.5.2 ULTRAFILTRATION

Ultrafiltration concentrates a protein solution using selectively permeable membranes. The function of the membrane is to let the water and small molecules pass through while retaining the protein (Fig. 17.14). The solution is forced against the membrane by mechanical pump, gas pressure, or centrifugation (Bennett, 2008).

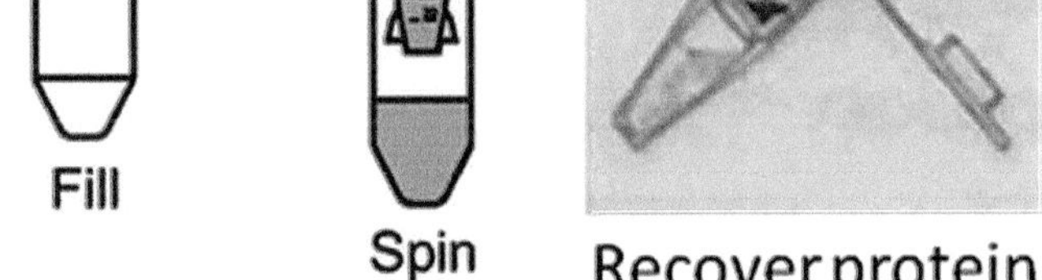

FIGURE 17.14 Centrifuge-based protein concentrator devices.

17.6 EVALUATING PURIFICATION YIELD

To evaluate the efficiency of multistep purification process, the amount of the specific protein has to be determined and compared to the amount of total protein. Protein estimation can be determined by the Bradford's assay or by absorbance at 280 nm. However, some reagents used during the purification process may interfere with the quantification. For example, imidazole (commonly used for purification of poly-histidine-tagged recombinant proteins) is an amino acid analogue and at low concentrations will interfere with the bicinchoninic acid (BCA) assay for total protein quantification. Impurities in low-grade imidazole will also absorb at 280 nm, resulting in inaccurate estimates of protein concentration from UV absorbance. Often ELISA technique is used for estimation of purified proteins; however, sensitivity of this technique is very high. Therefore, careful interpretation is needed at this point.

Besides knowing concentration of proteins, their impurity profile must be known. The most general method to monitor the purification process is by running an SDS-PAGE of the different steps. This method only gives a rough measure of the amounts of different proteins in the mixture, and it is not able to distinguish between proteins with similar apparent molecular weight. If the protein has a distinguishing spectroscopic feature or an enzymatic activity, this property can be used to detect and quantify the specific protein, and thus to select the fractions of the separation that contains the protein. If antibodies against the purified protein are available then western blotting and ELISA can specifically detect and quantify the amount of desired protein. Some proteins function as receptors and can be detected during purification steps by ligand-binding assays.

17.7 ANALYTICAL–QUALITY CONTROL ASPECTS

17.7.1 DENATURING-CONDITION ELECTROPHORESIS

Gel electrophoresis is a commonly used laboratory technique for protein analysis. The principle of electrophoresis relies on the movement of a charged ion in an electric field. In practice, the proteins are denatured in a solution containing a detergent (SDS). In these conditions, the proteins are unfolded and coated with negatively charged detergent molecules. The proteins in SDS-PAGE are separated on the sole basis of their size (Fig. 17.15). The protein migrates as bands based on size. Protein bands can be detected by stains such as Coomassie blue dye or silver stain. SDS PAGE gel stained with silver nitrate provides a very high sensitivity and detection of impurities (Rath et al., 2009).

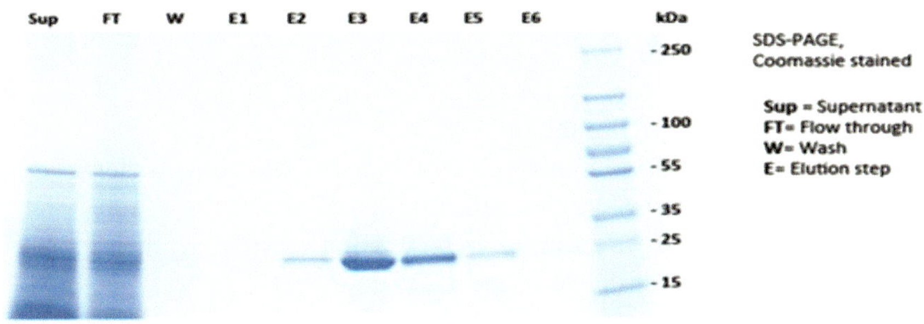

FIGURE 17.15 SDS-PAGE profile of typical purification process showing different fractions.

17.7.2 OTHER ANALYTICAL TECHNIQUES

- Protein concentration
  - UV/Vis, BCA, Bradford, Lowry even ELISA
- Protein purity/structure information
  - SDS PAGE—Coomassie or silver staining, HPLC, IEF, western blot
- Biological activity
  - ELISA, enzyme activity, receptor ligand interaction
- Spectroscopic characterization, protein native confirmation
  - CD spectroscopy, fluorescence spectroscopy
- Proteolytic digestion products, exact molecular weight determination
  - N-terminal sequencing, MALDI-TOF, mass spectrometry
- Impurities—allergens, immunogenic proteins, endotoxins, host cell proteins, viruses, etc.
  - For example, PCR for viruses or bacteria; LAL for endotoxin; western blots or ELISAs for protein contaminants

17.7.3 CONCEPT OF ORTHOGONALITY

It is understood that no single test can assure quality of the protein. In order to be sure about protein quality, one must use orthogonal testing concept for quality control purpose. That involves characterization of protein by series of parallel techniques rather than depending on a single assay, and collectively decides about quality of proteins. Also, requirement of protein purity depends on its application and from purpose to purpose.

17.8 APPLICATIONS OF PURIFIED PROTEINS

- Pure proteins are needed for large number and variety of applications, including but not limited to:
- Research and development on structure and function studies of individual proteins from given cell.
- Enzymes are needed in food processing, textile, leather industry, manufacturing of digestive aids, and detergent industry.
- Study protein–protein interactions.
- Industrial and biopharmaceutical applications.

17.8.1 EMERGENCE OF PROTEOMICS

Proteomics is the genome-wide study of proteins, especially their structures and functions. Proteins are important components of living cells, as they are the important components of the physiological and metabolic state of cells. The proteome is the entire set of proteins produced by a cell. This varies with time and distinct requirements, or stresses, that a cell or organism undergoes. Proteomics is an interdisciplinary domain formed on the basis of the research and development of the Human Genome Project. It is emerging area of research and exploration of proteomes as proteins is important part of functional elements of cells, metabolic enzymes, signaling molecules, disease as well as diagnostic markers. While proteomics generally refers to the large-scale experimental analysis of proteins, it often involves use of 2D-PAGE, protein purification, and mass spectrometry.

17.8.2 TWO-DIMENSIONAL GEL ELECTROPHORESIS

Two-dimensional gel electrophoresis, also abbreviated as 2-D electrophoresis/2-DE, is a form of gel electrophoresis commonly used to analyze entire set of proteins directly from cell or organ lysate without any prior purification step. Mixtures of proteins are resolved by two properties in two dimensions on 2-D gels. 2-D electrophoresis begins with 1-D electrophoresis and then separates the protein molecules in 90° direction from the first (Fig. 17.16). In 1-D electrophoresis, proteins and other molecules are separated in one dimension, so that all the proteins/molecules lie along the lane, later on the molecules are spread out across a 2-D gel (Fig. 17.16). As it is unlikely that two molecules will be similar in two distinct properties, molecules are effectively separated in 2-D electrophoresis than in 1-D electrophoresis. The 2-D separation in this technique is mainly based on isoelectric point complex mass in the native state, and protein molecular weight. Thereby, a gradient of pH is applied to a gel with the help of ampholytes and an electric potential gets created across the gel, making one end more positive than the other. Great resolution of proteins can be obtained in a single step. Well-resolved protein bands can be directly picked from gel or transferred to nitrocellulose membrane by electroblotting, then extracted and subjected for identification by N-terminal sequencing or mass determination by mass spectrometry.

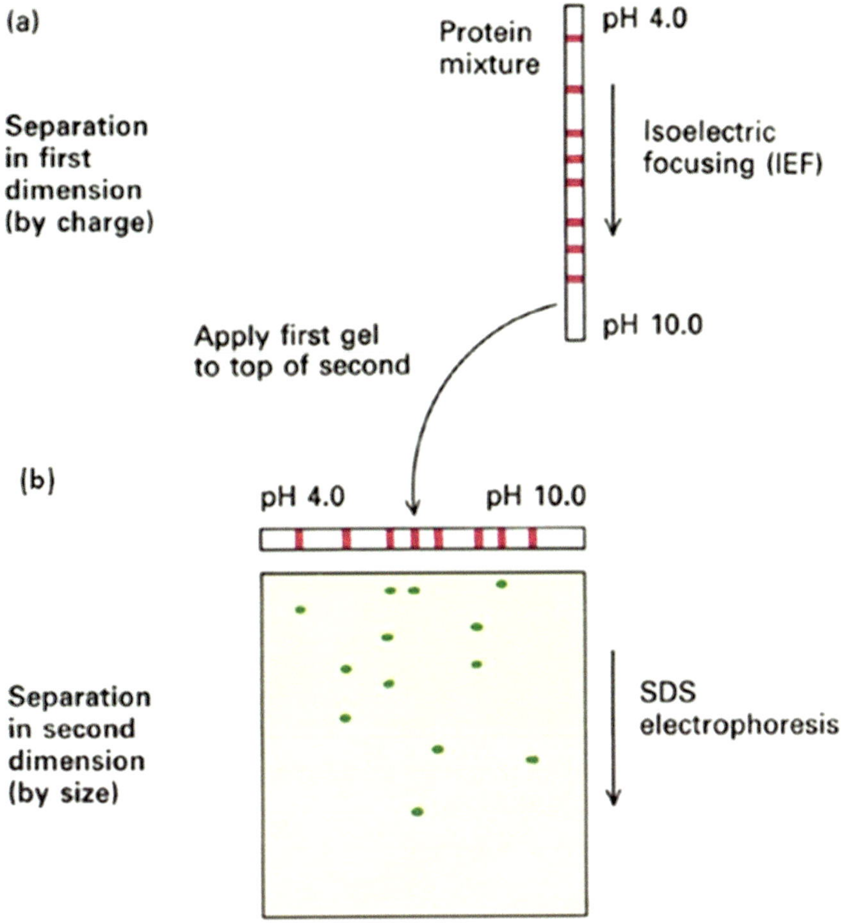

FIGURE 17.16 Principle of 2-D electrophoresis.

17.8.3 STUDYING PROTEIN–PROTEIN INTERACTIONS

Most proteins function in combination with other proteins, and one of the goals of proteomics is to identify which role of proteins is especially useful in determining potential partners in cell signaling cascades. Several methods are available to elucidate protein–protein interactions such as immunoaffinity chromatography, protein microarrays followed by mass spectrometry, and other experimental methods such as phage display. Protein functional analysis is used for the identification of protein–protein interactions (e.g., identification of protein complex members), protein–phospholipid interactions, small molecule targets, enzymatic substrates (mainly the substrates of kinases), and receptor ligands (Phizicky and Fields, 1995).

17.9 CONCLUSIONS

Protein separation science represents one of ever growing areas of science comprising combination of basic science and technology. Due to ever increasing demand of purified proteins for basic research and development as well as industrial applications, protein separation and purification processes have become one of the most important applied sciences in day-to-day life. Many protocols are available for small-scale to preparative-scale preparations of proteins. An array of techniques is available for analytical characterization and impurity profiling of proteins. Thus, protein science has become one of the most vibrant branches of research and development.

KEYWORDS

- **isolation**
- **purification**
- **chromatography**
- **centrifugation**
- **filtration**
- **gel filtration**
- **HPLC**
- **SDS-PAGE**

REFERENCES

Bennett, A. Membrane Technology: Developments in Ultrafiltration Technologies. *Filtr. +Sep.* **2012,** *49,* (6), 28–33.

Berg, J. M.; Tymoczko, J. L.; Stryer, L. The Purification of Proteins Is an Essential First Step in Understanding Their Function, **2002,** 5, 20.

Cole, J. L.; Lary, J. W.; Moody, T. P.; Laue, T. M. Analytical Ultracentrifugation: Sedimentation Velocity and Sedimentation Equilibrium. *Methods Cell Biol.* **2008,** *84,* 143–179.

Davis, R. A.; Vance, J. E. *Structure, Assembly and Secretion of Lipoproteins;* Elsevier Science B. V., **1996**.

Ehle, H.; Horn, A. Immunoaffinity Chromatography of Enzymes. *Bioseparation* **1990,** *1* (2), 97–110.

GE Healthcare. *Protein Purification Handbook,* **2009** (18-1142-75 AD 01/2009).

Jackson, P.; Haddad, R. *Ion Chromatography: Principles and Applications;* Elsevier: Amsterdam, **1990**.

Kennedy, R. M. Hydrophobic Chromatography. *Methods Enzymol.* **1990,** *182,* 339–343.

Klose, J. Protein Mapping by Combined Isoelectric Focusing and Electrophoresis of Mouse Tissues. A Novel Approach to Testing for Induced Point Mutations in Mammals. *Humangenetik* **1975,** *26* (3), 231–243.

Mohapatra, S. C.; Pattee, H. E. Lipid Metabolism in Dehydrating Peanut Kernels. *Physiol. Plant.* **1973,** *28* (2), 320.

Porath, J. Immobilized Metal Ion Affinity Chromatography. *Protein Expr. Purif.* **1992,** *3* (4), 263–281.

Phizicky, E. M.; Fields, S. Protein–Protein Interactions: Methods for Detection and Analysis. *Microbiol. Rev.* **1995,** *59,* 94–123.

O'Farrell, P. H. High Resolution Two-dimensional Electrophoresis of Proteins. *J. Biol. Chem.* **1975,** *250* (10), 4007–4021.

Rath, A.; Glibowicka, M.; Nadeau, V. G.; Chen, G.; Deber, C. M. Detergent Binding Explains Anomalous SDS-PAGE Migration of Membrane Proteins. *Proc. Natl. Acad. Sci. U. S. A.* **2009,** *106* (6), 1760–1765.

Regnier, F. E. High-performance Liquid Chromatography of Biopolymers. *Science* **1983,** *222* (4621), 245–252.

Ryan, B. J. Differential Precipitation and Solubilization of Proteins. *Protein Chromatogr. Methods Protoc.* **2011,** 203–213.

Scopes, R. K. *Protein Purification—Principles and Practice;* Springer-Verlag: New York, **1994**.

Smyth, C. J.; Jonsson, P.; Olsson, E.; Soderlind, O.; Rosengren, J.; Hjertén, S.; Wadström, T. Differences in Hydrophobic Surface Characteristics of Porcine Enteropathogenic *Escherichia coli* with or without K88 Antigen as Revealed by Hydrophobic Interaction Chromatography. *Infect. Immun.* **1978,** *22* (2), 462–472.

Tatjana, W.; Joachim, W. *Handbook of Ion Chromatography;* Wiley-VCH: Weinheim, **2005**.

Uhlén, M. Affinity as a Tool in Life Science. *Biotechniques.* **2008,** *44* (5), 649–654.

Urh, M.; Simpson, D.; Zhao, K. Affinity Chromatography: General Methods. *Methods Enzymol.* **2009,** *463,* 417–438.

Yang, Y.; Li-zhi, N.; Shao-nan, L. Purification and Studies on Characteristics of Cholinesterases from *Daphnia magna*. *J. Zhejiang Univ. Sci.* **2013,** *14* (4), 325–335.

CHAPTER 18

PROTEIN–PROTEIN INTERACTION DETECTION: METHODS AND ANALYSIS

VAISHALI SHARMA[1], TUSHAR RANJAN[2*], PANKAJ KUMAR[3], AWADHESH KUMAR PAL[3], VIJAY KUMAR JHA[4], SANGITA SAHNI[5], and BISHUN DEO PRASAD[3]

[1]*DOS in Biotechnology, University of Mysore, Mysore, Karnataka, India*

[2]*Department of Basic Science and Humanities Genetics, Bihar Agricultural University, Sabour, Bhagalpur, Bihar, India*

[3]*Department of Molecular Biology and Genetic Engineering, Bihar Agricultural University, Sabour, Bhagalpur, Bihar, India*

[4]*Department of Botany, Patna University, Patna, Bihar, India*

[5]*Tirhut College of Agriculture, Dholi, Dr. Rajendra Agricultural University, Pusa, Bihar, India*

[*]*Corresponding author. E-mail: mail2tusharranjan@gmail.com*

CONTENTS

ABSTRACT

Protein–protein interactions (PPIs) are ubiquitous to cellular process. Proteins are the workhorses that facilitate most biological processes in a cell, including gene expression, DNA replication, transcription, translation, splicing, secretion, cell cycle control, signal transduction, apoptosis, and intermediary metabolism, where PPI plays a crucial role to facilitate their smooth running. Implications about function of any protein can be made via PPI. These implications are based on the premise that the function of unknown proteins may be discovered if captured through their interaction with a protein target of known function. These interactions are critical for the interactomics system of the living cell (or organisms) and abnormal PPIs are the sole reason for several diseases such as Alzheimer's disease and cancer. Extracellular signal basically regulates the activities of cells. Transduction of signals inside the interior of cells depends critically on PPIs between the various signaling molecules and transmembrane proteins. These signals could be for either normal metabolism or even cell death. Not only this, but also PPIs help in carrying proteins from one place to targets (such as from cytoplasm to nucleus in case of the nuclear importin). In most of the cases, one protein carries another protein for proper functioning of cells. In almost all biosynthetic events in cells, enzymes interact or communicate with each other and generate other macromolecules. Regulation of all metabolic events (e.g., negative feedback) and biology of muscle contraction also involves several interactions.

18.1 INTRODUCTION

Protein–protein interactions (PPIs) are intrinsically ubiquitous to cellular process. Approximately 30,000 genes of the human genome are speculated to give rise to 1×10^6 proteins through a series of post-translational modifications and gene-splicing mechanisms. Although a population of these proteins can be expected to work in relative isolation, the majority are expected to operate in concert with other proteins in complexes and networks to orchestrate the myriad of processes that impact cellular structure and function. Proteins are the workhorses that facilitate most biological processes in a cell, including gene expression, DNA replication, transcription, translation, splicing, secretion, cell cycle control, signal transduction, apoptosis, and intermediary metabolism, where PPI plays a crucial role to facilitate their smooth running. Implications about function of any protein can be made via PPI. These implications are based on the premise that the function of unknown proteins may be discovered if captured through their interaction with a protein target of known function. PPI refers to enduring and specific physical interactions between two or more proteins as a result of biochemical events such as electrostatic forces, hydrophobic interactions, salt bridges, etc. In general, they are inferred as physical contacts with the molecular docking between proteins in a cell of a living organism in specific biomolecular circumstances (De Las Rivas and Fontanillo, 2010). Most of the proteins are heteromers and very rarely act alone. Intriguingly, many molecular events within a cell are conducted by molecular machines that comprise a large number of protein components basically organized or cross-talk by PPIs. These interactions are critical for the interactomics system of the living cell (or organisms) and abnormal PPIs are the sole reason for several diseases such as Alzheimer's disease and cancer. Extracellular signal basically regulates the activities of cells. Transduction of signals inside the interior of cells depends critically on PPIs between the various signaling molecules and transmembrane proteins. These signals could be for either normal metabolism or even cell death. Not only this, but also PPIs help in carrying proteins from one place to targets (such as from cytoplasm to nucleus in case of the nuclear importin). In most of the cases, one protein carries another protein for proper functioning of cells. In almost all biosynthetic events in cells, enzymes interact or communicate with each other and generate other macromolecules. Regulation of all metabolic events (e.g., negative feedback) and biology of muscle contraction also involves several interactions. Myosin filament which acts as molecular motors binds with the actin and enables sliding of filaments to each other (Cooper, 2000; Herce et al., 2013). Proteins with more than one subunit are found in many different classes of proteins. Some of the best characterized multisubunit proteins (including classical proteins such as hemoglobin, tryptophan synthetase, aspartate transcarbamylase, core DNA/RNA polymerase, replicase, helicase, and glycyl-tRNA synthetase) are those that, as originally purified, contains two or more different components. As most of the proteins purified were multisubunit complexes, their PPIs were apparent (Phizicky and Fields, 1995).

18.2 TYPES OF PPIs

Protein interactions radically could be characterized as either stable or transient. Both stable and transient interactions can be either strong or weak. Stable interactions could be seen in case of proteins that are purified as multisubunit complexes. These complexes can be either homomers or heteromers. Stable multisubunit complex interactions may be observed in case of hemoglobin and RNA polymerase. These stable interactions among complex proteins are best characterized by coimmunoprecipitation (coIP), pull-down, or far-Western methods. Transient interactions are temporary in nature and expected to control the majority of vast cellular processes (e.g. phosphorylation) (Rao et al., 2014). Transient interactions can be strong or weak and fast or slow depending upon requirement of the cells. Most of the transiently interacting proteins are thought to be involved in many cellular processes such as protein modification, transport, folding, signaling, and cell cycling once they come into contact with their binding partners. Scientists have captured several important transient interactions by crosslinking or label transfer methods. During the event of PPIs, proteins strictly bind to each other through a combination of interactions such as hydrophobic bonding, van der Waals forces, hydrogen bonding, and salt bridges situated at specific binding motif on each protein. These motifs could be either small binding clefts or large surfaces. The binding cleft could range from just a few peptides long or stretch hundreds of amino acids. A well-known binding motif that facilitates relatively stable PPIs is the leucine zipper, comprising α-helices on each protein that binds to each other in a parallel fashion through the hydrophobic interaction. This hydrophobic interaction operates among regularly-spaced leucine residues on each α-helix. Apart from leucine zipper, there are so many PPI motifs found in multisubunit complex, which facilitates stable interactions (Rao et al., 2014).

18.3 BIOLOGICAL EFFECTS OF PPIs/ADVANTAGES OF MULTISUBUNIT PROTEINS

Multisubunit proteins are much more efficient in terms of energy utilization relative to a single large protein with multiple sites. First, it is much more economical to build proteins from simpler subunits than to require multiple copies of the coding information to synthesize oligomers (actin filaments and virus coats are much more simply assembled from monomers than by translation of a large polyprotein of repeated domains). Similarly, it is much more convenient to have one gene encoding a protein with different interacting partners, such as some of the eukaryotic RNA polymerase subunits, than to have the gene for that subunit reiterated for each different polymerase. Second, translation of large proteins can cause a significant increase in errors in translation; if such errors cause a lack of activity, they are much more economically eliminated by preventing assembly of that subunit into the complex than by eliminating the whole protein. Third, multisubunit assemblies allow for synthesis at one locale, followed by diffusion and assembly at another locale; this allows for both faster diffusion (since the monomers are smaller) and compartmentalization of activity (if assembly is required for activity). Fourth, homo-oligomeric proteins, if they have an advantage over monomers, are easily selected in evolution if the oligomers interact in an antiparallel arrangement; in this case, a single amino acid change that increases interaction potential has effects at two such sites. Another advantage of multisubunit complexes is the ability to use different combinations of subunits to alter the magnitude or type of response. Thus, for example, adult hemoglobin (a2β2) and fetal hemoglobin (α2γ2) are each composed of hetero-oligomers with a common subunit; differences in the binding of oxygen in this hemoglobin allow oxygen to be readily passed from mother to fetus. Outcome of PPIs is very important for proper functioning of cells (Fields and Song, 1989). The significant effects of protein interactions have been listed below:

- Alters the kinetic properties of enzymes, which may be the result of subtle changes in substrate binding or allosteric effects;
- Allows for substrate channeling by moving a substrate between domains or subunits, resulting ultimately in an intended end product;
- Creates a new binding site, typically for small effector molecules;
- Inactivates or destroys a protein;
- Changes the specificity of a protein for its substrate through the interaction with different binding partners; e.g., demonstrates a new function that neither protein can exhibit alone;
- Serves a regulatory role in either an upstream or a downstream event.

18.4 COMMON METHODS TO ANALYZE PPIs

Various methods and approaches for studying of PPIs are reported, but description of all of these methods is beyond the scope of this chapter. Therefore, a summary of general methods to analyze PPIs and the types of interactions that can be studied using each method is presented in Table 18.1. Although, we tried to compile most of the techniques used for PPIs but in this chapter we will discuss only few important frequently used system. Yeast two-hybrid (Y2H) system and phage display are two most useful genetic approaches for studying PPIs.

TABLE 18.1 Different Methods for Detection of PPIs with Their Principles.

Approaches	Techniques	Highlights
In vivo	Yeast two-hybrid (Y2H)	Yeast two-hybrid is typically carried out by screening a protein of interest against a random library of potential protein partners.
	Synthetic lethality	Synthetic lethality is based on functional interactions rather than physical interaction
In vitro	Phage display	Phage-display approach originated in the incorporation of the protein and genetic components into a single phage particle
	Tandem affinity purification-mass spectroscopy (TAP-MS)	TAP-MS is based on the double tagging of the protein of interest on its chromosomal locus, followed by two steps purification process and mass spectroscopic analysis
	Affinity chromatography (Pull-down assay)	Affinity chromatography is highly responsive, can even detect weakest interactions in proteins, and also tests all the sample proteins equally for interaction
	Coimmunoprecipitation	Coimmunoprecipitation confirms interactions using a whole-cell extract where proteins are present in their native form in a complex mixture of cellular components
	Protein microarrays	Protein-fragment complementation assays (PCAs) can be used to detect PPI between proteins of any molecular weight and expressed at their endogenous levels
	X-ray crystallography	X-ray crystallography enables visualization of protein structures at the atomic level and enhances the understanding of protein interaction and function
	NMR spectroscopy	NMR spectroscopy can even detect weak protein–protein interactions
	Surface plasmon resonance	Relates binding information to small changes in refractive indices of laser light reflected from gold surfaces to which a bait protein has been attached. Changes are proportional to the extent of binding and analysis occurs in real time.
In silico	In silico 2 hybrid (I2H)	The I2H method is based on the assumption that interacting proteins should undergo coevolution in order to keep the protein function reliable
	Ortholog-based sequence approach	Ortholog-based sequence approach is based on the homologous nature of the query protein in the annotated protein databases using pairwise local sequence algorithm

TABLE 18.1 *(Continued)*

Approaches	Techniques	Highlights
	Domain-pairs-based approach	Domain-pairs-based approach predicts protein interactions based on domain–domain interactions
	Structure-based approaches	Structure-based approaches predict protein–protein interaction if two proteins have a similar structure (primary, secondary, or tertiary)
	Gene neighborhood	If the gene neighborhood is conserved across multiple genomes, then there is a potential possibility of the functional linkage among the proteins encoded by the related genes
	Gene fusion	Gene fusion, which is often called Rosetta stone method, is based on the concept that some of the single-domain containing proteins in one organism can fuse to form a multidomain protein in other organisms
	Phylogenetic tree	The phylogenetic tree method predicts the protein–protein interaction based on the evolution history of the protein

(Reprinted from Rao, V. S.; Srinivas, K.; Sujini, G. N.; Sunand, G. N. Protein–protein Interaction Detection: Methods and Analysis. Int. J. Proteomics 2014, 147648, 12. https://creativecommons.org/licenses/by/3.0/.)

18.4.1 YEAST TWO-HYBRID SYSTEM

Y2H system is an intriguing technique used in molecular biology to discover PPIs by examining physical interactions (such as binding) between two proteins or a single protein. This technique was originally devised by Stanley Fields and Ok-Kyu Song in 1989, while detecting PPIs using the GAL4 transcriptional activator of the yeast *Saccharomyces cerevisiae*. During galactose utilization, GAL4 activated transcription is required for the synthesis of galactose metabolizing enzymes, which formed the basis of selection (Hurt et al., 2003; Young, 1988). The basic funda behind this method is the activation of downstream reporter gene after the binding of a transcription factor onto the upstream regulatory region or upstream activator sequence (UAS). For two-hybrid system, the transcription factor is split into two separate domains, known as binding domain (BD) and activating domain (AD). The BD is mainly responsible for binding to the UAS and the AD is responsible for the activation of transcription. Because of these two different fragments of transcription factors explored, Y2H system is thus called a protein-fragment complementation assay. Figure 18.1 represents the overall working principle behind the operation of Y2H system. The key of the Y2H system is that most eukaryotic transcription factors comprises the ADs, and BDs are the modular and can perform function only in close proximity to each other instead of direct binding. This indicates that although the transcription factor is divided into two fragments, it still has potential to activate transcription once the two fragments are connected indirectly. Y2H system often explores a genetically engineered yeast strain in which the biosynthesis of certain nutrients (e.g., amino acids) is defective (Whipple, 1988). When grown on media that lack these amino acids, the yeast fails to survive. This mutant yeast strain can be engineered to incorporate foreign DNA in the form of plasmids. In Y2H screening, separate bait and prey plasmids are simultaneously introduced into the mutant yeast strain. Plasmids are designed to produce a protein product in which the DNA-BD fragment is fused onto a protein while another plasmid is engineered to produce a protein product in which the activation domain (AD) fragment is fused onto another protein. Generally, the protein fused to BD is referred as the bait protein, and is typically a known protein the investigator is using to identify new binding partners. The protein fused to the AD is referred as the prey protein and can be either a single known protein or a library of known or unknown proteins. In this context, a library may consist of a collection of protein-encoding sequences that represent all the proteins expressed in a particular organism or tissue, or may be generated by synthesizing random DNA sequences (Verschure et al., 2006).

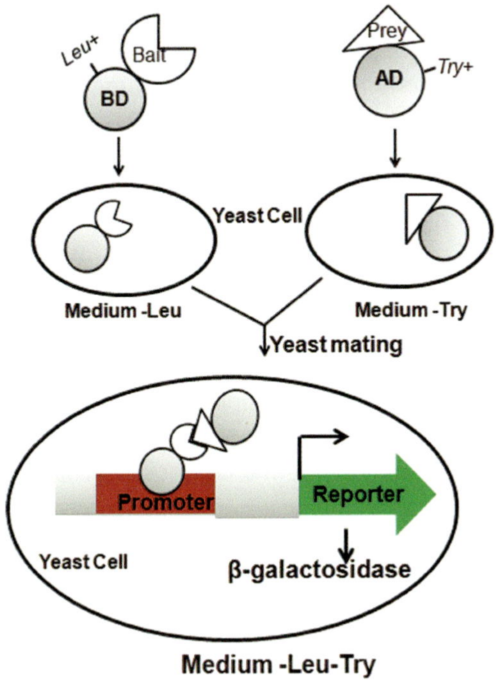

FIGURE 18.1 Working principle of Y2H system. The activation domain of yeast transcription factor is fused with a polypeptide called prey. Prey interacts with another polypeptide, bait, which in turn is fused with the DNA-binding domain of the same transcription factor. Successful transcription of an essential gene is possible only when the bait and prey interact, bringing together the DNA-binding and activation domains of the transcription factor.

Interaction between the two unknown proteins, i.e., baits and prey leads to the bringing of AD and BD of the transcription factor together. Bringing the AD in close proximity to the transcription initiation site, results in transcription of reporter gene. If the two unknown proteins do not interact, no transcription of the reporter gene will take place. Hence, a successful interaction between the fused proteins is linked to a change in the cell's phenotype (Gietz et al., 1997).

18.4.2 GST AND POLY HIS-TAGGED PULL-DOWN ASSAY

Pull-down assay basically works on the principle of binding affinity between biomolecules. Pull-down assays are akin in methodology to coIP because of the use of beaded support (stationary phase) to purify interacting proteins. Although the difference between these two approaches is that coIP uses antibodies to capture protein complexes, pull-down assays use a "bait" protein to purify any proteins in a lysate that bind to the bait. If antibody is not available, in that case pull-down assay is ideal for studying stable interactions. Glutathione S-transferase (GST) fusion proteins have had a range of applications. Pull-down approach has been greatly explored with expression and purification of novel proteins from bacterial lysate. Typically, GST pull-down experiments are used to identify interactions between a probe protein and unknown targets. The probe protein is a GST-fused protein, whose coding sequence is cloned under an IPTG-inducible expression vector. Now, this fusion protein is over expressed in bacteria and purified by affinity chromatography by applying on glutathione-agarose beads. The cell lysate and the GST fusion protein are incubated together with glutathione-agarose beads for their interactions. This incubation leads to the recovery of complexes from the beads and these complexes can be resolved by denaturing SDS-PAGE and further confirmed by the techniques such as western blotting, autoradiography (Jones and Thornton, 1996; Smith et al., 1988). Figure 18.2 describes the steps involved in pull-down assay. (1) A gene of interest (blue) is first subjected to a restriction digestion at restriction sites (D1 and D2) to produce sticky overhanging ends. Digested gene of interest is then inserted into a GST tagged vector (e.g., pGEX), which was also digested at the same restriction sites (D1 and D2). DNA ligase is used to ligate the desired gene or insert into the vector. The recombinant plasmid is then transformed into competent bacterial cells; (2) the bacteria cells harboring recombinant plasmid are then lysed and the lysate containing overexpressed protein is collected and incubated with GSH beads (yellow); (3) GSH beads bind to GST tags fused proteins with a high affinity. Therefore, the beads are strictly able to bind to the protein of interest in the lysate and form a large complex. The beads with large complex are then poured to a spin column; (4) this column contains inbuilt filter, which has pore size too small for the beads to be pass through. This allows anything that is not attached to the beads or the interaction large complex will pass through (pink). Column could be washed three–four times for the removal of any proteins that are not being bound to the bait protein. Finally, the SDS buffer is used to elute any proteins that have remained bound (green); (5) the flow through is then collected in fractionation tube; and (6) separated on SDS-PAGE. Any protein visualized here is an indicative of an interaction with the GST-fused bait protein (Fig. 18.2).

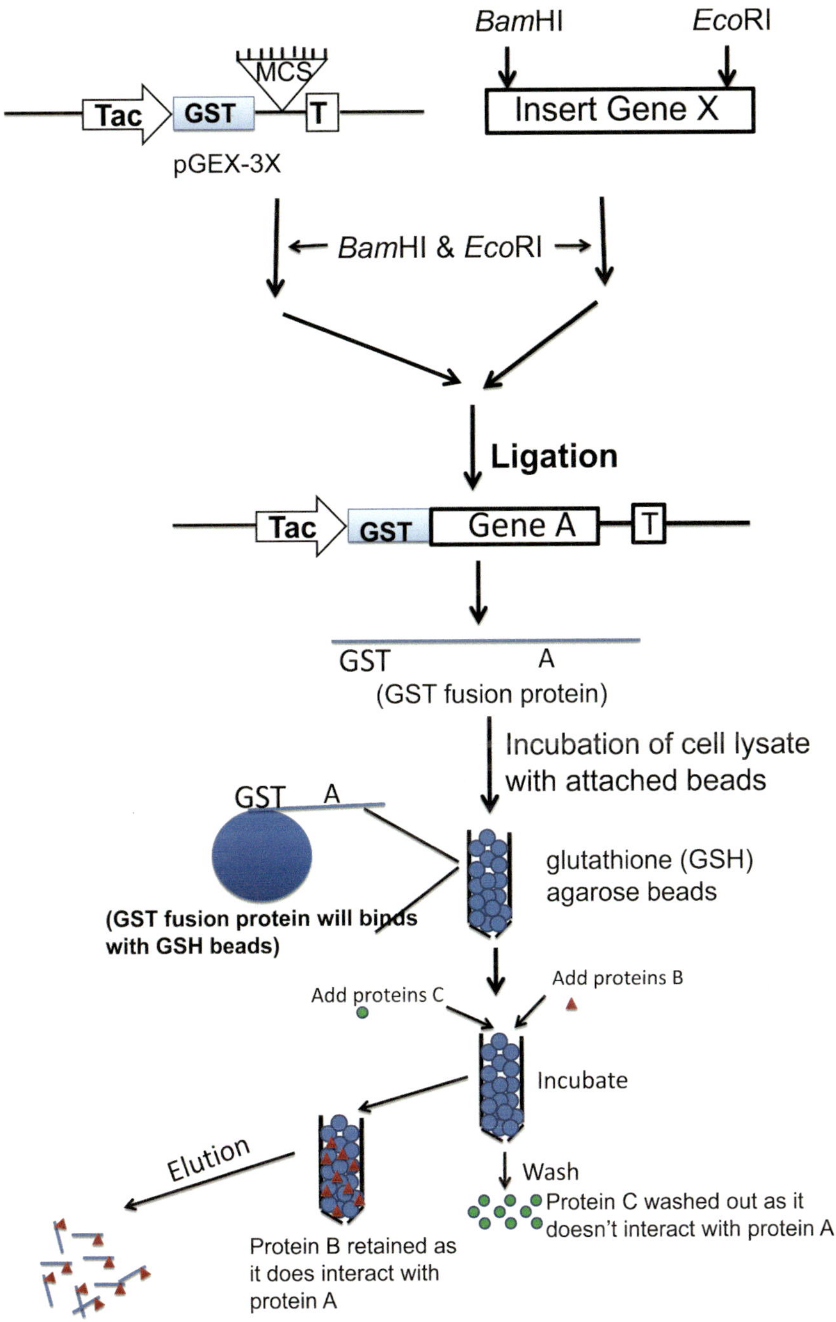

FIGURE 18.2 Working principle of GST pull-down assay.

18.4.3 COIMMUNOPRECIPITATION

CoIP is a classical method for the discovery of PPIs and has been used in literally thousands of experiments worldwide. CoIP is a powerful technique that is used routinely for analyzing PPIs. Immunoprecipitation (IP) is one of the oldest techniques of precipitating a protein antigen out of the solution containing thousands of different proteins using an antibody which strictly binds to target protein. In a simple way, this method can be used to remove and concentrate a particular protein from a sample containing many thousands of different proteins. IP technique basically involves the antibody to be coupled or adhered to a solid substratum. The basic experiment is simple. Cell lysates are generated, antibody is added, the antigen is precipitated and washed, and bound proteins are eluted and analyzed. CoIP works on the principle of selecting an antibody targeted against a known protein that is thought to be a member of a larger complex of proteins. By targeting this known member with an antibody, it may become possible to pull the entire novel protein complex out of the solution and thereby identify unknown members of the complex. Here, we also extract protein large complexes out of the solution mixture. This is the reason sometimes this method is also referred as a pull-down. By opting to these methods, many metabolic pathways have been deduced till date. During signal transduction, G-protein coupled receptors (GPRCs) are recognized by several hormones (e.g., glucagon). GPRCs are transmembrane proteins, helping in the activation (binding) of cytoplasmic downstream kinase to transfer the message from extracellular environment to intracellular environment. Cross-talk between GPCRs and different downstream kinases could be decoded with the help of this technique (Fig.18.3).

18.4.4 PHAGE DISPLAY METHOD

Similar to Y2H system, phage display is used for the high-throughput screening of protein interactions. Phage display is a most classical and reliable technique employing for the decoding of PPI and in some cases protein–DNA interactions as well. Phage display technique was first discovered by George P. Smith in 1985 while demonstrating the display or exposing of protein on filamentous phage by fusing the protein of interest onto gene III or VIII of filamentous phage M13. This method explores bacteriophage to attach target proteins with the genetic information that conceal them. This technique relies on the principle that the gene encoding a protein of interest

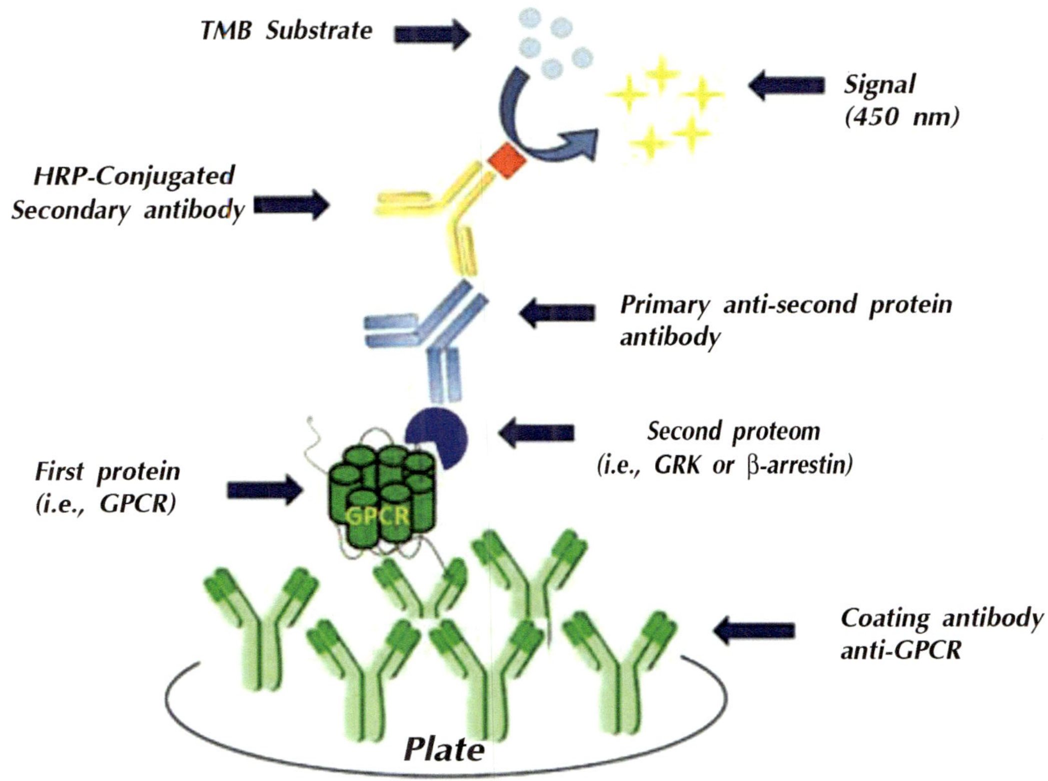

FIGURE 18.3 An immunoenzymatic assay to detect GPCRs and kinase interaction. (Reprinted from Zappelli, E, Daniele, S, Abbracchio, MP, Martini, C, Trincavelli, ML, A Rapid and Efficient Immunoenzymatic Assay to Detect Receptor Protein Interactions: G Protein-Coupled Receptors, Int. J. Mol. Sci. 2014, 15(4), 6252-6264. https://creativecommons.org/licenses/by/3.0/)

(unknown) is basically inserted (or tagged) with the phage coat protein gene, forcing the phage to display the desired protein on its outer surface. This display beautifully explains a connection between genotype and phenotype. These displaying phages can then be examined against other proteins in order to detect interaction between the displayed protein and other molecules. By exploring this way, large libraries of proteins can be screened and amplified. The method is particularly suited for the discovery of interactions between peptide BDs and their targets. This technique relies on the generation of library of millions of bacteriophages that have been genetically engineered to display different proteins on their outer surface. This is achieved by inserting a gene of interest encoding a protein into the phage's protein shell, which sets up a direct physical link between DNA sequences and their encoding proteins. Filamentous phages such as M13 are basically used for this kind of study, but T4, T7, and λ phages are also used (Kehoe and Kay, 2005; Smith, 1985).

The phage coat gene (pIII & VIII) and desired insert DNA (recombinant DNA) are transformed into *Escherichia coli* bacterial cells. Commonly used bacterial cells in this study are strains of *E. coli* such as TG1, SS320, ER2738, or XL1-blue. If only a phagemid is used (a simplified display construct vector) to infect the host, phage particles will not be excused out from the *E. coli* cells until and unless they are coinfected with the helper phage. This helper phage enables the packaging of phage genomic DNA and also helps in assembly of the mature virion particles with the relevant protein fragment (desired one) as part of their outer coat (Fig. 18.4). After immobilizing a target protein onto the surface of a microtiter plate well, only phage displaying a protein would bind to one of those targets. Other unwanted proteins would be removed while washing. The repeated cycle of above steps termed panning is necessary for the enrichment of a desired sample (displayed phage) by washing undesirable materials (Smith and Petrenko, 1997; Vidal et al., 2011).

18.4.5 SURFACE PLASMON RESONANCE

The recent development of a technique to monitor PPIs and ligand-receptor interactions by using changes in surface plasmon resonance (SPR) measured in real time spells the beginning of a minor revolution in biology. This method measures complex formation by monitoring changes in the resonance angle of light impinging on a gold surface as a result of changes in the refractive index of the surface up to 300 nm away. A ligand of interest

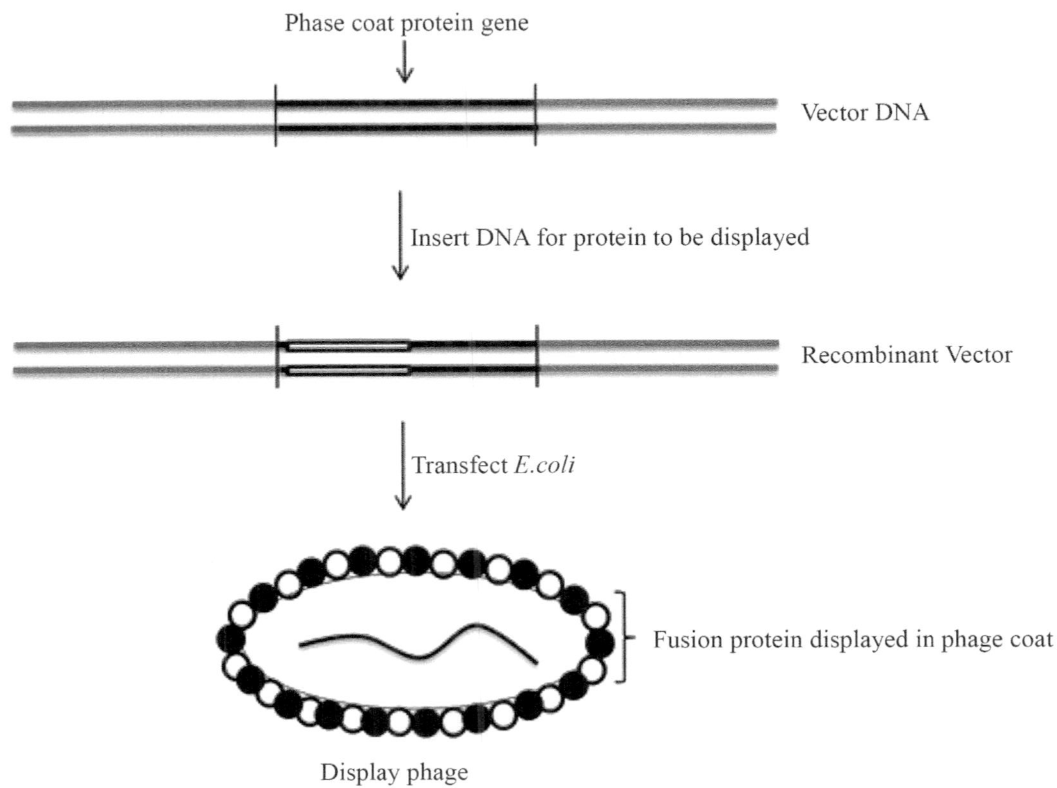

FIGURE 18.4 Phage display method: DNA coding for test protein is cloned into phage vectors in order to express hybrid protein (test protein fusion to the coat protein) on the phage surface. Phage particles produced by these transformed bacteria therefore display the test protein in their coats.

(peptide or protein in this case) is immobilized on a dextran polymer, and a solution of interacting protein is flowed through a cell, one wall of which is composed of this polymer. Protein that interacts with the immobilized ligand is retained on the polymer surface, which alters the resonance angle of impinging light as a result of the change in refractive index brought about by increased amounts of protein near the polymer (Fig. 18.5). Since all proteins have the same refractive index and there is a linear correlation between resonance angle shift and protein concentration near the surface, this allows one to measure changes in protein concentration at the surface due to protein–protein or protein-peptide binding. Furthermore, this can be done in real time, allowing direct measurement of kinetic of complex formation. This allows the monitoring in real time of the interactions between the immobilized ligand and the extracts. Interacting proteins from the extracts are then recovered, trypsinized, and identified using mass spectrometry. The data obtained are searched against a sequence database using the Mascot software. To exclude nonspecific interactors, control experiments using blank sensor chips and/or randomized peptides are performed.

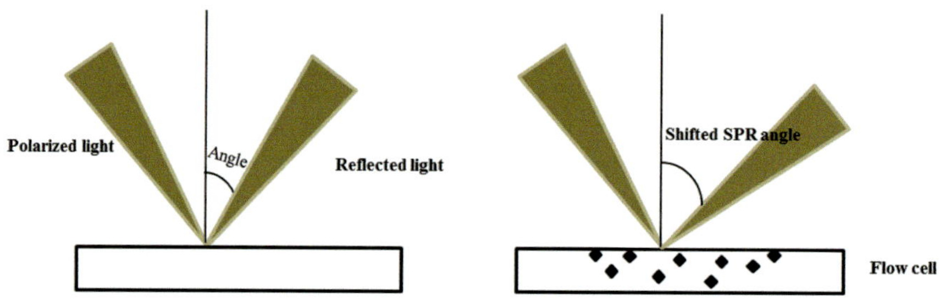

FIGURE 18.5 SPR working principle for studying of PPIs.

This method for detection of PPIs has several advantages. First, it requires very little material. Typically only 1–10 mg of protein has to be immobilized on a sensor chip, which can be reused up to 50 times after removal of adhering protein. Second, the method is very fast. A typical run for a given protein takes about 10 min. Third, no modifications of the proteins are required such as labeling or fluorescent tags. Fourth, interactions can be observed even in complex mixtures. Fifth, the system is useful over a wide range of protein concentrations. Sixth, the system is quite sensitive; the practical limit for association rates is 10^6 M/s, and off rates as low as 1.1×10^{-5}/s have been measured by recording for 6 h with buffer (Stenberg et al., 1991).

18.4.6 FLUORESCENCE RESONANCE ENERGY TRANSFER

Fluorescence resonance energy transfer (FRET) is a mechanism describing energy transfer between two light-sensitive molecules (chromophores) where a donor chromophore, initially in its electronic excited state, may transfer energy to an acceptor chromophore through nonradiative dipole–dipole coupling. The efficiency of this energy transfer is inversely proportional to the sixth power of the distance between donor and acceptor, making FRET extremely sensitive to small changes in distance. The process of resonance energy transfer (RET) can take place when a donor fluorophore in an electronically excited state transfers its excitation energy to a nearby chromophore, the acceptor. In principle, if the fluorescence emission spectrum of the donor molecule overlaps the absorption spectrum of the acceptor molecule, and the two are within a minimal spatial radius, the donor can directly transfer its excitation energy to the acceptor through long-range dipole–dipole intermolecular coupling. A theory proposed by Theodor Förster in the late 1940s initially described the molecular interactions involved in RET, and Förster also developed a formal equation defining the relationship between the transfer rates, inter chromophore distance, and spectral properties of the involved chromophores. FRET detects the proximity of fluorescently labeled molecules over distances >100 Å. When performed in a fluorescence microscope, FRET can be used to map PPIs *in vivo.* However, in order to understand the physical interactions between proteins partners involved in a typical biomolecular process, the relative proximity of the molecules must be determined more precisely than diffraction-limited traditional optical imaging methods permit (Kenworthy, 2001).

Although FRET has often been employed to investigate intermolecular and intramolecular structural and functional modifications in proteins and lipids, a major obstacle for implementation of FRET microscopy techniques in living cells has been the lack of suitable methods for labeling specific intracellular proteins with appropriate fluorophores. Cloning of the jellyfish green fluorescent protein (GFP) and its expression in a wide variety of cell types has become a critical key to developing markers for both gene expression and structural protein localization in living cells. Several spectrally distinct mutation variants of the protein have been developed, including a fluorescent protein that emits blue light (blue fluorescent protein, BFP). Both the excitation and emission spectra for the native GFP and BFP mutants are sufficiently separated in wavelength to be compatible with the FRET approach. Figure 18.6 illustrates the strategy for detection of PPIs using FRET and mutant fluorescent proteins. If two proteins, one labeled with

BFP (the donor) and the other with GFP (the acceptor), physically interact, then increased intensity at the acceptor emission maximum (510 nm) will be observed when the complex is excited at the maximum absorbance wavelength (380 nm) of the donor. Failure of the proteins to form a complex, results in no acceptor (GFP) fluorescence emission (Gadella, 2008; Truong and Ikura, 2001).

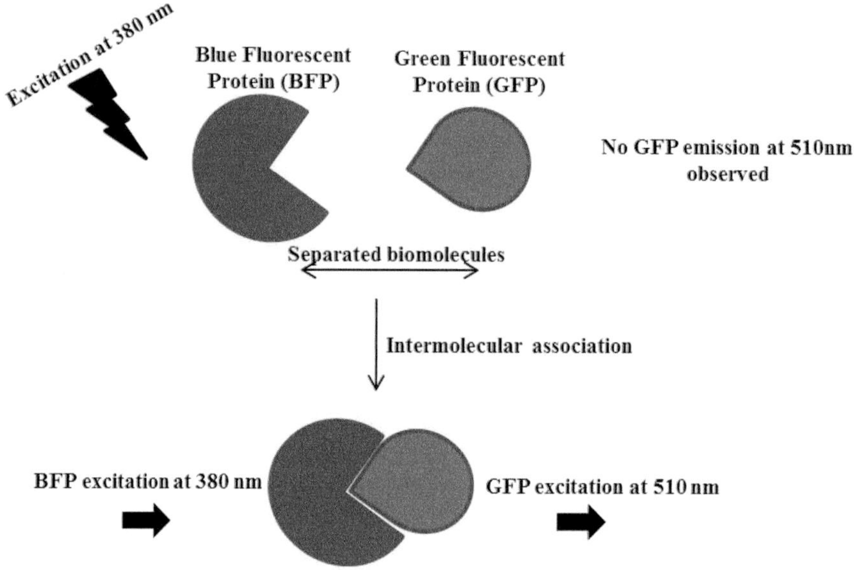

FIGURE 18.6 FRET working principle for studying of PPIs.

18.4.7 ISOTHERMAL TITRATION CALORIMETRY

Isothermal titration calorimetry (ITC) is considered as the most quantitative technique available for measuring the thermodynamic properties of PPIs and is becoming a necessary tool for protein–protein complex structural studies. This technique relies upon the accurate measurement of heat changes that follow the interaction of protein molecules in solution, without the need to label or immobilize the binding partners, since the absorption or production of heat is an intrinsic property of virtually all biochemical reactions (Fig. 18.7). ITC provides information regarding the stoichiometry, enthalpy, entropy, and binding kinetics between two interacting proteins. Measuring heat transfer during binding enables accurate determination of binding constants (K_D), reaction stoichiometry (n), enthalpy (ΔH), and entropy (ΔS).

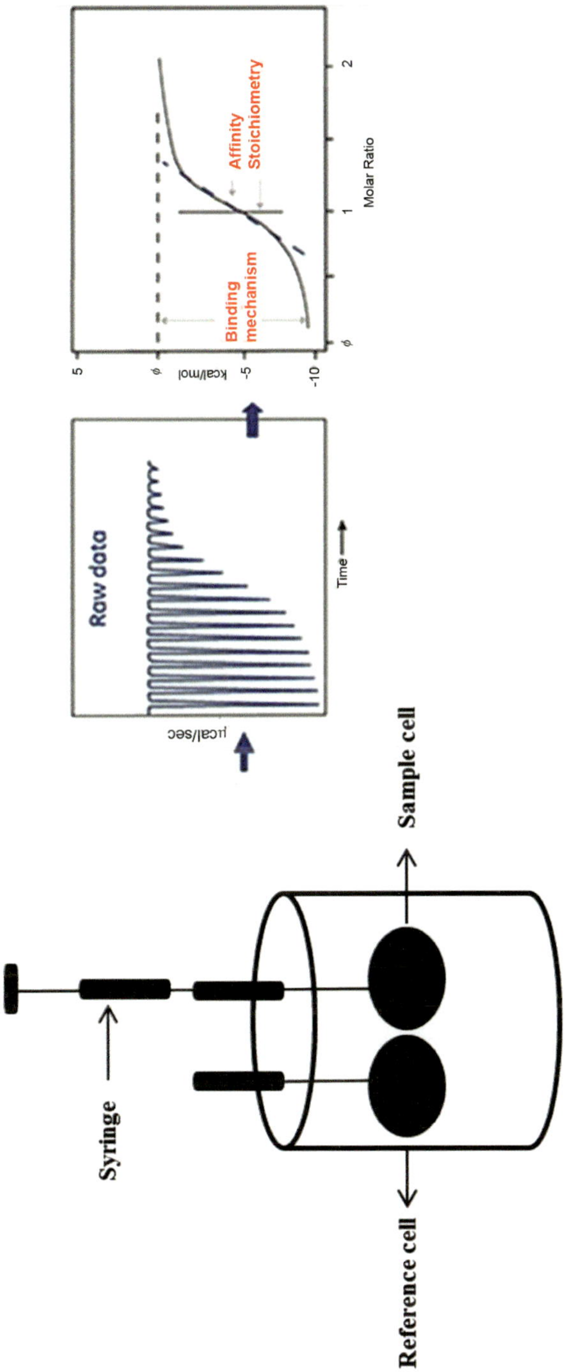

FIGURE 18.7 Working principle of ITC. (Reprinted from https://www.malvern.com/en/products/technology/isothermal-titration-calorimetry)

This provides a complete thermodynamic profile of the molecular interaction. ITC goes beyond binding affinities and can elucidate the mechanisms underlying molecular interactions (Bellucci, 2010).

Although there are many more important methods for detection of PPIs available, researchers use them according to their need and design of experiments. Most of these methods/techniques have their own limitation and sensitivity. Discussing or emphasizing all available methods for PPIs is not an easy task. By considering the above facts, here in this chapter, we tried to go into details of few of the well-known methods.

18.5 CONCLUSION

The study of protein interactions is vital to the understanding of how proteins function within the cell. Publication of the human genome and proteomics-based protein profiling studies catalyzed resurgence in protein interaction analysis. Characterizing the interactions of proteins in a given cellular proteome (now often referred as the "interactome") will be the next milestone along the road in understanding the biochemistry of the cell. There are many methods to investigate PPIs . Each of the approaches has its own strengths and weaknesses, especially with regard to the sensitivity and specificity of the method. A high sensitivity means that many of the interactions that occur in reality are detected by the screen. A high specificity indicates that most of the interactions detected by the screen are occurring in reality.

KEYWORDS

- **hydrophobic interaction**
- **electrostatic force**
- **leucin zipper**
- **multisubunit**
- **Y2H**
- **GAL4**
- **phage display**
- **FRET**
- **SPR**

REFERENCES

Bellucci, M. *Protein–Protein Interactions: A Tool Kit to Puzzle Out Functional Networks*; VDM Verlag Dr. Müller, **2010**.

Cooper, G. M. *The Cell: A Molecular Approach* (2nd Edn.); ASM Press: Washington, DC, **2000**.

De Las Rivas, J.; Fontanillo, C. Protein–protein Interactions Essentials: Key Concepts to Building and Analyzing Interactome Networks. *PLoS Comput. Biol.* **2010,** *6*(6), e1000807.

Fields, S.; Song, O. A Novel Genetic System to Detect Protein–protein Interactions (Abstract). *Nature,* **1989,** *340*(6230), 245–246.

Gadella, T. W. *FRET and FLIM techniques*. Elsevier, **2008**.

Gietz, R. D.; Triggs-Raine, B.; Robbins, A.; Graham, K.; Woods, R. Identification of Proteins that Interact with a Protein of Interest: Applications of the Yeast Two-hybrid System. *Mol. Cel. Biochem.* **1997,** *172*(1–2), 67–79.

Herce, H. D.; Deng, W.; Helma, J.; Leonhardt, H.; Cardoso, M. C. Visualization and Targeted Disruption of Protein Interactions in Living Cells. *Nat. Commun.* **2013,** *4*, 2660.

Hurt, J.; Thibodeau, S.; Hirsh, A.; Pabo, C.; Joung, J. Highly Specific Zinc Finger Proteins Obtained by Directed Domain Shuffling and Cell-based Selection. *Proc. Natl. Acad. Sci. U. S. A.* **2003,** *100*(21), 12271–12276.

Jones, S.; Thornton, J. M. Principles of Protein–protein Interactions. *Proc. Nat. Acad. Sci. U. S. A.* **1996,** *93*(1), 13–20.

Kehoe, J. W.; Kay, B. K. Filamentous Phage Display in the New Millennium. Chem. Rev. **2005,** *105(11), 4056–4072.*

Kenworthy, A. K. Imaging Protein–protein Interactions Using Fluorescence Resonance Energy Transfer Microscopy. *Methods* **2001,** *24*(3), 289–296.

Madeira, A.; Öhman, E.; Nilsson, A.; Sjögren, B.; Andrén, P. E.; Svenningsson, P. Coupling Surface Plasmon Resonance to Mass Spectrometry to Discover Novel Protein–protein Interactions. *Nat. Protoc.* **2009,** *4*, 1023–1037.

Phizicky, E. M.; Fields, S. Protein–protein Interactions: Methods for Detection and Analysis. *Microbiol. Rev.* **1995,** *59*, 94–123.

Rao, V. S.; Srinivas, K.; Sujini, G. N.; Sunand, G. N. Protein–protein Interaction Detection: Methods and Analysis. *Int. J. Proteomics* **2014,** *147648*, 12.

Smith, G. P. Filamentous Fusion Phage: Novel Expression Vectors that Display Cloned Antigens on the Virion Surface. *Science* **1985,** *228*(4705), 1315–1317.

Smith, D. B.; Johnson, K. S. Single-step Purification of Polypeptides Expressed in *Escherichia coli* as Fusions of Glutathione S-transferase. *Gene* **1988,** *67*, 31–40.

Smith, G. P.; Petrenko, V. A. Phage Display. *Chem. Rev.* **1997,** *97*(2), 391–410.

Truong, K.; Ikura, T. The Use of FRET Imaging Microscopy to Detect Protein–protein Interactions and Protein Conformational Changes In Vivo. *Curr. Opin. Struct. Biol.* **2001,** *11*(5), 573–578.

Verschure, P.; Visser, A.; Rots, M. Step Out of the Groove: Epigenetic Gene Control Systems and Engineered Transcription Factors. *Adv. Genet.* **2006,** *56*, 163–204.

Vidal, M.; Cusick, M. E.; Barabási, A. L. Interactome Networks and Human Disease. *Cell,* **2011,** *144*(6), 986–998.

Young, K. Yeast Two-hybrid: So Many Interactions, (in) so Little Time. *Biol. Reprod.* **1998,** *58*(2), 302–311.

Whipple, F. Genetic Analysis of Prokaryotic and Eukaryotic DNA-binding Proteins in *Escherichia coli. Nucleic Acids Res.* **1998,** *26*(16), 3700–3706.

PART IV
Molecular Markers and QTL Mapping

CHAPTER 19

MOLECULAR MARKERS IN PLANT BIOTECHNOLOGY

GAURAV V. SANGHVI[1*] and GAURAV S. DAVE[2]

[1]*Department of Botany, Faculty of Science, The Maharaja Sayajirao University of Baroda, Vadodara 390002, Gujarat, India.*

[2]*Department of Biochemistry, Saurashtra University, Rajkot 360005, Gujarat, India.*

[*]*Corresponding author. E-mail: gv.sanghvi@gmail.com*

CONTENTS

ABSTRACT

During the past few decades, significant progress has been made in plant breeding with advances in understanding of genetic variation, cytogenetics, quantitative genetics, and molecular biology. A molecular marker represents a handy tool for identification of polymorphism in field of plant biotechnology. Molecular markers can be classically divided into two types: Hybridization-based or non-PCR-based markers. Depending on type of studies, one can easily pick a particular marker from an assortment of various markers, viz. morphological, biochemical and DNA-based markers Allele mining, related hereditary qualities, and near genomics are promising new ways to deal with and get knowledge of the association and variety of genes that influence applicable phenotypic attributes. These DNA markers offer a favorable circumstances over conventional phenotypic markers, as they give information that can be investigated precisely. This chapter gives an inclusive use and types of various molecular markers used in various process in plant biotechnology.

19.1 INTRODUCTION

Plant breeding in combination with developments in agricultural technology has made notable advancement in increasing crop yields. However, plant breeders must constantly respond to many changes depending on type of soil, climate, etc. During the past century, there has been continuous upgradation and introduction of new techniques for improvement of crop quality to cope up with consumer's market demand as well as factors affecting crops. However, with increasing population and crop demand, some research points toward the decline in crop quality (Pingali and Heisey, 1999). The increase in crop demand has to cope with factors like pollution, inevitable emergence of new races and biotypes of pathogens and pests, and possible adverse effects of climate change (Collard and Mackill, 2008). Thus, to meet the present market demand with improved crop quality seems to be an unmatched challenge for plant breeders and agricultural scientists.

With advancements in the field of plant biotechnology and its associated technology, there are many research programs going on especially for crop improvement, yield, and quality by understanding the genetic mechanisms in cases ranging from detecting diseases in plant and its defense mechanisms to shelf life of fruits. In agriculture, biotechnology has found applications like cell and tissue culture engineering, diagnostics for detecting plant pests, and diseases based on the use of monoclonal antibodies and nucleic acid probes; and in genetic engineering of plant species, to introduce new traits, and in aiding conventional plant breeding programs use molecular markers (Persley, 1992).

19.1.1 HISTORY AND CURRENT SCENARIO OF PLANT BREEDING TECHNOLOGIES

Plant breeding has an extended history of assimilating the modern improvements in biology and genetics to enrich crop improvement. Prehistoric selection for visible phenotypes facilitated crop harvest and increased productivity led to the domestication of the first crop varieties (Jensen, 1992). Darwin outlined the scientific principles of hybridization and selection, and Mendel defined the fundamental association between genotype and phenotype, discoveries that enabled a scientific approach to plant breeding at the beginning of the 20th century (Shull, 1908). Subsequent advances in our understanding of plant biology, the analysis and induction of genetic variation, cytogenetics, quantitative genetics, molecular biology, biotechnology, and, most recently, genomics have been successively applied to further increase the scientific base and its application to the plant breeding process (Baenziger and Russell, 2006; Jauhar, 2006).

The plant biotechnology era began in 1980s with reports of generating transgenic plants by *Agrobacterium* (Bevan et al., 1983; Herrera-Estrella and Depicker, 1983). Molecular marker systems for crops were soon established for high-resolution genetic maps and genetic linkage between markers and key crop traits (Edwards et al., 1987; Tanksley et al., 1989). By 1996, the commercialization of transgenic crops confirmed integration of biotechnology into plant breeding and crop improvement programs (Koziel et al., 1993; Delannay et al., 1995). Conventional breeding is a dynamic area of applied science. It depends on genetic variation and choices to gradually improve plants for traits and features that are of interest for the grower and the consumer. Another important way of improvement is the introduction of new genetic material (e.g., genes for biotic and abiotic stress resistance) from other sources, such as gene bank accessions and related plant species. Although, current breeding practices have been successful in producing a continuous range of improved varieties, recent developments in the field of molecular biology can be employed to enhance plant breeding efforts and to speed up cultivar development.

Modern crop varieties are developed from plant population exhibition superior genetic variability and from developed hybrids which are produced by selective inbreeding and crossing (Hikmet et al., 2004). Crop improvement by selection process mainly depends on:

1. Discovery and generation of genetic variability in agronomic traits.
2. Precise selection of genotypes with favorable characteristics, as a product of a recombination among superior alleles at different loci (Sorrells and Wilson, 1997).

Modern biotechnology provides new tools that can facilitate development of improved plant breeding methods and augment our knowledge of plant genetics (Datta et al., 2011). During the past 25 years, the continued development and application of plant biotechnology, molecular markers, and genomics has established new tools for the creation, analysis, and manipulation of genetic variation and the development of improved cultivars (Sharma et al., 2002; Collard and Mackill, 2008). Methods for marker-assisted backcrossing were developed briskly for the development of new transgenic traits and reduction of linkage drag. The markers have been used over many years for classification of plants.

Markers are any trait of any organism that can be identified with confidence and relative ease which can be also followed in mapping population. In general, markers are heritable entities associated with economically important trait under the control of polygenes. In particular, molecular markers were used in genome scans to select those individuals that contained both the transgene and the greatest proportion of favorable alleles from the recurrent parent (RP) genome (Ragot et al., 1994; Johnson and Mumm, 1996). Molecular breeding is currently a standard practice in many crops, with the following sections briefly reviewing how molecular information and genetic engineering positively impacts the plant breeding paradigm. Molecular breeding may be defined as genetic manipulation performed at DNA molecular levels to develop characters of interest in plants, including genetic engineering or gene manipulation, molecular marker-assisted selection (MAS), genomic selection, etc. However, molecular breeding also implies molecular marker-assisted breeding (MAB) which may be defined as the novel application of molecular markers, in permutation with linkage maps and genomics, to alter and improve plant traits on the basis of genotypic assays.

In the current scenario, the DNA markers become the marker of choice for improving genetic diversity of crops and had revolutionized the plant biotechnology. Gradually, methods are developing more specifically, rapidly, and cheaply to assess genetic variation. Results obtained after successful usage of molecular markers indicated that when inbreed lines were unrelated, a measurement of relative relationship based on proportion of homomorphic marker loci was significantly correlated with a measure of relationship based on yield. Conversely, when lines were related the correlations were low (Dudley et al., 1991). With rise of DNA marker technology, several types of DNA markers and molecular breeding strategies are now available to plant breeders and geneticists, helping them to overcome many of the problems faced during conventional breeding. Moreover, molecular markers and more recently, high-throughput genome sequencing efforts, have dramatically increased knowledge of and ability to characterize genetic diversity in the germplasm pool for essentially any crop species (Moose and Mumm, 2008). In this chapter, basic qualities of molecular markers, their characteristics, usage, and their applications will be discussed. It is important to point out that no genetic approach is solely responsible for the development of crop trait; many methods complement each other. However, some techniques are clearly more appropriate than others for some specific applications likewise crop diversity and taxonomy studies (Kumar et al., 2009).

19.2 GENETIC MARKERS: CONCEPT AND TYPES

Genetic markers are the biological features that are determined by allelic forms of genes or genetic loci and can be transmitted from one generation to another, and thus they can be used as experimental probes or tags to keep track of an individual, a tissue, a cell, a nucleus, a chromosome, or a gene (Jiang, 2013). Genetic markers signify genetic variances between distinct organisms or species. Normally, they do not represent the target genes themselves but act as "signs" or "flags." Genetic markers that are positioned in near vicinity to genes may be stated as gene "tags." Such markers do not intrude the phenotype of the trait of interest as they are located only adjacent or "linked" to genes regulating the trait. All genetic markers occupy definite genomic positions within chromosomes (like genes) called "loci" (singular "locus"). The quality of a genetic marker is typically measured by its heterozygosity in population of interest. Polymorphism information content (PIC). PIC can be defined as probability of identifying one homologue of a given parent that transmitted and allele to given offspring, the other parent being genotyped as well. Prime properties defining genetic markers are: (1) It should be locus specific. (2) It should be polymorphic in the studied population. (3) It should be easily phenotyped.

Genetic markers may differ in key features like genomic abundance, level of polymorphism, locus specificity, and reproducibility. No marker is superior to all others for a wide range of applications. The selection of appropriate marker depends on specific application, presumed polymorphism level, economical variability (Kumar et al., 2009). Genetic markers were originally used in genetic mapping to determine the order of genes along the chromosomes. In 1913, Alfred Sturtevant generated the first genetic map using sex morphological traits (termed "factors") in fruit fly *Dorsophila melanogaster* (Sturtevant, 1913) and soon after, Karl Sax produced evidence for genetic linkage in common bean (Sax, 1923). Because of these pioneer studies, genetic markers have evolved from morphological markers through isozymes markers to DNA markers. Table 19.1 describes the comparison between the markers. Today, genetic markers are used in basic plant research for gene isolation, plant breeding for characterizations of germ plasm, for marker-assisted introgression of favorable alleles, and for variety protection (Henry, 2012). Especially, genetic DNA base marker has revolutionized the rational of genetic selection and genetic characterization in plants (Sergio and Gianni, 2005).

TABLE 19.1 Comparison of Markers.

Feature	Morphological Markers	Biochemical Markers	DNA Markers
Organisms scored	Phenotype	Enzymes	DNA
Plant material required for detection	Plant or plant organ	Tissues	DNA form tissues
Efforts required for detection	Easy	Medium	Medium to difficult
Dominance/codominance	Dominant	Codominant	Both depends on marker used
Reproducibility	High	High	Moderate to high

Genetic markers used in genetics and plant breeding can be classified into two categories: classical markers and DNA markers (Xu, 2010).

Classical markers are further classified as morphological markers, cytological markers, and biochemical markers.

DNA markers have been developed into many systems based on different polymorphism detecting techniques or methods (Southern blotting, polymerase chain reaction (PCR), and DNA sequencing) (Collard et al., 2005) such as restriction fragment length polymorphism (RFLP), amplified fragment length polymorphism (AFLP), random amplified polymorphic DNA

(RAPD), simple sequence repeats (SSRs), single nucleotide polymorphisms (SNP), etc.

19.2.1 CLASSICAL MARKERS

19.2.1.1 MORPHOLOGICAL MARKERS

Use of markers as a supportive tool for selection of plants with preferred traits had started in plant breeding long time ago. In early history of plant breeding, markers were used usually for visual phenotype character identification such as leaf shape, flower color, pubescence color, pod color, seed shape, hilum color, type and length, fruit shape, etc. (Jiang, 2013). These morphological markers mostly signify genetic polymorphisms which can be simple to identify and manipulate. Therefore, they are usually used in construction of linkage maps by classical two- and/or three-point tests. Some of these markers are linked with other agronomic traits and thus can be used as indirect selection criteria in practical breeding. In scientific terms, morphological markers are those traits that are scored visually, or morphological markers are those genetic markers whose inheritance can be followed with the naked eye (Bhat et al., 2010). The main advantages of using morphological markers are easy monitoring and economical viability. However, molecular markers are severely affected by the external environment. Such markers regularly cause major alternations in the phenotype which leads to failure of breeding programs. Morphological markers are also limited in number and appear late in plant development which makes scoring almost impossible. Also, dominant, recessive interactions mostly avoid differentiation of all genotypes linked with morphological traits. In addition, a given morphological marker can affect other morphological markers or traits of interest in breeding programs due to its pleiotropic gene action. A morphological marker masks the effect of linked minor gene, making it nearly impossible to identify desirable linkages for selection, influenced by environment and also specific stage of the analysis. Finally, the usefulness of morphological markers is restricted by their limited number (Jiang, 2013).

The best cited example in literature for successful usage of morphological marker that could be considered is selection of semidwarfism in rice and wheat leading to high-yield cultivation of crop. This could be regarded as a suitable illustration for effective use of morphological markers to modern breeding. In wheat breeding, the dwarfism directed by gene *Rht10* was introgressive hybridization into Taigu nuclear male-sterile wheat by backcrossing, with linkage was produced between *Rht10* and the male-sterility gene *Ta1*. Then resulted dwarfism was recognized and used as the marker for identification and selection of the male-sterile plants in breeding populations (Liu, 1991). This is predominantly supportive for execution of recurrent choice in wheat. However, limited availability and its close association with important economic crop traits (e.g., yield and quality), preferred less in modern plant breeding programs. It can even lead to some unexpected/undesirable effects on development, characterization, and growth of plants.

Cytological markers are used for the identification of structural features of chromosomes. Cytological markers can be revealed by chromosome karyotypes and bands. The distributional differences of euchromatin and heterochromatin in chromosomes including its color, width, order, and position in banding patterns can be studied using cytological markers. For example, Q bands are formed by quinacrine hydrochloride, G bands are visualized by Giemsa stain, and R bands are the reversed G bands. The applications of identified chromosome landmarks are not only for classification of normal chromosomes and uncovering of chromosome mutation but also broadly used in physical mapping and linkage group identification. The physical maps created using combination of morphological and cytological markers lay a perfect basis for genetic linkage mapping with the assistance of modern molecular techniques. However, direct usage of cytological markers has been very limited in genetic mapping and plant breeding (Jiang, 2013).

19.2.1.2 BIOCHEMICAL MARKERS

The first biochemical molecular markers used were the protein-based markers. Proteins are attractive molecules for direct genetic study since they are the primary yields of structural genes (Bhat et al., 2010). Biologically, any changes in the coding base sequence will lead to corresponding changes in primary structure of proteins. With change in single amino acid substitutions either by deletions or additions leads to mark change in protein migration under electric field during electrophoresis. One of the initial protein-based markers to be used was isozyme. Markert and Moller (1959) gave the term to define the multiple molecular forms of the same enzyme with the same substrate specificity. Isozymes are different forms of an enzyme exhibiting the same catalytic activity but differing in charge and electrophoretic mobility.

Isozyme analysis has been used for over more than 60 years for different research purposes for, for example, in phylogenetic relationships, to guess genetic variability, population and developmental biology, and in characterization of plant genetic resources management and improvement in plant breeding (Staub and Meglic, 1993). Isozymes were defined as structurally diverse molecular forms of an enzyme with, qualitatively, the same catalytic function. Isozymes start off through changes in amino acids, leading to change in net charge, spatial structure (conformation), and also electrophoretic mobility. After exact staining procedure, the isozyme profile of distinct samples can be observed (Hadačová and Ondřej 1972; Vallejos, 1983; Soltis, 1989).

Allozymes are allelic variations of enzymes determined by structural genes. Enzymes are proteins consisting of amino acids, some of which are electrically charged, resulting enzymes have a net electric charge, depending on the stretch of amino acids comprising the protein. During any of the situation leading to a mutation in the DNA results in a replacement of amino acid being substituted, then net electric charge of the protein may be modified, and overall shape (conformation) of the molecule may also be changed. Since, charge and conformation changes can affect the migration rate of proteins in an electric field, allelic variation can be detected by gel electrophoresis and consequent enzyme-specific stains that comprise substrate for the enzyme, cofactors, and an oxidized salt (e.g., nitroblue tetrazolium).

Usually minimum of two, or occasionally even more loci can be illustrious for an enzyme and these are called as isoloci. Hence, allozyme variation is often also called as isozyme variation (Kephart, 1990; May, 1992). Isozymes have been verified as consistent genetic markers in breeding and genetic studies of plant species (Heinz and Tew, 1987) due to uniformity in their expression, regardless of numerous environmental factors.

Advantages

The prime and most proven strength of allozymes is simplicity. Since, analysis of allozymes does not require DNA extraction, sequence information, primers, or probes, they are rapid and easy to use. A simple analytical protocol makes allozymes to work at low costs with consideration of enzyme staining and reagents. Isoenzyme markers are the oldest among the molecular markers. Isozymes markers have been successfully used in several crop improvement programs (Vallejos, 1983; Baes and Cutsem, 1993). Allozymes are codominant markers that have high reproducibility. Zymograms (the banding pattern of isozymes) can be readily understood in relationships of loci and alleles, or they may require separate analysis of progeny of known parental crosses for interpretation. However, zymograms can also present complex banding profiles from polyploidy or duplicated genes and the formation of intergenic heterodimers, which may lead to complex analysis.

Disadvantages

Low abundance and low level of polymorphism are viewed as main weaknesses of allozymes. Moreover, proteins with similar electrophoretic migration may not be same for distantly related germplasm. In addition, their specific neutrality may also be in question (Hudson et al., 1994; Berry and Kreitman, 1993; Ross, 2002). Importantly, often allozymes are regarded as molecular markers because they represent enzyme variants, and enzymes are molecules. Allozymes are in reality phenotypic markers, and as such they may be affected by environmental conditions. For example, the banding profile obtained for a particular allozyme marker may change depending on the type of tissue used for the analysis (e.g., root vs. leaf). This is because a gene that is being expressed in one tissue might not be expressed in other tissues. On the contrary, molecular markers, because they are based on differences in the DNA sequence, are not environmentally influenced, which means that the same banding profiles can be expected at all times for the same genotype.

Applications

Allozymes have found many application studies in population genetics studies, measurements of out crossing rates (Erskine and Muehlbauer, 1991), (sub) population structure, and population divergence (Fréville et al., 2001). Allozymes are useful specifically at the level of nonspecific populations and closely related species, and are thus useful to study diversity in crops. They have been used, often in concert with other markers, for fingerprinting purposes (Maass and Ocampo, 1995), and diversity studies (Ronning and Schnell, 1994), to study interspecific relationships (Garvin and Weeden, 1994), the mode of genetic inheritance (Warnke et al., 1998), and allelic frequencies in germplasm collections over serial increase cycles in germplasm banks, and to identify parents in hybrids (Parani et al., 1997).

19.3 MOLECULAR MARKERS

Molecular markers are fragments of DNA with changes like mutation and variations, which can be used to detect polymorphism in a gene for a particular sequence of DNA. These DNA fragments are associated with definite loci within the complete set of DNA and can be detected by specific molecular technologies (Henry, 2012).

Molecular marker detection system is moving progressively from first generation (i.e., RFLPs, RAPDs, SSRs, and AFLPs) (Lateef, 2015) to third generation [i.e., SNPs, KASper, DArT assays, and genotyping by sequencing (GBS); Paux et al., 2010; Paux et al., 2012]. Molecular markers are heritable variations in nucleotide arrangements of DNA at the comparing position on homologous chromosome of two distinct people, which take after a basic Mendelian example of legacy. In the course of the most recent two decades, the approach of molecular markers has changed the whole situation of biosciences. DNA-based markers are flexible apparatus in the fields of scientific classification, physiology, embryology, hereditary designing, and so forth (Schlotterer, 2004). The revelation of PCR was a milestone in this exertion and ended up being an exceptional procedure that realized another class of DNA profiling markers. Table 19.2 highlights important differences of molecular markers. This encouraged the improvement of marker-based gene identification, hereditary mapping, guide-based cloning of agronomically critical qualities, hereditary assorted qualities concentrates on, phylogenetic examination, and marker helped determination of attractive genotypes, and so on (Joshi et al., 2000). Along these lines, giving new measurements to reproducing and marker-based choice that can lessen the time compass of growing new and better assortments and the fantasy of super assortments work out. These DNA markers offer a few favorable circumstances over customary phenotypic markers, as they give information that can be broken down dispassionately.

19.3.1 COMMON TERMS

Dominant marker

A marker is called dominant if only one form of the trait (which is targeted to be marked) is associated with the marker, whereas the other form of the trait is not associated with any marker. Such markers cannot discriminate between heterozygote and homozygote marker allele.

TABLE 19.2 Comparison of Molecular Markers.

Feature	RFLP	RAPD	AFLP	SSRs	SNPs
DNA required (μg)	10	0.02	0.5–1.0	0.05	0.05
DNA quality	High	High	Moderate	Moderate	High
PCR based	No	Yes	Yes	Yes	Yes
No. of polymorph loci analyzed	1–3	1.5–50	20–100	1–3	1
Ease of use	Not easy	Easy	Easy	Easy	Easy
Amenable to automation	Low	Moderate	Moderate	High	High
Reproducibility	High	Unreliable	High	High	High
Development cost	Low	Low	Moderate	High	High
Type of probes/primers	Low copy DNA or cDNA clones	10 bp random nucleotides	Specific sequence	Specific sequence	Allele-specific PCR primers
Effective multiplex ratio	Low	Moderate	High	High	Moderate to high
Marker index	Low	Moderate	Moderate to high	High	Moderate
Genotyping throughput	Low	Low	High	High	High
Primary application	Genetics	Diversity	Diversity and genetics	All purposes	All purposes

Codominant marker

A marker is designated as codominant if both forms of the trait (which is targeted to be marked) are associated with the marker. It can discriminate between heterozygote and homozygote marker allele

19.3.1.1 CHARACTERISTICS OF IDEAL DNA MARKER

1. Highly polymorphic nature: It must be polymorphic as it is polymorphism that is measured for various genetic diversity studies.
2. Codominant inheritance: Determination of homozygous and heterozygous states of diploid organisms.
3. Frequent occurrence in genome: A marker should be evenly and frequently distributed throughout the genome.
4. Selective neutral behaviors: The DNA sequences of any organism are neutral to environmental conditions or management practices.
5. Easy access (availability): It should be easy, fast, and cheap to detect.
6. Easy and fast assay.
7. High reproducibility.
8. Easy exchange of data between laboratories.

It is extremely difficult to find a molecular marker, which would meet all the above criteria. A wide range of molecular techniques are available that detect polymorphism at the DNA level. Depending on the type of study to be undertaken, a marker system can be identified that would fulfill at least a few of the above characteristics (Weising et al., 1995). Various types of molecular markers are utilized to evaluate DNA polymorphism and are generally classified as hybridization-based markers and PCR-based markers. In the former, DNA profiles are visualized by hybridizing the restriction enzyme-digested DNA, to a labeled probe, which is a DNA fragment of known origin or sequence. PCR-based markers involve in vitro amplification of particular DNA sequences or loci, with the help of specifically or arbitrarily chosen oligonucleotide sequences (primers) and a thermostable DNA polymerase enzyme. The amplified fragments are separated electrophoretically and banding patterns are detected by different methods such as staining and autoradiography. PCR is a versatile technique invented during the mid-1980s (Saiki et al., 1985). Ever since thermostable DNA polymerase was introduced in 1988, the use of PCR in research and clinical laboratories has increased tremendously. The primer sequences are chosen to allow base-specific binding to the template in reverse orientation. PCR is extremely sensitive and operates at a very high speed. Its application for diverse purposes has opened up a multitude of new possibilities in the field of molecular biology.

19.3.2 RELATIVE VALUE OF MARKERS

Morphological, isozyme, and nuclear DNA markers are inherited in a Mendelian manner. Cytoplasmic markers are maternally inherited. In any genome, the number of morphological and isozyme markers is limited compared to DNA markers which are ubiquitous and numerous. To study the inheritance of chromosome segments, it is necessary to be able to distinguish the segments inherited from each of the parents. For this, each parent must possess different alleles of the gene in question. Natural variation of conventional genes is limited. If morphological or isozyme markers are to be used for this purpose, then variants will have to be produced by mutagenesis which often gives rise to lethal mutants. In DNA marker analysis, the natural variation in the DNA sequence is made use of and no mutagenesis is required. Essentially all DNA markers have no effect on the phenotype because they are reflections of the natural variation present in the DNA sequence. Thus, a detailed genetic linkage map can be constructed utilizing only one cross and one mapping population. In contrast, only a few phenotypic markers can be maintained in a single plant because some mutant phenotypes are harmful to the plant and it is time consuming and sometimes practically impossible to assemble all such viable mutations into a single plant. DNA markers are free of pleiotropic effects, thereby allowing any number of markers to be monitored in a single population. DNA marker analysis can be carried out at any stage of the life cycle of an organism and from almost any tissue including herbarium and mummified tissue. Morphological and biochemical markers depend upon the expression of certain genes which in turn are governed by environmental conditions, tissue specificity, and development stage. Isozymes and DNA markers with the exception of RAPDs, AFLPs, and DNA amplification fingerprinting (DAF) express codominance thereby allowing the genotype at any locus to be determined in any breeding scheme. Morphological markers have alleles that interact in a dominant/recessive manner. Moreover, PCR-based markers require only a few nanograms of DNA for analysis. These properties of DNA markers make them ideal candidates for their use in linkage mapping analysis.

19.3.3 RESTRICTION FRAGMENT LENGTH POLYMORPHISMS (RFLP)

RFLP markers are known as first generation molecular markers as they were mainly used in 1980s–1990s (Jones et al., 2009). Detection of changes like insertion and deletion in genetic code with one nucleotide leads to shift in fragment size of DNA by RFLP is based on generation/removal of restriction enzyme recognition site (Tanksley et al., 1989). Advantages of RFLP markers include codominance, high reproducibility, no need of prior sequence information, and high locus specificity (Lateef, 2015).

However, recently many direct uses of RFLP markers in genetic research and plant breeding started in practice. Moreover, RFLP is a lengthy procedure, requires relatively large amounts of pure DNA, tedious experimental procedure, and each point mutations have to be analyzed separately (Edwards and Batley, 2010; Wong, 2013).

In RFLP investigation, confinement chemical processed genomic DNA is acquired by gel electrophoresis and after that smeared on nitrocellulose layer. Particular banding examples are then envisioned by hybridization (Fig. 19.1). These tests are generally species-particular single or multi-locus tests of around 0.5–3.0 kb in size, acquired from a cDNA library or a genomic library. RFLP markers were utilized interestingly as a part of the development of hereditary maps. Being co-prevailing markers, RFLP can distinguish coupling period of DNA atoms, as DNA pieces from every single homologous chromosome are identified. The real quality of RFLP markers are their high reproducibility, co-overwhelming inheritance, and great transferability between research facilities which provides locus-particular markers that permit synteny considers. Still, there are a few impediments for RFLP examination: it requires the vicinity of high amount and nature of DNA (Young et al., 1992). RFLPs can be connected in assorted qualities and phylogenetic studies going from one sample within population or species, to firmly related species. It is generally utilized as a part of quality guide ping studies in view of their high genomic wealth because of the adequate accessibility of distinctive confinement compounds and arbitrary dispersion all through the genome. Essentially, it was utilized to examine connections of firmly related taxa (Miller and Tanksley, 1990), fingerprinting apparatuses for differing qualities contemplates, and for investigations of hybridization and introgression, and additionally investigations of quality stream in the middle of yields and weeds (Desplanque et al., 1999).

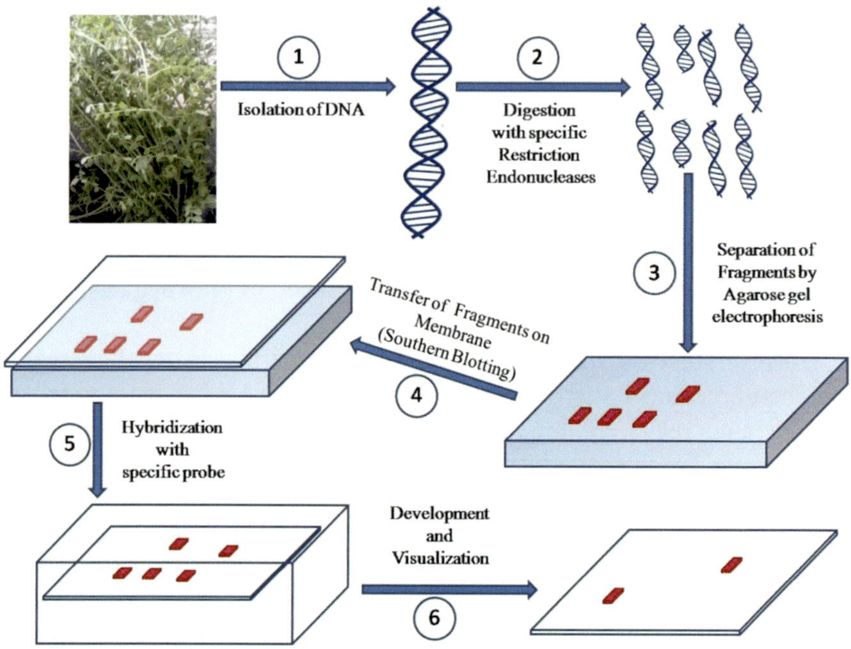

FIGURE 19.1 Overview of restriction fragment length polymorphism.

19.3.4 RANDOM AMPLIFIED POLYMORPHIC DNA (RAPD)

RAPD is based on the random amplification particularly DNA fragments by random primer set, which is a cheap and reproducible technique. Further, amplified products are analyzed and visualized by agarose gel electrophoresis (Williams et al., 1990; Varshney, 2013). The various steps of RAPD were represented in Figure 19.2.

RAPD markers have been employed to assess genetic variations and establish the phylogenetic relations between species and within species (Gupta et al., 1999). Main disadvantage of RAPD markers is it fails to detect allelic differences in heterozygotes and detect predominant markers only. Polymorphisms can be detected on the basis of band of amplified product of a certain molecular weight, without any information of heterozygosity (Edwards and Batley, 2010). Additionally, it does not carry sufficiency for whole genome mapping because of their random nature of amplification and short primer length. Moreover, these markers do not display reliable amplification patterns as well as fluctuate with the experimental conditions (Jones et al., 2009). Amplified pieces are generally from the selection

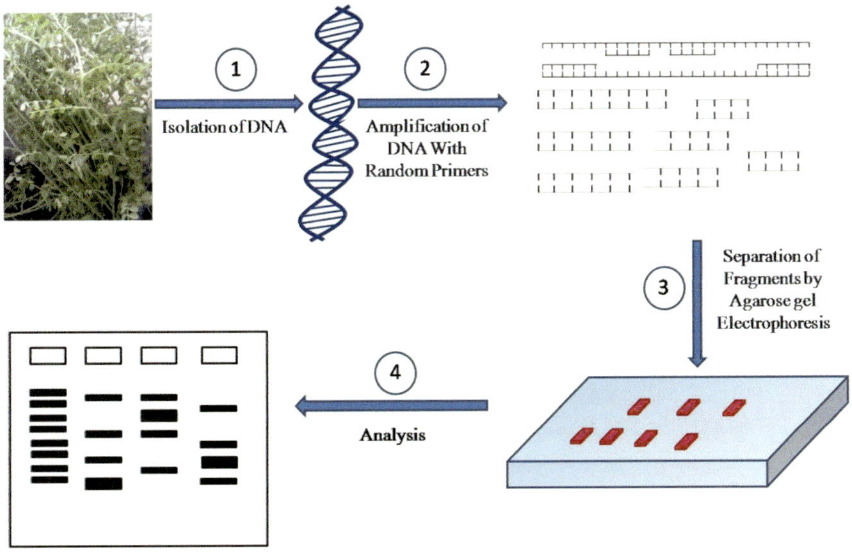

FIGURE 19.2 Steps of random amplified polymorphic DNA.

of 0.5–5 kb in size usually split up simply by agarose gel electrophoresis. Polymorphisms are usually accepted mainly on account of alternative inside the primer annealing sites, but it is produced simply by period variances inside the amplified series in between primer annealing sites (Williams et al., 1990). The main attributes of RAPDs are usually: less time consuming, easy assay, and also lower portions associated with template DNA are important, normally 5–50 ng for each reaction. As a result of professional accessibility to arbitrary primers, no sequence information with regard to primer building is needed. In addition, RAPDs employ an excessive genomic plethora and so are aimlessly dispersed through the genome. The main problem associated with RAPDs is lower reproducibility that is why remarkably standardized trial and error treatments are expected due to its tenderness for the reaction ailments. RAPD analyses normally involve purified, excessive molecular weight DNA, and also measures are expected to stop contamination associated with DNA examples as the quick random primers utilized may well increase DNA pieces in a number of microorganisms. Permanently, the purely natural troubles associated with reproducibility create RAPDs unsuitable with regard to transference or maybe assessment associated with results between research groups working in identical kinds and also subjects. Concerning most other multilocus strategies, RAPD markers will not be locus-specific, bands single profiles cannot be interpreted in terms of loci and also alleles (dominance associated with markers), and also similar sized pieces most likely are not homologous.

RAPDs at the individual level (e.g., hereditary identity) studies firmly related species and quality mapping studies to fill crevices not secured by different markers (Williams et al., 1990). Variations of the RAPD system incorporate arbitrarily primed (AP)-PCR which uses longer subjective preliminaries than RAPDs, and DAF that uses shorter, 5–8 bp ground works to create a bigger number of pieces. Multiple arbitrary amplicon profiling (MAAP) is the aggregate term for strategies utilizing single subjective preliminaries.

19.3.5 SIMPLE SEQUENCE REPEATS (SSRS)

SSRs also known as microsatellites were recognized and presented a preference for many genetic researchers for its low to high-throughput approaches. They are random tandem repeats of short nucleotide motifs (2–6 bp) (Edwards and Batley, 2010). SSRs are high-frequency polymorphic sequences present in animals and plants (Kalia et al., 2011b), and utilized to study relationship between inherited traits within a species (Dunn et al., 2005). Microsatellite markers are the sequences from noncoding regions of genome, that is, bacterial artificial chromosomes (BACs) and genomic survey sequences (GSSs).

Polymorphism is based on the variation in the number of repeats in different genotypes (Ellegren, 2000). In recent years, SSR markers can easily be developed in silico due to the availability of large-scale gene expressed sequence tag (EST) sequence information for many plant species. Since EST sequencing projects have provided sequence data that are available in online databases and can be scanned for identification of SSRs (Varshney et al., 2005).

The high degree of polymorphism as compared to RFLPs and RAPDs, their codominant nature, and locus specific make them the markers of choice for a diversity of purposes including practical plant breeding. Therefore, SSRs have become a marker of choice for an array of applications in plants due to extensive genome coverage and hyper variable nature (Kalia et al., 2011a).

19.3.6 INTER SIMPLE SEQUENCE REPEATS (ISSRS)

It includes the enhancement of DNA fragments present at an amplifiable separation in the middle of two indistinguishable microsatellite rehash locales situated in inverse bearings. The procedure utilizes microsatellites as basis as a part of a solitary preliminary PCR response focusing on numerous genomic loci to increase basically ISSR of diverse sizes. The microsatellite repeats utilized for ISSRs can be dinucleotide, trinucleotide, tetranucleotide, or pentanucleotide. This primer utilized can be either unanchored (Meyer et al., 1993; Gupta et al., 1994) or all the more for the most part secured at 3′ or 5′ end with one to four ruffian bases reached out into the flanking groupings (Zietkiewicz et al., 1994). ISSRs utilize longer primer (15–30 mers) when contrasted with RAPD preliminaries (10 mers), which allow the resulting utilization of a high annealing temperature prompting higher stringency. The annealing temperature relies on upon the GC substance of the preliminary utilized and ranges from 45°C to 65°C. The opened up items are generally 200–2000 bp long and can be identified by both agarose and polyacrylamide gel electrophoresis. ISSRs show the specificity of microsatellite markers; however do not need succession data for groundwork union getting a charge out of the benefit of arbitrary markers (Joshi et al., 2000). The system is straightforward, snappy, and the utilization of radioactivity is not vital. ISSR markers are randomly appropriated all through the genome and as a rule demonstrate high polymorphism in spite of the fact that the level of polymorphism has been appeared to change with the location strategy utilized. Detriments incorporate the likelihood of nonhomology of similar sized pieces. In addition, ISSRs, as RAPDs, may have reproducibility issues. ISSR examination can be connected in studies including hereditary character, parentage, clone and strain recognizable proof, and taxonomic investigations of firmly related species and in addition in quality mapping studies (Gupta et al., 1994; Zietkiewicz et al., 1994).

19.3.7 SINGLE-STRAND CONFORMATION POLYMORPHISM (SSCP)

This is an intense and fast system for quality examination especially for identification of point transformations and writing of DNA polymorphism. The technique depends on the way that the electrophoretic versatility of single-strand DNA relies on upon the auxiliary structure of the particle, which changed fundamentally because of transformation. In this way, SSCP gives a strategy to distinguish nucleotide variety among DNA tests without succession responses (Orita et al., 1989). Related strategies to SSCP are denaturing gradient gel electrophoresis (DGGE) taking into account twofold stranded DNA, changed over to single-stranded DNA in an undeniably denaturing physical environment amid gel electrophoresis and thermal gradient gel electrophoresis (TGGE) which utilizes temperature slopes to denature twofold-stranded DNA electrophoresis. Focal points of SSCP incorporate the codominance of alleles and the low amounts of format DNA required as the method is PCR based. However, the disadvantages are the need of succession information to outline PCR preliminaries, the need of profoundly institutionalized electrophoretic conditions keeping in mind the end goal to acquire reproducible results, and some of the time changes may stay undetected and thus nonappearance of transformation cannot be demonstrated. SSCP, a potential instrument for high throughput DNA polymorphism, valuable in the location of heritable human diseases, distinguish changes in qualities utilizing quality arrangement data for primer development. In plants, it is not very much created in spite of the fact that its application in separating descendants can be misused once suitable primers are intended for agronomically vital qualities (Kesawat and Das Kumar, 2009).

19.3.8 AMPLIFIED FRAGMENT LENGTH POLYMORPHISM (AFLPS)

AFLPs are PCR-based markers; merely RFLPs can be analyzed by selective PCR amplification of DNA restriction fragments. Such a marker could be a multilocus marker technique that mixes the techniques of selective PCR amplification of restriction and fragments restriction digestion and it is attainable to DNA from any source (Vos et al., 1995).

This method involves different steps which starts with digestion of DNA oligonucleotide adapters are ligated to each ends of the ensuing restriction fragments and genomic DNA is digestible. Afterwards, the fragments are united by selection amplified using the adapter and situation sequences as primer binding sites for following PCR reactions. As the 3′ ends of the primers extend into the restriction fragments by (1–4 bp), solely those fragments are amplified, whose ends are fully complementary to the 3’ ends of the selective primers. Therefore, solely an exact quantity of the restriction fragments is amplified. At the end, amplified fragments are subjected to gel electrophoresis and bands can be visualized by silver staining, autoradiography, or fluorescence, resulting in a unique reproducible fingerprint for each individual (Vos et al., 1995; Nicod and Largiader, 2003) (Fig. 19.3).

This method has many benefits such as cost-efficiency, since a single assay allows detection of a large number of coamplified restriction fragments and it requires moderate quantities of DNA. Moreover, higher levels of polymorphisms compared with RFLPs, can be detected also by AFLPs, have much higher multiplex ratio and better reproducibility than RAPDs (Gupta et al., 1999; Edwards and Batley, 2010).

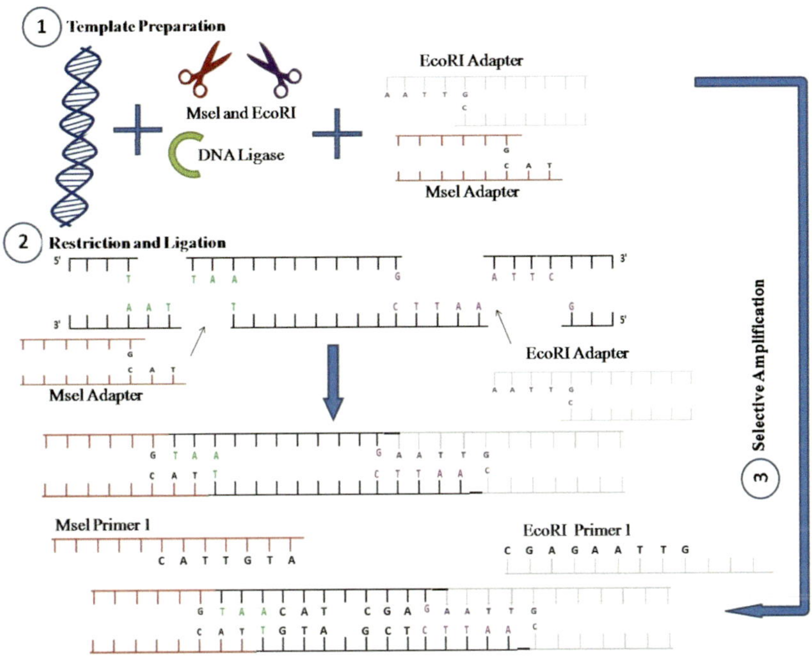

FIGURE 19.3 Steps of amplified fragments length polymorphism. (Reprinted from Mueller, Ulrich G. et al., AFLP Genotyping and Fingerprinting, Trends in Ecology & Evolution 14(10), 389–394. © 1999 with permission from Elsevier.)

However, AFLP assays have some limitations like, it requires the use of polyacrylamide gels for detection and it needs a greater technical skill. Moreover, because of multifarious banding arrangement most AFLP markers are dominant. In some cases, the scoring of AFLP polymorphisms as codominant marker *loci* is possible, for the reason that diploid homozygous individuals cause more intense peak than heterozygous individuals (Meudt and Clarke, 2007).

Specific fragment length amplification (SFLA) and selective restriction fragment amplification (SRFA) are equivalent words now and again used to allude to AFLPs. A variety of the AFLP system is known as selectively amplified microsatellite polymorphic locus (SAMPL). This innovation opens up microsatellite loci by utilizing a solitary AFLP preliminary as a part of combination with a groundwork reciprocal to compound microsatellite arrangements, which do not require earlier cloning and characterization (Morgante and Vogel, 1999). SAMPL is viewed as more relevant for intra-particular than between particular studies because of frequent invalid alleles. The quality of AFLPs incorporates its high genomic plenitude, significant reproducibility, the era of numerous instructive groups per response, extensive variety of functions, and the way that no arrangement in turn for foundation development are required (Saal and Wricke, 2002). AFLPs can be dissected on programmed sequencers; however, programming issues concerning the scoring of AFLPs are experienced on a few frameworks. For instance, on account of an insertion between two limitations locales, the opened up DNA part brings about expanded band size. This will be deciphered as the passing of a little band and in the meantime as the increase of a bigger band. It is vital for the examination of hereditary relations on the grounds that it would improve the heaviness of nonfree groups contrasted with alternate groups. AFLPs can be connected in studies including hereditary personality, fingerprinting, recognizable proof of clones and cultivars, and phylogenetic investigations of closely related species. Their high genomic plenitude and by and large irregular dissemination all through the genome make AFLPs a generally esteemed technology for quality mapping studies (Vos et al., 1995).

19.3.9 SINGLE NUCLEOTIDE POLYMORPHISMS (SNPS)

SNP is a single nucleotide polymorphism in which single base difference between two DNA sequences of samples, SNPs are typically biallelic and generated due to substitutions/point mutations (transversion and transition) or as a result of deletion/insertion of nucleotides (Xu, 2010).

SNPs provide the simplest and ultimate form of molecular markers as a single nucleotide base is the smallest unit of inheritance, and therefore they can provide a great marker density. High density of SNPs over other markers makes it advantageous to find more probability of polymorphisms in a target gene at best closely linked to a locus of interest (Ganal et al., 2009). Generally, linkages are lost when a marker is applied to other population with different recombination patterns. Typically, SNP frequencies are in a range of one SNP 100–300 bp in plants. SNPs are present within coding, noncoding regions or in the intergenic regions between genes at different frequencies in different chromosomal segments (Edwards et al., 2007; Edwards and Batley, 2010).

Most hybridization systems are gotten from the Dot Blot, in which DNA to be tried (either genomic, cDNA, or a PCR response) altered on a layer and is hybridized with a test, for the most part an oligonucleotide. Be that as it may, hybridization procedures are error inclined and require attentive test exactness and hybridization conventions. The most recent change strategies, utilization of DNA chips (gathering of infinitesimal DNA spots connected to a strong surface, for example, glass, plastic, or silicon chips) on which the tests are straightforwardly integrated utilizing a parallel technique including covers and photolithography (Pease and Solas, 1994). To exploit new ASO test groups for SNP typing, it is important to utilize recognition routines which genius vide high accuracy, affectability, and results. Groundwork expansion depends on the capacity of DNA polymerase to consolidate particular deoxy-ribonucleotides correlative to the grouping of the format DNA.

SNP discovery methods are broadly categorized into four segments: hybridization with allele-specific oligonucleotide probes, oligonucleotide ligation, enzymatic cleavage, and single nucleotide primer extension (Syvanen, 2005; Edwards et al., 2007; Gupta et al., 2008). In principle, the SNP methods show differences between a probe of known sequence and a target DNA containing the SNP site. The target DNA sections are typically PCR products and mismatches with the probe reveal SNPs within the amplified target DNA segment. The mismatched DNA can be sequenced to identify SNP polymorphisms (Weising et al., 2005).

Nowadays, it is very gainful and easier to quickly identify a large number of SNPs within limited time frame in any plant species. This is due to the emergence of the third generation DNA sequencing technologies. The advantages of this new sequence technology are expected to further reduction in sequencing costs to $1 per mega base compared to $60, $2, and $1 expected costs for sequences generated by next generation sequencing (Thudi et al., 2012).

19.3.10 THE KBIOSCIENCE COMPETITIVE ALLELE-SPECIFIC PCR (KASPAR)

KASPar enables researchers and breeders to analyze a small number of targeted SNPs in a large number of samples. This makes KASPar a cost-effective, simple, and flexible genotyping system; since the assays can be modified with a range of DNA samples, it additionally does not require a hybridization step; as an alternative it includes real-time detection of the product (McCouch et al., 2010). The chemistry of KASPar assays involves one common reverse primer and two competitive allele-specific tailed forward primers. This process involves SNP detection through fluorescence resonance energy transfer (FRET) to determine the alleles at a specific locus; this assay system relies on the discrimination allele-specific PCR for a given SNP (McCouch et al., 2010).

This assay for the target SNPs has been developed and used for genotyping commercially by Kbioscience UK (http://www.kbioscience.co.uk/). They have improved the performance of the detection platform by incorporating a 5′–3′ exonuclease cleaved *Taq* DNA polymerase and a homogeneous FRET detection system.

The two allele-specific primers of a SNP are designed so that they incorporate with a unique (18) bp tail to the respective allele-specific products, which in later cycles allow incorporation of allele-specific fluorescent labels to the PCR products (with the help of corresponding labelled primers) (http://www.kbioscience.co.uk/).

SNPs are not sequenced in KASPar assays, instead SNP flanking sequences are already known while developing different types of genotyping assays (for instance, illumina) can easily be used for primer design (one common and two allele-specific primers) (Gupta et al., 2008).

Though KASPar genotyping assays have been introduced very recently, it is used for large number of commercial species. In maize, a set of 695 SNPs from a total of 13,882 GG-validated SNPs were selected and converted into KASPar genotyping assay with a success rate of 98% (Mammadov et al., 2012), similarly, in wheat, to construct linkage map containing several hundred SNPs (Allen et al., 2011).

19.3.11 GENOTYPING BY SEQUENCING (GBS)

Resequencing for genome-wide surveys of genetic diversity became reasonable because of high throughput NGS platforms (Wheeler et al., 2008). Independent estimation of genetic diversity across the genome in both coding and noncoding regions can be determined through bioinformatics approach. Furthermore, it permits detection genetic variation; this detection capability contains SNPs, small indels and large mega-base scale indels (Kiani and Akhunov, 2013).

This assay enables the use of restriction enzymes for reducing the complexity of genomes and targeted sequencing of reduced proportions, so each marker can be sequenced at high coverage at low cost and high accuracy in different samples/individuals (Mir Hiremath et al., 2013).

The principle focus for building GBS libraries depended on lessening genome many-sided quality with restriction enzymes, which may reach critical locales of the genome that are inaccessible to succession catch approaches (Oeveren et al., 2011).

The system has been exhibited with grain (Oregon Wolfe Barley) at the recombinant ingrained lines populaces and maize (IBM) where around (25,000–200,000) succession labels were mapped separately.

With this technique, species that do not have a complete genome succession can have a reference guide settled around the confinement locales, which should be possible during the time spent example genotyping. This framework has been balanced for lessening missing information focuses and enhanced SNP calls (Poland et al., 2012).

The redundant and discretionary DNA markers are of decision in genotyping of cultivars. Microsatellites like $(CT)_{10}$, $(AAGG)_4$, $(AAT)_6$, $(GATA)_4$, $(CAC)_5$ (Gupta et al., 1994), and minisatellites (Ramakrishna et al., 1995) have been utilized in DNA fingerprinting for the discovery of hereditary variety, cultivar ID, and genotyping (Yang et al., 1994; Sant et al., 1999). These data are valuable for measurement of hereditary differing qualities, portrayal of increases in plant germplasm accumulations, and taxonomic studies (Morjane et al., 1994). In plants, the first use of microsatellites was for cultivar recognizable proof and was later used to genotype unequivocally different materials like rice, wheat, grapevine, soybean, and so forth. This is vital particularly for security of restrictive germplasm. Likewise, microsatellite markers have additionally been profitable in family examination as they speak to a solitary locus. The multiallelism of these markers encourages similar allelic variability identification dependably over an extensive variety of germplasm and permits individuals to be universally genotyped, with the goal that quality stream and paternity can be set up. It can be utilized as a symptomatic marker where male and female plants do not demonstrate any sex-particular morphological contrasts until blooming. Thus, previously a randomly enhanced (RAPD) DNA marker for pseudo-autosomal plant sex chromosome in *Silene dioica* (L.) was recognized (Di Stilio et al., 1998).

19.3.12 PYRAMIDING MULTIPLE LOCI AND FAVORABLE ALLELES

Gene pyramiding is characterized as a get together of numerous attractive genes which can be joined into a solitary genotype from different folks. This is eluded as one of the significant utilizations of marker collaborator choice, as gene pyramiding by means of traditional plant rearing is troublesome, if not incomprehensible (Gupta and Mir, 2010).

The systems for pyramiding great alleles can be utilized as a part of the same approach to amass QTL controlling diverse attributes. A primary distinction in the model is that alleles at diverse characteristic loci to be gathered may have diverse good bearings, for example, negative alleles are best for a few characteristics yet positive alleles are positive for others. Thus, to meet reproducing goals, one may need to consolidate the positive QTL alleles of a few attributes with the negative alleles of others (Yunbi et al., 2013). Determination for numerous qualities may be finished in one cycle if the populace size is sufficiently extensive to permit alluring samples to consolidate distinctive qualities. By and by, the quantity of characteristic loci that can be controlled in one cycle is limited on the grounds that the populace size required covering the recombinants increments exponentially with the increment of the quantity of characteristics/loci (Yunbi et al., 2013). To beat this constraint, a system was proposed by Bonnett and Rebetzke (Bonnett et al., 2005). In this technique, people have been chosen by all objective markers for both heterozygous and homozygous structures to pick up a subset of populace that contain higher frequencies of the objective alleles so that to get the homozygotes at the objective loci, a littler populace size is required in the accompanying era. In down to earth, what must be considered when applying such techniques pyramiding needs to be rehashed after every intersection, since the pyramided resistance qualities are isolating in the offspring (Werner et al., 2005).

19.4 MARKER-ASSISTED SELECTION (MAS)

MAS procedure and theoretical and practical considerations.

MAS refers to such a breeding procedure in which DNA marker detection and selection are integrated into a traditional breeding program. Taking a single cross as an example, the general procedure can be described as follows:

1. Select parents and make the cross, at least one (or both) possesses the DNA marker allele(s) for the desired trait of interest.
2. Plant F1 population and detect the presence of the marker alleles to eliminate false hybrids.
3. Plant segregating F2 population, screen individuals for the marker(s), and harvest the individuals carrying the desired marker allele(s).

4. Plant F2:3 plant rows, and screen individual plants with the marker(s). A bulk of F3 individuals within a plant row may be used for the marker screening for further confirmation in case needed if the preceding F2 plant is homozygous for the markers. Select and harvest the individuals with required marker alleles and other desirable traits.
5. In the subsequent generations (F4 and F5), conduct marker screening and make selection similarly as for F2:3s, but more attention is given to superior individuals within homozygous lines/rows of markers.
6. In F5:6or F4:5 generations, bulk the best lines according to the phenotypic evaluation of target trait and the performance of other traits, in addition to marker data.
7. Plant yield trials and comprehensively evaluate the selected lines for yield, quality, resistance, and other characters of interest.

In MAS, phenotypic evaluation and selection is still very helpful if conditions permit to do so, and even necessary in cases when the QTLs selected for MAS are not so stable across environments and the association between the selected markers and QTLs is not so close. Moreover, one should also take the impact of genetic background into consideration. The presence of a QTL or marker does not necessarily guarantee the expression of the desired trait. QTL data derived from multiple environments and different populations help a better understanding of the interactions of QTL × environment and QTL × QTL or QTL × genetic background, and thus help a better use of MAS. In addition to genotypic (markers) and phenotypic data for the trait of interest, a breeder often pays considerable attention to other important traits, unless the trait of interest is the only objective of breeding.

There are several indications for adoption of molecular markers in the selection for the traits of interest in practical breeding. The situations favorable for MAS include:

The selected character is expressed late in plant development, like fruit and flower features or adult characters with a juvenile period (so that it is not necessary to wait for the plant to become fully developed before propagation occurs or can be arranged).

- The target gene is recessive (so that individuals which are heterozygous positive for the recessive allele can be selected and/or crossed to produce some homozygous offspring with the desired trait).
- Special conditions are required in order to invoke expression of the target gene(s), as in the case of breeding for disease and pest

resistance (where inoculation with the disease or subjection to pests would otherwise be required), or the expression of target genes is highly variable with the environments.

- The phenotype of a trait is conditioned by two or more unlinked genes. For example, selection for multiple genes or gene pyramiding may be required to develop enhanced or durable resistance against diseases or insect pests.

19.4.1 MARKER-ASSISTED RECURRENT SELECTION (MARS)

Marker-helped intermittent determination was proposed in around 1990s, which utilizes markers at every era to target all attributes of significance and for which hereditary data can be accomplished. Folks contribute diverse good alleles when the QTL mapping is directed in view of a biparental populace. Along these lines, the ideal genotype is a mosaic of chromosomal fragments delivered by recombination between the two folks (Gupta et al., 2010).

MARS alludes to the change of a (F2) populace by one cycle of marker-helped determination (for case, taking into account information marker scores and phenotypic) took after ordinarily by a few cycles of marker-based example of QTLs, all QTLs conveying good alleles from different folks. After a few progressive eras of intersections, it may be conceivable to draw near to the perfect genotype. Furthermore, this framework can begin with no QTL data, and determination can be founded on huge marker-attribute affiliation set up amid the MARS process (Gupta et al., 2010).

Studies have uncovered that, in aggregating good alleles, MARS was better than phenotypic choice (Wang et al., 2009). The convenience of including former information of QTL under hereditary models has been examined that included QTL number, quality impacts, heritability, epistasis, and linkages. It is inferred that with known QTL, MARS is most great for qualities controlled by a vast number of QTL (Bernardo and Charcosset, 2006).

A substantial number of monogenic and polygenic loci for different qualities have been distinguished in numerous plants and are as of now being misused by reproducers and subatomic researchers together, in order to make the fantasy of marker-helped choice materialize. Labeling of valuable genes like the ones in charge of giving imperviousness to plant pathogens, combination of plant hormones, dry season resilience, and an assortment of other vital formative pathway qualities is a noteworthy target. Such labeled genes can likewise be utilized for recognizing the vicinity of helpful qualities in the new genotypes produced in a cross program or by different systems like change. The principal reports on gene labeling were from tomato (Williamson et al., 1994), benefiting the methods for ID of markers connected to genes included in a few characteristics like water use proficiency (Martin et al., 1989), imperviousness to *Fusarium oxysporum* (12 traits) (Sarfatti et al., 1989), leaf invulnerability to rust traits LR9 and LR24 (Schachermayr et al., 1995), and root hitch nematodes. Previous report (Xiao et al., 1998) has demonstrated the utility of RFLP markers in recognizing the attribute enhancing QTL alleles from wild rice relative *O. rufipogon*. Allele's specific related preliminaries have additionally shown their utility in genotyping of allelic variations of loci that result from both size contrasts and point transformations. A percentage of the bona fide cases of this are the waxy quality locus in maize (Shattuck-Eidens et al., 1991), the Glu D1 complex locus connected with bread making quality in wheat (D'Ovidio and Anderson, 1994), the Lr1 leaf imperviousness to rust locus in wheat (Feuillet et al., 1995), the Gro1, and H1 alleles presenting imperviousness to the root pimple nematode *Globodera rostochiensis* in potato (Niewöhner et al., 1995), and allele-particular enhancement of polymorphic locales for identification of fine mold resistance loci in grains. Various different attributes have been labeled utilizing ASAPs as a part of tomato, lettuce, and so forth (Paran and Michelmore, 1993). STMS markers are utilized as potential analytic markers for critical attributes in plant rearing projects, for example, (AT) 15 rehash situated inside of a soybean warmth stun protein quality which is around 0.5 cM from (Rsv), a quality giving imperviousness to soybean mosaic infection (Yu et al., 2002). A few resistance qualities including shelled nut mottle infection (Rpv), phytopthora (Rps3), and Japanese root tie nematode are bunched in this locale of the soybean genome. Like RFLPs, STMS, and ASAPs, subjective markers, RAPDs have additionally assumed imperative part in immersion of the hereditary linkage maps and quality labeling. Its utilization in mapping is particularly vital in frameworks, where RFLPs neglected to uncover much polymorphism. One of the first employments of RAPD markers in immersion of hereditary maps was accounted (Williams et al., 1990).

Apart from mapping and labeling of genes, an essential utility of RFLP markers has been seen in recognizing quality introgression in a backcross rearing and synteny mapping among firmly related animal types (Gale and Devos, 1998). Comparative utility of STMS markers has been watched for solid pre-s race in a marker-helped choice backcross plan. Other than particular markers, DAMD-based DNA fingerprinting in wheat has likewise been helpful for checking backcross-intervened genome introgression in hexaploid wheat (Somers et al., 1996).

19.4.2 MARKER-ASSISTED BACKCROSSING

Marker-assisted or marker-based backcrossing (MABC) is regarded as the simplest form of MAS, and at present it is the most widely and successfully used method in practical molecular breeding. MABC aims to transfer one or a few genes/QTLs of interest from one genetic source (serving as the donor parent (DP) and maybe inferior agronomically or not good enough in comprehensive performance in many cases) into a superior cultivar or elite breeding line (serving as the RP) to improve the targeted trait. Unlike traditional backcrossing, MABC is based on the alleles of markers associated with or linked to gene(s)/QTL(s) of interest instead of phenotypic performance of target trait. The general procedure of MABC is as follow, regardless of dominant or recessive nature of the target trait in inheritance:

1. Select parents and make the cross, one parent is superior in comprehensive performance and serves as RP, and the other one used as DP should possess the desired trait and the DNA markers allele(s) associated with or linked to the gene for the trait.
2. Plant F1population and detect the presence of the marker allele(s) at early stages of growth to eliminate false hybrids, and cross the true F1plants back to the RP.
3. Plant BCF1 population, screen individuals for the marker(s) at early growth stages, and cross the individuals carrying the desired marker allele(s) (in heterozygous status) back to the RP. Repeat this step in subsequent seasons for two to four generations, depending upon the practical requirements and operation situations as discussed below.
4. Plant the final backcrossing population (e.g., BC4 F1), and screen individual plants with the marker(s) for the target trait and discard the individuals carrying homozygous markers alleles from the RP. Have the individuals with required marker allele(s) selfed and harvest them.
5. Plant the progenies of backcrossing selfing (e.g., BC4 F2), detect the markers and harvest individuals carrying homozygous DP marker allele(s) of target trait for further evaluation and release.

The efficiency of MABC depends upon several factors, such as the population size for each generation of backcrossing, marker-gene association or the distance of markers from the target locus, number of markers used for target trait and RP background, and undesirable linkage drag. Based on simulations of 1000 replicates, Hospital (2003) presented the expected results of a typical MABC program, in which heterozygotes were selected at the target locus in each generation, and RP alleles were selected for two flanking markers on target chromosome each located 2 cm apart from the target locus and for three markers on nontarget chromosomes. As shown in Table 19.2, a faster recovery of the RP genome could be achieved by MABC with combined foreground and background selection, compared to traditional backcrossing. Therefore, using markers can lead to considerable time savings compared to conventional backcrossing (Frisch et al., 1999; Collard et al., 2005).

19.4.3 APPLICATION OF MABC

Success in integrating MABC as a breeding approach lies in identifying situations in which markers offer noticeable advantages over conventional backcrossing or valuable complements to conventional breeding effort. MABC is essential and advantageous.

1. Phenotyping is difficult and/or expensive or impossible.
2. Heritability of the target trait is low.
3. The trait is expressed in late stages of plant development and growth, such as flowers, fruits, seeds, etc.
4. The traits are controlled by genes that require special conditions to express.
5. The traits are controlled by recessive genes.
6. Gene pyramiding is needed for one or more traits.

19.5 CONCLUSIONS

In view of the above, as it is apparent from the discourse over, diverse levels of throughput are accessible. Along these lines, a suitable marker framework can be chosen in light of the need. The prior sorts of subatomic markers incorporate nonpartisan markers. RAPDs and AFLPs have likewise been generally utilized as a part of hereditary differences considers and quality mapping. Both innovations are especially valuable when there is a need to test loci over the whole genome. Their absence of reproducibility, predominant nature of RAPDs contrasted and AFLPs, and the need of specificity in both cases are restricting elements for their application in exact MAS reproducing methodologies (Edwards et al., 2007).

As of late, the accessibility of entire genome sequences of some crops and the arrangement data has likewise prompted the improvement of another era of markers, for example, KASpar, SNP measures. SNP markers which are transferable through diverse genotyping sciences will offer as adaptable choice devices for plant raisers in (MAS). Likewise, the center ought to be put in the distinguishing proof of SNPs in the same number of qualities as conceivable and the parallel investigation of a wide range of lines (Mammadov et al., 2012). It is likely that change of complex characteristics will rely upon the capacity to control qualities, which have minor impacts, and show association with one another.

Genotype by sequencing is by all accounts the best approach for development of complex characteristics. On the other hand, the parameters and conditions incorporated into reproductions may not completely mirror the complex circumstances of differing plant rearing projects, the hereditary increase per unit time and cost that has been accomplished in reproductions should be upheld by the long haul choice reaction by contrasting and other rearing methodologies. What is more, this methodology can actually prompt the revelation of a large number of SNPs in one single examination. In addition; it can be utilized of those plants that do not have the reference genome accessible (Poland et al., 2012; Yunbi et al., 2013).

SNPs appear to be extremely energizing markers yet costly, so they are unrealistic to be taken up by the national rural frameworks and colleges in creating nations. The utilization of EST-SSR and EST-RFLP are appropriate just for species which have been widely sequenced some time recently. In this manner, RFLP, SSR, RAPD, AFLP, and ISSR are the main markers that could be utilized for an extensive variety of uses in plants. The expanding accessibility of grouping information for more products by means of entire genome sequencing tasks, and access to EST databases, empowers the improvement of markers focusing on coding locales of the genome or even particular qualities (Kesawat and Das Kumar, 2009). Mechanical improvements keep on expanding by vast scale genotyping of hereditary assets. Allele mining, related hereditary qualities, and near genomics are promising new ways to deal with and get knowledge of the association and variety of genes that influence applicable phenotypic attributes. These improvements misused by consolidating mastery from a few orders, including subatomic hereditary genes, measurements, bioinformatics, and so forth (Kesawat and Das Kumar, 2009).

KEYWORDS

- **allele**
- **crops**
- **marker-assisted**
- **genotype**
- **DNA**

REFERENCES

Allen, A. M.; Barker, G. L.; Berry, S. T. Transcript-specific, Single-nucleotide Polymorphism Discovery and Linkage Analysis in Hexaploid Bread Wheat (*Triticum aestivum L.*). *Plant Biotechnol.* **2011,** *9*, 1086–1099.

Baenziger, P. S.; Russell, W. K.; Graef, G. L.; Campbell, B. T. Improving Lives. *Crop Sci.* **2006,** *46*, 2230–2244.

Baes, P.; Cutsem, P. Electrophoretic analysis of isozymes Eleven Systems and Their Possible use as Biochemical Markers in Breeding of Chicory (*Cichorium intybus L.*). *Plant Breed.* **1993,** *100*, 16–23.

Bernardo, R.; Charcosset, A. Usefulness of Gene Information in Marker-Assisted Recurrent Selection: A Simulation Appraisal. *Crop Sci.* **2006,** *46*, 614–621.

Berry, A.; Kreitman, M. Molecular Analysis of an Allozyme Cline: Alcohol Dehydrogenase in *Drosophila melanogaster* on the East Coast of North America. *Genetics* **1993,** *134*, 3869–3893.

Bevan, M.; Flavell, R.; Chilton, M. A Chimaeric Antibiotic Resistance Gene as a Selectable Marker for Plant Cell Transformation. *Nature* **1983,** *304*, 184–187.

Bhat, Z. A.; Dhillon, W. S.; Rashid, R.; Bhat, J.; Dar, W. A.; Ganaie, M. Y. The Role of Molecular Markers in Improvement of Fruit Crops. *Not. Sci. Biol.* **2010,** *2*, 22–30.

Bonnett, D. G.; Rebetzke, G. J.; Spielmeyer, W. Strategies for Efficient Implementation of Molecular Markers in Wheat Breeding. *Mol. Breed.* **2005,** *15*, 75–85.

Budak, H., Bölek, Y.; Tevrican Dokuyucu, A. A. Potential Uses of Molecular Markers in Crop Improvement. *KSU. J. Sci. Eng.* **2004,** *7*, 75–79.

Vallejos, C. E. Enzyme AcAtivity Staining. In *Isozymes in Plant Genetics and Breeding*. Tanksley, S.D. and Orton, T.J. Elsevier: Amsterdam, **1983.** 469–515.

Collard, B.; Jahufer, M.; Brouwer, J.; Pang, E. An Introduction to Markers, Quantitative Trait loci (QTL) Mapping and Marker-assisted Selection for Crop Imrovement: The Basic Concepts. *Euphytica* **2005,** *142*, 169–196.

Collard, B. C. Y.; Mackill, D. J. Marker-assisted Selection: An Approach for Precision Plant Breeding in the Twenty-first Century. *Philos. Trans. R. Soc. Lond. B. Biol. Sci.* **2008,** *363*, 557–572.

D'Ovidio, R.; Anderson, O. D. PCR Analysis to Distinguish between Alleles of a Member of a Multigene Family Correlated with Wheat Bread-making Quality. *Theor. Appl. Genet.* **1994,** *88*, 759–763.

Datta, D.; Gupta, S.; Chaturvedi, S. K.; Nadarajan, N. *Molecular Markers in Crop Improvement*. Indian Institute of Pulses Research: Kanpur. **2011,** pp 1–49.

Delannay, X.; Bauman, T.; Buettner, M. J.; Coble, H. D.; Defelice, M. S.; Derting, C. W.; Diedrick, T. J.; Padgette, S. R. Yield Evaluation of a Glyphosate-Tolerant Soybean Line After Treatment with Glyphosate. *Crop. Sci.* **1995,** *35*, 1461–1467.

Desplanque, B.; Boudry, P.; Broomberg, K. Genetic Diversity and Gene Flow Between Wild, Cultivated and Weedy Forms of Beta vulgaris L. (Chenopodiaceae), Assessed by RFLP and Microsatellite Markers. *Theor. Appl. Genet.* **1999,** *98*, 1194–1201.

Di Stilio, V. S.; Kesseli, R. V.; Mulcahy, D. L. A Pseudoautosomal Random Amplified Polymorphic DNA Marker for the Sex Chromosomes of Silene dioica. *Genetics* **1998,** *149*, 2057–2062.

Dudley, J.; Maroof, M.; Called, G. Molecular Markers and Grouping of Parents in Maize Breeding Programs. *Crop. Sci.* **1991,** *31*, 718–723.

Dunn, G.; Hinrichs, A. L.; Bertelsen, S. Microsatellites versus Single-nucleotide Polymorphisms in Linkage Analysis for Quantitative and Qualitative Measures. *BMC. Genet.* **2005,** *6*(Suppl 1):S122.

Edwards, D.; Batley, J. Plant Genome Sequencing: Applications for Crop Improvement. *Plant. Biotech.* **2010,** *8*, 2–9.

Edwards, D.; Forster, J. W.; Chagné, D.; Batley, J. What Are SNPs? In *Association Mapping in Plants*; Oraguzie, N. C., Rikkerink, E. H. A., Gardiner, S. E., Silva, H. N. De., Eds.; Springer-Verlag: New York, **2007**. 41–52.

Edwards, M.; Stuber, C.; Wendel, J. Molecular-marker-facilitated Investigations of Quantitative-trait loci in Maize. I. Numbers, Genomic Distribution and Types of Gene Action. *Gentics* **1987,** *116*, 113–125.

Ellegren, H. Microsatellite Mutations in the Germline: Implications for Evolutionary Inference. *Trends. Genet.* **2000,** *16*, 551–558.

Erskine, W.; Muehlbauer, F. Allozymes and Morphological Variability, Outcrossing Rate and Core Collection Formation in Lentil Germplasm. *Theor. Appl. Genet.* **1991,** *83*, 119–125.

Feuillet, C.; Messmer, M.; Schachermayr, G.; Keller, B. Genetic and Physical Characterization of the LR1 Leaf Rust Resistance locus in Wheat (*Triticum aestivum L.*). *Mol. Gen. Genet. MGG.* **1995,** *248*, 553–562.

Fréville, H.; Justy, F.; Olivieri, I. Comparative Allozyme and Microsatellite Population Structure in a Narrow Endemic Plant Species, Centaurea corymbosa Pourret (Asteraceae). *Mol. Ecol.* **2001,** *10*, 879–889.

Frisch, M.; Bohn, M.; Melchinger, A. Minimum Sample Size and Optimal Positioning of Flanking Markers in Marker-assisted Backcrossing for Transfer of a Target Gene. *Crop. Sci.* **1999,** *36*, 967–975.

Gale, M. D.; Devos, K. M. Comparative Genetics in the Grasses. *Proc. Natl. Acad. Sci. 95*, **1998,** 1971–1974.

Ganal, M. W.; Altmann, T.; Roder, M. S. SNP Identification in Crop Plants. *Curr. Opin. Plant. Biol.* **2009,** *12*, 211–217.

Garvin, D.; Weeden, N. Isozyme Evidence Supporting a Single Geographic Origin for Domesticated Tepary Bean. *Crop. Sci.* **1994,** *34*, 1390–1395.

Gupta. J.; Mir, R. R.; Kumar, A. P. Marker-Assisted Selection as a Component of Conventional Plant Breeding. In *Plant Breeding Reviews*; Janik, J., Ed.; John Wiley & Sons, Inc.: Hoboken, NJ, **2010**. 233–276.

Gupta, K.; Langridge, P.; Mir, R. R. Marker-assisted Wheat Breeding: Present Status and Future Possibilities. *Mol. Breed.* **2010,** *26*, 145–161.

Gupta, M.; Chyi, Y. S.; Romero-Severson, J.; Owen, J. L. Amplification of DNA Markers from Evolutionarily Diverse Genomes using Single Primers of Simple-sequence Repeats. *Theor. Appl. Genet.* **1994,** *89*, 998–1006.

Gupta, P. K.; Rustgi, S.; Mir, R. R. Array-based High-throughput DNA Markers for Crop Improvement. *Heredity* **2008,** *101*, 5–18.

Gupta, P. K.; Varshney, R. K.; Sharma, P. C.; Ramesh, B. Molecular Markers and their Applications in Wheat Breeding. *Plant Breed.* **1999,** *118*, 369–390.

Hadačová, V.; Ondřej, M. Isoenzymy. Biol. Listy. **1972,** *37*, 1–25.

Heinz, D.J.; Tew, T. Hybridisation Procedures. In *Sugarcane Improvement through Breeding*; Heinz, D.J., Ed.; Elsevier: Amsterdam, **1987.** 313–342.

Henry, R. J. *Molecular Markers in Plants*. Oxford Press: UK, **2012**.

Herrera-Estrella, L.; Depicker, A. Expression of chimaeric genes transferred into plant cells using a Ti-plasmid-derived vector. *Nature* **1987,** *303*, 209–213.

Hudson, R. R.; Bailey, K.; Skarecky, D.; Kwiatowski, J.; Ayala F. J. Evidence for Positive Selection in the Superoxide Dismutase (SOD) Region of *Drosophila melanogaster*. *Gentics* **1994,** *136*, 1329–1340.

Jauhar, P. Modern Biotechnology as an Integral Supplement to Conventional Plant Breeding: the Prospects and Challenges. *Crop. Sci.* **2006,** *46*, 1841–1859.

Jiang, G. Molecular MaMarkers and Marker-assisted Breeding in Plants. Intech Publishers, Croatia. **2013,** 45–83.

Johnson, G.; Mumm, R. Marker Assisted Breeding Maize. *Proc. 51. Annu. Corn Sorghum Res. Conf.* **1996,** 75–84.

Jones, N.; Ougham, H.; Thomas, H.; Pasakinskiene, I. Markers and Mapping Revisited: Finding your Gene. *New. Phytol.* **2009,** *183*, 935–966.

Joshi, S. P.; Gupta, V. S.; Aggarwal, R. K. Genetic Diversity and Phylogenetic Relationship as Revealed by Inter Simple Sequence Repeat (ISSR) Polymorphism in the Genus Oryza. *Theor. Appl. Genet.* **2000,** *100*, 1311–1320.

Jensen, R. H. *Crops and Man*. American Society of Agronomy and Crop Science Society of America: Madison, **1992**.

Kalia, R. K.; Rai, M. K.; Kalia, S. Microsatellite Markers: An Overview of the Recent Progress in Plants. *Euphytica* **2011,** *177*, 309–334.

Kephart, S. R. Starch Gel Electrophoresis of Isozymes Plans: A Comparative Analysis of Techniques. *Am. J. Bot.* **1990,** *77*, 693–712.

Kesawat, M.; Das Kumar, B. Molecular Markers: It's Application in Crop Improvement. *J. Crop. Sci. Biotechnol.* **2009,** *12*, 169–181.

Kiani, A.; Akhunov, E. S. Application of Next-Generation Sequencing Technologies for Genetic Diversity Analysis in Cereals. In *Cereal Genomics II*; Varshney, P. K. G., Ed.; Springer: Netherlands, Berlin, **2013**, 197–205.

Koziel, M.; Beland, G.; Bowman, C.; Evola S. V. Field Performance of Elite Transgenic Maize Plants Expressing an Insecticidal Protein Derived from *Bacillus thuringiensis*. *Nat. Biotech.* **1993,** *11*, 194–200.

Kumar, P.; Gupta, V.; Misra, A.; Modi, D. R.; Pandey, A. K. Potential of Molecular Markers in Plant Biotechnology. *Plant. Omic. J.* **2009,** *2*, 141–162.

Lateef, D. D. DNA Marker Technologies in Plants and Applications for Crop Improvements. *J. Biosci. Med.* **2015,** *3*, 7–18.

Liu, B. Development and Prospects of Dwarf Male—Sterile Wheat. *Chin. Sci. Bull.* **1991,** *36*, 306.

Maass, B.; Ocampo, C. Isozyme Polymorphism Provides Fingerprints for Germplasm ofArachis glabrata Bentham. *Genet. Resour. Crop. Evol.* **1995,** *42*, 77–82.

Mammadov Chen, W.; Mingus, J.; Thompson, S.; Kumpatla, S. J. Development of Versatile Gene-Based SNP Assays in Maize (Zea mays L.). *Mol. Breed.* **2012,** *29*, 779–790.

Markert, C.; Moller, F. Multiple Forms of Enzymes: Tissue, Ontogenetic, and Species Specific Patterns. *Proc. Natl. Acad. Sci.* **1959,** *45*, 753–763.

Martin, B.; Nienhuis, J.; King, G.; Schaefer, A. Restriction Fragment Length Polymorphisms Associated with Water Use Efficiency in Tomato. *Science* **1989,** *243*, 1725–1728.

May, B. Starch Gel Eectrophoresis of Allozymes. In *Molecular Genetic Analysis of Populations: A Practical Approach.* Hoelzel, A. R. Oxford University Press: Oxford, UK, **1992,** 1–27.

Meudt, H. M.; Clarke, A. C. Almost Forgotten or Latest Practice? AFLP Applications, Analyses and Advances. *Trends. Plant Sci.* **2007,** *12*, 106–117.

Meyer, W.; Mitchell, T.G.; Freedman, E. Z.; Vilgalys, R. Hybridization Probes for Conventional DNA Fingerprinting used as Single Primers in the Polymerase Chain Reaction to Distinguish Strains of Cryptococcus Neoformans. *J. Clin. Microbiol.* **1993,** *31*, 2274–2280.

Miller, J. C.; Tanksley, S. D. RFLP Analysis of Phylogenetic Relationships and Genetic Variation in the Genus Lycopersicon. *Theor. Appl. Genet.* **1990,** *80*, 437–448.

Mir Hiremath, P. J.; Riera-Lizarazu, O.; Varshney, R. K. Evolving Molecular Marker Technologies in Plants: From RFLPs to GBS. In *Diagnostics in Plant Breeding*. Springer: Netherlands, **2013,** 229–247.

Moose, S.; Mumm, R. Molecular Plant Breeding as the Foundation for 21st Century Crop Improvement. *Plant Physiol.* **2008,** *147*, 969–977.

Morgante, M.; Vogel, J. M. Compound Microsatellite Primers for the Detection of Genetic Polymorphisms. US5955276 A. Sep 21, **1999**.

Morjane, H.; Geistlinger, J.; Harrabi, M. Oligonucleotide fingerprinting detects genetic diversity among Ascochyta rabiei isolates from a single chickpea field in Tunisia. *Curr. Genetics* **1994,** *26*, 191–197.

Nicod, J. C.; Largiader, C. R. SNPs by AFLP (SBA): A Rapid SNP Isolation Strategy for Nonmodel Organisms. *Nucleic Acids Res.* **2003,** *31*, 19–21.

Niewöhner, J.; Salamini, F.; Gebhardt, C. Development of PCR Assays Diagnostic for RFLP Marker Alleles Closely Linked to Alleles gro1 and H1, Conferring Resistance to the Root Cyst Nematode Globodera rostochiensis in Potato. *Mol. Breed.* **1995,** *1*, 65–78.

Orita, M.; Suzuki, Y.; Sekiya, T.; Hayashi, K. Rapid and Sensitive Detection of Point Mutations and DNA Polymorphisms using the Polymerase Chain Reaction. G*enomic*s **1989,** *5*, 874–879.

Paran, I.; Michelmore, R. W. Development of Reliable PCR-based Markers Linked to Downy Mildew Resistance Genes in Lettuce. *Theor. Appl. Genet.* **1993,** *85*, 985–993.

Parani, M.; Lakshmi, M.; Elango, S. Molecular Phylogeny of Mangroves II. Intra-and Interspecific Variation in Avicennia Revealed by RAPD and RFLP Markers. *Genome* **1997,** *40*, 487–495.

Paux, E.; Faure, S.; Choulet, F. Insertion Site-based Polymorphism Markers Open New Perspectives for Genome Saturation and Marker-assisted Selection in Wheat. *Plant. Biotech.* **2010,** *8*, 196–210.

Paux, E.; Sourdille, P.; Mackay, I.; Feuillet, C. Sequence-based Marker Development in Wheat: Advances and Applications to Breeding. *Biotechnol. Adv.* **2012,** *30*, 1071–1088.

Pease, A.; Solas, D. Light-generated Oligonucleotide Arrays for Rapid DNA Sequence Analysis. *Proc. Nat. Aca. Sci.* **1994,** *91*, 5022–5026.

Persley, G. J. Beyond Mendel's Garden. Biotechnology in Agriculture, *Biotechnol. CTA/IITA Co-publication*, Nigeria, **1992**.

Pingali, P. L. and Heisey, P.W. Cereal Crop Productivity in Developing Countries. CIMMYT Econ Pap CIMMYT, Mex, D.F., 99–03, **1999**.

Poland, J. A.; Brown, P. J.; Sorrells, M. E.; Jannink, J. L. Development of High-density Genetic Maps for Barley and Wheat using a Novel Two-enzyme Genotyping-by-Sequencing Approach. *PLoS One.* **2012,** *7*, 322–353.

Ragot, M.; Biasiolli M.; Delbut M.; Gay, G. Marker-assisted Backcrossing: A Practical Example. Techniques et utiliation des marquereus molacularies. Montpelliar, Paris, **1994**.

Ramakrishna, W.; Chowdari, K. V.; Lagu, M. D. DNA Fingerprinting to Detect Genetic Variation in Rice using Hypervariable DNA Sequences. *Theor. Appl. Genet.* **1995,** *90*, 1000–1006.

Ronning, C.; Schnell, R. Allozyme Diversity in a Germplasm Collection of Theobroma cacao L. *J. Hered.* **1994,** *85*, 291–295.

Ross, M.; Warriors, K. Identification of a Major Genes Regulating Complex Social Behavior. *Science* **2002,** *295*, 328–332.

Saal, B.; Wricke, G. Clustering of Amplified Fragment Length Polymorphism Markers in a Linkage Map of Rye. *Plant. Breed.* **2002,** *121*, 117–123.

Saiki, R.; Scharf, S.; Faloona, F.; Mullis, K. Enzymatic Amplification of Beta-globin Genomic Sequences and Restriction Site Analysis for Diagnosis of Sickle Cell Anemia. Science **1985,** *230*, 1350–1354.

Sant, V. J.; Patankar, A. G.; Sarode, N. D. Potential of DNA Markers in Detecting Divergence and in Analysing Heterosis in Indian Elite Chickpea Cultivars. *Theor. Appl. Genet.* **1999,** *98*, 1217–1225.

Sarfatti, M.; Katan, J.; Fluhr, R.; Zamir, D. An RFLP Marker in Tomato Linked to the Fusarium Oxysporum Resistance Gene I2. *Theor. Appl. Genet.* **1989,** *78*, 755–759.

Sax, K. The Association of Size Differences with Seed-coat Pattern and Pigmentation in Phaseolus vulgaris. *Genetics* **1923,** *8*, 552–560.

Schachermayr, G. M.; Messmer, M. M.; Feuillet, C. Identification of Molecular Markers Linked to the Agropyron elongatum-derived Leaf Rust Resistance Gene Lr24 in Wheat. *Theor. Appl. Genet.* **1995,** *90*, 982–990.

Schlotterer, C. The Evolution of Molecular Markers [mdash] just a Matter of Fashion? Nat. Rev. Genet. **2004,** *5*, 63–69.

Sergio, L.; Gianni, B. *Molecular Markers Based Analysis for Crop Germplasm Preservation. The Role of Biotechnology*. Villa Gualino: Turin, Italy—5–7 March, **2005,** pp. 55–66.

Sharma, H.; Crouch, J.; Sharma, K.; Haas, C. T. Applications of Biotechnology for Crop Improvement: Prospects and Constraints. *Plant. Sci.* **2002,** *163*, 381–395.

Shattuck-Eidens, D. M.; Bell, R. N.; Mitchell, J. T.; McWhorter, V. C. Rapid Detection of Maize DNA Sequence Variation. *Genet. Anal. Tech. Appl.* **1991,** *8*, 240–245.

Shull G. A Pure-line Method in Corn Breeding. *Rep. Amer. Breed. Ass.* **1908,** *4*, 296–301.

Soltis, D. E. Isozymes in Plant Biology. Dioscorides Press: Portland, Oregon, USA, **1989,** 5–45.

Somers, D. J.; Zhou, Z.; Bebeli, P. J.; Gustafson, J. P. Repetitive, Genome-specific Probes in Wheat (Triticum aestivum L. em Thell) Amplified with Minisatellite Core Sequences. *Theor. Appl. Genet.* **1996,** *93*, 982–989.

Sorrells, M.E.; Wilson, W.A. Direct Classification and Selection of Superior Alleles for Crop Improvement. *Crop. Sci.* **1997,** *37*, 691–697.

Staub, J. E.; Meglic, V. Molecular Genetic Markers and Cultivar Discrimination : A Case Study in Cucumber. *Hort. Technol.* **1993,** *3*, 291–300.

Sturtevant, A. A Third Group of Linked Genes in Drosophila ampelophila. *Science* **1913,** *37*, 990–992.

McCouch, S. R.; Wright, M.; Tung, C. W.; Ebana, K.; Thomson, M.; Bustamante, K. Z. Development of Genome-wide SNP Assays for Rice. *Breed. Sci.* **2010,** *60*, 524–535.

Syvanen, A. C. Toward Genome-wide SNP Genotyping. *Nat. Genet.* **2005,** *37*, Suppl:S5–10.

Tanksley, N. D.; Paterson, A. H.; Bonierbale, M. W. RFLP Mapping in Plant Breeding: New Tools for an Old Science. *Nat. Biotechnol.* **1989,** *7*, 257–264.

Thudi, M.; Li, Y.; Jackson, S. A. Current State-of-art of Sequencing Technologies for Plant Genomics Research. *Br. Funct. Genom.* **2012,** *11*, 3–11.

Oeveren, J.; Ruiter, M.; Jesse, T. Sequence-based Physical Mapping of Complex Genomes by Whole Genome Profiling. *Genome. Res.* **2011,** *21*, 618–625.

Varshney, P. K.; Graner, A. *Cereal Genomics II*. Springer: Berlin, **2013,**161–179.

Varshney, R. K.; Graner, A.; Sorrells, M. E. Genomics-assisted Breeding for Crop Improvement. *Trends Plant Sci.* **2005,** *10*, 621–630.

Vos, P.; Hogers, R.; Bleeker, M. AFLP: A New Technique for DNA Fingerprinting. *Nucleic Acids Res.* **1995,** *23*, 4407–4414.

Wang, J.; Chapman, S.; Bonnett, D.; Rebetzke, G. Simultaneous Selection of Major and Minor Genes: Use of QTL to Increase Selection Efficiency of Coleoptile Length of Wheat (*Triticum aestivum* L.). *Theor. Appl. Genet.* **2009,** *119*, 65–74.

Warnke, S.; Douches, D.; Branham, B. Isozyme Analysis Supports Allotetraploid Inheritance in Tetraploid Creeping Bentgrass (Agrostis palustris Huds.). *Crop. Sci.* **1998,** *38*, 801–805.

Weising, K.; Atkinson, R.; Gardner, R. Genomic Fingerprinting by Microsatellite-primed PCR: A Critical Evaluation. *Genome. Res*. **1995,** *4*, 249–255.

Weising, K.; Nybom, H.; Wolff, K.; Kahl, G. *DNA Fingerprinting in Plants*. Taylor & Francis: U.K., **2005**, 338.

Werner, K.; Friedt, W.; Ordon, F. Strategies for Pyramiding Resistance Genes Against the Barley Yellow Mosaic Virus Complex (BaMMV, BaYMV, BaYMV-2). *Mol. Breed.* **2005,** *16*, 45–55.

Wheeler, D. A.; Srinivasan, M.; Egholm, M. The Complete Genome of an Individual by Massively Parallel DNA Sequencing. *Nature* **2008,** *452*, 872–876.

Williams, J. G. K.; Kubelik, A. R.; Livak, K. J. DNA Polymorphisms Amplified by Arbitrary Primers Are Useful as Genetic Markers. *Nucl. Aci. Res*. **1990,** *18*, 6531–6535.

Williamson, V. M.; Ho, J. Y.; Wu, F. F. A PCR-based Marker Tightly Linked to the Nematode Resistance Gene, Mi, in Tomato. *Theor. Appl. Genet*. **1994,** *87*, 757–763.

Wong, L. J. Next Generation Molecular Diagnosis of Mitochondrial Disorders. *Mitochondrion* **2013,** *13*, 379–387.

Xiao, J.; Li, J.; Grandillo, S. Identification of Trait-improving Quantitative Trait Loci Alleles from a Wild Rice Relative, Oryza rufipogon. *Genetics* **1998,** *150*, 899–909.

Xu, Y. Molecular lPlant Breeding. CAB International. Wallingford Oxfordshire: UK, **2010**, 151–194.

Yang, G. P.; Maroof, M. A.; Xu, C. G. Comparative Analysis of Microsatellite DNA Polymorphism in Landraces and Cultivars of Rice. *Mol. Gen. Genet*. **1994,** *245*, 187–194.

Young, M. H.; Fatokun, C. A.; Danesh, D. RFLP Technology, Crop Improvement, and International Agriculture. In *Biotechnology: Enhancing Research on Tropical Crops in Africa*; Thottappilly Monti, L. M., Moham, D. R., Moore, A. W. G., Eds.; **1992,** 841–846.

Yu, J.; Hu, S.; Wang, J. A. Draft Sequence of the Rice Genome (Oryza sativa L. ssp. indica). *Science* **2002,** *296*, 79–92.

Yunbi, Xu.; Jianmin, W.; Zhonghu, He.; Boddupalli, M.; Prasanna, C. X. Marker-assisted Selection in Cereals: Platforms, Strategies and Examples. In *Cereal Genomics II*; Varshney, P., Ed.; Springer: Netherlands, Berlin, **2013,** 375–411.

Zietkiewicz, E.; Rafalski, A.; Labuda, D. Genome Fingerprinting by Simple Sequence Repeat (SSR)-Anchored Polymerase Chain Reaction Amplification. *Genomics* **1994,** *20*, 176–183.

CHAPTER 20

DEVELOPMENT OF MAPPING POPULATIONS

ANAND KUMAR[1*], TUSHAR RANJAN[2], RAVI RANJAN KUMAR[3], KUMARI RAJANI[4], CHANDAN KISHORE[1] and JITESH KUMAR[3]

[1]*Department of Plant Breeding and Genetics, Bihar Agricultural University, Sabour, Bhagalpur 813210, Bihar, India*

[2]*Department of Basic Science and Humanities Genetics, Bihar Agricultural University, Sabour, Bhagalpur 813210, Bihar, India*

[3]*Department of Molecular Biology and Genetic Engineering, Bihar Agricultural University, Sabour, Bhagalpur 813210, Bihar, India*

[4]*Department of Seed Science and Technology, Bihar Agricultural University, Sabour, Bhagalpur 813210, Bihar, India*

Corresponding author. E-mail: anandpbgkvkharnaut@gmail.com

CONTENTS

ABSTRACT

Mapping populations consist of individuals of one species, or in some cases they derive from crosses among related species where the parents differ in the traits to be studied. These genetic tools are used to identify genetic factors or loci that influence phenotypic traits and determine the recombination distance between loci. The first step in producing a mapping population is selecting two genetically divergent parents, which show clear genetic differences for one or more traits of interest. In different organisms of the same species, the genes, represented by alternate allelic forms, are arranged in a fixed linear order on the chromosomes. Linkage values among genetic factors are estimated based on recombination events between alleles of different loci, and linkage relationships along all chromosomes provide a genetic map of the organism. The type of mapping population to be used depends on the reproductive mode of the plant to be analyzed. In this chapter we are going to discuss in detailed about the mapping population and their types.

20.1 INTRODUCTION

The development of genetics has been exponential with several milestones since Mendel discovered the laws of inheritance in 1865, which is a core component of biology to relate genetic factors to functions visible as phenotypes. During Mendel's time, genetic analysis was restricted to visual inspection of the plants. Pea (*Pisum sativum*) was already a model plant at that time, and Mendel studied visible traits such as seed and pod color, surface structure of seeds and pods (smooth versus wrinkled), and plant height. These traits are, in fact, the first genetic markers used in biology (Swiecicki et al., 2000). During last two decades, several linkage maps, genes/quantitative trait loci (QTLs)/traits have been discovered for important agronomic traits in several crop species. The prime requirement of construction of genetic linkage/QTL map in crop species is development of appropriate mapping population and selection of sample size from mapping populations. Mapping populations consist of individuals of one species, or in some cases they derive from crosses among related species where the parents differ in the traits to be studied. These genetic tools are used to identify genetic factors or loci that influence phenotypic traits and determine the recombination distance between loci (McCouch et al., 1988; Semagn et al., 2006). The first step in producing a mapping population is selecting two genetically divergent parents, which show clear genetic differences for one or more traits of interest. In different organisms of the same species, the genes, represented by alternate allelic forms, are arranged in a fixed linear order on the chromosomes. Linkage values among genetic factors are estimated based on recombination events between alleles of different loci, and linkage relationships along all chromosomes provide a genetic map of the organism. The type of mapping population to be used depends on the reproductive mode of the plant to be analyzed. In this respect, the plants fall into the main classes of self-fertilizers and self-incompatibles (McCouch et al., 1988). In 1912, Vilmorin and Bateson described the first work on linkages in *Pisum*. However, the concept of linkage groups representing chromosomes was not clear in *Pisum* until 1948, when Lamprecht described the first genetic map with 37 markers distributed on seven linkage groups (Swiecicki et al., 2000). Large collections of visible markers are today available for several crop species and for *Arabidopsis thaliana* (Koornneef et al., 1987; Neuffer et al., 1997). In the process of finding more and more genetic markers, the first class of characters scored at the molecular level was isoenzymes. These are isoforms of proteins that vary in amino acid composition and charge, and that can be distinguished by electrophoresis. The technique is applied to the characterization of plant populations and breeding lines and in plant systematics, but it is also used for genetic mapping of variants, as shown particularly in maize (Frei et al., 1986; Stuber et al., 1972). However, due to the small number of proteins for which isoforms exist and that can be separated by electrophoresis, the number of isoenzyme markers is limited. The advance of molecular biology provided a broad spectrum of technologies to assess the genetic situation at the DNA level. The first DNA polymorphisms described were restriction fragment length polymorphism (RFLP) markers (Botstein et al., 1980). This technique requires the hybridization of a specific probe to restricted genomic DNA of different genotypes. The whole genome can be covered by RFLP and, depending on the probe, coding or noncoding sequences can be analyzed. The next generation of markers was based on PCR: rapid amplified polymorphic DNA (RAPD) (Williams et al., 1990; Welsh and McClelland, 1990) and amplified fragment length polymorphism (AFLP) (Vos et al., 1995). Recently, methods have been developed to detect single nucleotide polymorphisms (Rafalski, 2002). Since these methods have the potential for automatization and multiplexing, they allow the establishment of high-density genetic maps. Whereas RAPD and AFLP analyses are based on anonymous fragments, RFLP and SNP analyses allow the choice of expressed genes as markers. Genes of a known sequence and that putatively influence the trait of interest can be selected and mapped. In this way, function maps can be constructed (Chen et al., 2001; Schneider et al., 2002).

The basis for genetic mapping is recombination among polymorphic loci which involves the reaction between homologous DNA sequences in the meiotic prophase. Currently, the double-strand-break repair model (Szostak et al., 1983) is acknowledged to best explain meiotic reciprocal recombination. In this model, two sister chromatids break at the same point and their ends are resected at the ends. In the next step, the single strands invade the intact homologue and pair with their complements. The single-strand gaps are filled in using the intact strand as template. The resulting molecule forms two Holliday junctions. Upon resolution of the junction, 50% of gametes with recombinant lateral markers and 50% nonrecombinants are produced. In the non-recombinants, genetic markers located within the region of strand exchange may undergo gene conversion which can result in nonreciprocal recombination, a problem interfering in genetic mapping. Figure 20.1 shows generation of recombinants by chiasma formation due to cross-over event. In plants, gene conversion events were identified by Buschges et al. (1997) when cloning the *Mlo* resistance gene from barley. The likelihood that recombination events occur between two points of a chromosome depends in general on their physical distance: the nearer they are located to each other, the more they will tend to stay together after meiosis. With the increase of the distance between them, the probability for recombination increases and genetic linkage tends to disappear (Buschges et al., 1997). This is why genetic linkage can be interpreted as a measure of physical distance. However, taking the genome as a whole, the frequency of recombination is not constant because it is influenced by chromosome structure. An example is the observation that recombination is suppressed in the vicinity of heterochromatin: here, the recombination events along the same chromatid appear to be reduced, an observation called positive interference (Szostak et al., 1983).

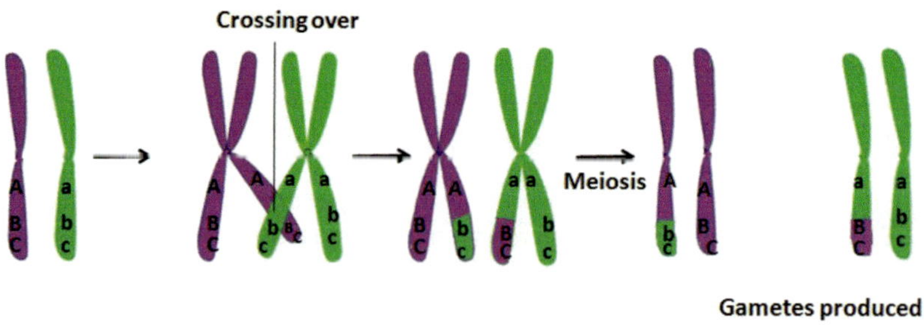

FIGURE 20.1 Generation of recombinants by chiasma formation due to cross-over event. In the meiotic prophase, two sister chromatids of each parent (labeled in maroon and blue, respectively) align to form a bivalent. A chiasma is formed by a physical strand exchange between two non-sister chromatids. Breakage and reunion of reciprocal strands leads to the generation of recombinants.

20.2 MAPPING POPULATION

Mapping populations are the tools used to identify the genetic loci controlling measurable phenotypic traits. For self-pollinating species, F_2 populations and recombinant inbred lines (RILs) are used; for self-incompatible, highly heterozygous species, F_1 populations are mostly the tools of choice. In short, population used for mapping the genes is commonly called mapping population and they are usually obtained from controlled crosses. Selection of parents is the first step for production of mapping population. Parents selected to develop mapping population should have sufficient variation for trait of interest both at DNA sequence level and at phenotypic level. Higher the variation, easier it is to find the recombination. Parents should not be so diverse that they are unable to cross. Backcross (BC) populations and doubled haploid lines (DHLs) are a possibility for both types of plants. The inheritance of specific regions of DNA is followed by molecular markers that detect DNA sequence polymorphisms. Recombination frequencies between traits and markers reveal their genetic distance, and trait-linked markers can be anchored, when necessary, to a more complete genetic map of the species. For map-based cloning of a gene, populations of a large size provide the resolution required (Szostak et al., 1983). Due to intensive breeding and pedigree selection, genetic variability within the gene pools of relevant crops is at risk. Interspecific crosses help to increase the size of the gene pool, and the contribution of wild species to this germplasm in the form of introgression lines is of high value, particularly with respect to traits like disease resistance. The concept of exotic libraries with near-isogenic lines (NILs), each harboring a DNA fragment from a wild species, implements a systematic scan of the gene pool of a wild species. To describe the complexity of genome organization, genetic maps are not sufficient because they are based on recombination, which is largely different along all genomes. However, genetic maps, together with cytogenetic data, are the basis for the construction of physical maps. An integrated map then provides a detailed view on genome structure and enforces positional cloning of genes, and ultimately the sequencing of complete genomes (Buschges et al., 1997).

20.3 TYPES OF MAPPING POPULATION

20.3.1 F_2 POPULATIONS

The simplest form of a mapping population is F_2 population. Parent 1 (P1) and parent 2 (P2) are two parents contrasting for trait of interest crossed to get F_1 population. Individual F_1 plant is then selfed to produce an F_2 population. F_2 populations are outcome of single meiotic cycle. The segregation ratio for codominant marker is 1:2:1 (homozygous like P1: heterozygous: homozygous like P2) while segregation ratio for each dominant marker is 3:1. Figure 20.2 represents the steps involved in generation of an F_2 population. This type of population was the basis for the Mendelian laws (1865) in which the foundations of classic genetics were laid. Two pure lines that result from natural or artificial inbreeding are selected as parents, P1 and P2. Alternatively, DHLs can be used to avoid any residual heterozygosity. If possible, the parental lines should be different in all traits to be studied. The degree of polymorphism can be assessed at the phenotypic level (e.g., morphology, disease resistance) or by molecular markers at the nucleic acid level. For inbreeding species such as soybean and the *Brassicaceae*, wide crosses between genetically distant parents help to increase polymorphism. However, it is required that the cross leads to fertile progeny. The progeny of such a cross is called the F_1 generation. If the parental lines are true homozygotes, all individuals of the F_1 generation will have the same genotype and

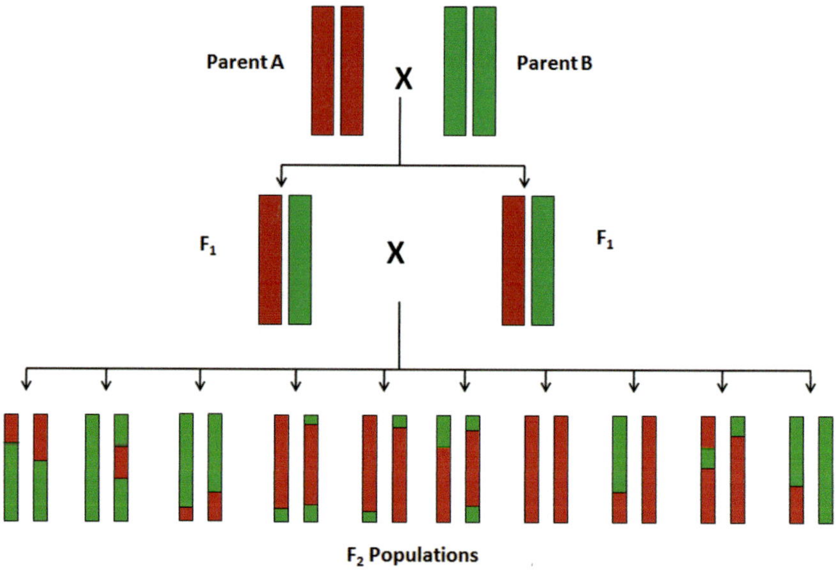

FIGURE 20.2 The generation of an F_2 population. Two chromosomes are shown as representatives of the diploid parental genome. In the parental generation, the genotypes are homozygous and in the F_1 generation, they are heterozygous. For gamete formation, the genetic material undergoes meiosis, leading to recombination events in F_1 gametes. Correspondingly, F_2 plants vary largely in their genetic constitution.

a similar phenotype. This is the content of Mendel's law of uniformity. An individual F_1 plant is then selfed to produce an F_2 population that segregates for the traits different between the parents. F_2 populations are the outcome of one meiosis, during which the genetic material is recombined. The expected segregation ratio for each codominant marker is 1:2:1 (homozygous like P1: heterozygous: homozygous like P2). It is a disadvantage that F_2 populations cannot be easily preserved, because F_2 plants are frequently not immortal, and F_3 plants that result from their selfing are genetically not identical. For species like sugar beet, there is a possibility of maintaining F_2 plants as clones in tissue culture and of multiplying and regrowing them when needed. A particular strategy is to maintain the F_2 population in pools of F_3 plants. Traits that can be evaluated only in hybrid plants, such as quality and yield parameters in sugar beet or maize, require the construction of test-cross plants by crossing each F_2 individual with a common tester genotype (Schneider et al., 2002). Ideally, different common testers should produce corresponding results to exclude the specific effects of one particular tester genotype. One of the limitations of this population is that it is unable to map the quantitative traits (QTs).

20.3.2 RECOMBINANT INBRED LINES

RILs are the homozygous selfed or sibmated progeny of the individuals of an F_2 population (Fig. 20.3). The RIL concept for mapping genes was originally developed for mouse genetics. In animals, approximately 20 generations of sibmating are required to reach useful levels of homozygosity. In plants, RILs are produced by selfing unless the species is completely self-incompatible. Because in the selfing process one seed of each line is the source for the next generation, RILs are also called single-seed descent lines. Self-pollination allows the production of RILs in a relatively short number of generations. In fact, within six generations, almost complete homozygosity can be reached. Along each chromosome, blocks of alleles derived from either parent alternate. Because recombination can no longer change the genetic constitution of RILs, further segregation in the progeny of such lines is absent. It is thus one major advantage that these lines constitute a permanent resource that can be replicated indefinitely and shared by many groups in the research community. A second advantage of RILs is that because they undergo several rounds of meiosis before homozygosity is reached, the degree of recombination is higher compared to F_2 populations. Consequently, RIL populations show a higher resolution than maps generated from F_2 populations (Burr and Burr, 1991), and the map positions of even tightly linked markers can be determined. In plants, RILs are available for many species including rice and oat (Wang et al., 1994). In *A. thaliana*, 300 RILs have become a public mapping tool (Lister and Dean, 1993). *Arabidopsis* RILs were constructed by an initial cross between the ecotypes *Landsberg erecta* and Columbia, and a dense marker framework was established. Every genomic fragment that displays a polymorphism between *Landsberg erecta* and Columbia can be mapped by molecular techniques (O'Donoughue et al., 1995).

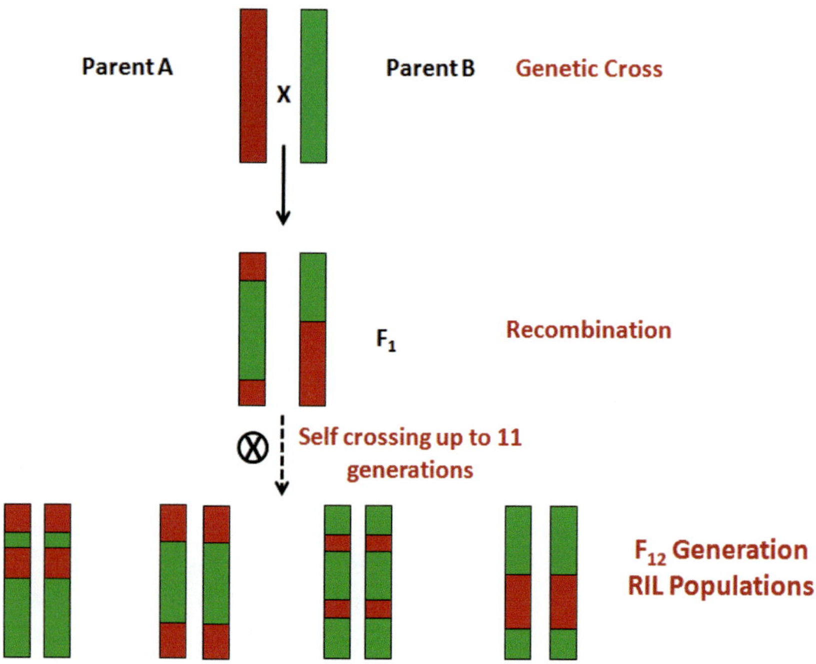

FIGURE 20.3 The generation of mapping population by RIL.

20.3.3 BACKCROSS POPULATIONS

To analyze specific DNA fragments derived from parent A in the background of parent B, a hybrid F_1 plant is backcrossed to parent B. In this situation, parent A is the donor of DNA fragments and parent B is the recipient. The latter is also called the recurrent parent. During this process two goals are achieved: unlinked donor fragments are separated by segregation and linked donor fragments are minimized due to recombination with the recurrent parent. To reduce the number and size of donor fragments, backcrossing is repeated and, as a result, so-called advanced BC lines are generated. With each round of backcrossing, the proportion of the donor genome is reduced by 50%. Figure 20.4 indicates the steps involved in generation of advanced BC lines. Molecular markers help to monitor this process and to speed it up. In an analysis of the chromosomal segments retained around the Tm-2 locus of tomato, it was estimated that marker-assisted selection (MAS) reduced the number of required BCs from 100 in the case of no marker selection to two (Young and Tanksley, 1989). The progeny of each BC is later screened for the trait introduced by the donor. In the case of dominant traits, the progeny can be screened directly; in the case of recessive traits, the selfed progeny of each BC plant has to be assessed. Lines that are identical, with the exception of a single fragment comprising one to a few loci, are called near-isogenic lines (NILs). The generation of NILs involves several generations of backcrossing assisted by marker selection. To fix the donor segments and to visualize traits that are caused by recessive genes, two additional rounds of self-fertilization are required at the end of the backcrossing process. If two NILs differ in phenotypic performance, this is seen as the effect of the alleles carried by the introgressed DNA fragment.

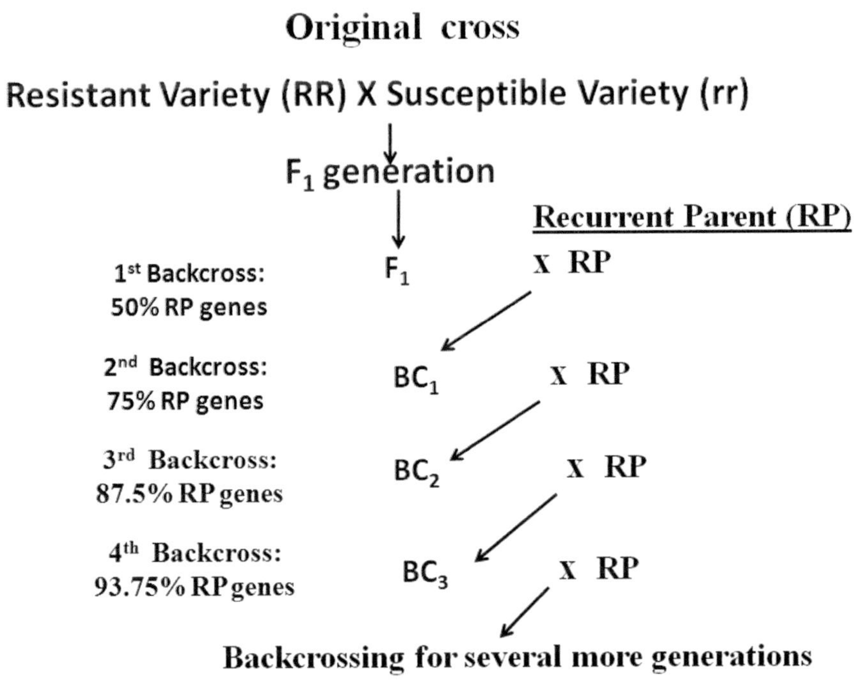

FIGURE 20.4 The generation of mapping population by advanced backcross lines.

The procedure is quite helpful in the functional analysis of the underlying genes. The strategy is particularly valuable for those species for which no transformation protocol is established to produce transgenics for the alleles of interest. A further advantage is that in NILs genomic rearrangements, which may happen during transformation, are avoided. BC breeding is an important strategy if a single trait, such as resistance, has to be introduced into a cultivar that already contains other desirable traits. The only requirement is that the two lines be crossable and produce fertile progeny. Lines incorporating a fragment of genomic DNA from a very distantly related species are called introgression lines whereas lines incorporating genetic material from a different variety are indicated as inter-varietal substitution lines (Young and Tanksley, 1989).

20.3.4 NEAR-ISOGENIC LINES

The development of NILs is a very important tool for both genetic and physiological dissection of drought resistance in rice. Two pairs of NILs differing for grain yield under drought stress were isolated and characterized for yield, yield-related traits, and several physiological traits in a range of contrasting environments. A polymorphism analysis study with 491 SSRs revealed that both NIL pairs are at least 96% genetically similar. The NILs show that small genetic differences can cause large difference in grain yield under drought stress in rice (Venuprasad et al., 2011). In both pairs, the drought-tolerant NILs showed a significantly higher assimilation rate at later stages both under stress and nonstress conditions. They also had a higher transpiration rate under non-stress condition. The most tolerant NIL (IR77298-14-1-2-B-10) had significantly higher transpiration rate and stomatal conductance in severe stress conditions. In one pair, the tolerant NIL had constitutively deeper roots than the susceptible NIL. In the second pair, which had higher mean root length than the first pair, the tolerant NIL had more roots, greater root thickness, and greater root dry weight than the susceptible NIL. Deeper root length may allow tolerant NILs to extract more water at deeper soil layers. The enhanced rooting depth is an important strategy for dehydration avoidance and rice adaptation to drought stress, but root architecture might not be the only mechanism causing the significant yield increase (Venuprasad et al., 2011). Figure 20.5 represents the construction and use of a NIL panel for identification of high-likelihood candidates for QTLs.

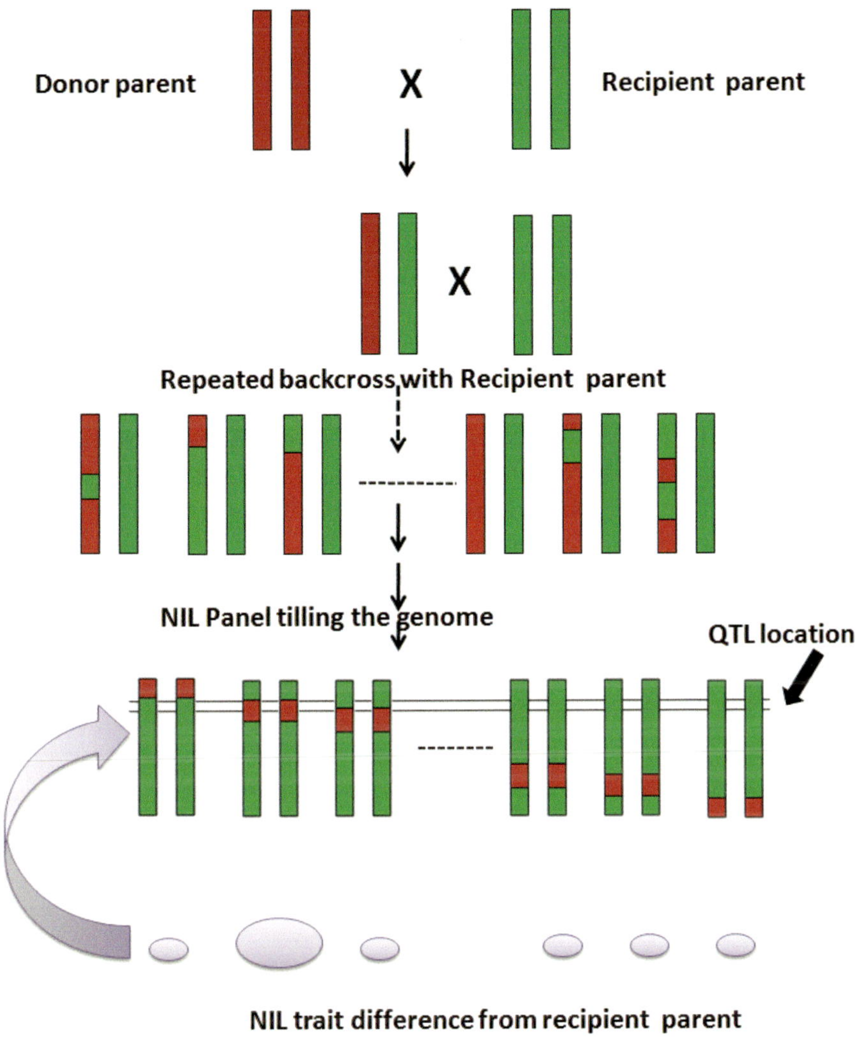

FIGURE 20.5 The construction and use of a near-isogenic line (NIL) panel for identification of high-likelihood candidates for QTLs. Initially, a donor and reference or recurrent parent are crossed and, subsequently, repeated backcrosses to the reference parent lead to a reduction of the donor genome contribution. With marker assisted selection (MAS), a panel of NILs that tile the genome can be constructed.

20.3.5 INTROGRESSION LINES

The breeding of superior plants consists of combining positive alleles for desirable traits on the elite cultivar. One source for such alleles conferring traits such as disease resistance or quality parameters is distantly related or even wild species. If the trait to be introduced is already known, the introgression can be performed in a direct way supported by MAS. However, the potential of wild species to influence QTs often is not yet assessed. In this case, BC breeding is a method to identify single genetic components contributing to the phenotype. NILs are constructed by an advanced BC program and their phenotypic effects are assayed. For example, in the work of Tanksley et al. (1996), loci from the wild tomato species *Lycopersicon pimpinellifolium* were shown to have positive effects on tomato fruit size and shape (Tanksley et al., 1996). To assess the effects of small chromosomal introgressions at a genome-wide level, a collection of introgression lines, each harboring a different fragment of genomic DNA, can be generated. Such a collection is called an exotic library which is achieved by advanced backcrossing. This corresponds to a process of recurrent backcrossing (ADB) and MAS for six generations and to the self-fertilization of two more generations to generate plants homozygous to the introgressed DNA fragments (Zamir 2001). An example is the introgression lines derived from a cross between the wild green-fruited species *L. pennellii* and the tomato variety M82 (Eshed and Zamir, 1995). The lines after the ADB program will resemble the cultivated parent, but introgressed fragments with even subtle phenotypic effects can be easily identified. In other words, phenotypic assessment for all traits of interest will reveal genomic fragments with positive effects on measurable traits. The introgressed fragments are obviously defined by the use of molecular markers (Eshed and Zamir, 1995).

20.3.6 DOUBLE HAPLOID LINES

DHLs contain two identical sets of chromosomes in each cell. They are completely homozygous as only one allele is available for all genes. Doubled haploids can be produced from haploid lines. Haploid lines either occur spontaneously, as in the case of rape and maize, or are artificially induced. Haploid plants are smaller and less vital than diploids and are nearly sterile (Chao et al., 1989). It is possible to induce haploids by culturing immature anthers on special media. Haploid plants can later be regenerated from the haploid cells of the gametophyte. A second option is microspore culture. In cultivated barley, it is possible to induce the generation of haploid embryos by using pollen from the wild species *Hordeum bulbosum*. During the first cell divisions of the embryo, the chromosomes of *H. bulbosum* are eliminated, leaving the haploid chromosomal set derived from the egg cell.

Occasionally in haploid plants the chromosome number doubles spontaneously, leading to doubled haploid (DH) plants. Such lines can also be obtained by colchicine treatment of haploids or their parts. Colchicines prevent the formation of the spindle apparatus during mitosis, thus inhibiting the separation of chromosomes and leading to doubled haploid cells. If callus is induced in haploid plants, a doubling of chromosomes often occurs spontaneously during endomitosis and DHLs can be regenerated via somatic embryogenesis (Heun et al., 1991). However, in vitro culture conditions may reduce the genetic variability of regenerated materials to be used for genetic mapping. DHLs constitute a permanent resource for mapping purposes and are ideal crossing partners in the production of mapping populations because they have no residual heterozygosity (Chao et al., 1989; Heun et al., 1991; McCouch et al., 1988). Figure 20.6 represents the development of DHL.

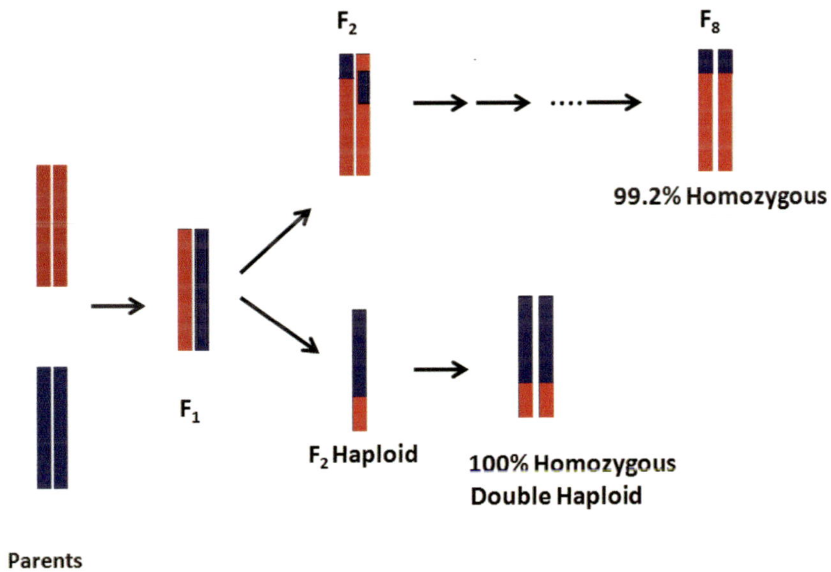

FIGURE 20.6 Development of double haploid lines.

20.3.6 MAGIC POPULATION

A major advantage for researchers in plant and animal genetics lies in the ability to create experimental populations. Such populations mix well-characterized founder genomes in controlled pedigrees, and facilitate the investigation of both the genome itself and its relationship with traits and the environment (Huang et al., 2015). Traditional experimental populations combine the genomes of two parents with contrasting phenotypes to identify regions of the genome affecting the trait. However, each of these populations captures only a small snapshot of the factors affecting the trait; due to the narrow genetic base, it is only possible to detect those genomic regions which differ between the two founders, and all alleles occur with high frequency in the population. The alternative of association mapping takes a panoramic view of the whole population by sampling distantly related individuals. It hence captures far greater diversity, but requires very large samples to have sufficient power to detect genomic regions of interest, and hence may have difficulty detecting rare alleles of importance. The weaknesses of existing designs have led to a new type of complex experimental design intermediate to biparental and association mapping designs in terms of power, diversity, and resolution. Multiparent advanced generation inter crosses (MAGIC) inter-mate multiple inbred founders for several generations prior to creating inbred lines, resulting in a diverse population whose genomes are fine-scale mosaics of contributions from all founders. Similar to biparental populations, alleles occur at relatively high frequencies due to the limited number of founders but the population encapsulates much higher diversity in polymorphisms. While a MAGIC population requires greater initial investment in capability and time than a biparental, careful selection of founders allows its generalizability to the wider breeding population and ensures relevance as a long-term genetic resource panel (Huang et al., 2015). Figure 20.7 represents the development of multitraits MAGIC population from monotypic characters.

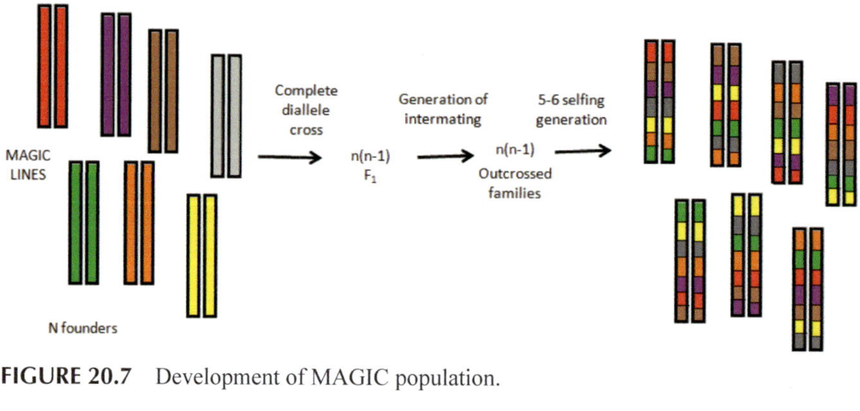

FIGURE 20.7 Development of MAGIC population.

20.4 GENETIC/QTL MAPPING IN POPULATION FOR AGRONOMIC TRAITS

With the development of molecular markers in the crops, genetics and physical maps have been constructed in many important crops. Gene or QTL conditioning important agronomic traits were mapped on to the chromosome or genetic linkage group through analysis of mapping population. QTLs mapping in crop plants has now become routine due to the progress made in this area during the last two decades. Although, initial QTL studies mainly focused on the identification of QTLs for only some important QTs in any individual crop, QTLs could later be identified for majority of the QTs in each of a number of crops, in many cases, leading to cloning of individual QTLs (Zargar et al., 2015). Recent advances in sequencing technology have brought several novel platforms for marker development and subsequent genotyping. The high-throughput and cost-effective marker techniques have changed the entire scenario of marker applications. The huge genotypic data obtained with next generation sequencing (NGS) also demands analytical tools, statistical advances, and comprehensive understanding to cope with breeding applications. In the present review, we discussed different available marker techniques, their strengths, and limitations. Emphasis was given on software tools, analytical pipelines, workbenches, and online resources available for marker development. Comparison of SNP genotyping involving complexity reduction techniques like GBS, RRL, RAD, and array-based platforms were presented in a view to describe suitability for specific purposes. The genotyping by whole genome resequencing has great potential and could be a routine application in the near future with continuously decreasing cost of sequencing. Microsatellites, still a valuable option for breeders, have also advanced with NGS. The most common applications of molecular marker are QTL mapping, genome-wide association mapping (GWAS), and genomic selection (Gupta et al., 2013).

MAS is most useful for traits that are difficult to select, e.g., disease resistance, salt tolerance, drought tolerance, heat tolerance, quality traits (aroma of Basmati rice, flavor of vegetables). This approach involves selecting plants at early generation with a fixed, favorable genetic background at specific loci, conducting a single large-scale marker-assisted selection while maintaining as much as possible the allelic segregation in the population and the screening of large populations to achieve the objectives of the scheme. No selection is applied outside the target genomic regions to maintain as much as possible the Mendelian allelic segregation among the selected genotypes. After selection with DNA markers, the genetic diversity at unselected loci may allow breeders to generate new varieties and hybrids through conventional breeding in response to targets set in breeding program (Datta et al., 2011). As more genomic DNA sequences became available in the crops, SSR or microsatellites were developed in wheat, chickpea, maize, rice, and soybean as well (Yadav et al., 2015; Varshney et al., 2015). Although SSR markers have multiple loci, their polymorphism is limited, especially in mapping population from parents with narrow genetic background. Further to increase the polymorphism, AFLP marker was proposed, which has advantage of both RFLP and PCR as well. Now, SNP markers have become powerful tools to detect polymorphism and are used as MAS in many crops. SNP could be used to generate numerous markers within a target region. SNP genotyping combining with allele-specific PCR has become valuable for genetic mapping, MAP-based cloning, and MAS as well. As more maps have been constructed in many important crops, breeder and geneticists can study the traits of interest using their own mapping population with references from the core maps (Hayashi et al., 2004). From past many years, MAS has been used in chickpea, wheat, maize, rice, and soybean for the resistance to biotic (bacteria, virus, fungi, nematodes, and insects) and abiotic stresses (cold, drought, and salinity) (Table 20.1).

The first step in defense is the recognition of the pathogen by the plant, which activates signal transduction cascades that subsequently trigger transcription of plant defense genes (Park et al., 2008). Gene expression studies provide information about the genes and metabolic pathways differentially regulated during plant-pathogen interactions and contribute to the identification of candidate resistance genes involved in each of these steps of the defense response (Rispail et al., 2013; Barilli et al., 2012). Knowledge of the genes involved in defense can be useful for MAS and also to select genes whose altered expression through transformation can result in an increased resistance (Rubiales et al., 2015). DNA-based markers have shown great promises in expediting plant breeding methods. The identification of molecular markers closely linked with resistance genes would facilitate expeditious pyramiding of major genes into elite background, making it more cost-effective. Once the resistance genes are tagged with molecular marker, the selection of resistant plant in the segregating generations becomes easy (Datta et al., 2011). Major success of identification of disease resistance genes is observed in chickpea and pigeon pea due to availability of genomic resources but still minor pulses like lentil, mung bean, and urd bean need an attention. Chickpea linkage map was established with help of 354 molecular surveyed among 130 RILs derived from a *Cicer arietinum* × *Cicer reticulatum* (Winter et al., 2000). DNA markers associated with two

TABLE 20.1 Genetic Mapping for Important Agronomic Traits in Cereals and Pulses.

Crops	Trait	Gene/QTL	Source
Chickpea	Fusarium wilt	*foc-0, foc-1, foc-2, foc-3, foc-4, foc-5*, QTL	Varshney et al. (2015)
	Ascochyta blight	QTL	Varshney et al. (2015)
	Rust	QTL	Varshney et al. (2015)
	Botrytis gray mold	QTL	Varshney et al. (2015)
	Pigeon pea		
	Fusarium wilt	BSA	Varshney et al. (2015)
	SMD resistance	QTL, BSA	Varshney et al. (2015)
	Lentil		
	Fusarium wilt	QTL	Hamweih et al. (2005)
	Aschochyta Blight	QTL	Rubeena et al. (2006)
Wheat	Stem rust	*Sr2*	Hyden et al. (2004)
	Green bug	*Gb3*	Zhu et al. (2005)
	Karnal bunt	QTL	Brooks et al. (2006)
	Spot blotch	QTL	Sharma et al. (2007)
	Drought	QTL	Kumar et al. (2007)
	Aluminum tolerance	QTL	Zhou et al. (2007)
Maize	Com earworm	QTL	Butron et al. (2001)
	Downy mildew	QTL	George et al. (2003)
	Gibberella ear rot	QTL	Ali et al. (2005)
	Bacterial leaf blight	*Xa23; Xa21*	Sanchez et al. (2000)
	Lysine	*O2, o16*	Yang et al. (2005)
Rice	Rice blast	*Pi-l, Pi-k, Pi-b*	Sharma et al. (2005)
	Brown plant hooper	*Bph 2, Bph 13, Bph 9*	Su et al. (2005)
	Root thickness	*brt5b*	Li et al. (2005)
	Drought tolerance	QTL	Babu et al. (2003)
	Cold tolerance	*Qsct-11*	Lou et al. (2007)
	Sheath blight resistance	*Qshb7.3, Qshb9.2*	Yadav et al. (2015)
	Grain quality	QTL	Xu et al. (2015)
Soybean	Soybean mosaic virus	*Rsv 3, Rsv 4*	Hwang et al. (2006)
	Soybean cyst nematode	*rhg 1*	Ferdous et al. (2006)
	Fatty acids contents	QTL	Hyten et al. (2004)
	Soybean rust	*Rpp1*	Hyten et al. (2007)
	Root and stem rot	*Rps 3, Rps 8*	Sandhu et al. (2005)

closely linked genes for resistance to Fusarium wilt race 4 and 5 in chickpea were also identified from a population of 131 RILs derived from a wide cross between *C. arietinum* and *C. reticulatum* (Benko-Iseppon et al., 2003). These markers will pave the way for MAS and searching other useful genes. Gowda et al. (2009) identified flanking markers for chickpea Fusarium wilt resistance genes in a RIL population. Reddy et al. (2009) performed bulk segregant analysis on a segregating population of ICPL 7035 × ICPL 8863 for identification of RAPD markers associated with pigeon pea sterility mosaic disease resistance. Dhanasekhar et al. (2010) identified two RAPD markers OPF04700 and OPA091375 were linked with the open and tall plant type gene in pigeon pea F_2 population of the cross between TT44-4 and TDI2004-1 through bulk segregant analyses. Kotresh et al. (2006) used bulk segregant analysis with 39 RAPD primers which led to identification of two markers (OPM03704 and OPAC11500) that were associated with Fusarium wilt susceptibility allele in a pigeon pea F_2 population derived from GS1 × ICPL87119. Saxena (2010) assessed the DNA polymorphism in a set of 32 pigeon pea lines screened with 30 SSR markers. Based on polymorphism of marker alleles, higher genetic dissimilarity coefficient and phenotypic diversity for Fusarium wilt and sterility mosaic disease resistance data, five parental combinations were identified for developing genetically diverse mapping populations suitable for the development of tightly linked markers for Fusarium wilt and sterility mosaic disease resistance. Tullu et al. (2003) tagged anthracnose resistance gene LCt-2 of lentil cultivar PI 320937 with RAPD and AFLP markers. Taran et al. (2003) identified two molecular markers associated with Ascochyta blight resistance in lentil, viz., UBC 2271290 linked with ral1 gene and RB18680 linked with AbR1 and a marker (OPO61250) linked with anthracnose resistance gene were utilized for identifying lines that possessed pyramided genes in a population of 156 RILs developed from a cross between 'CDC Robin' and a breeding line '964a-46.' These markers can be converted into more robust SCAR markers for routine use in marker-assisted selection. Basak et al. (2004) developed molecular marker linked to yellow mosaic virus (YMV) resistance gene in *Vigna* sp. from a population segregating for YMV disease resistance. Maiti et al. (2010) identified molecular markers CYR1 and YR4 in a F_2 population for screening of MYMIV resistance genes. CYR1 cosegregated with MYMV resistance gene in F_2 plants and F_3 progenies. These two markers can be used simultaneously with the help of a multiplex PCR reaction. Nguyen et al. (2001) converted a RAPD marker into a SCAR (SCARW19) for selecting Ascochyta blight resistance gene of lentil accession ILL5588. Rubeena et al. (2003) identified QTLs for Ascochyta blight resistance in lentil. Further validation is required to use these markers for MAS. Hamwieh et al. (2005) mapped microsatellite markers identified from a genomic library of lentil. The linkage spanning about 751cM, consisting of 283 marker loci was derived from 86 RILs derived from the cross ILL 5588 × L 692-16-1(s) using 41 microsatellite and 45 AFLP markers. The average marker distance was 2.6 cM. Two flanking markers (SSR marker SSR59-2B at 8.0 cM and AFLP marker p17m30710 at 3.5 cM) were linked with Fusarium resistance. Similar kind of study was also performed in chickpea, wheat, maize, rice, and soybean for the resistance to biotic (bacteria, virus, fungi, nematodes, and insects) and abiotic stresses (cold, drought, and salinity) (Table 20.1).

20.5 PROSPECTS OF GENETIC MAPPING

The recent advancement of genetic maps based on the markers that are simple to generate, highly reproducible, codominant, and specific for known linkage groups are highly desirable for their application in breeding and molecular geneticists as well. The transferability of maps constructed using AFLPs, RAPDs, and ISSRs is limited between populations and pedigrees within a species because each marker is primarily defined by its length (i.e., sequence information may be limited). Moreover, the same size band amplified across populations/species does not necessarily mean that bands possess the same sequence, unless proven by hybridization studies. In contrast, the development of high-density maps that incorporate EST-derived RFLP, SSR, and SNP markers will provide researchers with a greater arsenal of tools for identifying genes or QTLs associated with economically important traits. Furthermore, such EST-based markers mapped in one population can be used as probes and primers for characterizing other populations within the same species. The present chapter will be helpful for the effective utilization of available resources and the planning of crop improvement programs employing molecular marker techniques.

KEYWORDS

- **mapping population**
- **linkage**
- **cross-over**

- **F_2 population**
- **backcross population**
- **NIL**
- **RIL**
- **IL**
- **MAGIC population**
- **MAS**
- **QTL**

REFERENCES

Ali, M. L.; Taylor, J. H.; Jie, L.; Sun, G.; William, M.; Kasha, K. J.; Reid, L. M.; Pauls, K. P. Molecular Mapping of QTLs for Resistance to Gibberella Ear Rot, in Corn, Caused by *Fusarium graminearum*. *Genome* **2005,** *48*, 521–533.

Babu, R. C.; Nguyen, B. D.; Chamarerk, Y.; Shanmugasundaram, P.; Chezhian, P.; Jeyaprakash, P.; Ganesh, S. K.; Palchamy, A.; Sadasivam, S.; Sarkarung, S.; Wade, L. J.; Nguyen, H. T. Genetic Analysis of Drought Resistance in Rice by Molecular Markers. *Crop Sci. Soc. Am.* **2005,** *43*(4), 1457–1469.

Basak, J.; Kundagrami, S.; Ghoose, T. K.; Pal, A. Development of Yellow Mosaic Virus (YMV) Resistance Linked DNA Marker in *Vigna mungo* from Populations Segregating for YMV-reaction. *Mol. Breed.* **2004,** *14*, 375–382.

Benko-Iseppon, A.; Winter, M.; Huettel, P.; Staginnus, B.; Muehlbauer, C.; Kahl, F. G. Molecular Markers Closely Linked to Fusarium Resistance Genes in Chickpea Show Significant Alignments to Pathogenesis-related Genes Located on *Arabidopsis* Chromosomes 1 and 5. *Theor. Appl. Genet.* **2003,** *107*, 379–386.

Bergelson, J.; Roux, F. Towards Identifying Genes Underlying Ecologically Relevant Traits in *Arabidopsis thaliana*. *Nature Rev. Genet.* **2010,** *11*, 867–879.

Botstein, D.; White, R. L.; Skolnick, M.; Davis, R. W. Construction of a Genetic Linkage Map in Man Using Restriction Fragment Length Polymorphisms. *Am. J. Hum. Genet.* **1980,** *32*, 314–331.

Barilli, F.; Rubiales, D.; Castillejo, M. A. Comparative Proteomic Analysis of BTH and BABA-induced Resistance in Pea (*Pisum sativum*) Toward Infection with Pea Rust (*Uromycespisi*). *J. Proteomics* **2012,** *75*, 5189–5205.

Brooks, S. A.; See, D. R.; Brown-Guedira, G. SNP-based Improvement of a Microsatellite Marker Associated with Karnal Bunt Resistance in Wheat. *Crop Sci. Soc. Am.* **2005,** *46*(4), 1467–1470.

Butrón, A.; Li, R.; Guo, B. Z.; Widstrom, N. W.; Snook, M. E.; Cleveland, T. E. Molecular Markers to Increase Corn Earworm Resistance in a Maize Population. *Maydica* **2001,** *46*, 117–124.

Buschges, R.; Hollrichter, K.; Panstruga, R. The Barley Mlo Gene: A Novel Control Element of Plant Pathogen Resistance. *Cell* **1997,** *88*, 695–705.

Chen, X.; Salamini, F.; Gebhardt, C. A Potato Molecular-function Map for Carbohydrate Metabolism and Transport. *Theor. Appl. Genet.* **2001,** *102*, 284–295.

Chao, S.; Sharp, P. J.; Worland, A. J.; Warham, E. J.; Koebner, R. M. D.; Gale, M. D. RFLP-based Genetic Maps of Wheat Homoeologous Group 7 Chromosomes. *Theor. Appl. Genet.* **1989,** *78*, 495–504.

Dhanasekar, P.; Dhumal, K. N.; Reddy, K. S. Identification of RAPD Markers Linked Plant Type Gene in Pigeonpea. *Indian J. Biotech.* **2010,** *9*, 58–63.

Datta, D.; Gupta, S.; Chaturvedi, S. K.; Nadarajan, N. *Molecular Markers in Crop Improvement*; Indian Institute of Pulses Research: Kanpur, **2011,** 1–54.

Eshed, Y.; Zamir, D. An Introgression Line Population of *Lycopersicon pennellii* in the Cultivated Tomato Enables the Identification and Fine Mapping Of Yield-associated QTL. *Genetics* **1995,** *141*, 1147–1162.

Ferdous, S. A.; Watanabe, S.; Suzuki-Orihara, C.; Kamiya, M.; Yamanaka, N.; Harada, K. QTL Analysis of Resistance to Soybean Cyst Nematode Race 3 Toyomusume. *Breeding Sci.* **2006,** *56*, 155–163.

Frei, O. M.; Stuber, C. W.; Goodman, M. M. Use of Allozymes as Genetic Markers for Predicting Performance in maize single-cross hybrids. *Crop Sci.* **1986,** *26*, 37–42.

George, M. L. C.; Prasanna, B. M.; Rathore, R. S.; Setty, T. A. S.; Kasim, T.; Azrai, M.; Vasal, S.; Balla, O.; Hautea, D.; Canama, A.; Regalado, E.; Vargas, M.; Khairallah, M.; Jeffers, D.; Hoisington, D. Identification of QTLs Conferring Resistance to Downy Mildews of Maize in Asia. *Theor. Appl. Genet.* **2003,** *107*, 544–551.

Gowda, S. J. M.; Radhika, P.; Kadoo, N. Y.; Mhase. L. B.; Gupta, V. S. Molecular Mapping of Wilt Resistance Genes in Chickpea *Mol. Breed.* **2009,** *24*, 177–183.

Gupta, P. K.; Kulwal, P. L.; Mir, R. R. QTL Mapping: Methodology and Applications in Cereal Breeding. *Cereal Genom.* **2013,** *II*, 275–318.

Hamwieh, A.; Udupa, S. M.; Choumane, W.; Sarker, A.; Dreyer, F.; Jung, C.; Baum, M. A Genetic Linkage Map of Lens sp. Based on Microsatellite and AFLP Markers and the Localization of Fusarium Vascular Wilt Resistance. *Theor. Appl. Genet.* **2005,** *110*, 669–677.

Hayden, M. J.; Kuchel, H.; Chalmers, K. J. Sequence Tagged Microsatellites for the Xgwm533 Locus Provide New Diagnostic Markers to Select for the Presence of Stem Rust Resistance Gene *Sr2* in Bread Wheat (*Triticum aestivum* L.). *Theor. Appl. Genet.* **2004,** *109*, 1641–1647.

Heun, M.; Kennedy, A. E.; Anderson, J. A.; Lapitan, N. L. V.; Sorrells, M. E.; Tanksley, S. D. Construction of a Restriction Fragment Length Polymorphism Map for Barley (*Hordeum vulgare*). *Genome* **1991,** *34*, 437–447.

Huang, B. E.; Verbyla, K. L.; Verbyla, A. P.; Raghavan, C.; Singh, V. K.; Gaur, H.; Leung, H.; Varshney, R. K.; Cavanagh, C. R. MAGIC Populations in Crops: Current Status and Future Prospects. *Theor. Appl. Genet.* **2015,** *128*, 999–1017.

Hwang, T. Y.; Moon, J. K.; Yu, S.; Yang, K.; Mohankumar, S.; Yu, Y. H.; Lee, Y. H.; Kim, H. S.; Kim, M. A. H. M.; Maroof, S.; Jeong, S. N. Application of Comparative Genomics in Developing Molecular Markers Tightly Linked to the Virus Resistance Gene *Rsv*4 in Soybean. *Genome* **2006,** *49*(4), 380–388.

Hyten, D. L.; Pantalone, V. R.; Saxton, A. M.; Schmidt, M. E.; Sams, C. E. Molecular Mapping and Identification of Soybean Fatty Acid Modifier Quantitative Trait Loci. *J. Am. Oil Chem. Soc.* **2004,** *81*(12), 1115–1118.

Hyten, D. L.; Hartman, G. L.; Nelson, R. L.; Frederick, R. D.; Concibido, V. C.; Narvel, J. M.; Cregan, P. B. *Map Location of the Rpp1 Locus that Confers Resistance to Soybean Rust in Soybean*; Agronomy & Horticulture Faculty Publications: **2007,** Paper 784.

Koornneef, M.; Hanhart, C. J.; von Loenen Martinet, E. P.; Peeters, A. J. M.; van der Veen, J. H. Trisomicsin *Arabidopsis thaliana* and the Location of Linkage Groups. *Arabidopsis Inf. Serv. Frankfurt.* **1987,** *23*, 46–50.

Kotresh, H.; Fakrudin, B.; Punnuri, S. M.; Rajkumar, B. K.; Thudi, M.; Paramesh, H.; Lohithaswa, H.; Kuruvinashetti, M. S. Identification of Two RAPD Markers Genetically Linked to a Recessive Allele of a Fusarium Wilt Resistance Gene in Pigeon Pea (*Cajanus cajan* L. Millsp.). *Euphytica* **2006,** *149*, 113–120.

Kumar, N.; Kulwal, P. L.; Balyan, H. S.; Gupta, P. K. QTL Mapping for Yield and Yield Contributing Traits in Two Mapping Populations of Bread Wheat. *Mol. Breed.* **2007,** *19*, 163–177.

Li, Z.; Mu, P.; Li, C.; Zhang, H.; Li, Z.; Gao, Y.; Wang, X. QTL Mapping of Root Traits in a Doubled Haploid Population from a Cross Between Upland and Lowland *Japonica* Rice in Three Environments. *Theor. Appl. Genet.* **2005,** *110*(7), 1244–1252.

Lister, C.; Dean, C. Recombinant Inbredlines for Mapping RFLP and Phenotypic Markers in *Arabidopsis thaliana*. *Plant J.* **1993,** *4*, 745–750.

Lou, Q.; Chen, L.; Sun, Z.; Xing, Y.; Li, J.; Xu, X.; Mei, H.; Luo, L. A Major QTL Associated with Cold Tolerance at Seedling Stage in Rice (*Oryza sativa* L.). *Euphytica* **2007,** *158*(1), 87–94.

McCouch, S. R.; Kochert, G.; Yu, Z. H.; Wang, Z. Y.; Khush, G. S.; Coffman, W. R.; Tanksley, S. D. Molecular Mapping of Rice Chromosomes. *Theor. Appl. Genet.* **1988,** *76*, 815–829.

Neuffer, M. G.; Coe, E.; Wessler, S. *Mutants of Maize;* Cold Spring Harbor Laboratory: New York, **1997**.

O'Donoughue, L. S.; Kianian, S. F.; Rayapati, P. J.; Penner, G. A.; Sorrells, M. E.; et al. Molecular Linkage Map of Cultivated Oat (Avena Byzantina XA. sativa cv. Ogle). *Genome* **1995,** *38*, 368–380.

Park, C.; Peng, Y.; Chen, X.; Dardick, C.; Ruan, D.; Bart, R.; Canlas, P. E.; Ronald, P. C. Rice XB15, a Protein Phosphatase 2C, Negatively Regulates Cell Death and XA21-mediated Innate Immunity. *PLoS Biol.* **2008,** *6*, e231.

Rafalski, A. Applications of Single Nucleotide Polymorphisms in Crop Genetics. *Curr. Opin. Plant Biol.* **2002,** *5*, 94–100.

Reddy, L. P.; Reddy, B. V.; Rekha, R. K.; Sivaprasad, Y.; Rajeswari, T.; Reddy, K. R. RAPD and SCAR Marker Linked to the Sterility Mosaic Disease Resistance Gene in Pigeon Pea (*Cajanus cajan* L. Millsp.). *The Asian and Aust. J. Plant Sci. Biotech.* **2009,** *3*, 16–20.

Rispail, N.; Rubiales, D. Identification of Sources of Quantitative Resistance to *Fusarium oxysporum* f. sp. Medicaginis. *Med. Truncatula Plant Dis.* **2014,** *98*, 667–673.

Rubeena, P.; Taylor, W. J.; Ades, P. K. QTL Mapping of Resistance in Lentil (*Lens culinaris* ssp. *culinaris*) to Ascochyta Blight (*Ascochyta lentis*). *Plant Breed.* **2003,** *125*, 506–512.

Rubiales, D.; Fondevilla, S.; Chen, W.; Gentzbittel, L.; Higgins, T. J. V.; Castillejo, M. A.; Singh, K. B.; Nicolas, R. Achievements and Challenges in Legume Breeding for Pest and Disease Resistance. *Cr. Rev. Plant Sci.* **2015,** *34*(1–3), 195–236. DOI: 10.1080/07352689.2014.898445.

Sanchez, A. C.; Brar, D. S.; Huang, N.; Li, Z.; Khush, G. S. Sequence Tagged Site Marker-assisted Selection for Three Bacterial Blight Resistance Genes in Rice. *Crop Sci. Soc. Am.* **2000,** *40*(3), 792–797.

Sandhu, D.; Schallock, K. G.; Rivera-Velez, N.; Lundeen, P.; Cianzio, S.; Bhattacharyya, M. K. Soybean Phytophthora Resistance Gene Rps8 Maps Closely to the Rps3 Region. *J. Hered.* **2005,** *96*(5), 536–541.

Saxena, R. K.; Saxena, K. B.; Kumar, R. V.; Hoisington, D. A.; Varshney, R. K. SSR-based Diversity in Elite Pigeon Pea Genotypes for Developing Mapping Populations to Map Resistance to Fusarium Wilt and Sterility Mosaic Disease. *Plant Breed.* **2010,** *129*, 135–141.

Sharma, R. C.; Duveiller, E.; Jacquemin, J. M. Microsatellite Markers Associated with Spot Blotch Resistance in Spring Wheat. *J. Phytopathol.* **2007,** *155*(5); 316–319.

Schneider, K.; Weisshaar, B.; Borchardt, D. C.; Salamini, F. SNP Frequency and Allelic Haplotype Structure of *Beta vulgaris* Expressed Genes. *Mol. Breed.* **2001,** *8*, 63–74.

Semagn, K.; Bjornstad, A.; Ndjiondjop, N. Principles, Requirements and Prospects of Genetic Mapping in Plants. *Afr. J. Biotechnol.* **2006,** *5*(25), 2569–2587.

Stuber, C. W.; Goodman, M. M.; Moll, R. H. Improvement of Yield and Ear Number Resulting from Selection of Allozyme Loci in a Maize Population. *Crop Sci.* **1972,** *22*, 737–740.

Su, C. C.; Wan, J.; Zhai, H. Q.; Wang, C. M.; Sun, L. H.; Yasui, H.; Yoshimura, A. A New Locus for Resistance to Brown Planthopper Identified in the Indica Rice Variety DV85. *Plant Breed.* **2005,** *124*, 93–95.

Swiecicki, W. K.; Wolko, B.; Weeden, N. F. Mendel's Genetics, the *Pisum* Genome and Pea Breeding. *Vortr. Pflanzenzchtg.* **2000,** *48*, 65–76.

Szostak, J. W.; Orr-Weaver, T. L.; Rothstein, R. J.; Stahl, F. W. The Double-strand-break Repair Model for Recombination. *Cell* **1983,** *33*, 25–35.

Tanksley, S. D.; Grandillo, S.; Fulton, T. M.; Zamir, D.; Eshed, Y.; Petiard, V.; Lopez, J.; Beckbunn, T. Advanced Backcross QTL Analysis in a Cross Between an Elite Processing Line of Tomato and its Wild Relative L. pimpinellifolium. *Theor. Appl. Genet.* **1996,** *92*, 213–224.

Taran, B.; Buchwald, L.; Tullu, A.; Banniza, S.; Warkentin, T. D.; Vandenberg, A. Using Molecular Markers to Pyramid Genes for Resistance to Ascochyta Blight and Anthracnose in Lentil (*Lens culinaris* Medik). *Euphytica* **2003,** *134*, 223–230.

Tullu, A.; Buchwaldt, L.; Warkentin, T.; Taran, B.; Vandenberg, A. Genetics of Resistance to Anthracnose and Identification of AFLP and RAPD Markers Linked to the Resistance Gene in PI320937 Germplasm of Lentil (*Lens culinaris* Medikus). *Theor. Appl. Genet.* **2003,** *106*, 428–434.

Varshney, R. K.; Kudapa, H.; Pazhamala, L.; Chitikineni, A.; Thudi, M.; Bohra, A.; Gaur, P. M.; Janila, P.; Fikre, A.; Kimurto, P.; Ellis, N. Translational Genomics in Agriculture: Some Examples in Grain Legumes. *Cr. Rev. Plant Sci.* **2015,** *34*(13), 169–194.

Vos, P.; Hogers, R.; Bleeker, M.; Reijans, M.; vander Lee, T.; Fornes, M.; Frijters, A.; Pot, J.; Peleman, J.; Kuiper, M.; Zabeau, M. AFLP: A New Technique for DNA Fingerprinting. *Nucleic Acids Res.* **1995,** *23*, 4407–4414.

Wang, G. L.; Mackill, D. J.; Bonman, J. M.; McCouch, S. R.; Champoux, M. C.; Nelson, R. J. RFLP Mapping of Genes Conferring Complete and Partial Resistance to Blast in a Durably Resistant Rice Cultivar. *Genetics* **1994,** *136*, 1421–1434.

Welsh, J.; McClelland, M. Fingerprinting Genomes Using PCR with Arbitrary Primers. *Nucleic Acids Res.* **1990,** *18*, 7213–7218.

Williams, J. G. K.; Kubelik, A. R.; Livak, K. J.; RafalskiJ, A.; Tingey, S. V. DNA Polymorphisms Amplified by Arbitrary Primers are Useful as Genetic Markers. *Nucleic Acids Res.* **1990,** *18*, 6531–6535.

Winter, P.; Benko-Iseppon, A. M.; Huttel, B.; Ratnaparkhe, M.; Tullu, A.; Sonnante, G.; Pfaff, T.; Tekeoglu, M.; Santra, D.; Sant, V. J.; Rajesh, P. N.; Kahl, G.; Muehlbauer, F. J. A Linkage Map of the Chickpea (*Cicerarietinum* L.) Genome Based on Recombinant Inbred

Lines from *C. arietinum* × *C. reticulatum* Cross: Localization of Resistance Genes for Fusarium Wilt Race 4 and 5. *Theor. Appl. Genet.* **2000,** *101*, 115–1163.

Yadav, S.; Anuradha, G.; Kumar, R. R.; Vemireddy, L. R.; Sudhakar, R.; Donempudi, K.; Venkata, D.; Jabeen, F.; Narasimhan, Y. K.; Marathi, B.; Siddiq, E. A. Identification of QTLs and Possible Candidate Genes Conferring Sheath Blight Resistance in Rice (*Oryza sativa* L.). *SpringerPlus* **2015,** *4*(1), 1–2.

Yang, W.; Zheng, Y.; Zheng, W.; Feng, R. Molecular Genetic Mapping of a High-lysine Mutant Gene (opaque-16) and the Double Recessive Effect with Opaque-2 in Maize. *Mol. Breed.* **2005,** *15*, 257–269.

Young, N. D.; Tanksley, S. D. RFLP Analysis of the Size of Chromosomal Segments Retained Around the Tm-2 Locus of Tomato During Backross Breeding. *Theor. Appl. Genet.* **1989,** *77*, 353–359.

Zamir, D. Improving Plant Breeding with Exotic Genetic Libraries. *Nature Rev.* **2001,** *2*, 983–989.

Zargar, S. M.; Raatz, B.; Sonah, H.; Nazir, M.; Bhat, J. A.; Dar, Z. A.; Agrawal, G. K.; Rakwal, R. Recent Advances in Molecular Marker Techniques: Insight into QTL Mapping, GWAS and Genomic Selection in Plants. *J. Crop Sci. Biotechnol.* **2015,** *18*(5), 293–303.

Zhou, L. L.; Bai, G. H.; Ma, H.; Carver, B. F. Quantitative Trait Loci for Aluminum Resistance in Wheat. *Mol. Breed.* **2007**. DOI: 10.1007/s11032-006-9054.

Zhu, L. C.; Smith, C. M.; Fritz, A.; Boyko, E.; Voothuluru, P.; Gill, B. S. Inheritance and Molecular Mapping of New Greenbug Resistance Genes in Wheat Germplasms Derived from *Aegilops tauschii*. *Theor. Appl. Genet.* **2005**. DOI: 10.1007/s00122-005-0003-6.

CHAPTER 21

PRINCIPLES AND PRACTICES OF MAPPING QTLS IN PLANTS

SUNAYANA RATHI[1*], AKHIL RANJAN BARUAH[2], SUROJIT SEN[3], and SAMINDRA BAISHYA[1]

[1]*Department of Biochemistry and Agricultural Chemistry, Assam Agricultural University, Jorhat, Assam, India*

[2]*Department of Agricultural Biotechnology, Assam Agricultural University, Jorhat, Assam, India*

[3]*Department of Zoology, Mariani College, Mariani 785634, Jorhat, Assam, India*

[*]*Corresponding Author. Email: rathisunayana@yahoo.co.in*

CONTENTS

ABSTRACT

The major challenge in current biology is to understand the genetic basis of variation for quantitative traits. The genetic variation of a quantitative trait is controlled by the collective effects of numerous genes, known as quantitative trait loci (QTLs). Identification of QTLs of agronomic importance and its utilization in a crop improvement requires mapping of these QTLs in the genome of crop species using molecular markers. Mapping quantitative trait loci (QTL) for complex traits has become a routine tool in functional genomic research. This chapter will focus on the basic concepts, prerequisites of QTL mapping, approaches for understanding the function of a gene, a brief description of existing methodologies for QTL mapping and their advantages and disadvantages.

21.1 INTRODUCTION

The recent technology in the field of "omics" introduced us to the wealth of genetic information that has profound impact on the study of evolutionary biology, particularly on understanding one of the most enduring problems in evolution and molecular biology—the genetic basis of complex traits. The complexities of these phenotypic traits are mainly due to the involvement of environment and genetic effect adapted to particular situation, probably arises from segregation of alleles at many interacting loci (quantitative trait loci, or QTL). The advances in molecular genetics and statistical techniques make it possible to identify the chromosomal regions where these QTL are located and cloning of those. Ultimately, an understanding of adaptive evolution will require detailed knowledge of the genetic changes that accompany evolutionary change facilitated by QTL mapping.

A QTL is a section of DNA (the locus) that correlates with variation in a phenotype (the quantitative trait) (Miles and Wayne, 2008). The QTL typically is linked to, or contains, the genes that control that phenotype. QTLs are mapped by identifying which molecular markers [such as single nucleotide polymorphisms (SNPs) or amplified fragment length polymorphisms (AFLPs)] correlate with an observed trait. This is often an early step in identifying and sequencing the actual genes that cause the trait variation. QTL mapping is the statistical study of the alleles that occur in a locus and the phenotypes (physical forms or traits) that they produce. The process of constructing linkage maps and conducting QTL analysis to identify genomic regions associated with traits is known as QTL mapping (McCouch and Doerge, 1995). QTL mapping is based on the segregation of genes and markers via chromosome recombination (called crossing-over) during meiosis (i.e., sexual reproduction), thus allowing their analysis in the progeny (Paterson, 1996). The basic principle of determining whether a QTL is linked to a marker is to partition the mapping population into different genotypic classes based on genotypes at the marker locus, and to apply correlative statistics to determine whether the individuals of one genotype differ significantly with the individuals of other genotype with respect to the trait being measured.

21.2 PREREQUISITES FOR QTL MAPPING

Availability of a good linkage map (this can be done at the same time the QTL mapping) is a prerequisite for QTL mapping. A segregating population derived from parents that differ for the trait(s) of interest, and which allow for replication of each segregant, so that phenotype can be measured with precision [such as Recombinant Inbred Lines (RILs) or Double Haploids (DHs)].

21.2.1 SIZE OF MAPPING POPULATION USED IN QTL MAPPING

In large sample size, QTL with small effects cannot be observed but QTL with large effects can be observed. In small sample size also, QTL with small effects cannot be observed but QTL with major effects can be observed (Khan, 2015).

21.2.2 TYPE AND NUMBER OF MARKERS IN LINKAGE MAPS

If the number of markers used is more, amount of precision of estimation of both QTL position and effect will be more. Here, codominant marker shows three types of genetic differences while dominant marker shows two types of genetic differences. So, codominant marker provides more information than dominant marker regarding recombination within marker intervals (Khan, 2015).

21.2.3 PHENOTYPING OF MAPPING POPULATION AND SAMPLE SIZE

The target quantitative traits are measured as precisely as possible and limited amounts of missing data can be tolerated. The power to resolve the QTL location is confined first by sample size and then by genetic marker coverage of the genome. Generally, the number of individuals in a sample might appear to be large but missing data or skewed allele frequencies in the population cause the effective sample size to diminish, thus sacrificing the statistical power. Sometimes, it is must to sacrifice population size in favor of data quality and this trade-off means that only major QTL (with relatively large effect) can be detected. QTL data are typically pooled over locations and replications to obtain a single quantitative trait for the line. It is also preferred to measure the target trait(s) in experiments conducted in multiple (and appropriate) locations to have a better understanding of the QTL versus environment interaction, if any (Khan, 2015).

21.3 MAPPING QTL

The improvement of quantitative traits has been an important goal for many plant breeding programs. With a pedigree breeding program, the breeder will cross two parents and practice selection until advanced-generation lines with the best phenotype for the quantitative trait under selection are identified. These lines will then be entered into a series of replicated trials to further evaluate the material with the goal of releasing the best lines as a cultivar. It is assumed that those lines which performed best in these trials have a combination of alleles most favorable for the fullest expression of the trait.

This type of program, though, requires a large input of labor, land, and money. Therefore, plant breeders are interested in identifying the most promising lines as early as possible in the selection process. Another way to state this point is that the breeder would like to identify as early as possible those lines which contain those QTL alleles that contribute to a high value of the trait under selection. Plant breeders and molecular geneticists have joined efforts to develop the theory and technique for the application of molecular genetics to the identification of QTLs.

Molecular makers associated with QTLs are identified by first scoring members of a random segregating population for a quantitative trait. The molecular genotype (homozygous Parent A, heterozygous, or homozygous parent B) of each member of the population is then determined. The next step is to determine if an association exists between any of the markers and the quantitative trait.

The most common method of determining the association is by analyzing phenotypic and genotypic data by one-way analysis of variance and regression analysis. For each marker, each of the genotypes is considered a class, and all of the members of the population with that genotype are considered an observation for that class. Data is typically pooled over locations and replications to obtain a single quantitative trait value for the line. If the variance for the genotype class is significant, then the molecular marker used to define the genotype class is considered to be associated with a QTL. For those loci that are significant, the quantitative trait values are regressed onto the genotype. The R^2 value for the line is considered to be the amount of total genetic variation that is explained by the specific molecular marker. The final step is to take those molecular marker loci that are associated with the quantitative trait and perform a multiple regression analysis. From this analysis, you will obtain an R^2 value which gives the percentage of the total genetic variance explained by all of the markers. The two types of populations that have been used to identify markers linked to QTLs are F2*3 families (or F3 families from F2 plants) and recombinant inbred lines. Each population type has advantages and disadvantages. The primary advantage of F2*3 families is the ability to measure the effects of additive and dominance gene actions at specific loci because RI lines are essentially homozygous, only additive gene action can be measured. The advantage, though, of the RI lines is the ability to perform larger experiments at several locations and even in multiple years. For many crops, it is not possible to generate enough seed to perform a multilocation experiment with population of F2*3 families. The overview of QTL mapping is depicted in Figure 21.1.

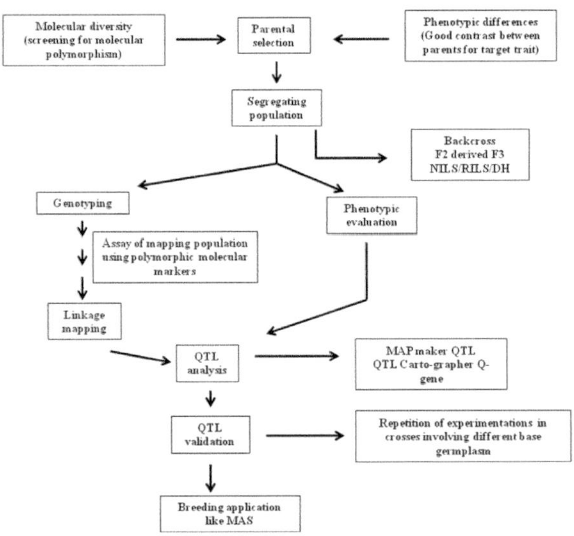

FIGURE 21.1 Overview of QTL mapping. (Reprinted from Neeraj Anand [2014], Concept of Molecular Marker: A Student's Approach. First Edition, Sharma Publications and Distributors, New Delhi, ISBN: 978-93-82310-06-8.)

21.4 APPROACHES FOR UNDERSTANDING THE FUNCTION OF A GENE

There are two general approaches to understanding the function of a gene: forward genetics and reverse genetics. Forward genetics begins with a well-characterized phenotype, such as resistance to a disease, and then works toward identifying the gene(s) responsible for the phenotype. Reverse genetics begins with a gene, for example a protein kinase, and works toward determining which phenotype(s) it determines (White et al., 2007). Reverse genetics is a particular approach in discovering the function of a gene that usually goes in the opposite direction of what is called as forward genetic screens associated with classical genetics. To put it simply, while forward genetics has the goal of trying to find the genetic basis of a phenotype or trait, reverse genetics is aimed at finding the possible phenotypes that may be derived from a specific genetic sequence that is detailed in DNA sequencing (Pekosz et al., 1999). Genetic mapping approaches such as QTL mapping and association mapping are also forward genetic approaches and are often used because gene transfer is not required (Tierney and Lamour, 2005). QTL mapping can be defined as the marker-facilitated genetic dissection of variation of complex phenotypes through appropriate experimental design and statistical analyses of segregating materials.

21.5 TYPES OF QTLS

To date, QTLs identified can be classified into two major types: main-effect QTLs (M-QTLs) and epistatic QTLs (E-QTLs), based largely on the presence or absence of epistasis. Distinction of the two types of QTLs is critical to our understanding of the genetic basis of quantitative trait variation (Li, 2000).

21.5.1 M-QTLS

M-QTLs are defined as single Mendelian factors at which effects (additive and/or dominance) on a given phenotype arise from allelic substitution and are detected by marker–trait associations using single-factor ANOVA or interval mapping models (Lander and Botstein, 1989; Li et al., 1997; Zeng, 1994). M-QTLs appear to include two groups of genes. The first group includes major genes of very large effects on highly heritable traits, which are typically detected with very large LOD (Logarithm of the Odds) scores (> 10.0), and each explains a large portion of the total trait variation in a mapping population. Examples of this type are sd-1 for semidwarf stature, *Xa4* for bacterial blight resistance, *Ta9* for tiller angle, *Hd*-1, *Hd*-3, and *QHd3* for heading date, etc. The second group includes the typical M-QTLs, which represent most (more than 90%) QTLs reported to date. These typical M-QTLs tend to have relatively small effects. There are two general results regarding the M-QTLs from previous studies (Newbury, 2003). First, the number of detected M-QTLs for a specific trait in a population evaluated in a specific environment is relatively small. Based on the 324 cases (trait/population/environment combinations) in the QTL mapping studies involving 46 mapping populations and 71 phenotypes, the average number of detected M-QTLs per trait/population/environment is 3.7 ± 1.2 and, surprisingly, this number does not differ between high- and low-heritability traits. For example, the average detectable number of M-QTLs per population/environment is 3.7 and 4.1 for two highly heritable traits, days to heading and plant height, respectively, and 3.3 and 3.4 for the low-heritability traits, grains per panicle and grain yield, respectively. However, when a comparison is made across mapping populations and environments, many more M-QTLs are detected for each trait. This underestimation of M-QTL number in most QTL mapping studies is due largely to epistasis and QE interactions (Li, 2000). Second, the accuracy of the estimated M-QTL genetic parameters, such as their effects, genomic locations, and gene actions, varies considerably, depending largely on errors in phenotyping and the statistical methods for parameter estimation. In the former case, the nature and size of mapping populations and use of replications in phenotyping play a key role in reducing phenotyping errors. In the latter case, control of background genetic variation and inclusion of epistasis and QE interactions in the statistical models are vital to obtaining more reliable parameter estimates (Zeng, 1994; Li et al., 1997; Li et al., 1999, Wang et al., 1999).

21.5.2 E-QTLS

The second type of QTLs is epistatic QTLs, or E-QTLs. E-QTLs are defined as loci at which trait values are determined by interactions between alleles at two or more loci and are detected by associations between trait values and multilocus marker genotypes using epistatic models (Li et al., 1997; Kao et al., 1999; Wang et al., 1999). In other words, trait values (phenotypes) are associated with specific alleles at single loci for M-QTLs, but with multilocus genotypes for E-QTLs. Three types of epistasis can be recognized. Type I is well described in classic quantitative genetics theory (Mather and Jinks, 1982), in which two M-QTLs are involved in epistasis and affect the same phenotype. Type II involves interactions between alleles at an M-QTL and a background (or modifying) locus. Type III represents epistasis between two complementary loci that do not have detectable main effects (Li et al., 1997).

21.6 APPROACHES TO MAPPING EXPERIMENTAL CROSS (SEGREGATING POPULATION)

Backcrosses, F_2 intercrosses, RIL, double haploids are used for mapping studies. Association studies (Linkage disequilibrium, LD mapping) with candidate genes (direct approach), localized association studies (chromosomal region) and whole-genome association studies are also regarded as approaches to mapping studies. There is a need of segregating population in natural population with consistent association between QTL and marker genotype which usually does not exist (except where marker is completely linked to QTL, which is very rare). So, to study the recombination between QTL and marker, segregating population is useful.

21.7 STATISTICAL METHODS FOR QTL MAPPING

21.7.1 SINGLE MARKER APPROACH

The single marker approach is also known as single factor analysis of variance or single point analysis. It is widely used method for quick scanning of whole genome to determine best QTLs. It is used for each marker locus which is free from other loci. Generally, this technique is unable to determine QTL position. F-test is used for determination of significant differences between various genotypes groups. Some major limitations of this approach are as follows: the method cannot determine whether the markers are associated with one or more QTLs; chance of QTL detection decreases with distance between marker and QTL; effect of QTL is underestimated due to confounding with recombination frequencies, and its accuracy is less compared with other methods.

Interval mapping uses the two flanking markers. A separate analysis is performed for each pair of adjacent marker loci. The use of such two-locus marker genotypes results in $n - 1$ separate tests of marker–trait associations for a chromosome with n markers (one for each marker interval). Both single-marker and interval mapping approaches are biased when multiple QTLs are linked to the marker/interval being considered. Methods simultaneously using three or more marker loci attempt to reduce or remove such bias. Interval mapping offers increased power of detection and more precise estimates of QTL effects and position. It evaluates the association between the trait values and the expected contribution of a QTL (the target QTL) at multiple analysis points between each pair of adjacent marker loci (the target interval). The expected QTL genotype is estimated from the genotypes of flanking marker loci and their distance from the QTL. Since there is usually uncertainty in the QTL genotype, the likelihood is a sum of terms, one for each possible QTL genotype, weighted by the probability of that genotype given the genotypes of the flanking markers. The analysis point that yields the most significant association may be taken as the location of a putative QTL.

21.7.2 SIMPLE INTERVAL MAPPING

Simple interval mapping (SIM) was first proposed by Lander and Botstein and it is based on linkage map. It can be called two-marker approach. Here, QTL is determined in interval generated between two markers at various points. It gives more accurate results compared with single-marker approach but less than composite interval mapping (CIM) and multiple interval mapping (MIM) techniques. In this technique, likelihood ratio test is used to determine every QTL position in interval created by both markers. SIM is mostly preferred as it can be easily performed through statistical packages such as MAPMAKER/QTL. Lander and Botstein (1989) developed formulae for significance levels appropriate for interval mapping when the genome size, number of chromosomes, number of marker intervals, and the overall false positive rate desired are given. However, when various QTLs are segregating in a cross, SIM will not take into consideration genetic variance due to other. In such a case, SIM is having same limitation as in single-marker analysis.

21.7.3 COMPOSITE INTERVAL MAPPING

CIM and MQM (Multiple QTL Mapping) techniques are developed by Jansen and Stam (1994). It is used to minimize effects of various linked QTLs. It is based on one QTL and other markers used as covariates. This technique gives more precise results and used to exclude bias due to other QTLs (nontarget QTLs) linked to target QTL. It is used to fit the parameters for a single QTL in one interval. The partial regression coefficient is used to determine genetic variance due to nontarget QTLs. It considers a marker interval and a few other selected single markers in each QTL analysis, so that $n - 1$ tests for interval–QTL associations are conducted on a chromosome with n markers. The merits of CIM are as follows: mapping of multiple QTLs can be carried out by the search in one dimension; by using linked markers as covariates, the test is not affected by QTL out of region, thereby increasing the precision of QTL mapping; and by eliminating as much as the genetic variance produced by other QTL, the residual variance is reduced, thereby increasing the efficiency of determination of QTL. CIM is more efficient than SIM, but not widely used in QTL mapping as in SIM.

21.7.4 MULTIPLE INTERVAL MAPPING

It is a recent method of QTL mapping. MIM is the extension of interval mapping to multiple QTLs, just as multiple regressions extend analysis of variance. It is used to map multiple QTLs. This method is a potential tool for detection of QTL versus QTL interaction.

21.8 METHODS TO DETECT QTLS

CIM evaluates the possibility of a target QTL at multiple analysis points across each intermarker interval (same as SIM). However, at each point, it also includes the effect of one or more background markers. Background markers have been shown to be associated with the trait and therefore lie close to other QTLs (background QTLs) affecting the trait (Jansen, 1993). The inclusion of a background marker in the analysis helps in one of two ways, based upon the linkage of background marker and the target interval if they are linked, inclusion of the background marker may help to separate the target QTL from other linked QTLs. If they are not linked, inclusion of the background marker makes the analysis more sensitive to the presence of a QTL in the target interval (Zeng, 1993, 1994). Interval mapping takes proper account of missing data, interpolates positions between markers, provides a support interval, provides more accurate estimate of QTL effect.

The power of a QTL detection experiment, defined as the probability of detecting a QTL at a given level of statistical significance, depends upon the strength of the QTL and the number of progeny in the population. If we consider the strength of the QTL in terms of the fraction of the total trait variance that it explains, we can define three categories of QTLs. Those which explain over 20% of the variance are strong QTLs; traits controlled by such QTLs can be considered almost Mendelian. Strong QTLs can be detected with a power greater than 80% even with the A×B/B×A set of recombinant inbred strains. At the other extreme, weak QTLs, which explain 1% or less of the trait variance, require at least a thousand progenies to detect them with high power. Detection of such QTLs is not routinely feasible. The number of progeny required to detect a QTL is, roughly speaking, proportional to the variance of the nongenetic (environmental) contributions and inversely proportional to the square of the strength of the QTL.

Different software is available for mapping QTLs (Table 21.1).

21.9 ADVANTAGES OF QTL MAPPING

There are four advantages of this mapping strategy. (1) First, by confining the test to one region at a time, it reduces a multidimensional search problem (for multiple QTLs) to a one-dimensional problem, and also estimates of locations and effects of individual QTLs are likely to be asymptotically unbiased. (2) Second, by conditioning on linked markers in the test, the precision of QTL mapping can be greatly improved. (3) Third, by selectively conditioning multiple markers in the test, the method simultaneously utilizes more information in the data to make inferences and should be more informative and efficient for mapping QTLs than the current methods. (4) Fourth, it can still use the QTL likelihood map (the likelihood profile) to present the strength of the evidence for QTLs at various positions along the entire genome, and preserves the feature of interval mapping. These advantages are brought about by the realization that a complete linkage map can be used not only to provide an anchor to fix a position to test for a QTL anywhere in a genome covered by markers (interval mapping), but also to provide a boundary condition for the test, and at the same time to control the residual genetic variation in the rest of the genome for the test (interval test).

Where mutant approaches fail to detect genes with phenotypic functions, QTL mapping can be good alternative when mutant screening is laborious and expensive, e.g., circadian rhythm screens can identify new functional alleles of known function genes, e.g., flowering time QTL, EDI was the CRY2 gene. Natural variation studies provide insight into the origins of plant evolution with identification of novel genes.

TABLE 21.1 Software Ava lable for Mapping QTLs.

Name	Platform	Remarks	References	URL
Mapmaker/QTL Version 1.1	Unix or DOS	Widely used program; interval mapping and composite interval mapping; requires Mapmaker/Exp for map construction; freely downloadable	Lincoln et al. (1992a, b)	ftp://genome.wi.mit.edu/pub/mapmaker3
QTL Cartographer V2.5	Unix; DOS; Mac; Windows	Offers several varieties of composite interval mapping; freely downloadable	Basten et al. (1997)	http://statgen.ncsu.edu/qtlcart/index.php
Map Manager QTX	Mac	Graphical user interface for data entry and display; designed as mapping program and data preparation program for other mapping programs	Manly et al. (2001)	http://www.mapmanager.org/
QGene 4.0	Mac	Graphical interface for displaying trait data; marker-trait relationship and marker genotype	Nelson (1997)	http://coding.plantpath.ksu.edu/qgene
MapQTL V6	Multiple OS	Mapping QTLs in populations of noninbred parents; offers nonparametric single locus association	van Ooijen and Maliepoard (1996)	http://www.kyazma.nl/index.php/mc.MapQ TL/
PLABQTL	DOS; AIX	QTL detection and environmental effect Estimation QTL x	Utz and Melchinger (1996)	https://www.unihohenheim.de/plantbreeding/software/
MQTL v 0.98	DOS; SUN	Simplified composite interval mapping estimation QTL × environment effect	Tinker and Mather (1995a, 1996b)	ftp://gnome.agrenv.megil.ca/pub/genetics/software/MQTL/
Multimapper	Unix	QTL mapping based on single linkage group; provision for comparing threshold statistic generated by other programs for QTL mapping	Sillanpaa (1998)	http://www.RNI. Helsinki.FI/~mjs/
Epistat	DOS	Analysis of interaction between QTL; does not perform interval mapping	Chase et al. (1997)	http://archives.math.utk.edu/software/msdos /statistics/epistat/.html
QTL mapper V 1.6	Windows 95, 98, NT, 2000, XP	A computer software for mapping quantitative trait loci (QTLs) with additive effects, epistatic effects and QTL environment interactions	Wang et al. (1999)	http://ibi.zju.edu.cn/software/qtlmapper/ind ex.htm

TABLE 21.1 *(Continued)*

Name	Platform	Remarks	References	URL
MultiQTL	Windows XP	Easy data input; fast bootstrapping and permutation test	MultiQTL LTD	http://www.multiqtl.com/
QTLNetwork 2.0	Windows XP	Software for mapping QTL with epistatic and QE interaction effects in experimental populations	Yang et al. (2007, 2008)	http://ibi.zju.edu.cn/software/qtlnetwork/
QTL	Windows XP	Express tools for permutation analysis to set significance levels and bootstrap analysis to estimate confidence regions for the QTL location are provided	Seaton et al. (2002)	http://qtl.cap.ed.ac.uk/
Grid QTL	Windows XP		Seaton et al. (2006)	http://www.gridqtl.org.uk/
The QTL Cafe	Multiple OS	The QTL café is a JAVA-based package which provides a user-friendly way to perform QTL analyses		http://www.biosciences.bham.ac.uk/labs/kea rsey/
QTLBIM	Windows XP	Provides a Bayesian model selection approach to map multiple interacting QTL		http://www.qtlbim.org/
R/qtl	Windows XP			http://www.rqtl.org/
MetaQTL	Windows XP	Java package designed to perform the integration of data from the field of gene mapping experiments		http://www.bioinformatics.org/mqtl/wiki/M ain/HomePage
MapQTL 6	Windows XP	Analyzing QTL experiments with interval mapping, with the powerful MQM mapping		http://www.kyazma.nl/
J/qtl	Windows XP		Jackson Laboratory	http://research.jax.org/faculty/churchill/rese arch/qtl/index.html

(Reprinted from M. Maheswaran, Mapping Quantitative Trait Loci in Plants: Approaches and Applications. Centre for Plant Breeding and Genetics, Tamil Nadu Agricultural University, Coimbatore, India.)

21.10 DISADVANTAGES OF QTL MAPPING

There are two major limitations of QTL mapping. First, only allelic diversity that segregates between the parents of the particular F2 cross or within the RIL population can be assayed (Borevitz and Nordborg, 2003), and second, the amount of recombination that occurs during the creation of the RIL population places a limit on the mapping resolution. Resolution can be dramatically improved with several generations of intercrossing when establishing the RIL population, e.g., advanced intercross RILs (Balasubramanian et al., 2009). Genome-wide association studies (GWAS) overcome the two main limitations of QTL analysis mentioned above, but introduce several other drawbacks as a trade-off (discussed below). Generally, after identifying a phenotype of interest, GWAS can serve as a foundation experiment by providing insights into the genetic architecture of the trait, allowing informed choice of parents for QTL analysis, and suggesting candidates for mutagenesis and transgenics. Thus, GWAS are often complementary to QTL mapping and, when conducted together, they mitigate each other's limitations (Brachi et al., 2010; Zhao et al., 2007). It mainly identifies loci with large effects, less strong ones can be hard to pursue, number of QTLs detected, their position and effects are subjected to statistical error, small additive effects/epistatic loci are not detected and may require further analyses, cloning can be challenging but not impossible.

21.11 QTL TO GENE IDENTIFICATION

QTL analysis, while lifting the 'statistical fog' surrounding conventional quantitative genetics (Mauricio, 2001), provides a powerful magnifying lens for deciphering the chromosome regions regulating complex traits. In this context, the introduction of QTL analysis in quantitative genetics can be compared with the introduction of the optical microscope in cell biology. However, the ultimate dissection of a phenotype can only be considered complete with a direct connection with a DNA sequence variation, and this cannot be obtained at present under a simple and unique QTL analysis framework. Several options are available to proceed from a supporting interval delimiting the QTL to the actual gene(s) responsible for the QTL effect.

Presently, positional cloning appears to be the main strategy toward QTL cloning. All the requirements for the positional cloning of Mendelian genes (Wicking and Williamson, 1991; Tanksley *et al.,* 1995) are also needed for the positional cloning of the gene underlying a QTL; additionally, a much larger effort is needed for the phenotypic scoring of the segregating progenies. The prerequisites are: (1) the confining of the QTL to a 2–3 (or preferably shorter) cM region and the availability of at least one pair of near-isogenic lines (NILs) carrying opposite alleles at the target region; (2) the availability of a large mapping population (> 2000 plants) derived from the cross of the NILs; (3) the presence of a high level of polymorphism between the NILs in order to allow for a level of marker density (usually AFLPs) in the target region suitable to chromosome walking or chromosome landing; (4) a high rate of recombination (i.e., a high ratio between genetic and physical distances) in the target region; (5) the availability of a BAC (bacterial artificial chromosome) and/or YAC (yeast artificial chromosome) genomic library covering the QTL region; and (6) a system (e.g., transformation) for proving the identity and testing the effects of the selected candidate gene.

21.12 CONCLUSION

QTL, otherwise described as hypothetical genes based on statistical inferences, have very little biological meaning. To date the knowledge on QTL mapping is enormous. Issues related to genetic mapping of QTL are well reviewed by Liu (1998), Lynch and Walsh (1998), and Paterson (1998). Mauricio (2001) and Doerge (2002) critically reviewed on various approaches adopted for QTL mapping and analysis of QTL in experimental populations. Price (2006) elaborated the success stories of QTL mapping in his article "Believe it or not, QTLs are accurate." However, the accrued knowledge does not have immediate solutions to the problems associated with QTL mapping. Detecting QTLs in biological systems has become a routine affair to make a way to understand the genetic architecture of complex traits. Still there is a kind of mysticism in QTL mapping to exploit the potential in practical plant breeding. QTL maps based on linkage studies and marker trait association can be effectively utilized for gene pyramiding, germplasm screening of diversified material for abiotic (salinity, cold, salt, drought) and biotic stresses (disease, pest), etc. The identification and location of specific genes mediating quantitative characters is having great importance in plant breeding. QTL analysis is helpful in assessing possible number of loci, their distribution in the genome, equality of effects and manner of their action. DNA markers are very useful for information about number and position of QTLs because they are highly polymorphic, abundant and codominant in nature. High resolution linkage maps based on various molecular markers are required for preparation of for QTL analysis. Proper development and understanding of the statistical background is essential for QTL mapping. The technique of marker-assisted selection and QTL mapping should be adopted at large scale for all major crops.

KEYWORDS

- **quantitative trait loci**
- **QTL mapping**
- **mapping population**
- **linkage maps**
- **association studies.**

REFERENCES

Balasubramanian, S.; Schwartz, C.; Singh, A.; Warthmann, N.; Kim, M. C.; Maloof, J. N.; Loudet, O.; Trainer, G. T.; Dabi, T.; Borevitz, J. O.; Chory, J.; Weigel, D. QTL Mapping in New *Arabidopsis thaliana* Advanced Intercross-recombinant Inbred Lines. *PLoS One* **2009,** *4* (2), e4318.

Basten, C.; Weir, B. S.; Zeng, Z. B. *QTL Cartographer. A Reference Manual and Tutorial for QTL Mapping;* Department of Statistics, North Carolina State University: Raleigh, N.C., **1997**.

Borevitz, J. O.; Nordborg, M. The Impact of Genomics on the Study of Natural Variation in *Arabidopsis*. *Plant Physiol*. **2003,** *132* (2), 718–725.

Brachi, B.; Faure, N.; Horton, M.; Flahauw, E.; Vazquez, A.; Nordborg, M.; Bergelson, J.; Cuguen, J.; Roux, F. Linkage and Association Mapping of *Arabidopsis thaliana* Flowering Time in Nature. *PLoS Genet.* **2010,** *6* (5), e1000940.

Chase, K.; Adler, F. R.; Lark, K. G. Epistat: A Computer Programme for Identifying and Testing Interactions between Pairs of Quantitative Trait Loci. *Theor. Appl. Genet.* **1997,** *94*, 724–730.

Doerge, R. W. Mapping and Analysis of Quantitative Trait Loci in Experimental Populations. *Nat. Rev.* **2002,** *3*, 43–52.

Jansen, R. C. Interval Mapping of Multiple Quantitative Trait Loci. *Genetics* **1993,** *135*, 205–211.

Jansen, R. C.; Stam, P. High Resolution of Quantitative Traits into Multiple Loci via Interval Mapping. *Genetics* **1994,** *136*, 1447–1455.

Kao, C. H.; Zeng, Z. B.; Teasdale, R. D. Multiple Interval Mapping for Quantitative Trait Loci. *Genetics* **1999,** *152*, 1203–1216.

Khan, S. QTL Mapping: A Tool for Improvement in Crop plants. *Res. J. Recent Sci.* **2015,** *4*, 7–12.

Lander, E. S.; Botstein, D. Mapping Mendelian Factors Underlying Quantitative Traits Using RFLP Linkage Maps. *Genetics* **1989,** *121*, 185–199.

Li, Z. K. QTL Mapping in Rice: A Few Critical Considerations. Rice Genetics IV, *Proceedings of the Fourth International Rice Genetics Symposium*, Oct 22–27, 2000, Los Baños, Philippines. Science Publishers, Inc.: New Delhi (India) and International Rice Research Institute: Los Baños (Philippines), **2000**.

Li, Z. K.; Pinson, S. R. M.; Paterson, A. H.; Park, W. D.; Stansel, J. W. Epistasis for Three Grain Yield Components in Rice (*Oryza sativa* L.). *Genetics* **1997,** *145*, 453–465.

Li, Z. K.; Shen, L. S.; Courtois, B.; Lafitte, R. Development of Near Isogenic Introgression Line (NIIL) Sets for QTLs Associated with Drought Tolerance in Rice. In *Proceedings of International Workshop for Molecular Approaches for the Genetic Improvement of Cereals for Stable Production in Water-Limited Environments*, CIMMYT, El Batán, Mexico, **1999**.

Lincoln, S.; Daly, M.; Lander, E. *Constructing Linkage Map with MAPMARKER/EXP;* Whitehead Institute Technical Report, **1992a**.

Lincoln, S.; Daly, M.; Lander, E. *Mapping Genes Controlling Quantitative Traits with MAPMARKER/QTL;* Whitehead Institute Technical Report, **1992b**.

Liu, B. H. *Statistical Genomics: Linkage, Mapping and QTL Analysis;* CRC Press: Boca Raton, New York, **1998**.

Lynch, M.; Walsh, B. *Genetics and Analysis of Quantitative Traits;* Sinauer Associates Inc.: Sunderland, **1998**.

Manly, K. F.; Cudmore, Jr., R. H.; Meer, J. M. Map Manager QTX, Cross-platform Software for Genetic Mapping. *Mamm. Genome* **2001,** *12*, 930–932.

Mather, K.; Jinks, J. L. *Biometrical Genetics* (3rd Edn.); Chapman and Hall: London, **1982**.

Mauricio, R. Mapping Quantitative Trait Loci in Plants: Uses and Caveats for Evolutionary Biology. *Nat. Rev.* **2001,** *2*, 370–381.

McCouch, S. R.; Doerge, R. W. QTL Mapping in Rice. *Trends Genet.* **1995,** *11*, 482–487.

Miles, C.; Wayne, M. Quantitative Trait Locus (QTL) Analysis. Nat. Educ. **2008,** *1 (1), 208.*

Nelson, J. C. Q GENE: Software for Marker Based Genomic Analysis and Breeding. *Mol. Breed.* **1997,** *3*, 239–245.

Newbury, H. J. *Plant Molecular Breeding;* Blackwell Publishing Ltd., U.K., 2003.

Paterson, A. H. Making Genetic Maps. In *Genome Mapping in Plants;* Paterson, A. H., Ed.; R. G. Landes Company: San Diego, California; Academic Press: Austin, Texas, **1996a;** pp 23–39.

Paterson, A. H. Mapping Genes Responsible for Differences in Phenotype. In *Genome Mapping in Plants;* Paterson, A. H., Ed.; R. G. Landes Company: San Diego, California; Academic Press: Austin, Texas, **1996b;** pp 41–54.

Paterson, A. H. (Ed.) Molecular Dissection of Complex Traits. CRC Press: Boca Raton, New York, **1998**.

Pekosz, A.; He, B.; Lamb, R. A. Reverse Genetics of Negative-strand RNA Viruses: Closing the Circle. *Proc. Natl. Acad. Sci. U. S. A.* **1999,** *96* (16), 8804–8806.

Price, A. H. Believe It or Not, QTLs Are Accurate. *Trends Plant Sci.* **2006,** *11*, 213–216.

Seaton, G.; Haley, C. S.; Knott, S. A.; Kearsey, M.; Visscher, P. M. QTL Express: Mapping Quantitative Trait Loci in Simple and Complex Pedigrees. *Bioinformatics* **2002,** *18*, 339–340.

Seaton, G.; Hernandez, J.; Grunchec, J. A.; White, I.; Allen, J.; De Koning, D. J.; Wei, W.; Berry, D.; Haley, C.; Knott, S. Grid QTL: A Grid Portal for QTL Mapping of Compute Intensive Datasets. In *Proceedings of the 8th World Congress on Genetics Applied to Livestock Production*, Aug 13–18, Belo Horizonte, Brazil, **2006**.

Sillanpaa, M. J. Multimapper Reference Manual, **1998**. http://www.rni.helsinki.fi/~mjs/. Accessed on 27.04.2017

Tanksley, S. D.; Ganal, M. W.; Martin, G. B. Chromosome Landing: A Paradigm for Map-based Gene Cloning in Plants with Large Genomes. *Trends Genet.* **1995,** *11*, 63–68.

Tierney, M. B.; Lamour, K. H. An Introduction to Reverse Genetic Tools for Investigating Gene Function. *The Plant Health Instructor,* **2005**. DOI: 10.1094/PHI-A-2005-1025-01.

Tinker, N. A.; Mather, D. E. Methods for QTL Analysis with Progeny Replicated in Multiple Environments. *J. Quant. Trait Loci.* **1995a,** *1*, 1–28.

Tinker, N. A.; Mather, D. E. MQTL: Software for Simplified Composite Interval Mapping of QTL in Multiple Environments. *J. Quant. Trait Loci.* **1995b,** *1*. 1–7.

Utz, H. F.; Melchinger, A. E. PLABQTL: A Program for Composite Interval Mapping of QTL. *J. Quant. Trait Loci.* **1996,** *2* https://wheat.pw.usda.gov/jag/papers96/paper196/utz.html

van Ooijen, T. W.; Maleipaard, C. *Map QTL Version 3.D Software for the Calculation of QTL Position on Genetic Maps;* CPRO–DLO: Wageningen, **1996**.

Wang, D. L.; Zhu, J.; Li, Z. K.; Paterson, A. H. Mapping QTLs with Epistatic Effects and Genotype x Environment Interactions by Mixed Linear Model Approaches. *Theor. Appl. Genet.* **1999,** *99*, 1255–1264.

White, T. L.; Adams, W. T.; Neale, D. B. *Forest Genetics;* CABI, U.K. **2007**.

Wicking, C.; Williamson, B. *From Linked Marker to Gene. Trends Genet.* **1991,** *7,* 288–*293.*

Yang, J.; Hu, C. C.; Hu, H.; Yu, R. D.; Xia, Z.; Ye, X. Z.; Zhu, J. QTL Network: Mapping and Visualizing Genetic Architecture of Complex Traits in Experimental Populations. *Bioinformatics* **2008,** *24*, 721–723.

Yang, J.; Zhu, J.; Williams, R. W. Mapping the Genetic Architecture of Complex Traits in Experimental Populations. *Bioinformatics* **2007,** *23*, 1527–1536.

Zeng, Z. B. Theoretical Basis for Separation of Multiple Linked Gene Effects in a Mapping Quantitative Trait Loci. *Proc. Natl. Acad. Sci. U. S. A.* **1993,** *90*, 10972–10976.

Zeng, Z. B. The Precision Mapping of Quantitative Trait Loci. *Genetics* **1994,** *136*, 1457–1468.

Zhao, K.; Aranzana, M. J.; Kim, S.; Lister, C.; Shindo, C.; Tang, C.; Toomajian, C.; Zheng, H.; Dean, C.; Marjoram, P.; Nordborg, M. An Arabidopsis Example of Association Mapping in Structured Samples. *PLoS Genet.* **2007,** *3* (1), e4.

CHAPTER 22

ASSOCIATION MAPPING: A TOOL FOR DISSECTING THE GENETIC BASIS OF COMPLEX TRAITS IN PLANTS

SWETA SINHA[1*], AMARENDRA KUMAR[2], RENU KUSHWAH[3], and RAVI RANJAN KUMAR[1]

[1]*Deptartment of Molecular Biology and Genetic Engineering, Bihar Agricultural University, Sabour, Bhagalpur, Bihar, India*

[2]*Deptartment of Plant Pathology, Bihar Agricultural University, Sabour Bhagalpur, Bihar, India*

[3]*Deptartment of Plant Molecular Biology & Biotechnology, IGKV, Raipur 492012, India*

[*]*Corresponding author. E-mail: bablysweta@gmail.com*

CONTENTS

ABSTRACT

Association mapping is an attractive approach used to detect complex quantitative trait loci (QTLs) and to identify candidate genes, based on linkage disequilibrium (LD) between genetic markers and genes controlling the phenotype of interest by exploiting the recombination events accumulating over many generations. Many factors affect linkage disequilibrium such as genetic drift, selection within populations, population admixture, recombination, genetic bottlenecks, founder effects and mating systems. Association mapping can be grouped into genome-wide association mapping and candidate-gene association mapping and the appropriate approaches can be selected based on the study of interest in different plant species. The success of association analysis depends on the choice of germplasm, quality of genotypic and phenotypic data, use of the appropriate statistical analysis for the detection of marker- phenotype associations and the validation of marker-phenotype associations. In recent years, it has continued to gain favorability in plant genetic research because of advances in high throughput genomic technologies, interests in identifying novel and superior alleles and improvements in statistical methods. In this chapter, we describe the opportunities and challenges of association mapping in plants for complex trait dissection.

22.1 INTRODUCTION

Diversity in the plant species is the gift of nature and arises due to geographical separation or due to genetic barriers to cross ability. Natural diversity is used for the identification and utilization of useful allelic variants for plant improvement. The common plants have a narrow genetic pool due to domestication. In contrast, their wild relatives, as a result of genetic history and selection pressure, are becoming reservoirs of natural genetic variation. A fundamental goal of evolutionary biology is to understand the genetic basis of adaptation in natural populations (Orr and Coyne, 1992) and it is therefore rather surprising that we still know relatively little about the genetic architecture of many adaptive traits (Mackay et al., 2009). The primary reason for this is that phenotypic variation in most adaptive traits in natural populations is caused by the action of many genes, each having only a small to moderate effect on the phenotype. Although Fisher (1918) showed that the underlying principles of Mendelian inheritance also explain segregating variation in quantitative traits, quantitative genetics has traditionally been used to partition phenotypic variation within and among individuals with known degrees of relatedness. However, as a result of the small contribution of individual genes, the methodology and technology required to dissect the genetic architecture of quantitative traits down to individual causal loci have long eluded us. Genes associated with desired agronomic traits such as higher yield or disease resistance that could be lost in the plant breeding process of a plant can be restored using these wild species. The problem for breeder is to find the genes and find an efficient way to trace the genes and to incorporate them in breeding populations (Abdurakhmonov and Abdukarimov, 2008).

The genetic basis of quantitative traits opened with the introduction of quantitative trait locus (QTL) mapping in the 1980s (Lander and Botstein, 1989). In QTL mapping, designed crosses are used to dissect quantitative variation that distinguishes the individuals making up the parental generation of the cross. Initially, it has been a challenge for breeders to work with QTL because they are controlled by polygene and are greatly dependent of genetic × environment interactions. However, with the advent of molecular marker technology that allowed for rapid and cost-effective genotyping in almost any plant, QTL mapping has proved to be extremely useful for identifying many genomic regions that influence complex traits in a large number of species (Mauricio, 2001; Doerge, 2002; Mackay et al., 2009). But QTL mapping has a number of drawbacks; for instance, genetic variation in the mapping population is usually quite restricted with only two parents used to initiate the QTL mapping population. Moreover, because a QTL mapping population usually consists of early-generation crosses (usually F_2), the number of recombination events per chromosome is small, which in turn limits the resolution of the genetic map. Finally, in many plants the generation of mapping populations through controlled crosses is either time consuming or not even possible, further restricting the utility of QTL mapping. Furthermore, when identified in a mapping population consisting of a few hundred individuals, a typical QTL region can span anywhere from between a few to tens of centimorgans, corresponding to genomic regions encompassing several megabases and which typically contain hundreds or even thousands of genes (Rae et al., 2009). Even when a QTL of large effect is identified, tracking down the causal gene is a tedious and time-consuming task. In addition, a single large-effect QTL often breaks down into multiple, closely linked QTLs of smaller, and sometimes opposite, effects on the phenotype (Doerge, 2002; Mackay et al., 2009). The phenotypic variation of many complex traits of agricultural or evolutionary importance is influenced by multiple QTL, their interaction, the environment, and the interaction between QTL and environment. Linkage analysis and association mapping are the two most commonly used tools for dissecting complex traits.

22.2 ASSOCIATION MAPPING

Association mapping has been used to identify QTLs by examining the trait–marker associations and enables researchers to use modern genetic technologies to exploit natural diversity and locate valuable genes in the genome. It takes advantage of historical recombination events, which have accumulated over thousands of generations in historical populations and thus enable high-resolution mapping for identification of possible target genomic regions for complex quantitative traits (Huang et al., 2010; Zhao et al., 2011). Association mapping makes it possible to eliminate many of the drawbacks of traditional, pedigree-based, QTL mapping such as limited sample sizes, low variation, and a lack of recombination within pedigrees. The key properties of QTL mapping and association mapping are summarized in Table 22.1. An association mapping study utilizes variation segregating in a diverse germplasm and therefore does not suffer from the lack of variation that characterizes many QTL mapping populations. The term association mapping and linkage disequilibrium (LD) have been used interchangeably, but association mapping is an application of LD. As a new alternative to traditional linkage mapping, association mapping offers three advantages: (1) increased mapping resolution, (2) reduced research time, and (3) greater allele number (Yu and Buckler, 2006). Association mapping refers to significant association of a molecular marker with a phenotypic traits; LD refers to nonrandom association between two markers or two genes/QTLs or between a gene/QTL and a marker locus. Thus, association mapping is actually one of the several uses of LD. In statistical sense, association refers to covariance of a marker polymorphism and a trait of interest, while LD represents covariance of polymorphisms exhibited by two molecular markers/genes. Thus, association mapping utilizes ancestral recombinations and natural genetic diversity within a population to dissect quantitative traits and is built on the basis of the LD concept.

Since its introduction to plants (Thornsberry et al., 2001), association mapping has continued to gain favorability in genetic research because of advances in high throughput genomic technologies, interests in identifying novel and superior alleles, and improvements in statistical methods. Association mapping originated in human genetics and it has been widely used in plant research, for example, Arabidopsis, maize, rice, barley, durum wheat, spring wheat, sorghum, sugarcane, sugar beet, soybean, tomato, grape, forest tree species and forage grasses in identifying phenotype-associated marker and trait-associated phenotypes, which show the versatility of this approach in identifying markers linked to genes and genomic regions associated with desirable traits.

TABLE 22.1 A Comparison of QTL Mapping and Association Mapping.

Attributes	QTL Mapping	Association Mapping
Detection aim	Quantitative trait locus	Quantitative trait nucleotide polymorphism, that is, SNPs
Experimental populations	Bi-parental mapping population (experimental cross required)	Genetically diverse individuals (no cross required)
Resolution	Low	High
Number of traits	Limited trait mapping	Multiple traits mapping
Familial relatedness	Minimized by controlled crossing	Minimized by kinship coefficient estimation and its use in association mapping
Population structure	Minimized by controlled crossing	Minimized by estimation of Q or P and its use in AM
Phenotypic data	Phenotypes to be collected	Phenotypic data can be already available
Time	Time consuming	Reduced time
Mapping based on	Recombination frequency between the loci	Linkage disequilibrium (LD) between the loci
Recombination events	Occurring after the crosses are made	Occurred since the LD was created
Linked marker	Few to several centimorgans (cM) away from gene/QTL	Much closer than those by linkage mapping
Markers required for genome coverage	10^2–10^3	10^5–10^9
QTL result confirmation/ validation	Confirmation as well as validation required	Often confirmation is done by replication studies

22.3 LD AND ASSOCIATION MAPPING

LD and association mapping have often been used interchangeably in the literature but have subtle differences. Association mapping refers to the significant association of a marker locus with a phenotype trait while LD refers to the nonrandom association between two markers or two genes/ QTLs (Gupta et al., 2005). Thus, association mapping is actually an application of LD. In other words, two markers in LD represent a nonrandom association between alleles, but do not necessarily correlate/associate with a particular phenotype, whereas association implies a statistical significance and refers to the covariance of a marker and a phenotype of interest.

Unlike QTL mapping, association mapping takes advantage of LD as well as historical recombinations present within the gene pool of an organism, thus utilizing a broader reference population (Breseghello and Sorrels, 2006; Myles et al., 2009). If two alleles from separate loci occur together more often than otherwise predicted, on the basis of their individual frequencies, that is, nonrandom association of alleles at separate loci, they are deemed to be in LD. Only those molecular markers that are tightly linked to the trait and located within the extent of LD decay will demonstrate significant marker–trait association. If markers are not tightly linked to a trait, they will be separated by recombination during meiosis throughout the evolutionary history of the plant. Accumulating meiotic events in a population will increase the statistical power and mapping resolution for detecting associations. However, it should be noted that the rate of LD decay should be sufficient enough to statistically identify associations, but not too high as it will make it difficult to narrow down the target genomic region.

22.3.1 QUANTIFICATION OF LD

Association mapping is quantified in terms of LD. The concept of LD was first described by Jennings in 1917, and its quantification (D) was developed by Lewontin in 1964 (Abdurakhmonov and Abdukarimov, 2008). The simplified explanation of the commonly used LD measure, D', is the difference between the observed gametic frequencies of haplotypes and the expected gametic frequencies of haplotype under linkage equilibrium.

$$D' = P_{AB} - P_A P_B.$$

In the absence of other forces, recombination through random mating breaks down the LD with $D_t = D_0 (1 - r)^t$, where D_t is the remaining LD between two loci after t generations of random mating from the original D_0 (Zhu et al., 2008). Several statistics have been proposed for LD, and these measurements generally differ in how they are affected by marginal allele frequencies and sample sizes. The two common measures of LD are r^2 and D'. Both D' (Lewontin, 1964) and r^2, the square of the correlation coefficient between two loci (Hill and Robertson, 1968), reflect different aspects of LD and perform differently under various conditions. D' only reflects the recombinational history and is therefore a more accurate statistic for estimating recombination differences, whereas r^2 summarizes both recombinational and mutational history (Flint-Garcia et al., 2003; Soto-Cerda and Cloutier, 2012). For two biallelic loci, D' and r^2 have the following formula:

$$D' = |D| / D_{mas}$$

$$D_{mas} = \min (P_A P_b, P_a P_B) \; if \, D > 0;$$

$$D_{mas} = \min (P_A P_B, P_a P_b) \; if \, D < 0$$

$$r_2 = D_2 / P_A P_a P_B P_b$$

Both the parameters D' and r^2 range from 0 to 1. D' is the standardized LD coefficient (Sorkheh et al., 2008). $D' < 1$, indicates the occurrence of recombination between loci, while, $D' = 0$, indicates the lack of recombination between the loci. Thus, at $D' = 1$ indicate complete LD, whereas $D' < 1$ indicate that compete LD has been disrupted probably due to continuous recombination resulting in all four possible haplotypes. The r^2 is essentially the correlation between the alleles at two loci (Sorkheh et al., 2008). $r^2 = 1$, when the alleles at two loci co-occur most frequently, while $r^2 = 0$, if they co-occur no more frequently than expected by chance. R^2 plots are used to illustrate the rate at which LD decays with physical distance and often form the basis for comparison between studies. r^2 is the more preferred measure of LD in comparison to D', because D' is affected by same samples unlike r^2. Thus, r^2 that can correlate the genetic markers with the QTL of interest is the preferential statistic for determining the resolution of association studies (Flint-Garcia et al., 2003). Another important measure is *P*-value. *P*-value is the probability of an association between a marker and a trait expected by chance. Hence, *P*-values less than 0.05 are 95% statistically significant, 0.01 are 99% statistically significant, and 0.001 are 99.9% statistically significant.

22.3.2 DEPICTION OF LD

There are two methods which have been widely used to visualize or depict the extent of LD between pairs of loci across a genomic region (1) LD decay plots or LD scatter plots, and (2) disequilibrium matrices or LD heat maps (Gupta et al., 2005; Flint-Garcia et al., 2003; Gaut and Long, 2003). LD scatter plots are used to estimate the rate at which LD declines with genetic or physical distance (Fig. 22.1a). These scatter plots are useful to determine the average effective distance threshold above which significant LD (commonly 0.5 for D' and 0.1 for r^2) is expected based on the curve of a nonlinear logarithmic trend drawn through the data points of the scatter plot

(Breseghello and Sorrells, 2006). Disequilibrium matrices or LD heat maps are also very useful for visualizing the linear arrangement of LD between polymorphic sites within a short physical distance such as a gene, along an entire chromosome or across the whole genome (Fig. 22.1b) (Flint-Garcia et al., 2003). LD heat maps are color-coded triangular plots where the diagonal represents ordered loci and the different intensity colored pixels depict significant pairwise LD level expressed as *P*-value or *D′* or r^2. The large red blocks of haplotypes along the diagonal of the triangle plot indicate the high level of LD between the loci in the blocks, meaning that there has been a limited or no recombination since LD block formations. There is freely available computer software, "graphical overview of linkage disequilibrium" (GOLD) (Abecasis et al., 2000), to depict the structure and pattern of LD. Some other software packages measuring LD such as "Trait Analysis by Association, Evolution and Linkage" (TASSEL) (Bradbury et al., 2007) and PowerMarker (Liu and Muse, 2005) have also similar graphical display features. These graphical representations enable us to determine the optimum number of markers to detect significant marker–trait associations and the resolution at which a QTL can be mapped.

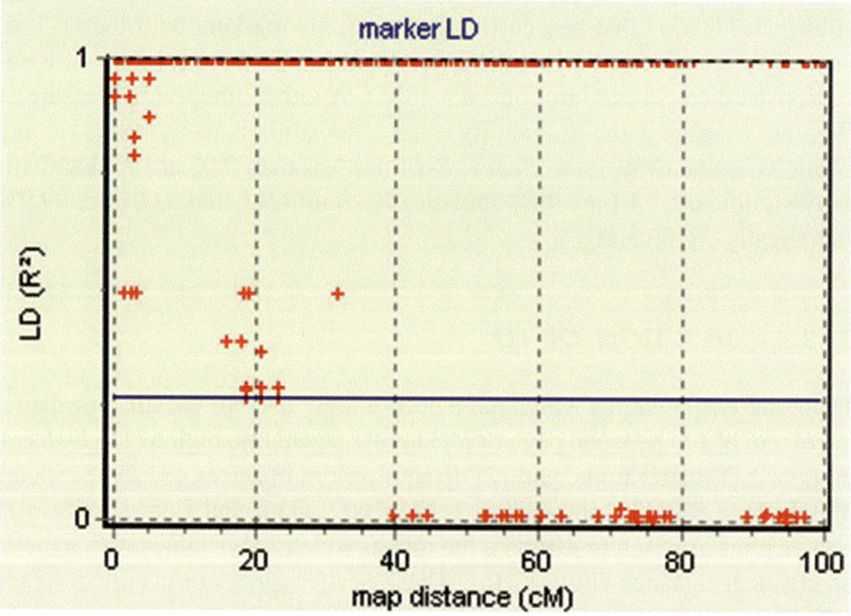

FIGURE 22.1a Scatter plot of LD decay (r^2) against genetic distance (cM). The decrease of the LD within the genetic distance indicates that the portion of LD is conserved with linkage and proportional to recombination.

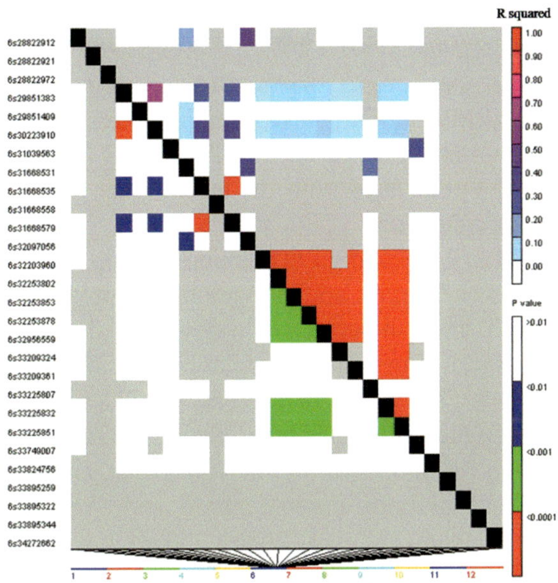

FIGURE 22.1b Disequilibrium matrices of LD variation between pairwise polymorphic loci. Pairwise calculations of LD (r^2) are displayed above the diagonal with the corresponding *P*-values for Fisher's exact test displayed below the diagonal. Each cell represents the comparison of two pairs of marker sites with the color codes for the presence of significant LD.

22.3.3 FACTORS AFFECTING LD AND ASSOCIATION MAPPING

There are many genetic and demographic factors that play a role in the shaping of the haplotypic LD blocks in a genome, out of which mutation and recombination are the key factors affecting LD significantly (Gupta et al., 2005; Stich et al., 2005, 2006, 2007). Generally, factors affecting LD can be grouped into two categories: (1) factors that increasing LD, and (2) factors that decreasing LD. Increased LD is the result of new mutations, population structure, autogamy, genetic isolation, admixture, genetic drift, small founder population size, epistasis, genomic rearrangement, selection and kinship, whereas higher rates of recombination and mutation, recurrent mutations, gene conversion, and outcrossing significantly decrease LD (Gupta et al., 2005). In presence of a high LD reduces the number of markers required for marker–trait association but lowers the mapping resolution and a whole genome scan approach may be performed. Conversely, with low LD, a large number of markers are required to cover the variation in the entire genome but increase mapping resolution and a candidate gene analysis may be conducted (Rafalski, 2002). The extent of LD over the genome is expected to vary according to the species, genome region, and population under study (Nordborg and Tavare, 2002). It is expected high variable levels of LD through the genome due to variations in recombination rates, presence of hot spots and selection, variation in recombination rate is a key factor that contributes to the variance observed in LD patterns (Salvi and Tuberosa, 2007). Selection of a population with LD level higher or lower depends on the objective of mapping study.

In order to appropriately apply LD mapping in crop plants, it is a prerequisite to characterize LD levels and patterns in a population analyzed. It is also important to distinguish between physical LD and the other different forces that can create LD in natural populations, to avoid the detection of spurious associations. The decay or decrease of LD with increasing map distance between markers in outcrossing plants is usually faster than that in inbreeding plants (Flint-Garcia et al., 2003). For example, LD decays rapidly within 1–5 kb in maize diverse inbred lines (Yan et al., 2009), 1.1 kb in cultivated sunflower (Liu et al., 2006), 300 bp in wild grapevine (Lijavetzky et al., 2007), whereas LD decays slowly within 250 kb in Arabidopsis (Nordborg et al., 2002), 212 kb in elite barley cultivars (Caldwell et al., 2006), 100–200 kb in rice diverse lines (McNally et al., 2009; Huang et al., 2010), and 250 kb in cultivated soybean (Lam et al., 2010). Also, the decay or decrease of LD in wild relatives is faster than that in modern varieties (Morrell et al., 2005; Song et al., 2009). The details of major factors responsible for the variation of LD are given below.

22.3.3.1 GERMPLASM

The variation in LD is influenced by the level of genetic diversity in the germplasm. In general, the larger the genetic variation, the faster the LD decays. Estimates of genome-wide average LD decay may not reflect LD patterns between different populations of the same species. Each of these populations should be explored independently for the extent of LD in order to conduct successful association mapping studies (Abdurakhmonov and Abdukarimov, 2008). Populations with narrow genetic diversity and long extent of LD are amenable to coarse mapping with fewer markers requiring fine mapping in more genetically diverse populations, assuming that the causal genetic factors are sufficiently similar across different germplasm groups. The population sample effect is evident in maize (*Zea mays*) where LD decays within 1 kb in landraces, approximately doubles (~2 kb) in diverse inbred lines and can extend up to several hundred kb in commercial elite inbred lines (Jung et al., 2004). In cotton (*Gossypium hirsutum*), the genome-wide average LD ($r^2 \leq 0.1$) declined to 10 cM in landraces, but was up to 30 cM in varieties (Abdurakhmonov et al., 2008). Myles et al. (2011) studied LD variation in over 1000 samples of domesticated grape (*Vitis vinifera*) and its wild relatives, reporting a rapid LD decay, even greater than in maize, as result of a weak domestication and widespread vegetative propagation.

22.3.3.2 MATING SYSTEM

Populations are termed as outcrossing or selfing populations on the basis of mating pattern prevalent in them. In outcrossing populations, a high level of heterozygosity is maintained so that opportunity for crossing over between pairs of loci is present in each generation while selfing populations consist primarily of homozygous genotypes, and any heterozygotes arising due to natural outcrossing or mutation are rapidly resolved into homozygous genotypes. The variation in LD was affected by mating system (Myles et al., 2009). The effective recombination reduces due to selfing than outcrossing species because individuals are more likely to be homozygous (Flint-Garcia et al., 2003). LD decay in self-pollinated and outcrossing species reviewed by Flint-Garcia et al. (2003) and Abdurakhmonov & Abdukarimov (2008). LD extends much further in self-pollinated species such as rice (*Oryza sativa*), Arabidopsis (*Arabidopsis thaliana*), and wheat (*Triticum aestivum*) (Garris et al., 2005; Nordborg, 2000; Zhang et al., 2010) as compared to outcrossing species such as maize (*Zea mays*), grapevine (*Vitis vinifera*), and rye (*Secale cereale*) (Li et al., 2011b; Myles et al., 2009; Tenaillon et al., 2001). In self-pollinated species, genetic polymorphisms tend to remain correlated, and LD is expected to be maintained over long genetic or physical distances (Gaut and Long, 2003). Higher resolution is expected in outcrossing species that enable more accurate fine mapping and potentially facilitating the cloning of candidate genes because LD declines more rapidly in outcrossing plant species than self-pollinated.

22.3.3.3 POPULATION STRUCTURE

Population structure is ubiquitous, and arises due to geographical isolation, and natural and artificial selections. Population structure signifies that individuals in a population do not form a single homogeneous group, but they are distributed in few to several distinct subgroups that show different gene frequencies. Population structure affects LD throughout the genome. Genome-wide patterns of LD can help to understand the history of changes in populations (Slatkin, 2008) and the power of AM can be strongly reduced as a consequence of population structure (Balding, 2006). Population structure occurs from the unequal distribution of alleles among subpopulations of different ancestries. When these subgroups are sampled to construct a panel of lines for AM, the intentional or unintentional mixing of individuals with different allele frequencies creates LD. Significant LD between unlinked loci results in false-positive associations between a marker and a trait. Thornsberry et al. (2001) reported significant associations between polymorphisms at the maize *Dwarf8* gene and variation in flowering time, but they also stated that up to 80% of the false-positive associations resulted from population structure. The occurrence of spurious associations is markedly higher in adaptation-related genes because they show positive correlations with the environmental variables under which they have evolved, and, as a result, the genomic regions carrying these genes could present stronger population differentiation.

22.3.3.4 SELECTION

Selection affects the genome and LD in locus-specific manner. Selection may be defined as differential reproduction rates for different genotypes. Ordinarily, selection operates on the phenotypes generated by various genotypes. Selection can generate LD between unlinked genes through epistatic selection as well as the "hitchhiking" effect. Domestication bottlenecks followed by strong selection for specific environments and end-use traits have modified the genome architecture in many crops reducing genetic diversity and creating population structure, which may be the main factor affecting the power of AM. If alleles at two loci are in LD and they both affect positively reproductive fitness, the response to selection at one locus might be accelerated by selection affecting the other (Slatkin, 2008). Thus, positive selection will increase LD between and in the vicinity of the selected loci, a phenomenon known as genetic hitchhiking (Slatkin, 2008). Even if the second locus is selectively neutral, the selection applied over the first will increase LD between them. The LD level between the two loci will remain constant over time depending on the genetic distance, the recombination rate, and the effective population size (N). In general, disease resistance genes in plants (*R*-genes) are affected by balancing selection with low intragenic LD and rapid decay (Yin et al., 2004), which could facilitate fine mapping of disease resistance genes providing high marker saturation. Artificial selection also has dramatic effects on LD. Mosaics of large LD blocks are observed, especially in regions carrying agronomic-related genes.

22.3.3.5 RECOMBINATION

It is generally accepted that different regions of the genome of a given species show different rates of recombination that may vary >10-fold. There is evidence that gene-rich genomic regions tend to have higher rates of recombination than gene-poor regions, and that regions having repetitive DNA and retroposons show little or no recombination. Thus, the rate of LD decay would be higher in such genomic regions that show higher recombination rates, and a higher marker density would be required for LD analysis in such regions. Several biological factors influence LD strength and its distribution across genomes. LD is strongly influenced by localized recombination rate and is correlated with other associated factors such as GC content and gene density (Dawson et al., 2002). In principle, local sequence features can affect LD directly and indirectly. For example, GC-rich sequences may be associated with higher rates of recombination and/or mutation, two phenomena that could directly lower surrounding levels of LD. Furthermore, in some protein-coding sequences, changes created by recombination or mutation may affect the fitness of an individual, and these sequences could be indirectly associated with unique patterns of LD as a consequence of natural selection (Smith et al., 2005). Because LD is broken down by recombination, and recombination is not distributed homogeneously across the genome, blocks of LD are expected. Also, differences in LD between micro chromosomes and macro chromosomes have been reported (Stapley et al., 2010) as well as intra-chromosomal variation, where centromeric regions showed higher levels of LD. Many regions of the human genome display rates of recombination that differ significantly from the genome average recombination rate of 1 cM/Mb (Arnheim et al., 2003). These regions have been called "hotspots" and "coldspots" for high and low recombination rates, respectively. Teo et al. (2009) conducted a comprehensive analysis of genomic regions with different patterns of LD to unravel the consequences of this patterning for AM in human populations. Plant genomes have revealed similar general conclusions with regard to LD distribution. Interchromosomal LD variation has been reported in barley (*Hordeum vulgare*), maize (*Zea mays*), tomato (*Solanum lycopersicum*), and bread wheat (*Triticum aestivum*) (Malysheva-Otto et al., 2006; Robbins et al., 2011; Yan et al., 2009; Zhang et al., 2010), where it varied between less than 1 cM to more than 30 cM ($r^2 > 0.1$).

22.3.3.6 GENETIC DRIFT, POPULATION BOTTLENECK, AND GENE FLOW

Factors such as genetic drift, population bottlenecks, and gene flow can contribute to generating artificial LD and negatively impact the ability to use LD in AM for the precise localization of QTL. In general, any biological or evolutionary forces that contribute to an increase of LD beyond that expected by chance in an "ideal" population will result in false positive associations (Gaut and Long, 2003). Genetic drift is the random change in gene frequency of a population due to random sampling of gametes that unite to produce a finite number of individuals in each generation. Genetic drift occurs in small populations and consistently leads to the loss of rare allelic combinations, leading to an upward bias in LD. The effect of genetic drift in a small population results in the consistent loss of rare allelic combinations which increases LD level (Flint-Garcia et al., 2003). Genetic drift can create LD between closely linked loci. The effect is similar to taking a small sample from a large population. Even if two loci are in linkage equilibrium, sampling only few individuals can create LD (Slatkin, 2008). LD can also be created in populations that have experienced a reduction in size (called a bottleneck) with accompanying extreme genetic drift (Flint-Garcia et al., 2003). In maize, ~80% of the allele richness has been lost as a consequence of domestication bottlenecks (Wright and Gaut, 2005) while this number is 40–50% in sunflower (Liu and Burke, 2006) and 10–20% in rice (Zhu et al., 2007). Gene flow introduces new individuals or gametes with different ancestries and allele frequencies among populations. If selection maintains differences in allele frequencies at two or more loci among subpopulations, LD in each subpopulation will persist (Slatkin, 2008), but generally when random mating and recombination take place, LD caused by gene flow eventually breaks down.

22.4 BASIC STEPS IN ASSOCIATION MAPPING

22.4.1 ASSEMBLY OF PLANT SAMPLE

Collection of unrelated genotypes representing wide gene pool variation is critical to the success of association analysis. Four main types of populations can be considered in a plant for association mapping study: (1) natural populations, (2) germplasm bank collections, (3) synthetic populations derived from a group of inbred lines, and (4) elite breeding material developed by breeding programs. These populations are expected to differ considerably with genetic diversity, the extent of genome-wide LD, as well as the level of population structure and relatedness (Stich and Melchinger, 2010).

22.4.2 PHENOTYPING

Precise phenotyping is a prerequisite for association mapping studies. An increase in the number of individuals/lines for phenotyping enhances the accuracy of AM in comparison to increase in the number of markers used for genotyping (Ingvarsson and Street, 2011). The selected sample is evaluated for the various traits of interest. Phenotyping should be preferably based on replicated trials conducted over locations and years to increase precision in phenotypic measurements by eliminating environmentally induced noise and measurement errors. The trials should be conducted using a suitable experimental design, which will provide robustness of positive associations across environments. An efficient field design with an incomplete block design (e.g., a-lattice) has the potential to increase the mapping power. Furthermore, in the case that unbalanced plant breeding trials are used as sources of phenotypic data, the proper statistical modelling of the experimental design and especially the consideration of genotype × environment as well as marker × environment interactions increases the mapping power.

22.4.3 GENOTYPING

Genotyping is required to know the population structure and relatedness as well as on marker–phenotype associations. The phenotyped sample is then genotyped with a set of molecular markers that are evenly distributed over the entire genome of the species and these markers should be unlinked. Random amplified polymorphic DNA (RAPD) and amplified fragment length polymorphism (AFLP) markers can be used, but as a result of their dominant inheritance demand special statistical methods if used to estimate population genetic parameters. Therefore, codominant simple sequence repeats (SSRs) and single nucleotide polymorphisms (SNPs) are more powerful in estimating population structure and familial relatedness. The polymorphism in candidate genes or regions is preferred for detection of marker-phenotype associations. If genome-wide association approaches are to be conducted to identify SNPs at a density that accurately reflects the genome-wide LD structure and haplotype diversity.

22.4.4 STRUCTURE AND KINSHIP ANALYSIS

Population structure and kinship present within the AM population is necessary take into consideration to avoid detecting spurious associations. The marker data are analyzed to detect and estimate the population structure of the sample using the STRUCTURE program and the extent of kinship among the individuals of the sample using the TASSEL program (Pritchard et al., 2000; Bradbury et al., 2007). The population structure and kinship matrix is calculated to determine relatedness among individuals. The estimates of population structure and kinship are used as covariates in the model to minimize false associations between the markers and the genes/QTLs of interest.

22.4.5 LD ANALYSIS

The average extent of LD decay at the whole genome level and for individual chromosomes will be quantified to understand the pattern of LD decay. The sample is genotyped with a sufficiently large number of molecular markers that cover the entire genome as densely as is feasible so that LD between markers and the loci of interest can be detected. The pattern of LD in the concerned genomic regions of the species and the extent of LD observed among different populations of the species would determine the number of markers required for adequate coverage of the whole genome. LD between markers on each chromosome is estimated with r^2 using TASSEL program.

22.4.6 ASSOCIATION ANALYSES

The genotypic and phenotypic data is used for association analysis. Generally, two approaches is used: a general linear model (GLM) that do not consider control for population structure (Q) and mixed linear model that considered both population structure and kinship (Q + K model) to reduce the number of false positive associations. A kinship matrix is calculated using the Van Raden method (K) to determine relatedness among individuals. The critical *P*-values for assessing the significance of markers are calculated based on a false discovery rate separately for each trait. Since these analyses are computationally intensive, suitable computer programs are used for their implementation.

22.5 TYPES OF ASSOCIATION MAPPING: GENOME-WIDE SCANS AND CANDIDATE GENES

Based on the scale and focus of a particular study, association mapping generally falls into two broad categories (1) genome-wide association mapping, which surveys genetic variation in the whole genome to find signals of association for various complex traits (Fig. 22.2), and (2) candidate-gene association mapping (Fig. 22.3), where selected candidate genes are sequenced and the sequence polymorphism is correlated to the phenotype variation (Risch and Merikangas, 1996).

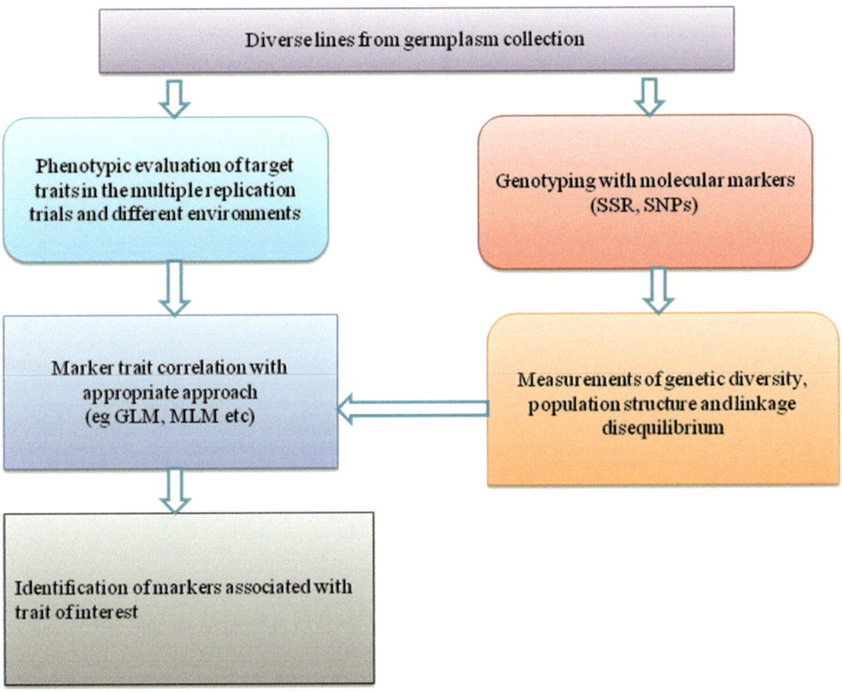

FIGURE 22.2 Steps for genome-wide association mapping.

Genome-wide association mapping is a systematic approach that requires saturating the genome with adequate marker coverage to test for associations with QTL of interest. It is preferred when the entire genome of the species in the study has been sequenced and no prior information regarding candidate genes is available (Zhu et al., 2008). This approach is employed when marker availability is a limiting factor or when LD extends to large distances, authorizing potential candidate regions associated with a trait of interest to

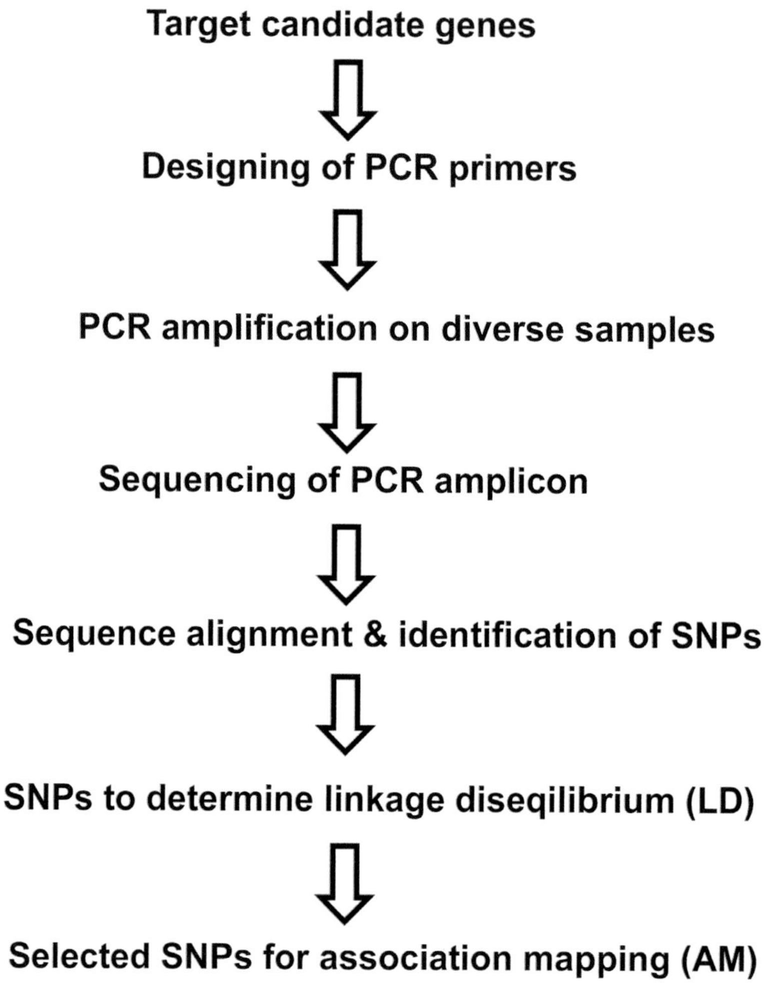

FIGURE 22.3 Steps for candidate gene association mapping.

be identified (Remington et al., 2001). However, the focus shifts to candidate-based gene association mapping to attain high resolution, when the LD decays too rapidly, significantly increasing the number of markers required to conduct genome-wide association mapping. Association mapping could be used to identify the causal molecular polymorphism responsible for trait differences for a trait of interest when the LD within the gene decays rapidly. An advantage of the genome-wide association mapping for the homozygous nature of crop is to employ a "genotype or sequence once and phenotype many times over" strategy, whereby once the lines are genomically characterized, the genetic data can be reused many times over across different phenotypes and environments (Zhao et al., 2011). It has tremendous significance for identification of genes particularly for complex quantitative traits.

Candidate-gene approach encompasses the use of candidate genes selected on the basis of knowledge for mutational analysis, biochemical pathways, or linkage analysis of traits of interest. An independent set of random background markers is screened for associations with the candidate genes (Abdurakhmonov et al., 2008). In maize, candidate gene association mapping identified SNPs in genes that affect abscisic acid levels in floral tissue during drought stress (Setter et al., 2011). Candidate-gene association mapping is selected over genome-wide association mapping to reduce research time accredited to the generation of lower marker number. Candidate gene approach is limited by the choice of candidate genes that are identified and hence always runs the risk of missing out on identifying causal mutations that are located in nonidentified candidate genes. However, the candidate-gene approach is a hypothesis driven, trait-specific approach. Hence, in contrast with genome-wide association mapping, candidate-gene mapping can overlook any other unknown loci partaking in the particular phenotypic trait of interest. Thus, appropriate approaches can be selected contingent for the study of interest.

22.6 APPROACHES FOR ASSOCIATION MAPPING

In recent years, association mapping is gaining popularity to identify genes of interest in plants for different traits with agricultural and evolutionary significance as the increase in availability and cost reduction of markers has made feasible. Taking the advantages of natural diversity coupled with the new decisive statistical analysis methods, association mapping has become a tempting and an affordable research program. Strategies of AM were initially developed for humans, and applied to plants without much modification. Subsequently, more precise and powerful methods for unbiased AM in plants were developed. There are several different approaches for the detection of significant LD, ranging from the simple chi-square test through analysis of variance to complex likelihood-based procedures. When the LD between a marker and a QTL is strong, the various methods would give comparable results. Generally, analysis of variance is not regarded as an effective procedure for AM. Therefore, AM for quantitative traits in plants is based on regression, maximum likelihood, and Bayesian approaches; a detailed treatment of some of these procedures can be found in Oraguzie et al. (2007). The following approaches are being widely used for AM.

22.6.1 MULTIPARENT ADVANCED GENERATION INTERCROSS (MAGIC)

The MAGIC was first proposed and applied to mice (Mott et al., 2000) as heterogeneous stock and later extended to plants by Mackay and Powell (2007), who also proposed the name MAGIC. MAGIC is an extension of the advanced intercross method in which multiple parents are intermated for several generations. An advantage of the method is that a population can be established containing lines that capture the majority of the variation available in the gene pool. Although it might take several years before these populations are suitable for fine mapping, they are cheap to set up and their value as mapping resources increases with each generation. In plants, MAGIC can be used to combine coarse mapping with low marker densities on lines derived from an early generation, with fine mapping using lines derived from a more advanced generation of crossing and a higher marker density.

22.6.2 CASE–CONTROL (CC)

This is the classical method of AM based on a group of unrelated individuals, called cases, carrying the allele of a gene responsible for a disease (or a mutant trait phenotype) and a sample of equal number of unrelated individuals, called control, lacking the disease. The allelic frequencies of the concerned gene and of the markers in the case and control groups are compared, and association between the gene and a marker can be detected by a suitable test. The case–control approach was developed for human populations to tag genes causing genetic diseases. Similar approaches have been used in some studies with plants to identify markers associated with qualitative traits, for example, identification of SNP and InDel (insertion–deletion) polymorphisms associated with the Y1 gene for endosperm color in maize (Palaisa et al., 2003). The chief limitations of this approach are the low frequency of "cases" in the population and a strong influence of population structure. If a mutation increases disease susceptibility, then we can expect it to be more frequent among affected individuals (cases) than among unaffected individuals (controls). The essential idea behind CC-based AM is that markers close to the disease mutation may also have allele frequency differences between cases and controls if there is LD between the marker locus and the "susceptibility" mutations (Schulze and McMahon, 2002). For accurate mapping, this design requires an equal number of unrelated and unstructured *case–control* samples. The Pearson $\chi 2$ test, Fisher's exact test, or Yates continuity correction can be used to compare allele frequencies and detect association between a phenotype and a marker (Abdurakhmonov and Abdukarimov, 2008). The CC tests are sensitive to overall population.

22.6.3 TRANSMISSION DISEQUILIBRIUM TEST

Transmission disequilibrium test (TDT) was the first family-based design developed to avoid limitations of the case–control approach (Spielman et al., 1993). The TDT design assumes linkage between the gene of interest and the marker under test. A chi-square test is used to compare transmission versus nontransmission of the marker to the affected progeny from one parent heterozygous and one parent homozygous for the concerned disease allele. TDT is widely used for unbiased mapping of genes with two alleles using biallelic markers. The TDT design requires genotyping of markers from three individuals: one heterozygous parent, one homozygous parent, and one affected offspring. In the absence of linkage between QTL and marker, the expected ratio of transmission to nontransmission is 1:1 (Nielsen and Zaykin, 2001). In the presence of linkage, it is distorted to an extent that depends on the strength of LD between the marker and the QTL. In addition, the power of the association will depend on the effectiveness of selection of extreme progeny in driving segregation away from expectation (Mackay and Powell, 2007). The initial TDT approach did not address the cases of multiallelic markers, multiple markers, missing parental information, large pedigrees, and complex quantitative traits (Schulze and McMahon, 2002).

22.6.4 STRUCTURED ASSOCIATION

Many association analysis models consider markers as linear fixed effects, in which each marker is individually examined for association with a QTL affecting the trait of interest. The structured association (SA) model was developed to tackle the problems due to population structure. Population structure describes the level of genetic differentiation among the different homogeneous groups present in the population, from which the sample was drawn for the AM study. Population structure generates LD between unlinked loci and tends to increase the likelihood of discovery of false positive associations. A Bayesian approach is used to detect population structure and to generate the clustering matrix Q, which is also known as "gross-level population structure". The SA model uses the Q matrix to correct, by logistic regression, the false associations due to population structure.

22.7 SOFTWARE FOR LD AND ASSOCIATION MAPPING

For LD studies and AM, a number of software packages have been developed. Most of these packages are available for free, but some of them are commercially available. The various packages provide a variety of functions useful for handling of animal and/or plant data, and some of them are specifically for particular design. The details of software packages used for AM based on Zhu et al. (2008), Gupta et al. (2014), and Singh and Singh (2015) are given in Table 22.2.

TABLE 22.2 Statistical Software Packages Generally Used for Association Mapping in Plants.

Software Package	Brief Description
Free packages	
TASSEL	LD statistic calculation and graphic visualization; sequence analysis; association mapping using logistic regression, GLM, MLM, and some other models; structure and kinship analyses; analysis of insertion/deletion, diversity estimation, etc. (http://sourceforge.net/projects/tassel; http://www.maizegenetics.net)
EMMAX	Fast computation, for large AM studies, corrects for population structure and kinship (http://genetics.cs.ucla.edu/emmax/)
GenAMap	Implements structured association mapping, employs various algorithms, good graphical presentation(http://sailing.cs.cmu.edu/genamap/)
GenABEL	GWAS for both quantitative and qualitative traits (http://www.genabel.org/packages/GenABEL
FaST-LMM	AM based on large samples of up to 120,000 individuals (http://fastlmm.codeplex.com/)
GAPIT	Implements CMLM, R-based, fast computation (http://www.maizegenetics.net/gapit

TABLE 22.2 *Continued)*

Software Package	Brief Description
STRUCTURE	Population structure analysis; generates Q matrix; computation intensive (http://pritch.bsd.uchicago.edu/structure.html)
SPAGeDI	Kinship analysis; generates K matrix (http://www.ulb.ac.be/sciences/ecoevol/spagedi.html)
EINGENSTRAT	Association analysis; PCA to generate P matrix to be used in the place of Q matrix (http://genepath.med/harvard.edu/~reich/software.html)
MTDFREML	MLM analysis of animal breeding data; can be used for plants (http://aipl.arsusda.gov/curtvt/mtdfreml.html
R	Generic package; convenient for simulation work; useful for researchers with good statistics and computer programming background (http://www.r-project.org/)
Commercial packages	
ASREML	MLM analysis for animal breeding data, can be used for plants (http://www.vsni.co.uk/products/asreml)
GenStat	Implements GLM and MLM, corrects for population structure (http://www.vsni.co.uk/software/ genstat)
JMP Genomics	Computation of population structure and kinship coefficient (marker based) (http://www.jmp.com/software/genomics/)
SAS	Standard statistical package used for data analysis and methodology work (http://www.sas.com)

(Adapted from Singh and Singh, 2015)

22.8 CONCLUSIONS

Advances in plant genomics will make it possible to scan a genome for polymorphisms associated with qualitative and quantitative traits. LD, the nonrandom association of alleles at different loci, plays an integral role in association mapping, and determines the resolution of an association study-Recently, association mapping has been exploited to dissect QTL. The potential high resolution in localizing a QTL controlling a trait of interest is the primary advantage of AM as compared to linkage mapping. AM has the potential to identify more and superior alleles and to provide detailed marker data in a large number of lines which could be of immediate application in breeding (Yu and Buckler, 2006). Furthermore, AM uses breeding populations including diverse and important materials in which the most relevant genes should be segregating. Complex interactions (epistasis) between alleles at several loci and genes of small effects can be identified, pinpointing the superior individuals in a breeding population (Tian et al., 2011). Genome-wide AM has not been able to identify the set of genes that together will explain the total phenotypic variation in any of the quantitative traits that have been extensively investigated. Generally, higher levels of LD are observed in newly founded populations. Therefore, younger populations should be used for initial detection of LD. Following this, older populations can be analyzed for fine mapping of the target locus/gene. One drawback of association mapping is that the underlying population stratification due to breeding history, selection, genetic drift, or founder effects can lead to false associations (Lander and Schork, 1994; Slatkin, 1991). However, this problem can be reduced by accounting for population structure using the relationship matrix or distance matrix among the lines (Pritchard and Rosenberg, 1999).

In view of the rapid decline in genotyping costs, the focus is likely to shift from candidate gene approach to genome-wide association studied based on complete genome sequencing of all the individuals in the sample. There will be an increasing trend of integrating gene expression data and even gene expression network information with genome-wide association studies. AM will be extended to nonmodel organisms, and efforts would be made to adequately address the issue like epistasis, G × E interaction effects and phenotypic plasticity. Integration of genome-wide association, eQTL, and molecular marker data is expected to yield valuable insights into the genetic architecture of quantitative traits and also to identify genes that are likely to have been the targets of natural selection. AM may provide a much clearer picture of the architecture of QTLs and to enable identification of individual causal mutations down to the nucleotide level changes.

KEYWORDS

- **association mapping**
- **genotyping**
- **linkage disequilibrium**
- **phenotyping**
- **population structure**

REFERENCES

Abdurakhmonov, I. Y.; Abdukarimov, A. Application of Association Mapping to Understanding the Genetic Diversity of Plant Germplasm Resources. *Int. J. Plant Genom.* **2008,** *574927*:,1–18.

Abecasis, G. R.; Cookson, W. O. C. GOLD-graphical Overview of Linkage Disequilibrium. *Bioinformatics* **2000,** *16*(2), 182–183.

Arnheim, N.; Calabrese, P.; Nordborg, M. Hot and Cold Spots of Recombination in the Human Genome: The Reason We Should Find Them and How This Can Be Achieved. *Am. J. Hum. Genet.* **2003,** *73*(1), 5–16.

Balding, D. A Tutorial on Statistical Methods for Population Association Studies. *Nat. Rev. Genet.* **2006,** *7*(10), 781–791.

Bradbury, P.; Zhang, Z.; Kroon, D.; Casstevens, T.; Ramdoss, Y.; Buckler, E. TASSEL: Software for Association Mapping of Complex Traits in Diverse Samples. *Bioinformatics* **2007,** *23*(19), 2633–2635.

Breseghello, F.; Sorrells, M. E. Association Analysis as a Strategy for Improvement of Quantitative Traits in Plants. *Crop Sci.* **2006,** *46*, 1323–1330.

Caldwell, K. S.; Russell, J.; Langridge, P.; Powell, W. Extreme Population Dependent Linkage Disequilibrium Detected in an Inbreeding Plant Species, *Hordeum vulgare*. *Genetics* **2006,** *172*, 557–567.

Dawson, E.; Abecasis, G.; Bumpstead, S.; Chen, Y.; Hunt, S.; Beare, D.; Pabial, J.; Dibling, T.; Tinsley, E.; Kirby, S.; Carter, D.; Papaspyridonos, M.; Livingstone, S.; Ganske, R.; Lõhmussaar, E.; Zernant, J.; Tõnisson, N.; Remm, M.; Mägi, R.; Puurand, T.; Vilo, J.; Kurg, A.; Rice, K.; Deloukas, P.; Mott, R.; Metspalu, A.; Bentley, D.; Cardon, L.; Dunham, I. A First-generation Linkage Disequilibrium Map of Human Chromosome 22. *Nature* **2002,** *418*(6897), 544–548.

Doerge, R. W. Mapping and Analysis of Quantitative Trait Loci in Experimental Populations. *Nat. Rev. Genet.* **2002,** *3*, 43–52.

Fisher, R. A. On the Correlation between Relatives on the Supposition of Mendelian Inheritance. *Philos. Trans. Royal Soc. Edinb.* **1918,** *52*, 399–433.

Flint-Garcia, S. A.; Thornsberry, J. M.; Buckler, S. E. Structure of Linkage Disequilibrium in Plants. *Ann. Rev. Plant Biol.* **2003,** *54*, 357–374.

Garris, A.; Tai, T.; Coburn, J.; Kresovich, S.; McCouch, S. Genetic Structure and Diversity in *Oryza sativa* L. *Genetics*. **2005,** *169*(3), 1631–1638.

Gaut, B. S.; Long, A. D. The Low Down on Linkage Disequilibrium. *Plant Cell.* **2003,** *15*, 1502–1506.

Gupta, P. K.; Kulwal, P. L.; Jaiswal, V. Association Mapping in Crop Plants: Opportunities and Challenges. In *Advances in Genetics*; Friedmann, T., Dunlap, J., Goodwin, S., Eds.; Academic, Elsevier: Waltham, MA, USA, **2014;** Vol. 85, pp 109–148.

Gupta, P. K.; Rustgi, S.; Kulwal, P. L. Linkage Disequilibrium and Association Studies in Higher Plants: Present Status and Future Prospects. *Plant Mol. Biol.* **2005,** *57*, 461–485.

Hill, W.; Robertson, A. Linkage Disequilibrium in Finite Populations. *Theor. Appl. Genet.* **1968,** *38*(6), 226–231.

Huang, X.; Wei, X.; Sang, T.; Zhao, Q.; Feng, Q. Genomewide Association Studies of 14 Agronomic Traits in Rice Landraces. *Nat Genet.* **2010,** *42*, 961–967.

Ingvarsson, P. K.; Street, N. R. Association Genetics of Complex Traits in Plants. *New Phytol.* **2011,** *189*, 909–922.

Jung, M.; Ching, A.; Bhattramakki, D.; Dolan, M.; Tingey, S.; Morgante, M.; Rafalski, A. Linkage Disequilibrium and Sequence Diversity in a 500-kbp Region Around the *adh*1 Locus in Elite Maize Germplasm. *Ther. Appl. Genet.* **2004,** *109*(4), 681–689.

Lam, H.; Xu, X.; Liu, X.; Chen, W.; Yang, G. Resequencing of 31 Wild and Cultivated Soybean Genomes Identifies Patterns of Genetic Diversity and Selection. *Nat Genet.* **2010,** *42*, 1053–1059.

Lander, E. S.; Botstein, D. Mapping Mendelian Factors Underlying Quantitative Traits using RFLP Linkage Maps. *Genetics* **1989,** *121*, 185–199.

Lander, E. S.; Schork, N. J. Genetic Dissection of Complex Traits. *Science* **1994,** *265*(5181), 2037–2048.

Lewontin, C. The Interaction of Selection and Linkage. I. General Considerations; Heterotic Models. *Genetics* **1964,** *49*(1), 49–67.

Li, Y.; Haseneyer, G.; Schön, C.; Ankerst, D.; Korzun, V.; Wilde, P.; Bauer, E. High Levels of Nucleotide Diversity and Fast Decline of Linkage Disequilibrium in Rye (*Secale cereale* L.) Genes Involved in Frost Response. *BMC Plant Biol.* **2011b,** *11*(6), 1–14.

Lijavetzky, D.; Cabezas, J. A.; Ibanez, A.; Rodriguez, V.; Martinez-Zapater, J. M. High Throughput SNP Discovery and Genotyping in Grapevine (*Vitis vinifera* L.) by Combining a Re-sequencing Approach and SNPlex technology. *BMC Genom.* **2007,** *8*, 424.

Liu, A.; Burke, J. M. Patterns of Nucleotide Diversity in Wild and Cultivated Sunflower. *Genetics* **2006,** *173*, 321–330.

Liu, K.; Muse, S. PowerMarker: An Integrated Analysis Environment for Genetic Marker Analysis. *Bioinformatics* **2005,** *21*(9), 2128–2129.

Mackay, I.; Powell, W. Methods for Linkage Disequilibrium Mapping in Crops. *Trends in Plant Science* **2007,** *12*(2), 57–63.

Mackay, T. F. C.; Stone, E. A.; Ayroles, J. F. The Genetics of Quantitative Traits: Challenges and Prospects. *Nature Reviews Genetics* **2009,** *10*, 565–577.

Malysheva-Otto, L.; Ganal, M.; Röder, M. Analysis of Molecular Diversity, Population Structure and Linkage Disequilibrium in a Worldwide Survey of Cultivated Barley Germplasm (*Hordeum vulgare* L.). *BMC Genet.* **2006,** *7*(6), 1–14.

Mauricio, R. Mapping Quantitative Trait Loci in Plants: Uses and Caveats for Evolutionary Biology. *Nat. Rev. Genet.* **2001,** *2*, 370–381.

McNally, K. L.; Childs, K. L.; Bohnert, R.; Davidson, R. M.; Zhao, K. Genome Wide SNP Variation Reveals Relationships among Landraces and Modern Varieties of Rice. *Proc. Natl. Acad. Sci. USA*. **2009,** *106*, 12273–12278.

Morrell, P. L.; Toleno, D. M.; Lundy, K.; Clegg, M. T. Low Levels of Linkage Disequilibrium in Wild Barley (*Hordeum vulgare* ssp. spontaneum) Despite High Rates of Self-fertilization. *Proc. Natl. Acad. Sci. USA.* **2005,** *102*, 2442–2447.

Mott, R.; Talbot, C. J.; Turri, M. G.; Collins, A. C.; Flint, J. A Method for Fine Mapping Quantitative Trait Loci in Outbred Animal Stocks. *Proc. Natl. Acad. Sci. USA.* **2000,** *97*, 12649–12654.

Myles, S.; Boyko, A.; Owens, C.; Brown, P.; Grassi, F.; Aradhya, M.; Prins, B.; Reynolds, A.; Chia, J.; Ware, D.; Bustamante, C.; Buckler, E. Genetic Structure and Domestication History of the Grape. *Proc. Natl. Acad. Sci. USA.* **2011,** *108*(9), 3530–3535.

Myles, S.; Peiffer, J.; Brown, P. J.; Ersoz, E. S.; Zhang, Z.; Costich, D. E.; Buckler, E. S. Association Mapping: Critical Considerations Shift from Genotyping to Experimental Design. *Plant Cell* **2009,** *21*, 2194–2202.

Nielsen, D.; Zaykin, D. Association Mapping: Where We've Been, Where We're Going. *Expert Rev. Mol. Diagn.* **2001,** *1*(3), 334–342.

Nordborg, M. Linkage Disequilibrium, Gene Trees and Selfing: An Ancestral Recombination Graph with Partial Self-fertilization. *Genetics* **2000,** *154*(2), 923–929.

Nordborg, M.; Tavare, S. Linkage Disequilibrium: What History has to Tell Us. *Trends Genet.* **2002,** *18*, 83–90.

Oraguzie, N.; Wilcox, P.; Rikkerink, H.; de Silva, H. Linkage Disequilibrium, In: *Association Mapping in Plants*; Oraguzie, N. C., Rikkerink, E. H. A., Gardiner, S. E., de Silve, H.N., Eds.; Springer: New York, USA, **2007;** pp 11–39.

Orr, H. A.; Coyne, J. A. The Genetics of Adaptation: A Reassessment. *Am. Nat.* **1992,** *140*, 725.

Palaisa, K.; Morgante, M.; Tingey, S.; Rafalski, A. Contrasting Effects of Selection on Sequence Diversity and Linkage Disequilibrium at Two Phytoene Synthase Loci. *Plant Cell* **2003,** *15*(8), 1795–1806.

Pritchard, J. K.; Rosenberg, N. A. Use of Unlinked Genetic Markers to Detect Population Stratification in Association Studies. *Am J. Hum. Genet.* **1999,** *65*(1), 220–228.

Rae, A. M.; Street, N. R.; Robinson, K. M.; Harris, N.; Taylor, G. Five QTL Hotspots for Yield in Short Rotation Coppice Bioenergy Poplar: The Poplar Biomass Loci. *BMC Plant Biol.* **2009,** *9*, 23.

Rafalski, A. Applications of Single Nucleotide Polymorphisms in Crop Genetics. *Curr. Opin. Plant Biol.* **2002,** *5*, 94–100.

Remington, D. L.; Thornsberry, J. M.; Matsuoka, Y.; Wilson, L. M.; Whitt, S. R.; Doebley, J.; Buckler, E. S. Structure of Linkage Disequilibrium and Phenotypic Associations in the Maize Genome. *Proc. Natl. Acad. Sci. USA.* **2001,** *98*, 11479–11484.

Risch, N.; Merikangas, K. The Future of Genetic Studies of Complex Human Diseases. *Science* **1996,** *273*, 1516–1517.

Robbins, M.; Sim, S.; Yang, W.; Deynze, A.; van der Knaap, E.; Joobeur, T.; Francis, D. Mapping and Linkage Disequilibrium Analysis with a Genome-wide Collection of SNPs that Detect Polymorphism in Cultivated Tomato. *J. Exp. Bot.* **2011,** *62*(6), 1831–1845.

Salvi, S.; Tuberosa, R. Cloning QTLs in Plants. In *Genomics-assisted Crop Improvement*; Varshney, R. K., Tuberosa, R., Eds.; Springer, Dordrecht, The Netherlands, **2007;** pp 207–225.

Schulze, T.; McMahon, F. Genetic Association Mapping at the Crossroads: Which Test and Why? Overview and Practical Guidelines. *Am. J. Med. Genet.* **2002,** *114*(1), 1–11

Schulze, T.; McMahon, F. Genetic Association Mapping at the Crossroads: Which Test and Why? Overview and Practical Guidelines. *Am. J. Med. Genet.* **2002,** *114*(1), 1–11.

Setter, T. L.; Yan, J. B.; Warburton, M.; Ribaut, J. M.; Xu, Y. B.; Sawkins, M.; Buckler, E. S.; Zhang, Z. W.; Gore, M. A. Genetic Association Mapping Identifies Single Nucleotide Polymorphisms in Genes That Affect Abscisic Acid Levels in Maize Floral Tissues During Drought. *J. Exp. Bot.* **2011,** *62*, 701–716.

Singh, B. D; Singh, A. K. *Marker-assisted Plant Breeding: Principles and Practices*. Springer: New Delhi, Heidelberg, New York, Dordrecht, London, **2015**.

Slatkin, M. Inbreeding Coefficients and Coalescence Times. *Genet. Res.* **1991,** *58*(2), 167–175.

Slatkin, M. Linkage Disequilibrium: Understanding the Evolutionary Past and Mapping the Medical Future. *Nat. Rev. Genet.* **2008,** *9*(6), 477–485.

Smith, A.; Thomas, D.; Munro, H.; Abecasis, G. Sequence features in regions of weak and strong linkage disequilibrium. *Genome Res.* **2005,** *15*(11), 1519–1534.

Song, B. H.; Windsor, A. J.; Schmid, K. J.; Ramos Onsins, S.; Schranz, M. E. Multilocus Patterns of Nucleotide Diversity, Population Structure and Linkage Disequilibrium in Boechera stricta, a Wild Relative of Arabidopsis. *Genetics* **2009,** *181*, 1021–1033.

Soto-Cerda, B. J.; Cloutier, S. Association Mapping in Plant Genomes. *Genetic Diversity in Plants* **2012,** Caliskan, M., Ed.; InTech (ISBN: 978-953-51-0185-7).

Sorkheh, K.; Malysheva otto, L. V.; Wirthensohn, M. G.; Tarkesh-esfahani, S.; Martínez gómez, P. Linkage Disequilibrium, Genetic Association Mapping and Gene Localization in Crop Plants. *Genet.Mol. Biol.* **2008,** *31*, 805–814.

Spielman, R.; McGinnis, R.; Ewens, W. Transmission Test for Linkage Disequilibrium: The Insulin Gene Region and Insulin-dependent Diabetes Mellitus (IDDM). *Am. J. Hum. Genet.* **1993,** *52*(3), 506–516.

Stapley, J.; Birkhead, T.; Burke, T.; Slate, J. Pronounced Inter- and intrachromosomal Variation in Linkage Disequilibrium Across the Zebra Finch Genome. *Genome Res.* **2010,** *20*(4), 496–502.

Stich, B; Melchinger, A. E. An Introduction to Association Mapping in plants. *CAB Rev. Perspectives Agric. Vet. Sci. Nutr. Nat. Resour.* **2010,** *5*(39), 1–9.

Stich, B.; Maurer, H. P.; Melchinger, A. E.; Frisch, M.; Heckenberger, M.; Van Der Voort, J. R.; Peleman, J.; Sørensen, A. P.; Reif, J. C. Comparison of Linkage Disequilibrium in Elite European Maize Inbred Lines using AFLP and SSR Markers. *Mol. Breed.* **2006,** *17*, 217–226.

Stich, B.; Melchinger, A. E.; Piepho, H. P.; Hamrit, S.; Schipprack, W.; Maurer, H. P.; Reif, J. C. Potential Causes of Linkage Disequilibrium in a European Maize Breeding Program Investigated with Computer Simulations. *Theor. Appl. Genet*. **2007,** *115*, 529–536.

Stich, B.; Melchinger, A. E.; Frisch, M.; Maurer, H. P.; Heckenberger, M.; Reif, J. C. Linkage Disequilibrium in European Elite Maize Germplasm Investigated with SSRs. *Theor. Appl. Genet*. **2005,** *111*, 723–730.

Tenaillon, M.; Sawkins, M.; Long, A.; Gaut, R.; Doebley, J.; Gaut, B. Patterns of DNA Sequence Polymosphism along Chromosome 1 of Maize (*Zea mays* ssp. *Mays* L.). *Proc. Natl. Acad. Sci. USA.* **2001,** *98*(16), 9161–9166.

Teo, Y.; Fry, A.; Bhattacharya, K.; Small, K.; Kwiatkowski, D.; Clark, T. Genome-wide Comparisons of Variation in Linkage Disequilibrium. *Genome Res.* **2009,** *19*(10), 1849–1860.

Thornsberry, J. M.; Goodman, M. M.; Doebley, J.; Kresovich, S.; Nielsen, D.; Buckler, E. S. *Dwarf 8* Polymorphisms Associate with Variation in Flowering Time. *Nat. Genet.* **2001,** *28*, 286–289.

Tian, F.; Bradbury, P.; Brown, P.; Hung, H.; Sun, Q.; Flint-Garcia, S.; Rocheford, T.; McMullen, M.; Holland, J.; Buckler, E. Genome-wide Association Study of Leaf Architecture in the Maize Nested Association Mapping Population. *Nat. Genet.* **2011,** *43*(2), 159–162.

Wright, S.; Gaut, B. Molecular Population Genetics and the Search for Adaptative Evolution in Plants. *Mol. Biol. Evol.* **2005,** *22*(3), 506–519.

Yan, J.; Shan, T.; Warburton, M.; Buckler, E.; McMullen, M.; Crouch, J. Genetic Characterization and Linkage Disequilibrium Estimation of a Global Maize Collection using SNP Markers. *PLoS One* **2009,** *4*, e8451.

Yu, J.; Buckler, E. S. Genetic Association Mapping and Genome Organization of Maize. *Curr. Opin. Biotechnol.* **2006,** *17*, 155–160.

Zhang, D.; Bai, G.; Zhu, C.; Yu, J.; Carver, B. Genetic Diversity, Population Structure, and Linkage Disequilibrium in U.S. Elite Winter Wheat. *Plant Genome* **2010,** *3*(2), 117–127.

Zhao, K.; Tung, C. W.; Eizenga, G. C. Genome-wide Association Mapping Reveals a Rich Genetic Architecture of Complex Traits in *Oryza sativa*. *Nat. Commun.* **2011,** *2*, 467.

Zhu, C.; Gore, M.; Buckler, E.; Yu, J. Status and Prospects of Association Mapping in Plants. *Plant Genome* **2008,** *1*(1), 5–20.

Zhu, Q.; Zheng, X.; Luo, J.; Gaut, B.; Ge, S. Multilocus Analysis of Nucleotide Variation of *Oryza sativa* and Its Wild Relatives: Severe Bottleneck During Domestication of Rice. *Mol. Biol. Evol.* **2007,** *24*(3), 875–888.

INDEX

A

B

C

F

G

H

I

K

L

M

N

P

T

For Product Safety Concerns and Information please contact our
EU representative GPSR@taylorandfrancis.com Taylor & Francis
Verlag GmbH, Kaufingerstraße 24, 80331 München, Germany